Ecological Studies

Analysis and Synthesis

Volume 250

Ecological Studies is Springer's premier book series treating all aspects of ecology. These volumes, either authored or edited collections, appear several times each year. They are intended to analyze and synthesize our understanding of natural and managed ecosystems and their constituent organisms and resources at different scales from the biosphere to communities, populations, individual organisms and molecular interactions. Many volumes constitute case studies illustrating and synthesizing ecological principles for an intended audience of scientists, students, environmental managers and policy experts. Recent volumes address biodiversity, global change, landscape ecology, air pollution, ecosystem analysis, microbial ecology, ecophysiology and molecular ecology.

Alessandra Fidelis • Vânia R. Pivello
Editors

Fire in the South American Ecosystems

Springer

Editors
Alessandra Fidelis
Departamento de Biodiversidade
Instituto de Biociências
Universidade Estadual Paulista
Rio Claro, Brazil

Vânia R. Pivello
Departamento de Ecologia
Instituto de Biociências
Universidade de São Paulo
São Paulo, Brazil

ISSN 0070-8356 ISSN 2196-971X (electronic)
Ecological Studies
ISBN 978-3-031-89371-1 ISBN 978-3-031-89372-8 (eBook)
https://doi.org/10.1007/978-3-031-89372-8

This work was supported by Fundação de Amparo à Pesquisa do Estado de São Paulo 2015/06743-0, 2023/16620-0
This work was supported by Conselho Nacional de Desenvolvimento Científico e Tecnológico 441968/2018-0 304559/2021-0

This Springer imprint is published by the registered company Springer Nature Switzerland AG
The registered company address is: Gewerbestrasse 11, 6330 Cham, Switzerland

We dedicate this book to all "brigadistas," firefighters, Indigenous Peoples, Traditional and Local Communities and pioneering scientists who, through their wisdom and dedication, teach us every day about the vital role of fire in managing and conserving our ecosystems.

Foreword

This book is a very welcome addition to understanding the role of fire in shaping the landscapes and biota of the South American continent. It also highlights the contributions of South American researchers to our global understanding of fire as an ecological and evolutionary process. The world has become familiar with media reports on catastrophic fires in South America, focused on the Amazon, but also fires in the Pantanal, the Cerrado, Chile, and elsewhere on the continent. We read speculation that these fires are a forerunner of the demise of the vast tropical forests and their conversion to "savannas" with major repercussions for regional (even global) earth system processes. Catastrophic fires have become widespread in recent decades causing loss of lives and property in all vegetated continents and even islands such as the Maui fires of 2023.

The perspective of fire as a disaster, comparable to earthquakes in its destructive force, is inevitably part of this book. However, most of the book focuses on the quite different subject of the role of fire as a natural process, of great antiquity, shaping the landscapes and biota of South America long before, and certainly after, human settlement. The worldwide recognition of fire as an ancient natural process, perhaps second only to climate in shaping the natural world, is quite recent in the scientific literature. But you would have been hard-pressed to write even a chapter on fire in South American ecosystems at the beginning of this century. This book shows the remarkable growth of knowledge on fire ecology since then. It has been written, with authority, by a new generation of South American researchers and their collaborators.

There is substantial literature on fire ecology and evolution from woody-fueled shrublands and forests of the northern hemisphere. However, such fire regimes are rare in South America. Instead, the most widespread fires are grass-fueled, with the dominant grasses belonging to the C4 clade of Andropogoneae. These are also the dominant grasses of the most frequently burnt savannas in the world in Africa and Australia. C4 grasses do not occur in Europe, so their unique properties as fuels for fire were unknown to European ecologists exploring southern ecosystems. Not surprisingly perhaps, the vast grassy ecosystems of the tropics in climates where forests also grew were interpreted as degraded secondary vegetation produced by anthropogenic fires. Research in Brazil has done much to dispel this notion. Molecular phylogenies were first used here to show the antiquity of woody species in Cerrado, their close relationships to forest sister taxa, and the evolution of adaptations that made life possible despite frequent burning. And it is not only woody trees and shrubs; there is a rich and diverse flora of sun-loving herbaceous species restricted to the flammable open ecosystems. Their diversity of fire responses seems unmatched in other flammable grasslands of the world.

Inevitably, a greater understanding of fire has led to pressures for changes in policy and land management practices. A good example is changes in legislation to allow burning after years of fire suppression. That policy caused the demise of sun-loving herbaceous species in nature reserves as trees thickened up. Changes in legislation permitting burning have followed, despite the global publicity on devastating fires in the Amazon. They must follow research-led attempts to change public perceptions and promote a more nuanced understanding of where fire is "good" and "bad" and for which objectives. The science of deciding whether a fire regime is "good" or "bad" is difficult enough. But changing mindsets of lawmakers and the public is another problem altogether. We look forward to innovation in how to avoid catastrophic fires in South America and worldwide. This has become the greatest technical challenge for managing fire regimes in the Anthropocene.

The excitement of new discoveries, the novelty of looking beyond the famous forests of the continent to its open ecosystems, and the willingness to share the science and insights as presented by this book represent a sort of coming of age of fire ecology in South America. The future looks very bright indeed.

Emeritus Professor, Biological Sciences
University of Cape Town,
Cape Town, South Africa
William Bond

Preface

For most people, fire is typically associated with disaster and destruction. However, for Traditional and Local Communities (LC) and Indigenous Peoples (IP), fire holds a different significance: it provides warmth on cold days, helps cook food, serves as a powerful agricultural tool, offers protection against undesirable animals and pests, and generally improves people's lives.

Long before humans arrived in South America, fire played a crucial ecological and evolutionary role in shaping ecosystems such as the Cerrado, Llanos, and Pantanal. With humans' arrival, fire regimes likely changed due to their activities. But the landscape underwent further significant transformations with the arrival of Europeans, who introduced a Eurocentric approach to land management that regarded fire as detrimental to conservation. As a result, for a long time (and still in some cases today), many ecosystems have been degraded due to the exclusion of fire or its overuse.

For decades, most knowledge about fire ecology and fire-prone ecosystems came from research focused on Mediterranean vegetation and African and Australian savannas. In contrast, there was limited literature on fire in South American ecosystems. The study of fire in these ecosystems is relatively recent, emerging primarily in the last 50 years. However, European naturalists of the eighteenth and nineteenth centuries, such as Eugenius Warming, had already recognized the importance of fire in South American ecosystems, noting both its positive and negative effects. South American researchers—for instance, Leopoldo Magno Coutinho, Guillermo Sarmiento and Ernesto Medina—have thoroughly documented the effects of fire on vegetation and its resulting consequences for the ecosystem. Most of this literature was written in Portuguese and Spanish, limiting its global accessibility. In the last 15 years, however, South American researchers have increasingly published their work internationally. This book aims to gather the essential knowledge on the role of fire in South American ecosystems, focusing on the main regions where fire plays a critical role—whether beneficial or detrimental—in shaping ecological dynamics.

Some South American countries have taken even further steps by incorporating local traditional knowledge into fire management practices. These efforts have impacted political decision-making, leading to changes in environmental laws and policies. Therefore, together with offering a comprehensive review of the knowledge accumulated over the past few decades regarding fire in South American ecosystems, this book highlights the advancements made in fire management practices and the valuable local knowledge about fire use, emphasizing the role of fire in shaping these ecosystems in the Anthropocene.

This book consists of 14 chapters authored mostly by South American researchers, with abstracts available in English, Spanish, and Portuguese. The authors provide a comprehensive review of available knowledge on fire in South American ecosystems—both published and unpublished—including topics where progress has been slower, such as the effects of fire on fauna, also providing a long list of references to be accessed by the readers. We hope that this book will draw the attention of the international scientific community to the rich body of knowledge and data accumulated by South American researchers and other scientists in the field over the past few decades.

Rio Claro, Brazil Alessandra Fidelis
São Paulo, Brazil Vânia R. Pivello

Contents

1 **Introduction: Fire and the South American Ecosystems** 1
Vânia R. Pivello and Alessandra Fidelis

2 **Fire as an Evolutionary Driver in South America** 9
Marcelo F. Simon and Vinicius L. Dantas

3 **Traditional Fire Uses by Indigenous Peoples and Local Communities in South America** 39
Bibiana A. Bilbao, Brian G. Ferrero, Rodrigo M. Falleiro, Livia Carvalho Moura, and Guilherme M. Fagundes

4 **Fire in Subtropical and Temperate Grasslands** 83
Luis López-Mársico, Luciana R. Podgaiski, Sergio M. Zalba, Claudia Rodríguez, and Gerhard E. Overbeck

5 **Fire in the Cerrado** 113
Vânia R. Pivello, Márcio Martins, Alexander V. Christianini, Ana Paula Carmignotto, Bruna França Gomes, and Alessandra Fidelis

6 **Fire in the Llanos of the Orinoco River** 157
Juan F. Silva and Bibiana A. Bilbao

7 **Fire Ecology on Megadiverse Montane Grasslands** 187
Soizig le Stradic, Swanni T. Alvarado, Orlando Vargas Ríos, and Fernando A. O. Silveira

8 **Fire in the Chaco Region: Ecological Aspects and Land Management** 213
Sandra Bravo, Roxana Ledesma, Darío Coria, and Dante Loto

9 **Fire in Temperate Forests and Shrublands** 243
Thomas Kitzberger, Mauro E. González, and Juan Paritsis

10 Fire in South American Wetlands 281
Geraldo Alves Damasceno Junior, Pia Parolin, Maxwell da Rosa Oliveira, Danilo Bandini Ribeiro, Thiago Silva Teles, Ieda Maria Bortolotto, Geraldo Wilson Fernandes, and Juan Jose Neiff

11 Fire in Mediterranean Ecosystems of South America 311
Susana Gómez-González, Susana Paula, Alejandro Miranda, Mabel Ortega, and Korina Ocampo-Zuleta

12 Fire in South American Tropical Forests: Examining the Past to Understand the Present and Predict the Future 347
Divino Vicente Silverio, Eddie Lenza, Letícia Gomes, Lucas Paolucci, Paulo Brando, and Marcia Macedo

13 Fire in the Anthropocene 369
Patrícia S. Silva, Renata Libonati, Jose A. Marengo, Mabel Calim Costa, Lincoln Muniz Alves, and Isabel Belloni Schmidt

14 Challenges for a Better Fire Management in South America 401
Livia Carvalho Moura, Vânia R. Pivello, Christian Niel Berlinck, Isabel Belloni Schmidt, and Alessandra Fidelis

Glossary of Terms Related to Fire Ecology 419

Editors and Contributors

About the Editors

Alessandra Fidelis is a Brazilian fire and savanna ecologist. She studied biology at the Universidade de São Paulo (USP) in Brazil and earned her Master's and PhD degrees from the Technische Universität München (Germany), working in collaboration with researchers from the Universidade Federal do Rio Grande do Sul. Her Master's and PhD's thesis dealt with fire effects on subtropical grasslands in Brazil. After completing her doctoral studies, she held postdoctoral positions at the Universidade de São Paulo (Brazil) and the Universidad de Alicante (Spain). In 2011, she was appointed to a professorship at the Universidade Estadual Paulista (UNESP). Currently, she is an Associate Professor at UNESP (Rio Claro, Brazil), where she leads the Lab of Vegetation Ecology. Her primary research interests include fire ecology, tropical grassland and savanna ecology, biological invasions, and seed ecology. She has been serving as the Vice President of the International Association for Vegetation Science since 2015 and is an honorary member with the Brigada Voluntária de Cavalcante (BRIVAC, Brazil).

Dr. Fidelis's research focuses on understanding the impact of fire on both above- and belowground plant communities in tropical and subtropical open ecosystems. She coordinates a long-term fire experiment in Central Brazil, which has been ongoing since 2013. With an extensive national and international research network, she collaborates with scientists and local stakeholders across Brazil and around the world. She and Vânia R. Pivello have been working together since 2009, mostly on topics related to fire in the Cerrado.

Vânia R. Pivello is a Brazilian ecologist in the field of fire ecology. She earned her undergraduate degree in biological sciences at the University of São Paulo (USP), Brazil, where she also completed her Master's degree. Her PhD in Environmental Sciences was pursued at Imperial College, University of London, UK. Both her Master's and PhD researches explored how fire influences vegetation dynamics and ecosystem processes in the Cerrado savannas. After completing her doctoral studies, Pivello continued her academic career at the USP, where she holds a full professorship at the Biosciences Institute. She is a senior researcher at the

Department of Ecology, where her work focuses on fire ecology, tropical grassland and savanna ecology, biological invasions, and conservation biology.

In addition to her academic and research achievements, Vânia R. Pivello has been actively involved in conservation initiatives and policy development related to her fields of expertise. She has contributed to shaping national fire policies and land management strategies aimed at balancing ecological conservation with human land use. She has contributed to the development of frameworks that integrate scientific knowledge with practical land management approaches, particularly in fire-prone ecosystems of the Cerrado.

Contributors

Swanni T. Alvarado Departamento de Biología, Facultad de Ciencias, Universidad Nacional de Colombia, Bogotá, Colombia

Lincoln Alves Instituto Nacional de Pesquisas Espaciais (INPE). São José dos Campos, Brazil

Christian Niel Berlinck Centro Nacional de Pesquisa e Conservação de Mamíferos Carnívoros (CENAP), Instituto Chico Mendes de Conservação da Biodiversidade (ICMBio), Atibaia, Brazil

Bibiana A. Bilbao Departamento de Estudios Ambientales, Universidad Simón Bolívar, Caracas, Venezuela

Université Paul Valéry-Montpellier III, Montpellier, France

Ieda Maria Bortolotto Instituto de Biociências, Universidade Federal de Mato Grosso do Sul, Campo Grande, Brazil

Paulo Brando Yale School of the Environment, Yale University, New Haven, CT, USA

Sandra Bravo Complejo Edilicio Jardín Botánico, Facultad de Ciencias Forestales, Universidad Nacional del Estero, Santiago del Estero, Argentina

Ana Paula Carmignotto Departamento de Biologia, Universidade Federal de São Carlos, Sorocaba, Brazil

Alexander V. Christianini Departamento de Ciências Ambientais, Universidade Federal de São Carlos, Sorocaba, Brazil

Darío Coria Estación Experimental Agropecuaria Santiago del Estero, Instituto Nacional de Tecnología Agropecuaria, Santiago del Estero, Argentina

Mabel Calim Costa Coordenação Geral de Pesquisa e Desenvolvimento, Centro Nacional de Monitoramento e Alertas de Desastres Naturais, São José dos Campos, Brazil

Geraldo Alves Damasceno Junior Laboratorio de Ecologia Vegetal, Instituto de Biociências, Universidade Federal de Mato Grosso do Sul, Campo Grande, Brazil

Vinicius L. Dantas Instituto de Geografia, Universidade Federal de Uberlândia, Uberlândia, Brazil

Guilherme M. Fagundes Departamento de Antropologia, Universidade de São Paulo, São Paulo, Brazil

Rodrigo M. Falleiro Ministério dos Povos Indígenas, Brasília, Brazil

Geraldo Wilson Fernandes Departamento de Genetica, Ecologia & Evolucao, Instituto de Ciências Biológicas, Universidade Federal de Minas Gerais, Belo Horizonte, Brazil

Brian G. Ferrero Centro de Investigación y Transferencia Rafaela, CONICET-UNRaf, Rafaela, Argentina

Alessandra Fidelis Departamento de Biodiversidade, Instituto de Biociências, Universidade Estadual Paulista, Rio Claro, Brazil

Bruna França Gomes Departamento de Ecologia, Instituto de Biociências, Universidade de São Paulo, São Paulo, Brazil

Letícia Gomes Universidade do Estado de Mato Grosso, Nova Xavantina, Brazil

Susana Gómez-González Departamento de Biologia-IVAGRO, Universidad de Cadiz, Puerto Real, Spain

Centro de Ciencia del Clima y la Resiliencia, (CR)2, Santiago, Chile

Mauro E. González Laboratorio de Ecología de Bosques, Instituto de Conservación, Biodiversidad y Territorio, Facultad de Ciencias Forestales y Recursos Naturales, Universidad Austral de Chile, Valdivia, Chile

Center for Climate and Resilience Research (CR2), Santiago, Chile

Thomas Kitzberger INIBIOMA-Universidad Nacional del Comahue, CONICET, Bariloche, Argentina

Roxana Ledesma Estación Experimental Agropecuaria Santiago del Estero, Instituto Nacional de Tecnología Agropecuaria, Santiago del Estero, Argentina

Eddie Lenza Universidade do Estado de Mato Grosso, Nova Xavantina, Brazil

Renata Libonati Departamento de Meteorologia, Universidade Federal do Rio de Janeiro, Rio de Janeiro, Brazil

Luis López-Mársico Instituto de Ecología y Ciencias Ambientales, Universidad de la República, Montevideo, Uruguay

Dante Loto Complejo Edilicio Jardín Botánico, Facultad de Ciencias Forestales, Universidad Nacional del Estero, Santiago del Estero, Argentina

Consejo Nacional de Investigaciones Científicas y Técnicas, El Zanjón, Argentina

Marcia Macedo Woodwell Climate Research Center, Falmouth, USA

Jose A. Marengo Coordenação Geral de Pesquisa e Desenvolvimento, Centro Nacional de Monitoramento e Alertas de Desastres Naturais, São José dos Campos, Brazil

Márcio Martins Departamento de Ecologia, Instituto de Biociências, Universidade de São Paulo, São Paulo, Brazil

Alejandro Miranda Centro de Ciencia del Clima y la Resiliencia, (CR)2, Santiago, Chile

Laboratorio de Ecologia del Paisaje y Conservacion, Departamento de Ciencias Forestales, Universidad de La Frontera, Temuco, Chile

Livia Carvalho Moura Instituto Sociedade, População e Natureza, Brasília, Brazil

Juan Jose Neiff Centro de Ecología Aplicada del Litoral (CECOAL), CONICET-UNNE, Corrientes, Argentina

Korina Ocampo-Zuleta Instituto de Ecología y Biodiversidad (IEB), Concepción, Chile

Mabel Ortega Departamento Desarrollo e Investigacion, Gerencia Proteccion contra Incendios Forestales, Corporacion Nacional Forestal, Santiago, Chile

Gerhard E. Overbeck Instituto de Biociências, Universidade Federal do Rio Grande do Sul, Porto Alegre, Brazil

Lucas Paolucci Universidade Federal de Viçosa, Viçosa, Brazil

Juan Paritsis INIBIOMA-Universidad Nacional del Comahue, CONICET, Bariloche, Argentina

Pia Parolin Max-Planck-Institute of Animal Behaviour, Radolfzell am Bodensee, Germany

Susana Paula Facultad de Ciencias, Instituto de Ciencias Ambientales y Evolutivas, Universidad Austral de Chile, Valdivia, Chile

Instituto de Ecología y Biodiversidad (IEB), Concepción, Chile

Vânia R. Pivello Departamento de Ecologia, Instituto de Biociências, Universidade de São Paulo, São Paulo, Brazil

Luciana R. Podgaiski Instituto de Biociências, Universidade Federal do Rio Grande do Sul, Porto Alegre, Brazil

Danilo Bandini Ribeiro Instituto de Biociencias, Universidade Federal de Mato Grosso do Sul, Campo Grande, Brazil

Orlando Vargas Ríos Departamento de Biología, Facultad de Ciencias, Universidad Nacional de Colombia, Bogotá, Colombia

Claudia Rodríguez Instituto de Ecología y Ciencias Ambientales, Universidad de la República, Montevideo, Uruguay

Maxwell da Rosa Oliveira Instituto de Ciências Biológicas, Universidade Federal de Minas Gerais, Belo Horizonte, MG, Brazil

Laboratório de Ecologia Vegetal, Instituto de Biociências, Universidade Federal de Mato Grosso do Sul, Campo Grande, Brazil

Isabel Belloni Schmidt Departamento de Ecologia, Instituto de Ciências Biológicas, Universidade de Brasília, Campus Universitário Darcy Ribeiro, Brasília, Brazil

Juan F. Silva Instituto de Ciencias Ambientales y Ecológicas ICAE, Universidad de Los Andes, Mérida, Venezuela

Fondo Verde Institute, Huancayo, Peru

Patrícia S. Silva Instituto Dom Luiz, Faculdade de Ciências, Universidade de Lisboa, Lisboa, Portugal

Yale School of the Environment, Yale University, New Haven, CT, USA

Fernando A. O. Silveira Departamento de Genética, Ecologia e Evolução, Universidade Federal de Minas Gerais, Belo Horizonte, Brazil

Divino Vicente Silverio Universidade Federal Rural da Amazônia, Capitão Poço, Brazil

Marcelo F. Simon Embrapa Recursos Genéticos e Biotecnologia, Brasília, Brazil

Soizig le Stradic Biodiversité, Génes & Communautés-BIOGECO, UMR INRAE/ Université de Bordeaux, Pessac Cedex, France

Thiago Silva Teles Instituto de Biociências, Universidade Federal de Mato Grosso do Sul, Campo Grande, Brazil

Sergio M. Zalba Departamento de Biología, Bioquímica y Farmacia, Universidad Nacional del Sur, Bahía Blanca, Argentina

Introduction: Fire and the South American Ecosystems

1

Vânia R. Pivello and Alessandra Fidelis

Abstract

Fire has played a crucial role as an evolutionary and ecological factor across the globe, and many South American ecosystems have been shaped by fire for thousands of years. However, human activity has significantly altered fire regimes over time. This book explores key ecosystems of South America and their historical and contemporary relationships with fire. It also highlights the traditional use of fire by Indigenous Peoples and their influence on environmental shaping. A growing concern is the impact of climate change on fire dynamics, which is increasingly relevant in today's world. Ultimately, this book aims to deepen our understanding of fire as a powerful ecological force in South America, offering essential insights into fire dynamics that can inform effective conservation strategies.

Fire is an important evolutionary and ecological factor globally (Bond et al. 2005), including several ecosystems in South America. The occurrence of fire in the planet is prior to the presence of humans (Pausas and Keeley 2021). However, humans have changed fire regimes for different reasons, mostly by increasing the number of ignitions, altering fuel load and fire timing (Bowman et al. 2011, 2020). Despite the

V. R. Pivello (✉)
Departamento de Ecologia, Instituto de Biociências, Universidade de São Paulo,
São Paulo, Brazil
e-mail: vrpivel@usp.br

A. Fidelis
Departamento de Biodiversidade, Instituto de Biociências, Universidade Estadual Paulista,
Rio Claro, Brazil
e-mail: alessandra.fidelis@unesp.br

A. Fidelis, V. R. Pivello (eds.), *Fire in the South American Ecosystems*,
Ecological Studies 250, https://doi.org/10.1007/978-3-031-89372-8_1

importance of fire in several ecosystems worldwide, including in South America, its role is often misunderstood and mistakenly regarded as always destructive (Fidelis 2020).

Hardesty et al. (2005) classified regions of the world based on the adaptability of their vegetation to fire, a classification further highlighted by Pivello (2011) for South America (Fig. 1.1). This adaptability is directly correlated with the frequency of fire exposure that the vegetation has experienced throughout its evolutionary history, which can be reconstructed through fossil plants and paleocharcoal records. These fossil records indicate that fire events were quite frequent on Earth during the Carboniferous geological period, with rarer occurrences dating back to the Devonian and even the Silurian periods, over 400 million years ago (Myr BP), coinciding with the emergence of the first terrestrial plants and atmospheric oxygen levels sufficient to support combustion (Scott 2000; Scott and Grasspool 2006). Fire-prone ecosystems persisted during past climatic changes (Keeley et al. 2011), with the diversification of angiosperms promoted by the Cretaceous fires (Bond and Scott 2010), and the rise of C_4 grasses occurring at the end of the Miocene, under a fire-prone climate (Edwards et al. 2010; Keeley and Rundel 2005). As a result, vegetation types that have had more frequent exposure to fire events exhibit greater adaptation to fire compared to those with limited fire contact, as seen in subtropical and tropical savannas and grasslands.

But what determines the burning of vegetation? First, the minimum conditions for fire to occur must be present. Fire results from a combustion reaction, and the essential requirements to start a combustion are fuel, an igniting source and oxygen. The igniting source provides the necessary heat to initiate combustion, which then continues as long as there is sufficient fuel and oxygen. This process releases light, heat and other reaction products, transferring energy to the surrounding environment. These three elements—oxygen, heat, and fuel—form the **fire triangle** (Fig. 1.2). While this physical reaction occurs at a localized scale and takes seconds, fire propagation and behavior are influenced by additional drivers on broader spatial and temporal scales (see Cochrane and Ryan 2009 for details). **Fire behavior** is shaped by environmental conditions, including variations in fuel types, terrain characteristics (e.g., slope, aspect, and barriers) and weather features (e.g., temperature, humidity, and wind) (Fig. 1.2). Over decades, the recurring pattern of fires within a specific ecosystem or biome defines its **fire regime**. This regime is influenced by the typical regional vegetation, landforms and climate. Thus, the fire regime combines the season and frequency that fire normally occurs, as well as the intensity, severity and extent of the area burned (Fig. 1.2).

The concept of a fire regime assumes that fires occur naturally (without human influence), whether ignited by lightning, rock friction, or even spontaneous combustion. It further presupposes that these fire events take place under relatively stable conditions, shaped by long-term evolutionary processes (Kobziar et al. 2024). Such natural patterns are inferred from the reconstruction of the regional fire history, which relies on evidence from fossil records. At present, however, fire regimes have greatly changed, as human activity has significantly altered natural fire regimes over time. As a result, a wide range of plants and animals are now threatened and at high

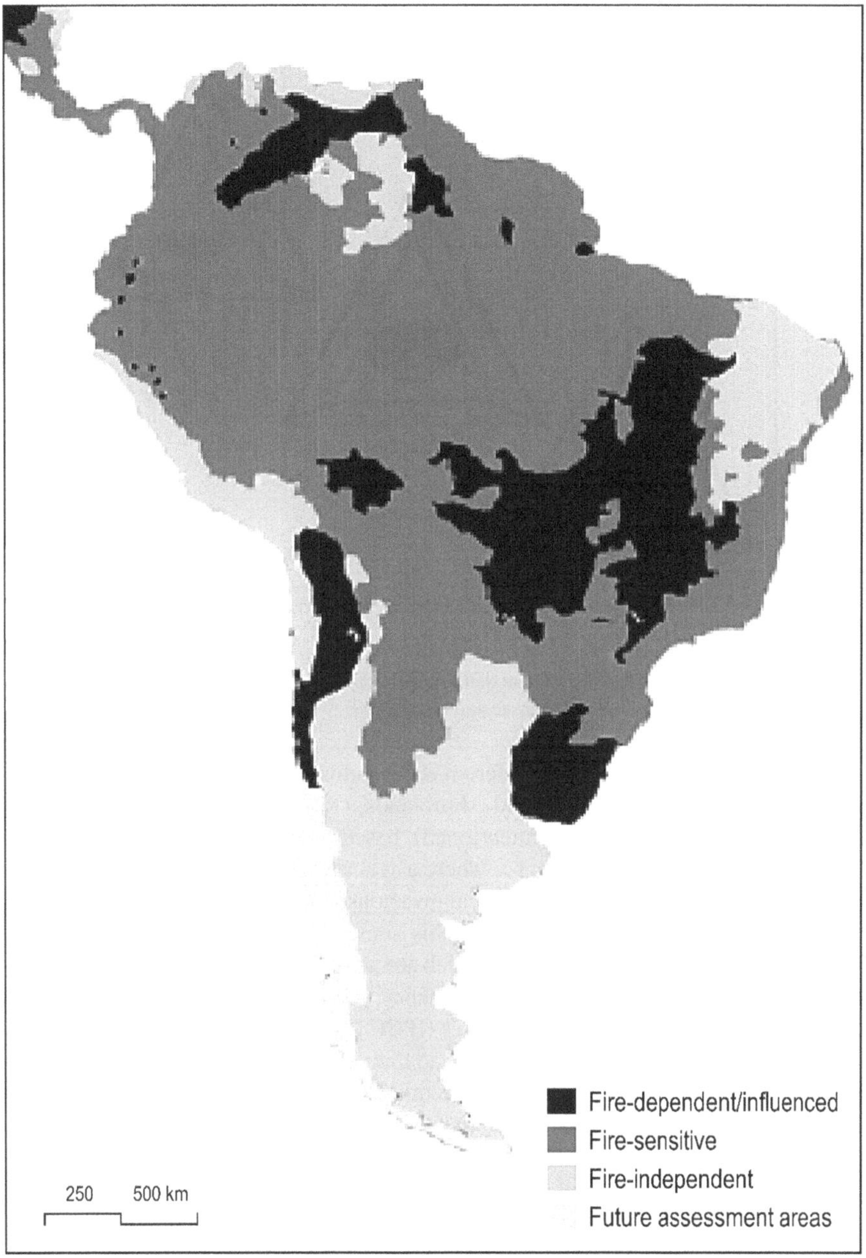

Fig. 1.1 Broad classification of South American vegetation based on fire sensitivity. (Modified from Pivello 2011)

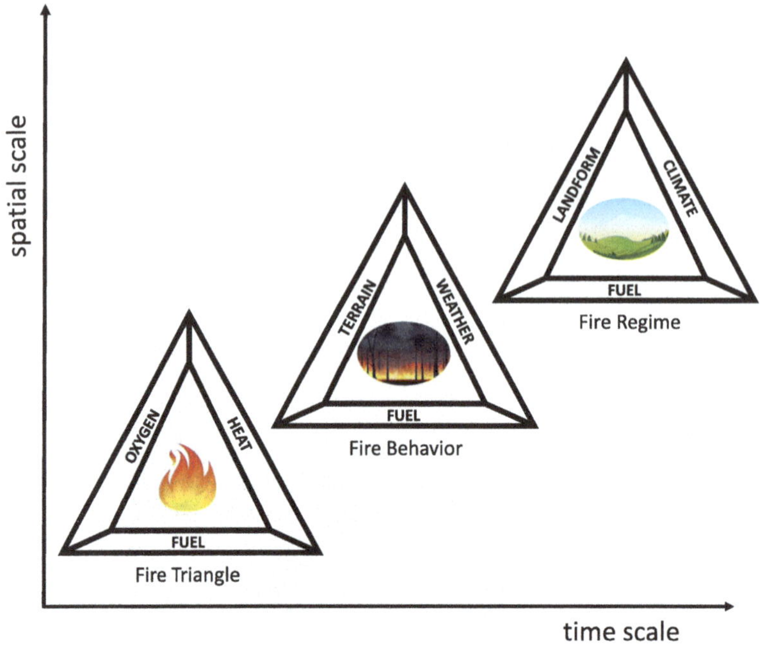

Fig. 1.2 The scales and main factors that determine the fire triangle, fire behavior and fire regime. (Based in Moritz et al. 2005 and Cochrane and Ryan 2009)

risk of fire-driven extinctions in different ecosystems in the world, with the greatest risks in savannas, grasslands and shrublands (Kelly et al. 2020). Therefore, Hardesty's classification (above mentioned), based on natural fires, no longer accurately reflects the current landscape, where a synergy among various anthropogenic land uses, climate change and biological invasions have profoundly altered both the type and distribution of vegetation. In this sense, trends that scientists had been inferring in the last two decades, and which are now being experienced by the global population, indicate increasing fire frequency, extent of burned areas, and often greater intensity and severity of wildfires (IPCC 2023; UNEP 2022). Thus, ecosystems that are naturally fire-sensitive, due to their evolutionary history of infrequent exposure to fire, are now subjected to frequent fires, leading to drastic impacts in both biota and the environment. The most representative example of this situation can be found in tropical rainforests.

South America is home for a great variety of ecosystems, from the fire-sensitive rainforests to the fire-prone grasslands and savannas. Each of them has different fire histories, fire regimes and biota responses to fire. Fire has been present for thousands of years in the South-American lands, acting as an evolutionary force on the vegetation and fauna, as shown in Chap. 2. In the fire-prone ecosystems, various adaptations—primarily morphological in plants and behavioral in animals—can be found in response to fire. For example, plants in these ecosystems have traits that allow their persistence and survival after fire (Keeley et al. 2011), such as thick bar on trees, bud protection (Charles-Dominique et al. 2015; Chiminazzo et al. 2021, 2023), belowground bud-bearing organs with reserves (Fidelis et al. 2014; Pausas et al. 2018), resprouting capacity (Clarke et al. 2013), post-fire flowering (Pyke 2017; Lamont et al. 2019; Fidelis and Zirondi 2021), seed dispersal (Coutinho 1990) and germination (Lamont et al. 2019; Moreira et al. 2010). At the ecosystem level, fire adaptations are also found, for example, directed to nutrient use and cycling. In contrast, a variety of South American forests and humid regions have had little interaction with fire along their history and are sensitive to periodical fires, showing little to no fire adaptations in their biotas.

The relationships between fire and key South American ecosystems are explored in Chaps. 4–12, respectively related to subtropical and temperate grasslands, the Cerrado, the Orinoco Llanos, montane grasslands, the Chaco, temperate forests and shrublands, wetlands, Mediterranean ecosystems and tropical rainforests (Fig. 1.3). There, the natural and human-modified fire regimes and their effects on the vegetation and fauna are presented in detail. A deep understanding of the natural role of fire and its dynamics in these diverse ecosystems is essential for guiding policymakers, land managers and conservationists in using fire as a tool for biodiversity conservation, ecosystem restoration and the maintenance of ecosystem services while also minimizing the negative impacts of wildfires.

Traditional knowledge is recognized as another important source of information for managing fire in natural and semi-natural ecosystems. Indigenous Peoples and Local Communities of South America have long employed fire for land management, agriculture, and cultural practices. As highlighted in Chap. 3, these traditional practices have historically supported the sustainable management of native environments. Together with the scientific knowledge, these practices can be very relevant in managing fire in a modern context, offering strategies that balance human needs with ecological health.

Still, an emerging concern is related to the changing world climate and the replacement of native vegetation by anthropogenic land uses, which are profoundly affecting the South American ecosystems. Chapter 13 explores the current shifts in fire patterns driven by these factors, and also projects future tendencies driven by climate and land use changes. Finally, Chap. 14 brings an overview on how the different countries are using (or not) fire as a management tool and the challenges they have faced to implement policies related to the use of fire in different ecosystems in South America.

This book brings to the public valuable information that has not been widely disseminated through mainstream scientific channels. Relevant studies and findings on fire in South American ecosystems are restricted to technical reports, internal documents, or specialized research papers that are not easily accessible to the broader scientific community, environmental managers, decision- and policy-makers or the general public. Additionally, a great deal of information is not processed in an integrated way and synthesized for practical use. By doing this, knowledge gaps are detected for future research. Therefore, this book aims to consolidate all this information and bridge the gap between academic scientists and practitioners.

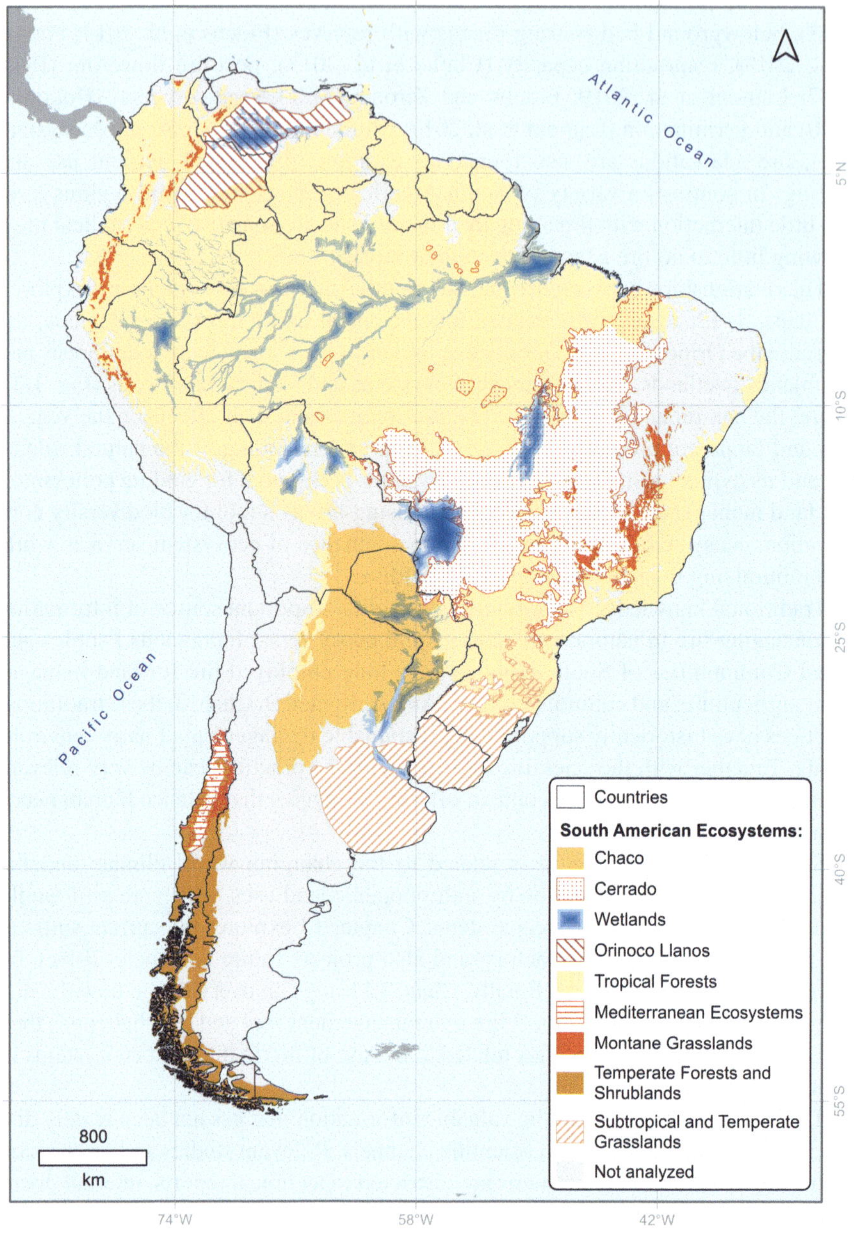

Fig. 1.3 South American ecosystems addressed in this book. Olson et al. (2001) served as the primary source for most of these ecosystems, however, other specific sources were consulted according to the vegetation type, which are indicated in the respective chapters. (Produced by Júlia Abrantes Rodrigues)

In essence, this book aims to deepen understanding of fire as an ecological force in South America, providing critical insights into fire dynamics that can support conservation strategies. These strategies are especially vital in light of the challenges posed by climate change and land-use transformations, benefiting both scientific research and practical land management efforts.

References

Bond WJ, Scott AC (2010) Fire and the spread of flowering plants in the Cretaceous. New Phytol 188:1137–1150. https://doi.org/10.1111/j.1469-8137.2010.03418.x

Bond WJ, Woodward FI, Midgley GF (2005) The global distribution of ecosystems in a world without fire. New Phytol 165:525–537. https://doi.org/10.1111/j.1469-8137.2004.01252.x

Bowman DMJS, Balch J, Artaxo P, Bond WJ, Cochrane MA et al (2011) The human dimension of fire regimes on Earth. J Biogeogr 38:2223–2236. https://doi.org/10.1111/j.1365-2699.2011.02595.x

Bowman DMJS, Kolden CA, Abatzoglou JT, Johnston FH, van der Werf GR, Flannigan M (2020) Vegetation fires in the Anthropocene. Nat Rev Earth Environ 1:500–515. https://doi.org/10.1038/s43017-020-0085-3

Charles-Dominique T, Beckett H, Midgley GF, Bond WJ (2015) Bud protection: a key trait for species sorting in a forest-savanna mosaic. New Phytol 207:1052–1060. https://doi.org/10.1111/nph.13406

Chiminazzo MA, Bombo AB, Charles-Dominique T, Fidelis A (2021) Your best buds are worth protecting: variation in bud protection in a fire-prone Cerrado system. Funct Ecol 35:2424–2434. https://doi.org/10.1111/1365-2435.13907

Chiminazzo MA, Bombo AB, Charles-Dominique T, Fidelis A (2023) Bark production of generalist and specialist species across savannas and forests in the Cerrado. Ann Bot 131:613–621. https://doi.org/10.1093/aob/mcad014

Clarke PJ, Lawes MJ, Midgley JJ, Lamont BB, Ojeda F et al (2013) Resprouting as a key functional trait: how buds, protection and resources drive persistence after fire. New Phytol 197:19–35. https://doi.org/10.1111/nph.12001

Cochrane MA, Ryan KC (2009) Fire and fire ecology: concepts and principles. In: Cochrane MA (ed) Tropical fire ecology: climate change, land use and ecosystem dynamics. Springer, Chichester, UK

Coutinho LM (1990) Fire in the ecology of Brazilian Cerrado. In: Goldamer JG (ed) Fire in the tropical biota: ecosystem processes and global challenges, Ecological studies, vol 84. Springer, Berlin, pp 81–105

Edwards E, Osborne C, Strömberg C, Smith S, Consortium CG (2010) The origins of C4 grasslands: integrating evolutionary and ecosystem science. Science 328:587–591. https://doi.org/10.1126/science.11772

Fidelis A (2020) Is fire always the 'bad guy'? Flora 268:151611. https://doi.org/10.1016/j.flora.2020.151611

Fidelis A, Zirondi HL (2021) And after fire, the Cerrado flowers: a review of post-fire flowering in a tropical savanna. Flora 280:151849. https://doi.org/10.1016/j.flora.2021.151849

Fidelis A, Appezzato-da-Glória B, Pillar VDVD, Pfadenhauer J (2014) Does disturbance affect bud bank size and belowground structures diversity in Brazilian subtropical grasslands? Flora 209:110–116. https://doi.org/10.1016/j.flora.2013.12.003

Hardesty J, Myers RL, Fulks W (2005) Fire ecosystems and people: a preliminary assessment of fire as a global conservation issue. George Wright Forum 22:78–87

IPCC (2023) Sections. In: Core Writing Team, Lee H, Romero J (eds) Climate change 2023: synthesis report. Contribution of Working Groups I, II and III to the sixth assessment report of the Intergovernmental Panel on Climate Change. IPCC, Geneva, Switzerland, pp 35–115. https://doi.org/10.59327/IPCC/AR6-9789291691647

Keeley JE, Rundel PW (2005) Fire and the Miocene expansion of C4 grasslands. Ecol Lett 8:683–690. https://doi.org/10.1111/j.1461-0248.2005.00767.x

Keeley JE, Pausas JG, Rundel PW, Bond WJ, Bradstock RA (2011) Fire as an evolutionary pressure shaping plant traits. Trends Plant Sci 16:406–411. https://doi.org/10.1016/j.tplants.2011.04.002

Kelly LT, Giljohann KM, Duane A, Aquilué N, Archibald S, Batllori E et al (2020) Fire and biodiversity in the Anthropocene. Science 370:6519. https://doi.org/10.1126/science.abb0355

Kobziar LN, Hiers JK, Belcher CM, Bond WJ, Enquist CA et al (2024) Principles of fire ecology. Fire Ecol 20:39. https://doi.org/10.1186/s42408-024-00272-0

Lamont BB, He T, Yan Z (2019) Evolutionary history of fire-stimulated resprouting, flowering, seed release and germination. Biol Rev 94:903–928. https://doi.org/10.1111/brv.12483

Moreira B, Tormo J, Estrelles E, Pausas JG (2010) Disentangling the role of heat and smoke as germination cues in Mediterranean Basin flora. Ann Bot 105:627–635. https://doi.org/10.1093/aob/mcq017

Moritz MA, Morais ME, Summerell LA, Carlson JM, Doyle J (2005) The fire regime of an ecosystem is the collective outcome of multiple drivers, such as ignition patterns, climate, and vegetation characteristics (Wildfires, complexity, and highly optimized tolerance). Proc Natl Acad Sci USA 102:17912–17917. https://doi.org/10.1073/pnas.0508985102

Olson D M, Dinerstein E, Wikramanayake ED et al 2001. Terrestrial ecoregions of the world: a new map of life on Earth. Bioscience 51:933-938. https://doi.org/10.1641/0006-3568(2001)051[0933:TEOTWA]2.0.CO;2

Pausas JG, Keeley JE (2021) Wildfires and global climate change. Front Ecol Environ. https://doi.org/10.1002/fee.2359

Pausas JG, Lamont BB, Paula S, Appezzato-da-Glória B, Fidelis A (2018) Unearthing belowground bud banks in fire-prone ecosystems. New Phytol 217:1435–1448. https://doi.org/10.1111/nph.14982

Pivello VR (2011) The use of fire in the Cerrado and Amazonian Rainforests of Brazil: past and present. Fire Ecol 7:24–39. https://doi.org/10.4996/fireecology.0701024

Pyke GH (2017) Fire-stimulated flowering: a review and look to the future. Crit Rev Plant Sci 36:179–189. https://doi.org/10.1080/07352689.2017.1364209

Scott AC (2000) The Pre-Quaternary history of fire. Palaeogeogr Palaeoclimatol Palaeoecol 164:281–329. https://doi.org/10.1016/S0031-0182(00)00192-9

Scott AC, Grasspool IJ (2006) The diversification of Paleozoic fire systems and fluctuations in atmospheric oxygen concentration. Proc Natl Acad Sci USA 103(29):10861–10865. https://doi.org/10.1073/pnas.0604090103

UNEP—United Nations Environment Programme (2022) Spreading like wildfire: the rising threat of extraordinary landscape fires. UNEP, Nairobi. Fire Ecol 20:39

Fire as an Evolutionary Driver in South America

2

Marcelo F. Simon and Vinicius L. Dantas

Abstract

Ancient fire regimes have been a key element in terrestrial ecosystems, and their influence on the evolution of fire-related traits spans many millions of years. We classified 115 South American ecoregions into three major Pyromes (areas with similar wildfire regimes) and identified key plant fire-related traits. The "Rare" Pyrome included rainforests and deserts where natural fires are unusual, and most species are fire-sensitive. The "Intense" Pyrome was characterized by infrequent high intensity fires associated with high interannual rainfall variability and anthropogenic activity, occurring in seasonal tropical/temperate forests and semi-arid shrublands, where post-fire regeneration generally occurs via basal resprouting or reseeding. The "Frequent" Pyrome comprised mostly tropical grasslands and savannas experiencing regular and frequent fires. Fire-related traits such as thick/corky bark, resprouting from underground organs and aerial bud banks, as well as fire induced flowering, are commonly found in the Frequent Pyrome. Plant traits evolved over millions of years in response to particular fire regimes. However, modern fire regimes are increasingly affected by human influence, with negative consequences for species and ecosystems. We advocate that the legacy of ancient fire regimes on the evolution of the South America ecosystems should be considered to keep them biologically and functionally diverse.

M. F. Simon (✉)
Embrapa Recursos Genéticos e Biotecnologia, Brasília, Brazil
e-mail: marcelo.simon@embrapa.br

V. L. Dantas
Instituto de Geografia, Universidade Federal de Uberlândia, Uberlândia, Brazil
e-mail: viniciusdantas@ufu.br

A. Fidelis, V. R. Pivello (eds.), *Fire in the South American Ecosystems*,
Ecological Studies 250, https://doi.org/10.1007/978-3-031-89372-8_2

Resumen

Los antiguos regímenes de incendios han sido fundamentales en los ecosistemas terrestres, influyendo en la evolución de rasgos relacionados al fuego durante millones de años. Hemos clasificado 115 ecorregiones sudamericanas en tres grandes Piromas (áreas con regímenes de incendios forestales similares), identificando rasgos vegetales clave relacionados al fuego. El Piroma "Raro" incluye selvas tropicales y desiertos con incendios naturales inusuales, donde la mayoría de las especies son sensibles al fuego. El Piroma "Intenso" presenta incendios infrecuentes, pero de alta intensidad, asociados a alta variabilidad de lluvias interanuales y actividad antropogénica, ocurriendo en bosques tropicales/templados estacionales y matorrales semiáridos, con regeneración mediante rebrote basal o resiembra. El Piroma "Frecuente" abarca pastizales y sabanas tropicales con incendios regulares. Rasgos como corteza gruesa, rebrota a partir de bancos de yemas subterráneas o aéreas y floración inducida por el fuego son comunes en plantas de este Piroma, que han evolucionado durante millones de años en respuesta a estes regímenes de incendios. Sin embargo, los regímenes modernos están cada vez más afectados por la influencia humana, con consecuencias negativas para especies y ecosistemas. Abogamos por considerar el legado de los antiguos regímenes de incendios en la evolución de los ecosistemas de Sudamérica para mantener su diversidad biológica y funcional.

Resumo

Regimes ancestrais de incêndios têm desempenhado um papel-chave nos ecossistemas terrestres e sua influência na evolução de adaptações de plantas ao fogo compreende milhões de anos. Classificamos 115 ecorregiões sulamericanas em Piromas (áreas com regimes de incêndio semelhantes) e identificamos as principais adaptações. O Piroma "Raro" inclui florestas tropicais úmidas e desertos, onde incêndios são incomuns e as espécies são sensíveis ao fogo. O Piroma "Intenso" é caracterizado por incêndios infrequentes, de alta intensidade, associados à variabilidade interanual da pluviosidade e atividade antropogênica, ocorrendo em florestas tropicais/temperadas estacionais e vegetações arbustivas semiáridas, onde a regeneração pós-fogo ocorre predominantemente via rebrota basal ou banco de sementes. O Piroma "Frequente" compreende campos e savanas tropicais sujeitos a incêndios regulares e frequentes. Casca espessa/corticosa, rebrota por órgãos subterrâneos ou bancos de gemas aéreos, e floração induzida pelo fogo são adaptações encontradas neste Piroma. As características das plantas evoluíram em resposta a milhões de anos na presença de regimes de fogo. Porém, esses regimes têm sido cada vez mais alterados pelo homem, com consequências negativas para espécies e ecossistemas. Defendemos que o legado dos regimes ancestrais de incêndios sobre a evolução dos ecossistemas sulamericanos seja considerado, a fim de mantê-los biológica e funcionalmente diversos.

2.1 The Long History of Fire on Earth's Terrestrial Ecosystems

The association of vegetation and fire on Earth has a long history that spans many millions of years. It involves different plant assemblages and ecosystems subject to a range of climatic conditions over the geological time. The presence of fire in terrestrial ecosystems is widely documented from fossilized charcoal (Scott 2000), which provides unequivocal evidence of the ancient nature of wildfires. During hundreds of millions of years, fire has been operating as a key selective force driving the evolution of numerous adaptations in plants and animals.

Early evidence of fires in the fossil record dates back from the colonization of land by plants in the Silurian Period around 400 million years ago (Myr BP; Wellman et al. 2003; Glasspool et al. 2004; Scott and Glasspool 2006). As plants spread and dominated the terrestrial system during the Devonian (c. 420–360 Myr BP), biomass production and available fuels increased dramatically, allowing large wildfires to take place (Belcher et al. 2013). The occurrence of wildfires over geological time scales and in the present depends on three factors: an ignition source, fuels, and sufficiently high amounts of oxygen in the atmosphere. Ignition sources have been present during the entire Earth's history (Scott and Glasspool 2006). Although there are potentially numerous sources of ignition such as volcanic eruption, friction, and even spontaneous combustion, most fires are ignited by lightning strikes (Janssen et al. 2023; Song et al. 2024). Fuel availability is fundamentally linked to plant biomass and flammability, which have been present since the colonization of land by plants. While atmospheric oxygen concentration was already high enough (> 20%) to ignite frequent fires since the Cambrian (c. 540 Myr BP), charcoal only became abundant in the fossil record after the formation of the first forest in the Early Carboniferous (350 Myr BP; He and Lamont 2018). Ecosystem flammability over geological times was largely influenced by variation in the atmospheric oxygen concentration. For example, oxygen concentrations higher than 30% (c. 50% higher than current levels) in the Late Carboniferous-Early Permian (c. 300 Myr BP) facilitated ignitions and are linked to a widespread period of high fire activity in Earth's ecosystems (Scott and Glasspool 2006; Belcher et al. 2013; He and Lamont 2018).

Interactions between fire and vegetation involved diverse evolutionary lineages of land plants that once dominated Earth's ecosystems, including major plant groups that are currently extinct. These ancient fire-prone ecosystems looked quite different in terms of physiognomy and species composition compared to those that we see today. Forests dominated by tree-sized lycopsid plants from the genus *Lepidodendron* during the Carboniferous (360–300 Myr BP), and the flammable "prairies" dominated by ferns in the Late Cretaceous (c. 100 Myr BP) are examples of ancient fire-prone systems that no longer exist (Falcon-Lang 2000; Scott 2000). Several other fire-adapted plant lineages evolved to colonize fire-prone ecosystems, some of which persist until the present. One remarkable example is pine trees (*Pinus*), which have been dominating fire-prone temperate and boreal forests since the mid Cretaceous (120 Myr BP; He et al. 2012). Different *Pinus* species display strategies to both tolerate and induce fire depending on the fire regimes under which they evolved (Keeley 2012).

A key event that strongly influenced terrestrial ecosystems and fire regimes on Earth was the origin of the angiosperms in the Cretaceous. Angiosperms became the major group of land plants by the Cretaceous-Paleogene boundary around 66 Myr BP, largely replacing formerly dominant gymnosperms and ferns, particularly in tropical and subtropical regions (Jaramillo 2023). Higher leaf photosynthetic/transpiration efficiency and associated fast-growth rates are thought to have given angiosperms an advantage over other plant groups, enabling them to colonize different ecosystems. The rise to dominance of angiosperms, which include ecologically diverse lineages with a suite of adaptive traits, created novel global fire regimes, including frequently burned pyrophytic grasslands, as well as closed broad-leaved forests that exclude fire (Bond and Scott 2010; Belcher et al. 2021).

Among the modern fire-prone ecosystems dominated by angiosperms, shrublands and heathlands associated with mediterranean-type climate regions (currently found in the Mediterranean Basin, California, central Chile, the Cape Region of South Africa, and southwestern Australia) have been well studied (Rundel et al. 2016). The origin of the unique vegetation found in mediterranean-type climates can be traced back to the Paleocene (60 Myr BP; Crisp and Cook 2013; Linder 2014) and is tightly linked to fire regimes. The mild winters, and the hot and dry summers typical of these ecosystems are highly conducive to high-intensity crown fires. A number of plant genera that diversified in mediterranean-type climate ecosystems such as *Arctostaphylos* (Ericaceae), *Banksia, Leucadendron, Protea* (Proteaceae), *Cistus* (Cistaceae), *Eucalyptus* (Myrtaceae), and *Quercus* (Fagaceae), evolved an array of fire-adaptive traits that enabled them to survive periodic fires (Rundel et al. 2018). Most notably, is serotiny, a fire-adaptive trait in which cones and fruits are kept closed attached to the plant crown for many years until a fire occurs, after which cones/fruits are opened, and seeds are released (Lamont et al. 2020). Although adult plants with serotinous cones/fruits generally die after fire, seeds germinate readily following post-fire dispersal, allowing a pulse of recruitment and population recovery (Keeley and Pausas 2022).

More recently in the geological time scale, a remarkable evolutionary event was the rise to ecological dominance of C_4 grasses around the Late Miocene-Pliocene (10–5 Myr BP), which culminated with the origin of the tropical grasslands and savannas worldwide, the most flammable modern ecosystems (Keeley and Rundel 2005; Beerling and Osborne 2006; Strömberg 2011). The rise of biomes dominated by C_4 grasses was possibly driven by climate changes (lower precipitation and increased seasonality) in the Late Miocene, coupled with lower atmosphere CO_2 concentration, which favor plants with C_4 physiology in the tropics (Edwards et al. 2010). Such combination of climate conditions allowed high biomass production by grasses during the wet season, followed by grass fuel curing in the dry season, producing highly flammable surface fuels that further increased fire activity (Keeley and Rundel 2005; Hoetzel et al. 2013), and promoting the gradual replacement of fire-sensitive woodlands by C_4 grasslands. The emergence of these new open ecosystems characterized by frequent surface fires led to the independent evolution of fire-related adaptations in a number of plant lineages inhabiting different continents (Simon and Pennington 2012; Dantas and Pausas 2013; Maurin et al. 2014; Pellegrini et al. 2017; Pausas et al. 2018; Azevedo et al. 2020; Pilon et al. 2021).

In addition to the well-documented occurrence of fire in paleoecosystems, based on fossil records (e.g., Bond 2015; Archibald et al. 2018), there is a growing body of evidence showing an early origin of fire-related traits in different plant lineages. Evolutionary analysis of dated phylogenies has shown that the first appearance of fire-related traits enabling extant plant groups to tolerate fire regimes, dates back many millions of years. Such traits include fire-induced resprouting, fire-stimulated flowering, serotiny, thick/corky bark, fire-induced seed germination, and the geoxyle habit (Simon et al. 2009; Crisp et al. 2011; He et al. 2012; Simon and Pennington 2012; Maurin et al. 2014; Pausas et al. 2018; Lamont et al. 2019; Keeley and Pausas 2022), which arose independently in different plant lineages in response to fire regime changes. These findings, based on the reconstruction of the evolutionary history of typical fire-tolerant plant groups, provide robust evidence for the role of fire as a major driver in the evolution and diversification of land plants.

There are a number of authoritative reviews on the role of fire as an evolutionary driver at the global scale (Keeley et al. 2011; Pausas and Keeley 2014; Bond 2015; Pausas 2015; He and Lamont 2018; Lamont et al. 2019; Keeley and Pausas 2022) and we also recommend a booklet for the general public published in Spanish about this subject (Pausas 2012). Here, we will focus on the role of fire as an evolutionary driver in South American ecosystems.

2.2 Fire Regimes and Fire-Related Plant Traits

Fire regimes in natural ecosystems depend essentially on fuel loads that result from an ecosystem primary productivity, on fuel moisture, which largely depends on the occurrence and frequency of dry climatic periods, and on favorable weather conditions to fire ignition and spread (Pausas et al. 2017). Fire regimes also depend on the fuel traits of organisms and how they influence flammability (Simpson et al. 2016).

Flammability, rather than a single trait, is the result of multiple non-independent dimensions, such as ignitability (probability of ignition), combustibility (rate of combustion), sustainability (burning time) and consumability (proportion of fuel consumed), each of which may be related to different morphological and physiological traits (Simpson et al. 2016). Flammability results from a set of plant attributes and can be characterized in different scales, such as the organ, the individual and the community. At the interspecific level, plant syndromes related to flammability can be characterized into three broad types (Pausas et al. 2017): non-flammable, fast-flammable, and hot-flammable. Non-flammable plants are those whose traits are linked to reduced overall plant flammability, preventing the ignition and consumption of plant biomass. The fast-flammable strategy is the one in which individual plants (and/or organs) promote fires that release little heat but spread fast, resulting in little damage to vital plant tissues, but that may cause damage to fire-sensitive species. Traits associated with the fast-flammable strategy include thin long leaves with a high surface-volume ratio that die back in the dry season while remaining attached to the plant. Finally, hot-flammable species (and/or organs) are linked to plants promoting hot and slow-moving fires, causing severe damage to plant aboveground tissues. Each syndrome is not necessarily associated with a different fire regime, and in many cases, they co-occur. However, the dominant strategy at the community scale largely influences the fire regime of fire-prone ecosystems (Simpson et al. 2016). In savannas, for example, non-flammable woody plants with sparsely branched, thick and corky stems and thick, large, leaves coexist with fast-flammable grasses with thin highly flammable leaves. Nevertheless, because savannas are dominated by fast-flammable grasses whereas non-flammable woody species occur scattered, these ecosystems are characterized by frequent low intensity surface fires. Despite growing evidence that flammability traits are under selection by fire (Archibald et al. 2018), more studies are necessary to demonstrate the extent to which flammability is adaptive under different fire regimes.

In addition to flammability, a number of fire-adaptive traits are the hallmark of plants that evolved in fire-prone environments. Some of these traits evolved to promote post-fire survival, enabling plants to persist after fire. These include the ability to resprout, which is frequent in ecosystems under a variety of fire regimes (Clarke et al. 2013; Lamont et al. 2019; Pilon et al. 2021). Basal resprouting occurs when aerial parts are consumed or killed by fire (top-kill), and regrowth occur via buds located at the plant base or underground organs. Resprouting can also occur from stem buds that remained alive after fire passage (epicormic resprouting). Stem buds of trees and shrubs that inhabit fire-prone vegetation are often protected by a thick bark (Charles-Dominique et al. 2015; Pausas 2015; Dantas et al. 2024), which is another trait that evolved in response to low intensity surface fires to ensure post-fire survival in many plant lineages. Allocation of plant biomass into underground woody structures, where buds are protected from fire by the soil, is an evolutionary strategy strongly associated to fire-prone systems, enabling plants to quickly resprout after fire (Clarke et al. 2013; Pausas and Keeley 2014). Some plants, especially in crown fire ecosystems with low fire frequency but high fire intensity, invest in traits related to post-fire recruitment, which includes fire-triggered seed dispersal from fruits or cones retained on the plant (serotiny), fire or smoke-induced germination from soil seed bank, and fire-stimulated flowering (Lamont et al. 2020; Keeley and Pausas 2022). All these traits involve fire-related cues that would enhance the chances of plant reproduction and establishment after fire, when competition is generally reduced.

It is important to note that adaptation to fire does not evolve in response to fire *per se*. Instead, the success of different plant strategies in fire-prone ecosystems depends on the fire regimes under which plants evolved. Thus, plant traits that are suitable for persistence in a particular ecosystem may not be suitable for another with a different fire regime. For instance, while serotiny is adaptive in crown fire ecosystems with fire intervals longer than a decade (e.g., some mediterranean-climate type ecosystems), it is not common in plant species from ecosystems subjected to frequent surface fires (e.g., tropical savannas). In ecosystems with very frequent fires, such as tropical savannas, serotiny tend to be absent in woody plants because the time for these plants to reach maturity from seeds is often longer than the typical fire return interval (Pausas and Keeley 2014). On the other hand, a thick protective bark on the base of the trunk is more effective in ecosystems characterized by low-intensity surface fires, but less effective against high-intensity crown fires.

In some cases, the association between traits and fire regimes is not evident and there are mismatches between plant traits and current fire regimes. This occurs because, while plant traits are generally believed to result from current ecological pressures, some traits are anachronistic, that is, they result from past environmental conditions that no longer exist. That explains the occurrence of species with typical fire-adapted traits in ecosystems that are currently not fire-prone, as well as the lack of fire adapted species in ecosystems that currently experience frequent fires. Moreover, it is important to mention that some traits such as resprouting ability could represent adaptations that were selected by or evolved in response to different factors, such as large mammal herbivory, drought, or frost, but that are also favored under frequent fire regimes, representing, therefore exaptations. Although distinguishing between true fire adaptations (traits that originated in response to fire) and exaptations is sometimes complicated (Bradshaw et al. 2011; Keeley et al. 2011), it is clear that many fire-related traits evolved in response to the selective pressures imposed by fire regimes (Simon and Pennington 2012).

2.3 Fire Regimes in Modern South America

Major biomes in South America are distributed in a broad range of environmental and primary productivity conditions (e.g., Raich et al. 1991; Eva et al. 2004), and are currently submitted to different degrees of anthropogenic influence. Thus, fire regimes vary widely across the diverse South American landscapes and vegetation types (e.g., Pivello 2011; Cochrane 2013; Table 2.1). Fire tends to be rare in environments with low primary productivity (deserts; fuel load-limited) or that lack annual dry periods (rainforests; fuel moisture-limited). On the other hand, high amounts of fine grass fuel produced during the wet season that dries out during the pronounced dry season followed by frequent lightning storms, particularly those in the transition from dry to wet season, make tropical savannas and grasslands ecosystems to burn frequently (Pausas and Ribeiro 2013). In seasonally dry and moist forests and shrublands, in contrast, fires do not burn as frequently, but unusual extreme weather events, as well as anthropogenic activity, may eventually cause fires, especially as fuels accumulate after many fire-free decades.

Surface fires, where the fuel consumed is mostly at the ground level (graminoids, subshrubs, small shrubs and litter), are generally of low to medium intensity and occur mostly in South American savannas and grasslands, but also in forests (Table 2.1). Because recovery of ground layer biomass after fire in open ecosystems is fast, surface fires tend to be frequent (every few years). Conversely, crown fires are associated with high fire intensity, burning most of the vegetation (shrubs and tree crowns), and are generally associated with tropical and temperate forests and

Table 2.1 Characteristics associated with fire regimes in major South American biomes

Biome	Rainfall seasonality	Aboveground biomass productivity	Dry fuel load	Fire type	Fire frequency	Fire intensity	Fire-related traits (woody)	Fire-related traits (herbs and subshrubs)
Tropical rainforests	Low/ intermediate	High	Low	Surface/ ground (rare)	Very low	Low/ medium	Absent	Absent
Tropical/ subtropical dry forests and shrublands	High	Intermediate	Low/ intermediate	Crown/ surface	Low	Medium/ high	Basal resprouting	Absent
Tropical savannas and grasslands	High	Intermediate	High	Surface	High	Low/ medium	Corky bark; woody underground organs; fire induced flowering; basal and epicormic resprouting	High flammability (grasses); fire induced flowering; resprouting from underground organs
Temperate forests	Low	High	Intermediate/ high	Crown	Low	High	Absent	Absent
Mediterranean vegetation	High	High	High	Crown	Low	High	Resprouting	Absent
Temperate grasslands and savannas	Low	Intermediate	High	Surface	Low	Low	Woody underground organs; resprouting	Basal resprouting
Floodplain grasslands and savannas	High	Intermediate	High	Surface/ ground	High	Low/ medium	Corky bark (rare); basal and epicormic resprouting	High flammability (grasses); resprouting; fire induced flowering (rare)
Deserts	High	Low	Low	Surface	Very low	Low	Absent	Absent

shrublands. Since most woody biomass is consumed or killed in crown fires, and its recovery is slow, fire return interval in these ecosystems can take many decades, specially if primary productivity is low (Keeley and Pausas 2022). Another type of fire observed in South American ecosystem is ground (turf) fire, associated with ecosystems in which the water table lies close to or above the soil surface (e.g., floodplain grasslands and savannas; Table 2.1). In these ecosystems, low oxygen concentrations resulting from soil waterlogging prevents organic matter decomposition and this material can fuel underground fires in unusual dry years, when water table depth increases (e.g., Damasceno-Junior et al. 2022). Plants inhabiting South American biomes where fire is frequent display a range of traits and strategies that evolved to cope with these fire regimes (Table 2.1), a topic that will be discussed later in this chapter.

In the next session we attempted to classify major South American ecoregions into Pyromes, reflecting current wildfire regime categories in order to discuss the extent to which traits and strategies of plant species in these ecoregions reflect modern or past fire regimes.

2.4 Classification of South American Ecoregions into Pyromes

Understanding natural fire regimes is fundamental to recognize the selective forces that have been favoring different levels and types of adaptations. Unfortunately, detailed information on historical fire frequency and intensity at broad spatial and temporal scales is generally lacking.

Previous studies have identified the main existing fire regimes at the global scale based on fire data derived from remote sensing (Archibald et al. 2013; Bowman et al. 2020; Pais et al. 2023). However, such approaches have some limitations to describe natural fire regimes. One limitation is the short period provided by fire data derived from satellite observations (c. 20 years), which may be insufficient to define historical fire regimes more precisely, particularly in ecosystems with longer fire return intervals. Another limitation is that these fire regime classifications are largely unable to distinguish natural ignitions (lightning) from man-made fires. Because fire regimes based on such framework have large anthropogenic influence, they are likely different from past fire regimes under which species evolved. Yet, these approaches can be used as a starting point to evaluate the extent to which plant traits observed in different regions are in equilibrium with current fire regimes or likely represent anachronistic features from past fire regimes.

Here, we used remote sensing data from 2000 to 2019 to identify the main large-scale fire regimes observed in South America ecoregions. An ecoregion is defined as infra biome regions characterized by distinct assembly of species and communities whose boundaries reflect their distribution prior to anthropogenic land use changes (Olson et al. 2001). Thus, the ecoregion likely represents true plant assemblages dominated by a particular vegetation type that evolved in equilibrium with a particular fire regime. Here we used the revised version Ecoregions2017©Resolve

(Dinerstein et al. 2017) that includes 115 South American ecoregions. We classified ecoregions into Pyromes, that is, large geographic regions showing similar fire regimes.

To reconstruct wildfire regimes at the ecoregion scale, we obtained data on fire incidence (number of hotspots per area) and intensity for all wildfire records in South America between 2000 and 2019 using the MODIS product MCD14ML (Giglio et al. 2016). To reduce anthropogenic influence, we excluded fires in areas with direct human influence using land use data from ESA Land Cover CCI project team and Defourny (2019), which comprises a land use map for 2015 at 300 m spatial resolution. Specifically, we excluded from our analyses fires in which more than 10% of the land cover in a circular 1 km buffer surrounding the fire consisted of human land uses, including the following classes: no data (class 0), croplands (classes 10 to 40), urban areas (class 190) and water bodies (class 210). Water bodies were excluded because in many areas (e.g. Amazon), human activity is especially high along river margins, and also because the presence of water bodies reduces fire activity per unit area. All other classes in this land cover product consisted of natural terrestrial ecosystems. Fire incidence per ecoregion was standardized by dividing the number of remaining fires (after excluding those in areas with substantial anthropogenic influence and water) by the area of the ecoregion that corresponded to natural vegetation (using the same criteria).

Fire intensity was calculated as the ecoregion level mean of the Fire Radiative Power of all the retained fires within the ecoregion (see Dantas and Pausas 2022 for more detail). We subsequently used hierarchical clustering on principal component axes in the ecoregion level fire data to classify South American ecoregions into Pyromes. Whenever possible, we compared the results with fire data from Mapbiomas Fogo v.2.0 (https://plataforma.brasil.mapbiomas.org/fogo; Alencar et al. 2022), available for Brazil only, and corrected eventual misclassification resulting from detection problems associated with the MODIS product. To be able to classify all ecoregions, we set fire intensity to zero in areas that did not burn during the period.

While the exclusion of fires near areas with direct anthropogenic activity likely contributed to reducing human influence in our ecoregion analysis, fire regimes presented here are likely to be substantially influenced by humans. Small scale anthropogenic activity (e.g., logging), intentional/criminal fires, as well as fires spreading from distant locations, are unlikely to have been completely excluded from our analysis. In fact, distinguishing natural ignitions (lightning) from man-made fires is a major challenge on large scale fire analyses that rely on satellite data. Some recent attempts to disentangle natural and anthropogenic-induced wildfires at the global (Janssen et al. 2023) and regional (Menezes et al. 2022) scales have revealed an overall reduced proportion of lightning ignited fires. Such approaches are promising since they can reveal the real influence and extent of natural wildfires on terrestrial ecosystems and provide a better understanding of ancient fire regimes.

2.4.1 Pyromes of Modern South America

We classified 115 South American ecoregions into three Pyromes (Fig. 2.1) based on fire incidence and intensity data from 2000 to 2019, as described in the previous section.

2.4.1.1 The Rare Pyrome

The Rare Pyrome represents ecosystems in which wildfires are extremely rare and with low intensity (Fig. 2.1). This generally corresponded to well-preserved rainforests and desert/xeric biomes, but also included some grassy ecosystems and dry forests. In this Pyrome, a lack of seasonality and either a high fuel moisture (rainforests) or low fuel load (dry ecosystems) result in very infrequent fires. Thus, fires, if any, in this Pyrome are likely to be of anthropogenic origin. The majority of ecoregions belonging to this class are lowland and upland moist forests, as well as flooded forests and mangroves, where persistence of high fuel-moisture throughout the year prevents fire ignition and spread. In addition to moist forests, the Rare Pyrome also includes large South American deserts (Atacama and Sechura), xeric shrublands on the coastal areas in the Pacific and northern S. America, where the

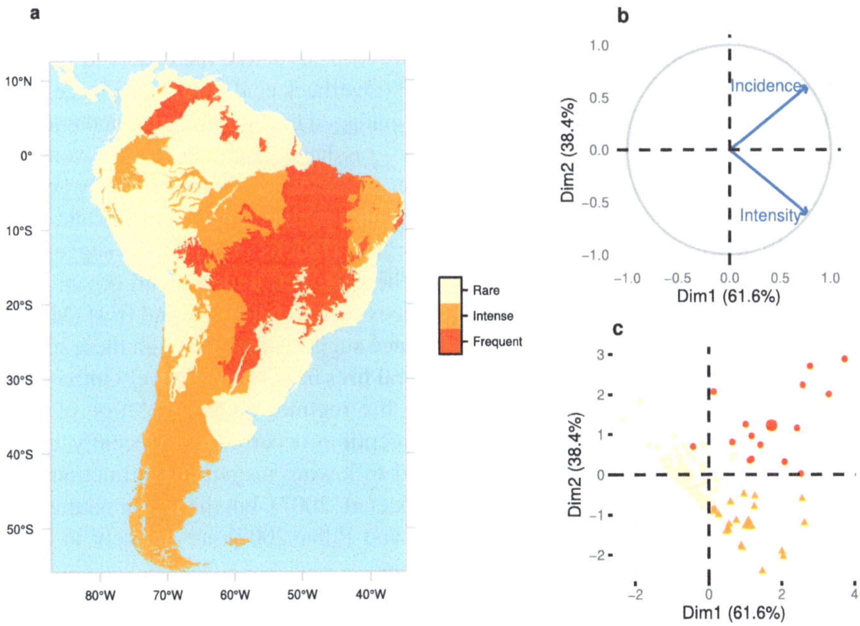

Fig. 2.1 (**a**) South American ecoregions classified into three major Pyromes, based on incidence and intensity of wildfire records between 2000 and 2019 using the product MODIS (MCD14ML; Giglio et al. 2016). (**b, c**) Hierarchical clustering of principal component axes (PCA) on ecoregion level fire data. Axes represent the principal components of fire regime variation (based on fire incidence and intensity) across South American ecoregions. Fires in areas with direct human influence were not considered in the analysis

lack of fuels strongly limits fires, as well as dry forests located in intermountain dry valleys (e.g., Marañón, Cauca, Magdalena, and Patía valleys).

Plants living in ecosystems in which fire is extremely rare, such as deserts or rainforests, generally lack fire-resistant traits. For instance, estimates of fire return intervals for Northern and Western Amazon forests suggest intervals of several hundreds of years (Bush et al. 2008; Feldpausch et al. 2022) and, as a consequence, most tree species are fire-sensitive. Accordingly, tree mortality after a low intensity surface fire in these Amazon forests is large enough to reduce tree biomass by more than 50% within 3 years after fire (Barlow et al. 2003). In these ecosystems, adaptations to deal with surface fires, such as thick bark protecting the base of the trunk (Staver et al. 2020) and post-fire resprouting are generally lacking.

Ecoregions corresponding to subtropical and temperate wet grasslands and savannas located in southern South America were also included in this Pyrome. Some of these grass-dominated ecosystems are special cases as they might represent relicts from past fire and herbivory regimes associated with the now extinct Neotropical megafauna (large mammal herbivores with over 45 kg). These ecosystems developed during the Pleistocene last glacial, a period characterized by drier and cooler climates that favored fire (Overbeck et al. 2007; Dantas and Pausas 2022). Thus, this current fire-free vegetation is not in equilibrium with current climatic conditions (absence of rainfall seasonality). For example, most subtropical and temperate grasslands in Southern South America (see Chap. 4) are currently characterized by the lack of a pronounced dry season, unlike the observed in fire-prone subtropical grasslands across the world (Overbeck et al. 2007). Evidence suggests that human intervention (livestock and managed fire) is currently fundamental to prevent forest encroachment within these grasslands (Behling 2002; Overbeck and Pfadenhauer 2007). Nevertheless, these ecosystems are characterized by herbaceous species with high resprouting capacity, including perennial grasses and subshrubs/forbs with developed underground organs (Fidelis et al. 2014), suggesting that fire could have been more frequent in the past. Yet, underground organs also seem to evolve in response to other disturbances such as herbivory and frost (Meller et al. 2022). Overall, palaeoecological evidence suggests that although these grasslands persisted for at least 40,000 years, natural fires have been relatively infrequent (Behling and Pillar 2007), so the impact of fire regimes on the evolution of fire-resistance traits in this region is uncertain. Exceptions occurred more recently, in the Holocene, in which fire frequency increased following megafauna extinction and human arrival during drier periods (Overbeck et al. 2007), but the recent nature and short-term duration of these fires (Behling and Pillar 2007) are unlikely to have allowed the *in situ* evolution of fire adaptations.

2.4.1.2 The Intense Pyrome

This Pyrome represents ecoregions that experience low/mid frequency of high intensity fires, generally corresponding to closed canopy ecosystems with periodic drought. It also included semi-arid open shrublands and woodlands located in both temperate and tropical climates. In these ecosystems, climatic conditions and competition for light with trees do not favor the development of a flammable herbaceous layer and, as a result, fires tend to be infrequent, especially compared to grass-dominated and highly flammable savannas. This Pyrome corresponds to regions experiencing substantial anthropogenic activity that account for most of the fires (Veblen et al. 1999; Bravo et al. 2010; Aragão et al. 2018). Deforestation, driving the spread of uncontrolled fires, especially during extreme climatic conditions, is likely to be the main cause of wildfires in this Pyrome. While in semi-arid forests and woodlands these fires occur after unusual wet periods, after fine fuels build up (e.g., Bravo et al. 2010), in moist forests, these fires occur under unusually dry conditions in which plant-derived fuels are dry and fires can easily spread. Under these extreme climatic conditions, many fires can also be ignited by lightning strikes (see Fig. 2.1c in Janssen et al. 2023). Moreover, many parts of this Pyrome are in close proximity with fire-prone savannas, which rather than forming a single sharp boundary with forest biomes, form intricate mosaics, implying that savanna fire can spread more easily to the forest under such conditions. In fact, a long dry season do not seem to be necessary for fires to occur, and even relatively moist forests with a short dry season experience natural fires when in close proximity with flammable vegetation (e.g., Flores et al. 2016; Pontes-Lopes et al. 2021). Regardless of the ignition source, the extreme climate required for fires to spread in these regions and the high accumulation of fuel loads caused by a low fire frequency imply that fires are often intense and, thus, we called this Pyrome "Intense" (Fig. 2.1).

The Intense Pyrome comprises seasonally dry tropical and subtropical forests and shrublands (Dry Chaco, Caatinga, and Espinal), tropical moist forests under a seasonal climate (Madeira-Tapajós and Tapajós-Xingu moist forests), and temperate vegetation, including the mediterranean-climate type ecosystems of Chile (Fig. 2.1; Chap. 11). Unexpectedly, two rainforest ecoregions were assigned to this Pyrome (Caqueta and Napo moist forests). This could be explained by the presence of savanna patches in these ecoregions, and their close proximity to the Llanos savanna ecoregion, from which fires could eventually spread in dry years, as well as the occurrence of anthropogenic fires in more isolated forest areas.

In the Intense Pyrome, plants generally lack specialized fire resistance mechanisms, with numerous trees relying on basal resprouting as a post-fire regeneration mechanism. Yet, while many trees are top-killed and resprout from basal buds after fire (Sampaio et al. 1993; Hoffmann et al. 2009; Brando et al. 2012; De Faria et al. 2021), the levels of adult tree stem survival are much higher than in the Rare Pyrome (Gurvich et al. 2005; Hoffmann et al. 2009; Balch et al. 2011; Mostacedo et al. 2022). Accordingly, there is evidence that trees from Amazonian regions classified here to be part of the Intense Pyrome (Southeastern Amazon), are characterized by relatively thicker bark (compared to rainforests of the Rare Pyrome, but thinner than in savannas of the Frequent Pyrome), therefore, reducing the probability of top-kill (Hoffmann et al. 2009; Staver et al. 2020). This pattern is consistent with the idea that at least some natural fires occurred in the region, possibly spreading from nearby savannas. Evidence that forest species can develop thicker barks when living in close proximity to savannas exist for riparian forests found in South American savanna regions (Hoffmann et al. 2009). The thicker barks found in Southeastern Amazon could also be a legacy from the Pleistocene, when evidence suggest that a large fraction of the Intense Pyrome used to be savannas, before megafauna extinction (Dantas and Pausas 2022). The extent to which relatively thicker barks are restricted to Southeastern Amazon or are a general feature of the Intense Pyrome needs to be further investigated.

The Intense Pyrome also included regions with mediterranean-type climate in Central Chile. Mediterranean-type climate regions in the world are characterized by dry summers and experience low frequency (10–100 years) crown-fires, which have been recognized as an important selection driver in assembling their modern floras (Rundel et al. 2016). However, central Chile is an exception to mediterranean-type climate biomes since natural fires are far less frequent in that region than elsewhere. In contrast to other mediterranean-type climate regions, the flora of central Chile lacks fire-related traits, such as serotiny, present in other Mediterranean regions, although plants have resprouting capacity (Rundel et al. 2018) and smoke-stimulated germination (in a limited number of plant groups; Keeley and Pausas 2022). Current fire activity in Central Chile is strongly related to tree plantations and most ignitions derive from anthropogenic sources, although lightning strikes also play a minor role (Gómez-González et al. 2019; Úbeda and Sarricolea 2016). A similar pattern is found in the Dry Chaco, a South American seasonally dry forest that is not currently subject to frequent natural fires, experiencing mostly human-ignited fires (Bravo et al. 2010; Argañaraz et al. 2015; Vidal-Riveros et al. 2023). Evidence based on a representative group of plant species from the Chaco shows limited representation of fire-related traits, except for species with high basal resprouting capacity and heat-tolerant seeds (Jaureguiberry and Díaz 2023; Chap. 8). In addition, bark thickness seems to be involved in fire resistance in some species (Bravo et al. 2014), although barks of Chaco trees are not as thick as those found in the Frequent Pyrome. While climate in the Chaco is seasonal, and evidences suggest that the Dry Chaco might have been a grass-dominated savanna in the past, extinct large herbivores may have had a more prominent role in maintaining the vegetation open compare to fire (Dantas and Pausas 2022), explaining the paucity of more specialized fire adaptations in the woody flora. In fact, traits of woody species in the Chaco are more similar to those found in savannas maintained by large mammal herbivores in Africa, in which plants are more reliant on basal resprouting as a fire escape strategy (Dantas and Pausas 2013, 2020).

2.4.1.3 The Frequent Pyrome

The third and last Pyrome comprises ecoregions that experience frequent low/medium intensity surface fires. More localized ground fires can also occur in seasonally flooded grasslands and swamps, environments that accumulate great amounts of non-decomposed dead biomass belowground and burn in unusually dry years. This Pyrome corresponds to the main tropical seasonally dry grasslands and savannas in the continent, like the Brazilian Cerrado, the Humid Chaco, and Llanos in Venezuela and Colombia, including some floodplain grasslands and savannas (e.g., Beni savannas and the Pantanal) and transitional zones. A combination of abundant grass cover (high fuel load and continuity, and low fuel bulk density) and seasonal climate (low fuel moisture) make the vegetation of these ecosystems highly flammable and, thus, fire frequency tends to be high. For example, in large savanna reserves, natural fires ignited by lightning are frequent, particularly during the beginning of rainy season (Ramos-Neto and Pivello 2000; Medeiros and Fiedler 2004; Schumacher et al. 2022). These fires are, however, of low intensity (Haas et al. 2022) as grasses, the main fuel of these surface fires, incinerate quickly while the woody flora has very low flammability (Zanzarini et al. 2022), resulting in low heat release, low fire residence time and short flame heights (Castro and Kauffman 1998). However, many if not most of the fires in these savannas are currently ignited by humans and, thus, occur in the dry season, rather than under the moister conditions that often characterize lightning ignited fires (Pivello 2011). While this change in ignition source likely increased fire intensity, whether fire frequency was also affected is less clear. Because of the typically high fire frequency of this regions, we called this Pyrome "Frequent" (Fig. 2.1). The Guianan savannas, which are isolated patches of Amazonian savannas with low anthropogenic activity surrounded by rainforests, although originally grouping within the Intense Pyrome in our ordination analysis (red spot on the left of Fig. 2.1c), were considered here as belonging to the Frequent Pyrome based on large and frequent burned areas, as revealed in the Mapbiomas fire dataset (Alencar et al. 2022).

The maintenance of grasslands and savannas in this Pyrome, in most cases, occurs because frequent fires prevent forest dominance, despite of a mesic climate, restricting forest distribution to small patches associated with water bodies, fertile soils and topographic complexity (Dantas et al. 2016; Bernardino et al. 2022). Thus, coexisting forests and savannas in South America are largely considered to be alternative biome states across tropical landscapes (e.g., Staver et al. 2011; Dantas et al. 2016), where the prevalence of one vegetation type or another depends on the local fire regime as influenced by vegetation, and conditions favoring forest recovery (e.g., soil fertility). As climate becomes more seasonal in the tropics, both fire frequency and intensity tend to increase. At a given point, fire frequency increases to such an extent that forest trees become unable to recover biomass between fire events, resulting in increasing gap formation and forest dieback, allowing invasion by shade intolerant flammable C_4 grasses (Hoffmann et al. 2012; De Faria et al. 2021). This positive feedback between grasses and fire increases the probability of a transition to a savanna-type fire regime, characterized by frequent low intensity grass-fueled surface fires. As a result, savanna-forest mosaics become widespread, with increasing dominance of savannas as rainfall seasonality increases. Savannas will become the dominant vegetation up to the point that aridity becomes a constraint to grass spread. However, the exact threshold after which grasses become sparse differs among continents, with grassy vegetation being common under much drier conditions in other tropical continents compared to South America (Lehmann et al. 2011).

Under the mesic, but seasonally dry conditions that characterize the climatic envelope of Neotropical savannas, fire becomes highly frequent. As a result, shrubs and tree species that have a thick corky bark on the main stem and branches are favored and become dominant in the woody flora (Hoffmann et al. 2009; Dantas and Pausas 2013). Tree barks are formed by two main parts with distinct functions. The inner bark part consists of the secondary phloem, is formed by living cells, and is involved in physiological functions and resource storage, whereas the outer bark is formed by dead cells and is mainly involved in temperature insulation. It is the thickening of the outer bark that differentiates savannas from forest species (Dantas et al. 2024), supporting the idea that selection by fire is a main driver of allocation to bark in savanna species. A thick outer bark protects the cambium and buds from heating, allowing for epicormic resprouting after fire (Souchie et al. 2017), greatly enhancing the probability of survival in the flammable grass matrix. As a result, post-fire mortality among woody species is generally low (4–22.5% of the individuals; Medeiros and Miranda 2008, Souchie et al. 2017). Evidence suggests that shade is a fundamental constraint on allocation to bark, because combining fire and shade tolerance implies high costs to plants (Dantas et al. 2024). Thus, savanna tree species are generally shade-intolerant and, therefore, depend on recurrent fires to prevent wood encroachment and canopy development (Dantas et al. 2016).

The prevalence of open canopy conditions in the Frequent Pyrome allowed the spread of numerous shade-intolerant herbs (including grasses and sedges) and subshrubs, all of which have a particular strategy to deal with fires (Simon and Pennington 2012; Dantas and Pausas 2020; Pilon et al. 2021). The herbaceous-subshrub stratum includes many resprouters that keep their bud banks and reserve organs protected underground (rhizomes, xylopodia and lignotubers; Pausas et al. 2018) or at the soil level, which allow for post-fire resprouting (Fig. 2.2). Moreover, some herbs and shrubs, such as those from the Eriocaulaceae and Velloziaceae families, protect their aerial stems by retaining the base of thick lignified leaves growing around them, which play an analogous role to the thick insulating bark of woody species. Some Velloziaceae species also show vegetative (clonal) spread, regrowing towards the ground and rooting to produce new genets (Maracahipes et al. 2024). A subset of herbs and subshrubs, while having fire-sensitive aerial parts, can resprout and reproduce quickly after fires (Pilon et al. 2018; Damasceno-Junior et al. 2022; Fidelis and Zirondi 2021) and, therefore, allow their descendants to colonize the post-fire environment (Fig. 2.3). In savannas, soil temperature increases very little during fires and, therefore, soil represents an important insulating mechanism. This strategy is different from the seeder strategy with serotine cones found in many mediterranean-type climate ecosystems, in which plants have fire-stimulated release of dormant seed stored in the canopy.

A remarkable evolutionary trend observed is the savanna flora is the tendency for a change in growth form associated with habitat shift from fire-free forests into fire-prone savanas. This pattern is evident in evolutionary reconstructions of disparate plant lineages such as *Andira* (Fabaceae), Cocoseae (Palmae), *Jacaranda* (Bignoniaceae), *Myrcia* (Myrtaceae), *Pradosia* (Sapotaceae), *Simaba* (Simaroubaceae), *Stryphodendron* (Fabaceae), where fire-adapted geoxylic (or acaulescent in palms) habit evolved from tree ancestrals to withstand the frequent fire regime of the Cerrado savannas (Gottsberger and Silberbauer-Gottsberger 2006; Simon and Pennington 2012; Terra-Araújo et al. 2015; Simon et al. 2016; Santos et al. 2017; Devecchi et al. 2018; Ragsac et al. 2019; Cássia-Silva et al. 2022). Similarly, the acquisition of a thick bark, a common fire-tolerance strategy in a

Fig. 2.2 Diversity of woody underground structures in subshrubs/forbs from fire-prone South American savannas. These structures allow resprouting after aerial parts are consumed by fire. (**a**) *Manihot porphyrantha* M.Mend. & T.B.Cavalc., (**b**) *Mimosa diminuta* Marc.F.Simon & C.E.Hughes, (**c**) *Myrcia tocantinensis* Glaz. ex P.O.Rosa & Proença, (**d**) *Stryphnodendron heringeri* Occhioni, (**e**) *Ctenodon simplicifolius* (G.P.Lewis) D.B.O.S.Cardoso & H.C.Lima, (**f**) *Casselia chamaedryfolia* Cham., (**g**) *Esenbeckia pumila* Pohl. Scale bar: 10 cm. (Photos: Marcelo Simon)

Fig. 2.3 Mass flowering of *Paepalanthus canastrensis* Silveira after a wildfire in Serra da Canastra National Park, Brazil, in September 2010. Plants were able to synchronously resprout and flower only a few days after fire. (Photo: Vinícius Dantas)

number of unrelated lineages of South American savanna trees is also associated with evolutionary habitat shift from forest to savanna. Thick, deeply fissured outer barks of savanna trees largely contrasts with the thin barks of their close relatives that inhabit fire-free rainforests (Hoffmann et al. 2012; Dantas et al. 2013a, Fig. 2.4).

The presence of attributes such as a thick corky bark and a geoxylic growth form in savannas is more pronounced in South America than in other tropical savannas in the world (Lawes et al. 2013; Dantas and Pausas 2013, 2020; Pellegrini et al. 2017). Differences in relation to Africa are probably explained by the different megafauna histories of the savannas in each continent (i.e., a lack of megafauna in South America in the last 10,000 years), resulting in a greater dominance of fire-maintained savannas in South America in detriment of the herbivory-maintained savannas that are widespread in Africa (Dantas and Pausas 2020, 2022). In the latter, plants tend to escape fires mainly via basal resprouting and canopy escape (high stem height-to-diameter ratios), rather than epicormic resprouting, as this strategy protects from both fire and herbivory (Dantas and Pausas 2013). In contrast, differences between South American and Australian savanna woody species probably result from a distinct mechanism of bud protection among Australian eucalypts (dominant woody elements), in which buds are hidden within the wood, allowing plants to reduce allocation to bark thickness and maintain tall trunks (Lawes et al. 2013; Dantas and Pausas 2013). Despite of these differences, savanna plants generally display thicker barks compared to forest species in all tropical continents (Pellegrini et al. 2017), allowing for greater post-fire survival.

Fig. 2.4 Five congeneric pairs illustrating the difference in bark thickness between fire-prone savanna and fire-free rainforest trees. Top row (savanna): *Couepia grandiflora* (Mart. & Zucc.) Benth., *Eschweilera nana* (O.Berg) Miers, *Machaerium opacum* Vogel, *Miconia ferruginata* DC., *Vochysia thyrsoidea* Pohl. Bottom row (rainforest): *Couepia guianensis* Aubl., *Eschweilera obversa* (O.Berg) Miers, *Machaerium aculeatum* Raddi, *Miconia poeppigii* Triana, *Vochysia ferruginea* Mart. (Photos: Marcelo Simon, except for *Miconia poeppigii* by Glocimar Pereira-Silva)

2.5 Modern Fire Regimes and Their Potential Impacts on Different Ecosystems

As discussed in this chapter, plant traits and strategies are the result of millions of years of evolution in response to particular fire regimes. For much of this time, humans had no or limited influence on these fire regimes. Instead, these regimes were primarily shaped by atmospheric oxygen concentration, ignition sources, climate, and vegetation. Nowadays, terrestrial ecosystems are strongly influenced by human activities and, in many ecosystems, especially in highly managed landscapes, fire regimes are likely different from historical benchmarks (Bowman et al. 2011; Archibald et al. 2013). In addition, recent trends associated with climate change indicate an increased flammability across most ecosystems in tropical South America (Feron et al. 2024). Such recent changes in fire intensity and frequency may have profound negative consequences for species that became dependent on historical fire regimes. Here, we discuss the extent to which anthropogenic fire regimes have been negatively affecting native species in some South American ecosystems by changing the selective pressure towards different plant traits.

Human intervention on fire regimes can result in either increasing or inhibiting wildfires. Perhaps contrary to the general perception, there is a trend of decline in the incidence of wildfires on a global scale over the last two decades (Archibald et al. 2013; Doerr and Santín 2016). This means that current human activities are reducing total burned area and fire intensity on a broad scale. Anthropogenic factors, such as density of human settlements and land use change, which reduce extension and connectivity of flammable vegetation, are likely to reduce fire activity (Forkel et al. 2019; Bowman et al. 2020). This is especially the case of some fire-prone ecosystems, such as tropical savannas (Criado et al. 2020), and can also be observed in South American savannas (e.g., the Brazilian Cerrado ecoregion; Fig. 2.5).

Declining fire frequencies in South America have been suggested as an important driver of woody plant encroachment within the Frequent Pyrome. For instance, evidence suggests that 19% of the native Brazilian Cerrado vegetation (including transitions within the surrounding forest ecoregions) have experienced woody encroachment in the last 15 years, with reduced fire frequency explaining 34% of

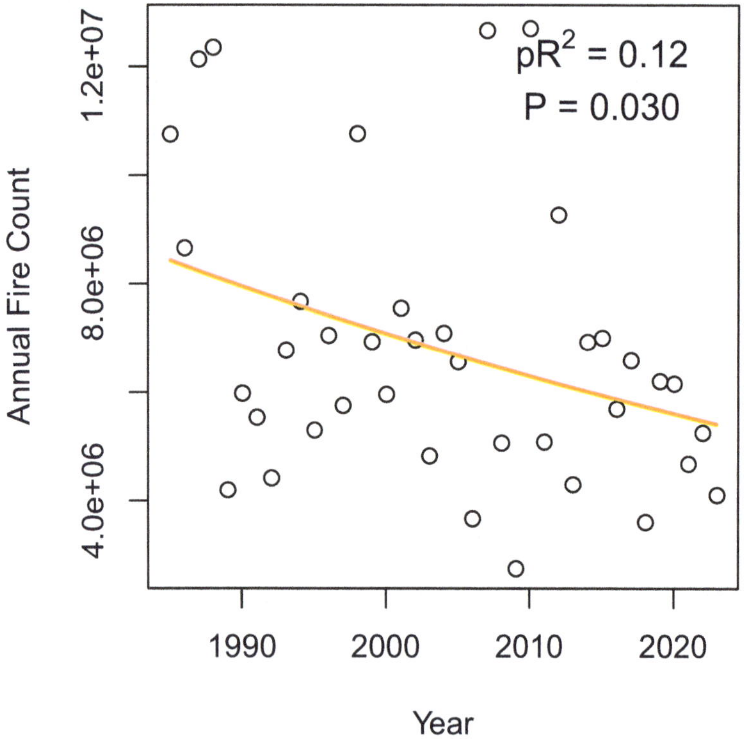

Fig. 2.5 Annual fire counts in savanna vegetation within the Brazilian Cerrado ecoregion between 1985 and 2023, according to data from the Mapbiomas dataset (Collection 3; 30 m spatial resolution). The results are based on generalized linear models (family quasipoisson) and show a trend towards declining fire frequency. pR^2 means pseudo R^2

the variability (Rosan et al. 2019). In managed savanna areas in Brazil, several decades of fire suppression have resulted in increased tree cover, leading to reduced light availability for shade-intolerant ground layer plants, such as grasses and forbs, as well as endemic savanna woody species (Abreu et al. 2017). This process of woody encroachment reduces savanna species richness, changes species composition and, ultimately, ecosystem function, resulting in the exclusion of savanna endemic species that depend on high light incidence maintained by regular fires to survive (see Chap. 5).

Conversely, increases in fire activity are expected for South American savannas in response to climate change (Silva et al. 2019). Evidence suggests that increasing the frequency of fire to an annual burning regime reduces tree basal area and tree height among savanna woody species (Dantas et al. 2013b). Increased fire frequency also reduces bark thickness and leaf nutrient content of the plants at the intraspecific level, suggesting that savanna woody plants may become increasingly more vulnerable to fire as fire frequency increases (thinner bark and lower storage of resources on plant tissues). Eventually, the reserve tissues that fuel post-fire resprouting may become insufficient to allow future resprouting, resulting in an increase in treeless landscapes. On the other hand, species tend to produce leaves with lower specific leaf areas under increasing fire frequency (Dantas et al. 2013b), a trait associated with reduced leaf flammability (Pausas et al. 2017). Moreover, increased fire frequencies tend to reduce fuel loads and continuity, resulting in fires of lower intensity and residence time (Rodrigues et al. 2021).

Changes in fire activity in response to climate change can also affect fire-sensitive vegetation within the savanna matrix, such as moist gallery forests. Under normal conditions, fire spreading from savannas does not penetrate the forest, given a lack of flammable herbaceous fuel loads and the high litter moisture content. However, reduced precipitation and more frequent droughts may cause fires to spread from grasslands/savannas to nearby gallery forests, resulting in highly destructive effects in this fire-sensitive vegetation (Pivello et al. 2021). Considering the complexity associated with the contrasting effects of climate change (increasing fire activity because extreme weather events) and land use change (reducing it) in savanna fire regimes, the fate of the savanna biodiversity under climate and land use changes remains uncertain.

While the recent trend in some savannas of the Frequent Pyrome seems to have been of decreasing fire activity (Fig. 2.5), largely due to fire suppression policies and habitat fragmentation (e.g., Abreu et al. 2017), fire activity has been increasing in many fire-sensitive rainforest regions in the Amazon. This increase has several drivers, such as logging, forest degradation, increasing edge effects and ignitions, invasion by flammable grasses, and increased drought frequency and severity (Aragão et al. 2018; De Faria et al. 2021; Alencar et al. 2022; Feron et al. 2024). Moreover, abandoned pastures in rainforest areas may not spontaneously return to a forest state, but rather, shift to a fire-maintained state dominated by thick-barked savanna trees and a dense grass cover that increase fire spread to the surrounding forests (Sansevero et al. 2020). Recurrent fires, in turn, tend to promote fire adapted and flammable species, further propagating forest degradation, an effect that is expected to become especially relevant with climate change (De Faria et al. 2021). In many areas, fire return intervals are already shorter than the time required for forest recovery, making these areas highly vulnerable to regime shifts resulting from the triggering of endogenous vegetation-fire feedbacks (De Faria et al. 2021). Considering the broad dependence of the hydrologic cycle on forest cover, this process could have cascading effects by creating climatic conditions that are more conductive to natural fires. Overall, climate change is predicted to directly increase both fire frequency and intensity in tropical and temperate rainforests (De Faria et al. 2017; Fonseca et al. 2019; Kitzberger et al. 2022; Feron et al. 2024). The increased fire frequency and intensity predicted for these forests will impose an important threat to endemic forest species, as most of these species lack fire adaptations.

It is clear that recent changes in fire regimes are affecting the biota of natural ecosystems by selecting particular traits that confer higher survival. In some regions, individuals and species are maladapted to current fire regimes, given that they evolved during pre-human fire regimes, which means that traits selected for in the past are no longer advantageous in the current conditions. For example, species that depend on fire for flowering, such as some herbs and subshrubs frequently found in South American tropical savannas and grasslands (e.g., Fidelis and Zirondi 2021) would have their reproductive output severely affected if fire is artificially suppressed. Overall, a key aspect that must be considered is that the persistence of novel fire regimes might select for different plant traits over time, and eventually alter the functioning of an entire ecosystem.

One critical point involving the conservation of remaining natural environments and associated biota are fire management and policy. Although there are many aspects that must be considered in fire management strategies and decision making (human health and safety, economic losses, air pollution, atmospheric emissions, etc.; Bowman et al. 2020), managers and politicians must recognize that plants and fire have an ancient history on Earth. Fire regimes that are more similar to natural/historical regimes would be required to maintain biologically and functionally diverse ecosystems.

2.6 Conclusion

The high climatic, topographic and vegetation heterogeneity of South America in time and space (e.g., Mayle et al. 2009; Jaramillo 2023) resulted in a variety of natural fire regimes. Some of these fire regimes have occurred for sufficient time to allow the evolution of an associated fire-adapted flora, whereas others seem to be more recent and there is a mismatch between the fire regime and the flora. While fire regimes have likely changed from historical benchmarks as anthropogenic pressure increased, South American ecoregions, as classified in our Pyrome framework, generally differ in the frequency of species and individuals bearing fire-related traits. Ecosystems where fire has been a recurrent element over evolutionary time scale (millions of years) tend to have more plants with specialized traits. In these ecosystems, fire is fundamental to maintain their high diversity of fire-adapted and fire-dependent plants. On the other hand, ecosystems in which fire is infrequent but often intense, plants tend to have less specialized adaptations that are useful against fire as well as other disturbances (e.g., basal resprouting). Linking the origin of these traits with fire is more difficult especially as there is uncertainty on the extent to which recent fire regimes are similar to ancient fire regimes. Finally, ecosystems where fire has been historically absent or in which fire is the result of recent (few thousand years or less) anthropogenic influence tend to have a limited proportion of fire related traits in their biota, because of insufficient time for plant evolution. Recent anthropogenic modifications of fire regimes may represent important threats to both fire-prone and fire-sensitive species across biomes, which means that efforts to understand natural fire regimes and associated fire-adaptations are fundamental to guide conservation and management efforts.

References

Abreu RC, Hoffmann WA, Vasconcelos HL, Pilon NA, Rossatto DR, Durigan G (2017) The biodiversity cost of carbon sequestration in tropical savanna. Sci Adv 3(8):e1701284. https://doi.org/10.1126/sciadv.1701284

Alencar AA, Arruda VL, Silva WVD, Conciani DE, Costa DP et al (2022) Long-term Landsat-based monthly burned area dataset for the Brazilian biomes using deep learning. Remote Sens 14:2510. https://doi.org/10.3390/rs14112510

Aragão LE, Anderson LO, Fonseca MG, Rosan TM, Vedovato LB et al (2018) 21st Century drought-related fires counteract the decline of Amazon deforestation carbon emissions. Nat Commun 9:536. https://doi.org/10.1038/s41467-017-02771-y

Archibald S, Lehmann CE, Gómez-Dans JL, Bradstock RA (2013) Defining pyromes and global syndromes of fire regimes. Proc Natl Acad Sci USA 110:6442–6447. https://doi.org/10.1073/pnas.1211466110

Archibald S, Lehmann CE, Belcher CM, Bond WJ, Bradstock RA et al (2018) Biological and geophysical feedbacks with fire in the Earth system. Environ Res Lett 13:033003. https://doi.org/10.1088/1748-9326/aa9ead

Argañaraz JP, Pizarro GG, Zak M, Landi MA, Bellis LM (2015) Human and biophysical drivers of fires in Semiarid Chaco mountains of Central Argentina. Sci Total Environ 520:1–12. https://doi.org/10.1016/j.scitotenv.2015.02.081

Azevedo JA, Collevatti RG, Jaramillo CA, Strömberg CA, Guedes TB et al (2020) On the young savannas in the land of ancient forests. In: Rull V, Carnaval AC (eds) Neotropical diversification: patterns and processes. Springer Nature, Cham, pp 271–298. https://doi.org/10.1007/978-3-030-31167-4_12

Balch JK, Nepstad DC, Curran LM, Brando PM, Portela O et al (2011) Size, species, and fire behavior predict tree and liana mortality from experimental burns in the Brazilian Amazon. For Ecol Manag 261:68–77. https://doi.org/10.1016/j.foreco.2010.09.029

Barlow J, Lagan BO, Peres CA (2003) Morphological correlates of fire-induced tree mortality in a central Amazonian forest. J Trop Ecol 19:291–299. https://doi.org/10.1017/S0266467403003328

Beerling DJ, Osborne CP (2006) The origin of the savanna biome. Glob Change Biol 12:2023–2031. https://doi.org/10.1111/j.1365-2486.2006.01239.x

Behling H (2002) South and southeast Brazilian grasslands during Late Quaternary times: a synthesis. Palaeogeogr Palaeoclimatol Palaeoecol 177:19–27. https://doi.org/10.1016/S0031-0182(01)00349-2

Behling H, Pillar VD (2007) Late Quaternary vegetation, biodiversity and fire dynamics on the southern Brazilian highland and their implication for conservation and management of modern Araucaria forest and grassland ecosystems. Philos Trans R Soc Lond Ser B Biol Sci 362:243–251. https://doi.org/10.1098/rstb.2006.1984

Belcher CM, Collinson ME, Scott AC (2013) A 450-million-year history of fire. In: Belcher CM (ed) Fire phenomena and the Earth system: an interdisciplinary guide to fire science. Wiley, Chichester, pp 229–249. https://doi.org/10.1002/9781118529539.ch12

Belcher CM, Mills BJ, Vitali R, Baker SJ, Lenton TM, Watson AJ (2021) The rise of angiosperms strengthened fire feedbacks and improved the regulation of atmospheric oxygen. Nat Commun 12:503. https://doi.org/10.1038/s41467-020-20772-2

Bernardino PN, Dantas VL, Hirota M, Pausas JG, Oliveira RS (2022) Savanna-forest coexistence across a fire gradient. Ecosystems 25:279–290. https://doi.org/10.1007/s10021-021-00654-4

Bond WJ (2015) Fires in the Cenozoic: a late flowering of flammable ecosystems. Front Plant Sci 5:749. https://doi.org/10.3389/fpls.2014.00749

Bond WJ, Scott AC (2010) Fire and the spread of flowering plants in the Cretaceous. New Phytol 188:1137–1150. https://doi.org/10.1111/j.1469-8137.2010.03418.x

Bowman DM, Balch J, Artaxo P, Bond WJ, Cochrane MA et al (2011) The human dimension of fire regimes on Earth. J Biogeogr 38:2223–2236. https://doi.org/10.1111/j.1365-2699.2011.02595.x

Bowman DM, Kolden CA, Abatzoglou JT, Johnston FH, Werf GR, Flannigan M (2020) Vegetation fires in the Anthropocene. Nat Rev Earth Environ 1:500–515. https://doi.org/10.1038/s43017-020-0085-3

Bradshaw SD, Dixon KW, Hopper SD, Lambers H, Turner SR (2011) Little evidence for fire-adapted plant traits in Mediterranean climate regions. Trends Plant Sci 16:69–76. https://doi.org/10.1016/j.tplants.2010.10.007

Brando PM, Nepstad DC, Balch JK, Bolker B, Christman MC, Coe M, Putz FE (2012) Fire-induced tree mortality in a neotropical forest: the roles of bark traits, tree size, wood density and fire behavior. Glob Change Biol 18:630–641. https://doi.org/10.1111/j.1365-2486.2011.02533.x

Bravo S, Kunst C, Grau R, Aráoz E (2010) Fire-rainfall relationships in Argentine Chaco savannas. J Arid Environ 74:1319–1323. https://doi.org/10.1016/j.jaridenv.2010.04.010

Bravo S, Kunst C, Leiva M, Ledesma R (2014) Response of hardwood tree regeneration to surface fires, western Chaco region, Argentina. For Ecol Manag 326:36–45. https://doi.org/10.1016/j.foreco.2014.04.009

Bush MB, Silman MR, McMichael C, Saatchi S (2008) Fire, climate change and biodiversity in Amazonia: a Late-Holocene perspective. Philos Trans R Soc B Biol Sci 363(1498):1795–1802. https://doi.org/10.1098/rstb.2007.0014

Cássia-Silva C, Oliveira RS, Sales LP, Freitas CG, Jardim L et al (2022) Acaulescence promotes speciation and shapes the distribution patterns of palms in neotropical seasonally dry habitats. Ecography 2022:06072. https://doi.org/10.1111/ecog.06072

Castro EA, Kauffman JB (1998) Ecosystem structure in the Brazilian Cerrado: a vegetation gradient of aboveground biomass, root mass and consumption by fire. J Trop Ecol 14:263–283. https://doi.org/10.1017/S0266467498000212

Charles-Dominique T, Beckett H, Midgley GF, Bond WJ (2015) Bud protection: a key trait for species sorting in a forest-savanna mosaic. New Phytol 207:1052–1060. https://doi.org/10.1111/nph.13406

Clarke PJ, Lawes MJ, Midgley JJ, Lamont BB, Ojeda F et al (2013) Resprouting as a key functional trait: how buds, protection and resources drive persistence after fire. New Phytol 197:19–35. https://doi.org/10.1111/nph.12001

Cochrane MA (2013) Current fire regimes, impacts and the likely changes—V: tropical South America. In: Goldammer JG (ed) Vegetation fires and global change: challenges for concerted international action. Kessel Publishing House, Remagen-Oberwinter, pp 101–114

Criado MG, Myers-Smith IH, Bjorkman AD, Lehmann CE, Stevens N (2020) Woody plant encroachment intensifies under climate change across tundra and savanna biomes. Glob Ecol Biogeogr 29:925–943. https://doi.org/10.1111/geb.13072

Crisp MD, Cook LG (2013) How was the Australian flora assembled over the last 65 million years? A molecular phylogenetic perspective. Annu Rev Ecol Evol Syst 44:303–324. https://doi.org/10.1146/annurev-ecolsys-110512-135910
Crisp MD, Burrows GE, Cook LG, Thornhill AH, Bowman DM (2011) Flammable biomes dominated by eucalypts originated at the Cretaceous-Palaeogene boundary. Nat Commun 2:193. https://doi.org/10.1038/ncomms1191
Damasceno-Junior GA, Pereira ADMM, Oldeland J, Parolin P, Pott A (2022) Fire, flood and Pantanal vegetation. In: Damasceno-Junior GA, Pott A (eds) Flora and vegetation of the Pantanal wetland. Springer International Publishing, Cham, pp 661–688. https://doi.org/10.1007/978-3-030-83375-6_18
Dantas VL, Pausas JG (2013) The lanky and the corky: fire-escape strategies in savanna woody species. J Ecol 101:1265–1272. https://doi.org/10.1111/1365-2745.12118
Dantas VL, Pausas JG (2020) Megafauna biogeography explains plant functional trait variability in the tropics. Glob Ecol Biogeogr 29:1288–1298. https://doi.org/10.1111/geb.13111
Dantas VL, Pausas JG (2022) The legacy of the extinct neotropical megafauna on plants and biomes. Nat Commun 13:129. https://doi.org/10.1038/s41467-021-27749-9
Dantas VL, Batalha MA, Pausas JG (2013a) Fire drives functional thresholds on the savanna-forest transition. Ecology 94:2454–2463. https://doi.org/10.1890/12-1629.1
Dantas VL, Pausas JG, Batalha MA, Paula Loiola PD, Cianciaruso MV (2013b) The role of fire in structuring trait variability in neotropical savannas. Oecologia 171:487–494. https://doi.org/10.1007/s00442-012-2431-8
Dantas VL, Hirota M, Oliveira RS, Pausas JG (2016) Disturbance maintains alternative biome states. Ecol Lett 19:12–19. https://doi.org/10.1111/ele.12537
Dantas VL, Oliveira LCS, Marcati CR, Sonsin-Oliveira J (2024) Coordination of bark and wood traits underlies forest-to-savanna evolutionary transitions. J Biogeogr. https://doi.org/10.1111/jbi.14850
De Faria BL, Brando PM, Macedo MN, Panday PK, Soares-Filho BS, Coe MT (2017) Current and future patterns of fire-induced forest degradation in Amazonia. Environ Res Lett 12:095005. https://doi.org/10.1088/1748-9326/aa69ce
De Faria BL, Staal A, Silva CA, Martin PA, Panday PK, Dantas VL (2021) Climate change and deforestation increase the vulnerability of Amazonian forests to post-fire grass invasion. Glob Ecol Biogeogr 30:2368–2381. https://doi.org/10.1111/geb.13388
Devecchi MF, Thomas WW, Plunkett GM, Pirani JR (2018) Testing the monophyly of *Simaba* (Simaroubaceae): evidence from five molecular regions and morphology. Mol Phylogenet Evol 120:63–82. https://doi.org/10.1016/j.ympev.2017.11.024
Dinerstein E, Olson D, Joshi A, Vynne C, Burgess ND et al (2017) An ecoregion-based approach to protecting half the terrestrial realm. Bioscience 67:534–545. https://doi.org/10.1093/biosci/bix014
Doerr SH, Santín C (2016) Global trends in wildfire and its impacts: perceptions versus realities in a changing world. Philos Trans R Soc B Biol Sci 371:20150345. https://doi.org/10.1098/rstb.2015.0345
Edwards EJ, Osborne CP, Strömberg CA, Smith SA, C4 Grasses Consortium et al (2010) The origins of C4 grasslands: integrating evolutionary and ecosystem science. Science 328:587–591. https://doi.org/10.1126/science.1177216
ESA Land Cover CCI project team, Defourny P (2019) ESA Land Cover Climate Change Initiative (Land_Cover_cci): global land cover maps, version 2.0.7. Centre for Environmental Data Analysis. https://catalogue.ceda.ac.uk/uuid/b382ebe6679d44b8b0e68ea4ef4b701c. Accessed 21 Sept 2022
Eva HD, Belward AS, Miranda EE, Bella CM, Gond V et al (2004) A land cover map of South America. Glob Change Biol 10:731–744. https://doi.org/10.1111/j.1529-8817.2003.00774.x
Falcon-Lang HJ (2000) Fire ecology of the Carboniferous tropical zone. Palaeogeogr Palaeoclimatol Palaeoecol 164:339–355. https://doi.org/10.1016/S0031-0182(00)00193-0

Feldpausch TR, Carvalho L, Macario KD, Ascough PL, Flores CF et al (2022) Forest fire history in Amazonia inferred from intensive soil charcoal sampling and radiocarbon dating. Front For Glob Change 5:815438. https://doi.org/10.3389/ffgc.2022.815438

Feron S, Cordero RR, Damiani A, MacDonell S, Pizarro J et al (2024) South America is becoming warmer, drier, and more flammable. Commun Earth Environ 5:501. https://doi.org/10.1038/s43247-024-01654-7

Fidelis A, Zirondi HL (2021) And after fire, the Cerrado flowers: a review of post-fire flowering in a tropical savanna. Flora 280:151849. https://doi.org/10.1016/j.flora.2021.151849

Fidelis A, Appezzato-da Glória B, Pillar VD, Pfadenhauer J (2014) Does disturbance affect bud bank size and belowground structures diversity in Brazilian subtropical grasslands? Flora 209:110–116. https://doi.org/10.1016/j.flora.2013.12.003

Flores BM, Fagoaga R, Nelson BW, Holmgren M (2016) Repeated fires trap Amazonian blackwater floodplains in an open vegetation state. J Appl Ecol 53:1597–1603. https://doi.org/10.1111/1365-2664.12687

Fonseca MG, Alves LM, Aguiar APD, Arai E, Anderson LO et al (2019) Effects of climate and land-use change scenarios on fire probability during the 21st century in the Brazilian Amazon. Glob Change Biol 25:2931–2946. https://doi.org/10.1111/gcb.14709

Forkel M, Dorigo W, Lasslop G, Chuvieco E, Hantson S et al (2019) Recent global and regional trends in burned area and their compensating environmental controls. Environ Res Commun 1:051005. https://doi.org/10.1088/2515-7620/ab25d2

Giglio L, Schroeder W, Justice CO (2016) The collection 6 MODIS active fire detection algorithm and fire products. Remote Sens Environ 178:31–41. https://doi.org/10.1016/j.rse.2016.02.054

Glasspool IJ, Edwards D, Axe L (2004) Charcoal in the Silurian as evidence for the earliest wildfire. Geology 32:381–383. https://doi.org/10.1130/G20363.1

Gómez-González S, González ME, Paula S, Díaz-Hormazábal I, Lara A, Delgado-Baquerizo M (2019) Temperature and agriculture are largely associated with fire activity in Central Chile across different temporal periods. For Ecol Manag 433:535–543. https://doi.org/10.1016/j.foreco.2018.11.041

Gottsberger G, Silberbauer-Gottsberger I (2006) Life in the Cerrado, a South American tropical seasonal ecosystem. Vol 1. Origin, structure, dynamics and plant use. Reta Verlag, Ulm

Gurvich DE, Enrico L, Cingolani AM (2005) Linking plant functional traits with post-fire sprouting vigour in woody species in Central Argentina. Austral Ecol 30:868–875. https://doi.org/10.1111/j.1442-9993.2005.01529.x

Haas O, Prentice IC, Harrison SP (2022) Global environmental controls on wildfire burnt area, size, and intensity. Environ Res Lett 17:065004. https://doi.org/10.1088/1748-9326/ac6a69

He T, Lamont BB (2018) Baptism by fire: the pivotal role of ancient conflagrations in evolution of the Earth's flora. Natl Sci Rev 5:237–254. https://doi.org/10.1093/nsr/nwx041

He T, Pausas JG, Belcher CM, Schwilk DW, Lamont BB (2012) Fire-adapted traits of *Pinus* arose in the fiery Cretaceous. New Phytol 194:751–759. https://doi.org/10.1111/j.1469-8137.2012.04079.x

Hoetzel S, Dupont L, Schefuß E, Rommerskirchen F, Wefer G (2013) The role of fire in Miocene to Pliocene C4 grassland and ecosystem evolution. Nat Geosci 6:1027–1030. https://doi.org/10.1038/ngeo1984

Hoffmann WA, Adasme R, Haridasan M, Carvalho MT, Geiger EL et al (2009) Tree topkill, not mortality, governs the dynamics of savanna-forest boundaries under frequent fire in Central Brazil. Ecology 90:1326–1337. https://doi.org/10.1890/08-0741.1

Hoffmann WA, Geiger EL, Gotsch SG, Rossatto DR, Silva LC et al (2012) Ecological thresholds at the savanna-forest boundary: how plant traits, resources and fire govern the distribution of tropical biomes. Ecol Lett 15:759–768. https://doi.org/10.1111/j.1461-0248.2012.01789.x

Janssen TA, Jones MW, Finney D, Werf GR, Wees D, Xu W, Veraverbeke S (2023) Extratropical forests increasingly at risk due to lightning fires. Nat Geosci 16:1136–1144. https://doi.org/10.1038/s41561-023-01322-z

Jaramillo C (2023) The evolution of extant South American tropical biomes. New Phytol 239:477–493. https://doi.org/10.1111/nph.18931

Jaureguiberry P, Díaz S (2023) A three-dimensional approach to general plant fire syndromes. Funct Ecol 37:2143–2158. https://doi.org/10.1111/1365-2435.14272
Keeley JE (2012) Ecology and evolution of pine life histories. Ann For Sci 69:445–453. https://doi.org/10.1007/s13595-012-0201-8
Keeley JE, Pausas JG (2022) Evolutionary ecology of fire. Annu Rev Ecol Evol Syst 53:203–225. https://doi.org/10.1146/annurev-ecolsys-102320-095612
Keeley JE, Rundel PW (2005) Fire and the Miocene expansion of C4 grasslands. Ecol Lett 8:683–690. https://doi.org/10.1111/j.1461-0248.2005.00767.x
Keeley JE, Pausas JG, Rundel PW, Bond WJ, Bradstock RA (2011) Fire as an evolutionary pressure shaping plant traits. Trends Plant Sci 16:406–411. https://doi.org/10.1016/j.tplants.2011.04.002
Kitzberger T, Tiribelli F, Barberá I, Gowda JH, Morales JM et al (2022) Projections of fire probability and ecosystem vulnerability under 21st century climate across a trans-Andean productivity gradient in Patagonia. Sci Total Environ 839:156303. https://doi.org/10.1016/j.scitotenv.2022.156303
Lamont BB, He T, Yan Z (2019) Evolutionary history of fire-stimulated resprouting, flowering, seed release and germination. Biol Rev 94:903–928. https://doi.org/10.1111/brv.12483
Lamont BB, Pausas JG, He T, Witkowski ET, Hanley ME (2020) Fire as a selective agent for both serotiny and nonserotiny over space and time. Crit Rev Plant Sci 39:140–172. https://doi.org/10.1080/07352689.2020.1768465
Lawes MJ, Midgley JJ, Clarke PJ (2013) Costs and benefits of relative bark thickness in relation to fire damage: a savanna/forest contrast. J Ecol 101:517–524. https://doi.org/10.1111/1365-2745.12035
Lehmann CE, Archibald SA, Hoffmann WA, Bond WJ (2011) Deciphering the distribution of the savanna biome. New Phytol 191:197–209. https://doi.org/10.1111/j.1469-8137.2011.03689.x
Linder HP (2014) The evolution of African plant diversity. Front Ecol Evol 2:38. https://doi.org/10.3389/fevo.2014.00038
Maracahipes L, Faggiani APS, Campos BH, Oliveira RS, Pilon N (2024) Shedding light on the relevance of asexual reproduction in the savannas: cloning in Velloziaceae. Biotropica 56:E13343. https://doi.org/10.1111/btp.13343
Maurin O, Davies TJ, Burrows JE, Daru BH, Yessoufou K et al (2014) Savanna fire and the origins of the 'underground forests' of Africa. New Phytol 204:201–214. https://doi.org/10.1111/nph.12936
Mayle FE, Burn MJ, Power M, Urrego DH (2009) Vegetation and fire at the Last Glacial Maximum in tropical South America. In: Vimeux F, Sylvestre F, Khodri M (eds) Past climate variability in South America and surrounding regions. Developments in paleoenvironmental research, vol 14. Springer, Dordrecht, pp 89–112. https://doi.org/10.1007/978-90-481-2672-9_4
Medeiros MB, Fiedler NC (2004) Incêndios florestais no Parque Nacional da Serra da Canastra: desafios para a conservação da biodiversidade. Cienc Florest 14(2):157–168. https://doi.org/10.5902/198050981815
Medeiros MB, Miranda HS (2008) Post-fire resprouting and mortality in cerrado woody plant species over a three-year period. Edinb J Bot 65:53–68. https://doi.org/10.1017/S0960428608004708
Meller P, Stellmes M, Fidelis A, Finckh M (2022) Correlates of geoxyle diversity in Afrotropical grasslands. J Biogeogr 49(2):339–352. https://doi.org/10.1111/jbi.14305
Menezes LS, Oliveira AM, Santos FL, Russo A, Souza RA et al (2022) Lightning patterns in the Pantanal: untangling natural and anthropogenic-induced wildfires. Sci Total Environ 820:153021. https://doi.org/10.1016/j.scitotenv.2022.153021
Mostacedo B, Viruez A, Varon Y, Paz-Roca A, Parada V, Veliz V (2022) Tree survival and resprouting after wildfire in tropical dry and subhumid ecosystems of Chiquitania, Bolivia. Trees For People 10:100327. https://doi.org/10.1016/j.tfp.2022.100327
Olson DM, Dinerstein E, Wikramanayake ED, Burgess ND, Powell GV, Underwood EC, Kassem KR (2001) Terrestrial ecoregions of the world: a new map of life on Earth. Bioscience 51:933–938. https://doi.org/10.1641/0006-3568(2001)051[0933:TEOTWA]2.0.CO;2
Overbeck GE, Pfadenhauer J (2007) Adaptive strategies in burned subtropical grassland in southern Brazil. Flora 202:27–49. https://doi.org/10.1016/j.flora.2005.11.004

Overbeck GE, Müller SC, Fidelis A, Pfadenhauer J, Pillar VD et al (2007) Brazil's neglected biome: the South Brazilian Campos. Perspect Plant Ecol Evol Syst 9:101–116. https://doi.org/10.1016/j.ppees.2007.07.005

Pais C, Gonzalez-Olabarria JR, Elimbi Moudio P, Garcia-Gonzalo J, González MC, Shen ZJM (2023) Global scale coupling of pyromes and fire regimes. Commun Earth Environ 4:267. https://doi.org/10.1038/s43247-023-00881-8

Pausas JG (2012) Incendios forestales. Catarata/CSIC, Madrid

Pausas JG (2015) Bark thickness and fire regime. Funct Ecol 29:315–327. https://doi.org/10.1111/1365-2435.12372

Pausas JG, Keeley JE (2014) Evolutionary ecology of resprouting and seeding in fire-prone ecosystems. New Phytol 204:55–65. https://doi.org/10.1111/nph.12921

Pausas JG, Ribeiro E (2013) The global fire–productivity relationship. Glob Ecol Biogeogr 22:728–736. https://doi.org/10.1111/geb.12043

Pausas JG, Keeley JE, Schwilk DW (2017) Flammability as an ecological and evolutionary driver. J Ecol 105:289–297. https://doi.org/10.1111/1365-2745.12691

Pausas JG, Lamont BB, Paula S, Appezzato-da-Glória B, Fidelis A (2018) Unearthing belowground bud banks in fire-prone ecosystems. New Phytol 217:1435–1448. https://doi.org/10.1111/nph.14982

Pellegrini AF, Anderegg WR, Paine CT, Hoffmann WA, Kartzinel T, Rabin SS, Pacala SW (2017) Convergence of bark investment according to fire and climate structures ecosystem vulnerability to future change. Ecol Lett 20:307–316. https://doi.org/10.1111/ele.12725

Pilon NA, Hoffmann WA, Abreu RC, Durigan G (2018) Quantifying the short-term flowering after fire in some plant communities of a cerrado grassland. Plant Ecol Divers 11:259–266. https://doi.org/10.1080/17550874.2018.1517396

Pilon NA, Cava MG, Hoffmann WA, Abreu RC, Fidelis A, Durigan G (2021) The diversity of post-fire regeneration strategies in the cerrado ground layer. J Ecol 109:154–166. https://doi.org/10.1111/1365-2745.13456

Pivello VR (2011) The use of fire in the Cerrado and Amazonian rainforests of Brazil: past and present. Fire Ecol 7:24–39. https://doi.org/10.4996/fireecology.0701024

Pivello VR, Vieira I, Christianini AV, Ribeiro DB, Silva Menezes L, Berlinck CN, Overbeck GE (2021) Understanding Brazil's catastrophic fires: causes, consequences and policy needed to prevent future tragedies. Perspect Ecol Conserv 19:233–255. https://doi.org/10.1016/j.pecon.2021.06.005

Pontes-Lopes A, Silva CV, Barlow J, Rincón LM, Campanharo WA et al (2021) Drought-driven wildfire impacts on structure and dynamics in a wet Central Amazonian forest. Proc R Soc B Biol Sci 288:20210094. https://doi.org/10.1098/rspb.2021.0094

Ragsac AC, Farias-Singer R, Freitas LB, Lohmann LG, Olmstead RG (2019) Phylogeny of the neotropical tribe Jacarandeae (Bignoniaceae). Am J Bot 106:1589–1601. https://doi.org/10.1002/ajb2.1399

Raich JW, Rastetter EB, Melillo JM, Kicklighter DW, Steudler PA et al (1991) Potential net primary productivity in South America: application of a global model. Ecol Appl 1:399–429. https://doi.org/10.2307/1941899

Ramos-Neto MB, Pivello VR (2000) Lightning fires in a Brazilian savanna national park: rethinking management strategies. Environ Manag 26:675–684. https://doi.org/10.1007/s002670010124

Rodrigues CA, Zirondi HL, Fidelis A (2021) Fire frequency affects fire behavior in open savannas of the Cerrado. For Ecol Manag 482:118850. https://doi.org/10.1016/j.foreco.2020.118850

Rosan TM, Aragão LE, Oliveras I, Phillips OL, Malhi Y et al (2019) Extensive 21st-century woody encroachment in South America's savanna. Geophys Res Lett 46:6594–6603. https://doi.org/10.1029/2019GL082327

Rundel PW, Arroyo MT, Cowling RM, Keeley JE, Lamont BB, Vargas P (2016) Mediterranean biomes: evolution of their vegetation, floras, and climate. Annu Rev Ecol Evol Syst 47:383–407. https://doi.org/10.1146/annurev-ecolsys-121415-032330

Rundel PW, Arroyo MT, Cowling RM, Keeley JE, Lamont BB et al (2018) Fire and plant diversification in Mediterranean-climate regions. Front Plant Sci 9:851. https://doi.org/10.3389/fpls.2018.00851

Sampaio EVSB, Salcedo IH, Kauffman JB (1993) Effect of different fire severities on coppicing of caatinga vegetation in Serra Talhada, PE, Brazil. Biotropica 25:452–460. https://doi.org/10.2307/2388868

Sansevero JB, Garbin ML, Sánchez-Tapia A, Valladares F, Scarano FR (2020) Fire drives abandoned pastures to a savanna-like state in the Brazilian Atlantic Forest. Perspect Ecol Conserv 18:31–36. https://doi.org/10.1016/j.pecon.2019.12.004

Santos MF, Lucas E, Sano PT, Buerki S, Staggemeier VG, Forest F (2017) Biogeographical patterns of *Myrcia* sl (Myrtaceae) and their correlation with geological and climatic history in the Neotropics. Mol Phylogenet Evol 108:34–48. https://doi.org/10.1016/j.ympev.2017.01.012

Schumacher V, Setzer A, Saba MM, Naccarato KP, Mattos E, Justino F (2022) Characteristics of lightning-caused wildfires in Central Brazil in relation to cloud-ground and dry lightning. Agric For Meteorol 312:108723. https://doi.org/10.1016/j.agrformet.2021.108723

Scott AC (2000) The Pre-Quaternary history of fire. Palaeogeogr Palaeoclimatol Palaeoecol 164:281–329. https://doi.org/10.1016/S0031-0182(00)00192-9

Scott AC, Glasspool IJ (2006) The diversification of Paleozoic fire systems and fluctuations in atmospheric oxygen concentration. Proc Natl Acad Sci USA 103:10861–10865. https://doi.org/10.1073/pnas.0604090103

Silva PS, Bastos A, Libonati R, Rodrigues JA, DaCamara CC (2019) Impacts of the 1.5 C global warming target on future burned area in the Brazilian Cerrado. For Ecol Manag 446:193–203. https://doi.org/10.1016/j.foreco.2019.05.047

Simon MF, Pennington T (2012) Evidence for adaptation to fire regimes in the tropical savannas of the Brazilian Cerrado. Int J Plant Sci 173:711–723. https://doi.org/10.1086/665973

Simon MF, Grether R, Queiroz LP, Skema C, Pennington RT, Hughes CE (2009) Recent assembly of the Cerrado, a neotropical plant diversity hotspot, by in situ evolution of adaptations to fire. Proc Natl Acad Sci USA 106:20359–20364. https://doi.org/10.1073/pnas.0903410106

Simon MF, Pastore JFB, Souza AF, Borges LM, Scalon VR et al (2016) Molecular phylogeny of *Stryphnodendron* (Mimosoideae, Leguminosae) and generic delimitations in the Piptadenia group. Int J Plant Sci 177:44–59. https://doi.org/10.1086/684077

Simpson KJ, Ripley BS, Christin PA, Belcher CM, Lehmann CE et al (2016) Determinants of flammability in savanna grass species. J Ecol 104:138–148. https://doi.org/10.1111/1365-2745.12503

Song Y, Xu C, Li X, Oppong F (2024) Lightning-induced wildfires: an overview. Fire 7(3):79. https://doi.org/10.3390/fire7030079

Souchie FF, Pinto JRR, Lenza E, Gomes L, Maracahipes-Santos L, Silvério DV (2017) Post-fire resprouting strategies of woody vegetation in the Brazilian savanna. Acta Bot Bras 31:260–266. https://doi.org/10.1590/0102-33062016abb0376

Staver AC, Archibald S, Levin SA (2011) The global extent and determinants of savanna and forest as alternative biome states. Science 334:230–232. https://doi.org/10.1126/science.1210465

Staver AC, Brando PM, Barlow J, Morton DC, Paine CT, Malhi Y, Aguila Pasquel J (2020) Thinner bark increases sensitivity of wetter Amazonian tropical forests to fire. Ecol Lett 23:99–106. https://doi.org/10.1111/ele.13409

Strömberg CA (2011) Evolution of grasses and grassland ecosystems. Annu Rev Earth Planet Sci 39:517–544. https://doi.org/10.1146/annurev-earth-040809-152402

Terra-Araújo MH, Faria AD, Vicentini A, Nylinder S, Swenson U (2015) Species tree phylogeny and biogeography of the Neotropical genus *Pradosia* (Sapotaceae, Chrysophylloideae). Mol Phylogenet Evol 87:1–13. https://doi.org/10.1016/j.ympev.2015.03.007

Úbeda X, Sarricolea P (2016) Wildfires in Chile: a review. Glob Planet Change 146:152–161. https://doi.org/10.1016/j.gloplacha.2016.10.004

Veblen TT, Kitzberger T, Villalba R, Donnegan J (1999) Fire history in northern Patagonia: the roles of humans and climatic variation. Ecol Monogr 69:47–67. https://doi.org/10.1890/0012-9615(1999)069[0047:FHINPT]2.0.CO;2

Vidal-Riveros C, Souza-Alonso P, Bravo S, Laino R, Bieng MAN (2023) A review of wildfires effects across the Gran Chaco region. For Ecol Manag 549:121432. https://doi.org/10.1016/j.foreco.2023.121432

Wellman CH, Osterloff PL, Mohiuddin U (2003) Fragments of the earliest land plants. Nature 425:282–285. https://doi.org/10.1038/nature01884

Zanzarini V, Andersen AN, Fidelis A (2022) Flammability in tropical savannas: variation among growth forms and seasons in Cerrado. Biotropica 54:979–987. https://doi.org/10.1111/btp.13121

Traditional Fire Uses by Indigenous Peoples and Local Communities in South America

3

Bibiana A. Bilbao, Brian G. Ferrero, Rodrigo M. Falleiro, Livia Carvalho Moura, and Guilherme M. Fagundes

Abstract

Indigenous Peoples and Local Communities (IPLCs) of South America have developed complex fire use systems for over 10,000 years that go beyond basic needs like warmth, cooking, and socializing. These practices evolved into sophisticated techniques for creating pyrobiocultural landscapes, polyculture agriculture, plant domestication, resource acquisition, soil enhancement, territory protection and cleaning, and spiritual and aesthetic purposes. Drawing from scientific literature and fieldwork, we demonstrate that these communities have used fire in diverse and complex ways to achieve multiple benefits across a wide

B. A. Bilbao (✉)
Departamento de Estudios Ambientales, Universidad Simón Bolívar, Caracas, Venezuela

Université Paul Valéry-Montpellier III, Montpellier, France
e-mail: bbilbao@usb.ve

B. G. Ferrero
Centro de Investigación y Transferencia Rafaela, CONICET-UNRaf, Rafaela, Argentina
e-mail: brianferrero@conicet.gov.ar

R. M. Falleiro
Ministério dos Povos Indígenas, Brasília, Brazil
e-mail: rodrigo.falleiro@povosindigenas.gov.br

L. C. Moura
Instituto Sociedade, População e Natureza, Brasília, Brazil
e-mail: livia@ispn.org.br

G. M. Fagundes
Departamento de Antropologia, Universidade de São Paulo, São Paulo, Brazil
e-mail: gmfagundes@usp.br

A. Fidelis, V. R. Pivello (eds.), *Fire in the South American Ecosystems*, Ecological Studies 250, https://doi.org/10.1007/978-3-031-89372-8_3

range of ecosystems—from savannas and grasslands to fire-sensitive tropical and template forests that have not evolved with high fire exposure. Paradoxically, despite South America having an outstanding cultural diversity of Indigenous groups and substantial evidence supporting traditional fire management practices, current policies often ignore or prohibit them. Instead, "zero-fire" approaches have been promoted. Some of these policies persist despite their limited effectiveness, controversial outcomes in conservation and ecology, and increased socio-environmental risks, especially under climate and land-use changes. We argue that developing more holistic, effective, and equitable fire management policies requires the inclusion of diverse perspectives and revitalizing Indigenous and local traditional fire knowledge and adaptive practices that have sustainably managed South American landscapes for millennia.

Resumen

Los Pueblos Indígenas y Comunidades Locales tradicionales de Sudamérica han desarrollado sistemas complejos de uso del fuego durante más de 10.000 años, que van más allá de necesidades básicas como calor, cocción y socialización. Estas prácticas evolucionaron en técnicas sofisticadas de agricultura policultivo, domesticar plantas, obtener recursos, mejorar suelos, limpieza y protección, fines espirituales y estéticos resultando en paisajes pirobioculturales de alta diversidad. Con base en la literatura científica y trabajo de campo, demostramos que estas comunidades han usado el fuego de maneras diversas y complejas en sabanas y pastizales hasta bosques tropicales y templados sensibles al fuego. Paradójicamente, a pesar de que Sudamérica posee una gran diversidad cultural indígena y la evidencia que respalda la sustentabilidad de estas prácticas, las políticas actuales frecuentemente las ignoran o prohíben, promoviendo enfoques "cero fuego". Estas políticas todavía persisten a pesar de su limitada efectividad, resultados controvertidos en la conservación y mayores riesgos socioambientales especialmente en condiciones de cambio climático y uso del suelo. Sostenemos que desarrollar políticas de manejo del fuego más holísticas, efectivas y justas requiere incluir diversas perspectivas y revitalizar el conocimiento y las prácticas adaptativas indígenas y locales tradicionales que han gestionado estos paisajes de manera sostenible durante milenios.

Resumo

Os Povos Indígenas e Comunidades Locais tradicionais da América do Sul desenvolveram sistemas complexos de uso do fogo por mais de 10.000 anos, que vão além das necessidades básicas, como aquecimento, cozimento e socialização. Essas práticas evoluíram para técnicas sofisticadas de criação de paisagens pirobioculturais, policultura, domesticação de plantas, obtenção de recursos, enriquecimento do solo, proteção e limpeza das terras, além de propósitos espirituais e estéticos. Com base na literatura científica e trabalho de campo, demonstramos que essas comunidades utilizaram o fogo de maneiras diversas e complexas para obter múltiplos benefícios em savanas e pastagens, até mesmo em florestas tropicais, sensíveis ao fogo. Paradoxalmente, apesar da grande diversidade cultural indígena e das evidências que sustentam a sustentabilidade dessas práticas, as políticas frequentemente as proíbem, promovendo abordagens de "fogo zero". Essas políticas persistem, apesar de sua eficácia limitada, com resultados controversos na conservação e maiores riscos socioambientais, especialmente em condições de mudança climática e uso da terra. Argumentamos que o desenvolvimento de políticas de manejo do fogo mais holísticas, eficazes e justas exige a inclusão de diversas perspectivas e a revitalização do conhecimento e práticas adaptativas indígenas e locais tradicionais, que têm gerido essas paisagens de forma sustentável por milênios.

3.1 Introduction

South America is one of the world's most biologically and culturally diverse regions. It contains 40% of the planet's biodiversity, 25% of its forests and 26% of its freshwater resources (UNEP-WCMC and IUCN 2016). It is also the continent with the highest diversity of Indigenous Peoples globally, enriched by multiple traditional Local Communities (IPLCs), including small collective groups such as Afro-descendent communities (Maroons, Quilombolas—descendants of escaped slaves who formed independent settlements) and rural criollo (Creole) mixed descent (e.g. mestizos, *caboclos*), fishermen, riverside and island communities, etc. This plurality of cultures has developed multiple forms of traditional life based on cultural heritage and adaptive practices deployed in close connection with the natural environment. Indigenous Peoples are recognized as biological and cultural diversity protectors and promoters, beyond principal actors in implementing sustainable development and territorial management (Trauernicht et al. 2015; Mistry et al. 2016; Bilbao et al. 2019; IPBES 2019; Moura et al. 2019; UNESCO 2024). There is growing scientific evidence that fire practices of Indigenous Peoples are achieving multiple benefits in a high diversity of ecosystems, from savanna and grasslands—which are fire-dependent biomes- to the fire-sensitive tropical and temperate forests that have not evolved with high exposure to fire (Fig. 3.1) (Myers 2006; Piperno 2011; Pivello 2011; Piperno et al. 2019; Pivello et al. 2021; Bilbao et al. 2022; Ponce-Calderón et al. 2022; MacLean et al. 2023; UNESCO 2024).

Ethnographic and archaeological evidence from South America reveals Indigenous Peoples have used fires for various purposes for millennia (32,000 to 11,400–10,000 years BP, Guidon and Delibrias 1986; Roosevelt 1994). It is estimated that the most primitive uses of fire were for warmth, cooking food, and socializing (including paintings), and it evolved into very complex forms such as an engineering tool in the creation of pyrobiocultural landscapes, domestication of plants, raising animals, obtaining resources, improving soils, protection and defense, and aesthetic and spiritual reasons. Thus, according to previous studies, Indigenous Peoples have developed a sophisticated and controlled use of fire for agriculture,

Fig. 3.1 Some Indigenous and Local Communities (IPLCs) fire practices, where we can notice the coexistence of the traditional practices with the presence of forests and woodlands. (**a**) Traditional fire Pemón Arekuna Indigenous practice in *conuco* (shifting cultivation plot) in Canaima National Park, Northern Amazonia, Venezuela (Photo: Ruth Salazar-Gascón); (**b**) Traditional fire practice from local islander in wetlands of Paraná river in Santa Fe, Argentina, for cattle ranching purposes (source: Brián Ferrero); (**c**) Krahô Indigenous woman using fire to scare the bees away so she can collect honey (Photo: Peter Caton, Institute Society, Population and Nature's (ISPN) collection); (**d**) Ceremonial Paresi Indigenous fire in Brazil (Photo: André Dib, Instituto Sociedade, População e Natureza-ISPN collection)

where biomass burning contributes to nutrient cycling, soil quality, and soil pH level improvement, allowing crops to be sown in the weathered and impoverished soils of tropical forests in South America. Indigenous Peoples have also used fire for hunting and fishing, to induce fruiting of edible wild plants, communication, visibility and control high levels of savanna and grassland plant fuels, preventing the spread of wildfires into adjacent forests between others (Figs. 3.1 and 3.2) (Bilbao et al. 2019).

Thus, considering ancient pre-Columbian human occupation, the most significant evidence of the sustainability of these Amerindian practices is the extensive forest mass on the continent that was home to an estimated 20–30 million pre-Columbian Indigenous Peoples before the arrival of Europeans (Sutter 2021). However, cultural and political approaches to the use of fire, introduced by Europeans in the 1600's, produced a dramatic change in South American landscapes (Moura et al. 2019; Bilbao et al. 2022). The new settlers disregarded Indigenous

Fig. 3.2 Indigenous and local traditional communities' cultural burns. (**a**) In the Quilombola communities of Rio Preto, located in the Jalapão region of the Brazilian Cerrado, burned areas are utilized as "natural cattle fences" within the traditional agropastoral system known as "na solta" (fenceless grazing) (Photo: Guilherme Moura Fagundes).; (**b**) Controlled burn in Chapada das Mesas National Park in Brazil for the cattle ranching for the native vegetation to resprout, and the older man appearing is the land owner who received help from the fire brigade to burn, within the IFM perspective (Photo: Leonardo Milano); (**c**) Patch mosaic burning in the transition savanna—forest to protect the forest from the risk of uncontrolled and dangerous wildfires that could arise in fire-prone savanna areas in Canaima National Park, Northern Amazonian Basin, Guiana Shield, Venezuela (Photo: Ruth Salazar-Gascón); (**d**) The photo captures the traditional ritual of Pemón Arekuna community burning a *conuco* (swidden cultivation) after asking for permission from God Wacarampü and have a good and controlled burn, in Kavanayén, Gran Sabana, Canaima National Park, Venezuela (Photo: Leonardo Gutiérrez)

practices and introduced agricultural and livestock production systems based on monocultures and pastures, following deforestation and the indiscriminate use of fire, drastically opposed to those of the original inhabitants. Pre-Columbian Amerindian practices, maintained for millennia, suffered after European colonizers' practices, leading to a marked deterioration in natural and sociocultural systems, together with the disappearance of many populations of Indigenous communities, their culture, and practices.

Until today, Indigenous Peoples' participation in policymaking determining the management even of their own territories in South America is very limited. Furthermore, the use of fire by Indigenous and other local non-Indigenous communities has been perceived as a threat to biodiversity conservation and natural resource management by Protected Areas (PA) administrators, fire-fighters bodies and policymakers. There is no real understanding of the benefits of ancestral Indigenous fire use practices for management and sustainability practices. On the contrary, often, the Indigenous Peoples have been considered "tenacious arsonists" who caused the large fires. These nicknames have been fueled by the misconception that all fires are catastrophic wildfires (severe events that spread uncontrollably) and reveal the unawareness of the variety of types of fires carried out by Indigenous Peoples and Local Communities (IPLCs), which are carried out in a controlled manner, defining function, where, how and when to carry out the fires (Fagundes 2019a, b, c; Bilbao et al. 2020).

Consequently, fire suppression and firefighting have historically been the dominant approaches to managing fire on the continent. Thus, most countries adopted 'zero-fire' policies to avoid and control virtually any fires by focusing on firefighting techniques such as fire brigades, technical support, predictive fire risk modelling, and environmental education programs to dissuade Indigenous and local people from burning (Mistry et al. 2016; Eloy et al. 2019; Kreider et al. 2024).

Critiques of widespread fire suppression policies underlined the unique role fire plays in the ecosystems and cultures in many parts of the world, as well as highlighting the costly and unaffordable for South American countries and the ineffectiveness of these policies. Evidence also showed that the exclusion of fire, especially in fire-prone ecosystems such as tropical savannas, promotes fires of greater intensity and extension due to the accumulation of combustible material (Barradas 2017; Bilbao et al. 2010; Pereira et al. 2014; Fidelis et al. 2018; Pivello et al. 2021; Kreider et al. 2024). This situation is expected to be exacerbated as climate change and sociocultural and demographic dynamics extend the periods and spaces of fire occurrence, as has been reported in recent megafire events, even the most sophisticated techniques and fire combat equipment cannot control (Bilbao et al. 2020; UNEP 2022).

Co-developing new visions and capacities for Integrated Fire Management instead of just fighting them requires the inclusion of multiple perspectives and actors and revitalizing the knowledge and adaptive practices of Indigenous Peoples and local and traditional communities that inhabit natural spaces. Other winds of change are blowing in South America. There are some pioneering initiatives for Integrated Fire Management policies with an intercultural and participative approach, joining multiple perspectives (Indigenous, academic and institutional) to enhance wildfire risk reduction through sustainable, effective, Integrated Fire Management strategies. Some long-term initiatives such as the Intercultural and Participatory Fire Management Network, endorsed by Indigenous Peoples, some academics, and public servants in Venezuela, Brazil, and Guyana, The National Center to Prevent and Combat Wildfires (Prevfogo/IBAMA) Program have promoted cross-cultural knowledge integration for fire management of Indigenous territories. The drafting of two new bills in South American countries in 2019 (one submitted to the Brazilian National Congress—recently approved—and another to the Colombian National Congress), changed laws (Paraguay), and the Memorandum of Understanding in 2019 between the Amazon Co-operation Treaty Organisation countries are also examples of these efforts aiming to enact new national or regional policies based on Integrated Fire Management to replace the "zero fire" provisions still in force today in most of these countries.

Understanding the changing dynamics and geographic patterns of human-induced landscape burning is critical. The survival of numerous species and ecosystems relies on comprehending the historical range of variability in fire activity. This perspective not only influences the scientific foundation of fire management but also extends into cultural values. It sheds light on how various cultures perceive and manage fire, revealing the interconnectedness of ecological and cultural perspectives.

Although Indigenous Peoples and Local Communities (IPLCs) can show some notable differences related to recognition and identity and historical context and might not have the same level of recognition as distinct cultural or ethnic entities, they share several commonalities, primarily rooted in their close connection to specific lands, shared cultural practices, and a deep sense of identity. In this chapter, we will focus on analyzing, compiling and highlighting the value of the uses of fire by both types of communities, considering the knowledge that supports these traditional practices. We adopted the definition of Local and Indigenous knowledge provided by LINKS, which refers to the understandings, skills and philosophies developed by societies with long histories of interaction with their natural surroundings. For Indigenous Peoples and local traditional communities, local knowledge informs decision-making about fundamental aspects of day-to-day life. Their traditional knowledge and practices, developed over centuries, are invaluable for understanding ecological processes related to fire management and mitigating wildfire impacts.

The present chapter aims to identify the current status of the primary uses of fire by Indigenous Peoples and local traditional communities in South America, their main actors, the socio-environmental benefits of the cultural legacy of fire use, the threats that these peoples face, and the contribution to the construction of new paradigms of fire management with an inclusive vision of our continent's cultural diversity.

3.2 Indigenous Peoples and Local Communities (IPLCs): Traditional Fire Users in South America

As previously mentioned in the introduction, there is evidence that pre-Columbian communities used to fire at various geographical regions in South America. However, the drastic changes brought about by European colonization, accompanied by the decimation of Indigenous populations, raise questions about the current situation on the continent and who the heirs of these traditional fire practices are. In the context of this chapter, and after the drastic impact of European colonization since the fifteenth century, it is relevant to ask how Indigenous Peoples are defined, what are the present populations, where they are currently located, and whether they continue to use fire. This chapter will also define local traditional communities, a new population of inhabitants, and fire users that emerged after colonization due to their mixing of European new settlers with the local Indigenous inhabitants.

Who are Indigenous Peoples? While there is not universally unique and accepted definition of Indigenous groups, the concept of Indigenous Peoples has been outlined on the main commonly recognized characteristics. Thus, Indigenous Peoples refer to the distinct social and cultural groups that share ancestral ties to the lands and natural resources where they live, occupy, or are displaced. These groups maintain social, cultural, economic, and political characteristics distinct from those of the dominant societies in which they live. Indigenous Peoples are known for their rich cultural heritage, deep-rooted cosmologies, spiritual beliefs, languages, customs, traditions, and laws passed down and recreated through generations. They have a special relationship with their land and environment, seeing themselves as integral to the ecosystem.

In addition, most South American nations sign the ILO Convention No. 169 (1989) on Indigenous and Tribal Peoples in Independent Countries. Articles 1 and 2 of this convention define Indigenous Peoples based on their descent from the populations that inhabited the country or geographical region during conquest, colonization, or when current state boundaries were established. This definition holds regardless of their legal status and acknowledges that these groups maintain some or all their distinct social, economic, cultural, and political institutions.

Estimating the total Indigenous population in South America can be challenging due to differences in data collection methods, definitions, and recognition of Indigenous identity across countries. However, according to the Economic Commission for Latin America and the Caribbean (ECLAC 2022) and the International Work Group for Indigenous Affairs (IWGIA 2019), the estimated Indigenous population in South America is approximately 42–50 million people. This represents roughly 8–10% of the continent's total population (FAO and FILAC 2021). The populations vary significantly by country, with some countries having a much higher percentage of Indigenous Peoples in proportion to their total populations than others (UNICEF 2009; IWGIA 2019; FAO and FILAC 2021; ECLAC 2022) (Fig. 3.3). For example, Brazil has the most significant diversity of Indigenous groups in South America across over 305 different ethnicities, according to the Instituto Brasileiro de Geografia e Estatística (IBGE 2010) and Agência Brasil (2023). However, this country has one of the lowest Indigenous populations and a percentage of the total population in the South American continent. In contrast, Bolivia and Peru show the largest Indigenous population and proportion (especially Bolivia with 48% of the total population), and Chile follows with around 12 million Indigenous populations, constituting 2.2% of its population (mainly Mapuche) (INE 2018; IWGIA 2019; ECLAC 2022) (Fig. 3.3).

According to the Atlas Sociolingüístico de pueblos Indígenas en América Latina (UNICEF 2009) and the Economic Commission for Latin America and the Caribbean (ECLAC 2022), in South America, Amerindian groups belong to 400–500 distinct Indigenous linguistic families, representing one of the continents with the highest Indigenous group diversity globally. These groups speak hundreds of different languages, many of which are not mutually intelligible. Regarding the

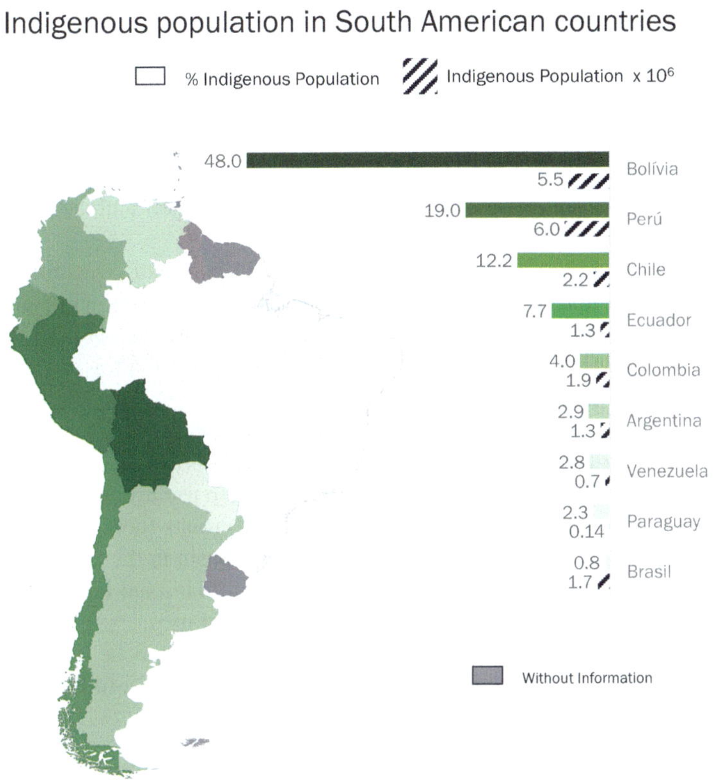

Fig. 3.3 Total and percentage of Indigenous population in South American countries. Information collected and prepared by authors based on data reported by FAO and FILAC (2021), Economic Commission for Latin America and the Caribbean (ECLAC 2022), Brazil Demographic Census 2022 (Agência Brasil 2023) and The International Work Group for Indigenous Affairs (IWGIA 2024)

number of Indigenous groups per country, Brazil hosts the most, with approximately 305, followed by Colombia with 102, and Peru with 85; the lowest numbers are found in Uruguay and Chile with two and nine groups, respectively.

Indigenous Peoples have also diverse histories of colonization under Spanish (Argentina, Bolivia, Colombia, Chile, Ecuador, Paraguay, Peru, Uruguay, Venezuela), Portuguese (Brazil), British (Guyana), Dutch (Suriname), French (French Guiana) rules, different languages and worldviews, and distinct political and institutional developments about Indigenous issues (Huber and Zent 1995; Halbmayer 2013; Bilbao et al. 2019; UNESCO 2024).

The diverse uses of fire among Indigenous Peoples in South America illustrate its importance in various ecosystems and cultural contexts. The major Indigenous Peoples, their geographical distribution and their main livelihood related to fire activities include, for example, (Fig. 3.1 and 3.2) (UNESCO 2024)

- The Caribs and Arawaks, who are widespread throughout the Guiana Shield, regularly use fire for swidden agriculture and to manage community savannas for cattle grazing. These practices are crucial for their economy and helps reduce bush encroachment, which increases the risk of dangerous and uncontrolled wildfires.
- In the Andes mountains, the Quechuas and Aymaras use fire for agricultural purposes, such as burning old crop residues to fertilize the soil with ash and preparing fields for the next planting season.
- The Guarani, spread across Paraguay, Brazil, Bolivia, and Argentina, employ swidden agriculture to cultivate staples like cassava (manioc), while also using fire to manage pasturelands for hunting and clear paths through dense vegetation. Controlled burns stimulate the growth of new vegetation, attracting game animals and facilitating hunting.
- In the Amazon Basin, the Yanomami Peoples, located in northern Brazil and southern Venezuela, use controlled fires to clear underbrush in forests, enhancing access to nuts, fruits, and hunting grounds. They also use fire to shift cultivation to crops like cassava. Similarly, the Xingu ethnic groups (composed by around 16 distinct ethnic groups), located in the Amazon Basin in the northeastern part of the Mato Grosso state in Brazil, utilize fire as a traditional land management tool for shifting cultivation, hunting and a diversity of uses.
- Meanwhile, in Chilean and Argentine Patagonia, the Mapuche use fire to burn shrubs and grasses, promoting new growth and supporting hunting and gathering activities.

On the other hand, traditional "Local Communities" refer to groups of people who share common cultural practices and customs and often reside in a specific geographic area (IPBES 2016). These communities typically have a strong sense of identity and continuity as they pass down traditions, ways of life, and values from one generation to the next. The term "Local" emphasizes the geographical proximity of the community members, indicating that their physical location closely connects them. "Traditional" underscores the continuity of customs and practices passed down through generations, rooted deeply and emerge in a way of life that is ecologically grounded (Mauro and Hardison 2000; Ingold, 2004; Athayde et al. 2021). However, Local Communities form a very heterogeneous group, including communities that hold collective knowledge and have their livelihoods tightly connected to their territories and natural resources. They might or might not have formal recognition of specific rights over their lands, territories, and cultural identity. In the Amazon and other regions of South America, local or traditional communities are generally descendants of migrants who intermarried with local Indigenous Peoples. They often exchange practices, objects and knowledge with members of Indigenous communities (Levis et al. 2018).

The traditional communities we refer to here comprise highly diverse social groups: riverine communities, artisanal freshwater fishers; coastal-based communities such as artisanal fishers, crab collectors etc.; forest-based livelihoods: rubber tappers, coconut breakers, açaí palm extractors, etc.; afro-descendant traditional populations and communities, for example *Quilombolas*, *caboclos*; peasants: family farmers/ smallholders etc.; urban-based Local Communities (social groups) whose livelihoods are tightly connected to biodiversity and natural resources (Athayde et al. 2021; Barretto Filho et al. 2021).

Certain common characteristics are linked to the fact that they are communities that carry out agricultural, livestock, and forestry production activities in rural areas where fire is part of their cultural practices, land management strategies, and daily subsistence activities (e.g., Fig. 3.1a).

In conclusion, Indigenous Peoples and Local Communities (IPLCs) of South America exhibit a vast diversity in language, culture, livelihood and historical backgrounds, contributing significantly to the continent's rich cultural heritage. Their presence and traditional knowledge are critical in maintaining the continent's biodiversity and managing natural resources sustainably. Understanding and respecting this diversity is crucial for their preservation against threats such as deforestation, climate change, and industrialization and for their active participation in regional development (IPBES 2016; Bilbao et al. 2019; IPBES 2016; Athayde et al. 2021; Barretto Filho et al. 2021; UNESCO 2024).

This chapter highlights the diverse fire uses of various Indigenous Peoples and Local Communities (IPLCs) groups and their importance to each community's subsistence and cultural practices.

3.3 Traditional Fire Practices and Uses by Indigenous Peoples and Local Communities (IPLCs)

Fire practices are deeply ingrained in the lives of Indigenous Peoples, serving various purposes, from agriculture, hunting, and other productive practices and preserving cosmological traditions (Figs. 3.1 and 3.2). Many of these practices developed during the pre-Columbian period continue in a great diversity of ecosystems in South America, from tropical forests to temperate forests, savannas, grasslands, floodplains and palm ecosystems. However, the largest management burns are carried out in areas occupied by savanna formations. Yet, forests are burned in smaller areas to establish crops, scare the bees away to collect honey, perform rituals and celebrations and protect villages or communities from pests (Fig. 3.1a, c, d; Fig. 3.2d). Protecting fire-sensitive vegetation is widespread in different cultures since these ecosystems' fauna and flora species are of great nutritional, sociocultural and economic interest (IBAMA 2018; Welch and Coimbra Jr. 2021).

It is crucial to recognize that Indigenous uses of fire are regarded as cultural burnings. However, what is the meaning of "cultural burnings"? Cultural burning, also known as traditional or controlled burning (Bilbao et al. 2020; Maclean et al. 2023), refers to the intentional use of fire by Indigenous and Local Communities to manage landscapes for ecological and cultural purposes. This practice is deeply embedded in the cultural heritage of many Indigenous Peoples worldwide, including those in South America. Their controlled nature and specific purposes characterize these, requiring detailed knowledge of what, where, when, how, and why to burn and the skill and knowledge of the person conducting burns (Bilbao et al. 2010; Fagundes 2019b; Steffensen 2020; Ponce-Calderón et al. 2022; Fagundes 2021; UNESCO 2024). Cultural burnings are conducted based on sophisticated and diversified fire techniques profoundly intertwined with Indigenous and Local Communities' intricate traditional ecological knowledge and sustainable practices, revealing multiple functions that benefit the environment, enhance safety, and sustain the community's way of life (Fagundes 2019a, b, c, 2021; UNESCO 2024). Also, fire activities among Indigenous Peoples and traditional communities are often oriented to protect the production or sacred sites from wildfires (Leonell 2000).

The burnings are carried out in previously defined areas. The time, extent and frequency are determined according to the management objective and families' or communities' needs (UNESCO 2024). These definitions are usually made collectively, allowing greater control over fire, mainly once most fires are carried out during the dry season. However, the fire started by management is not always controlled throughout the process. Often, these burnings continue until the fire is suppressed on its own and may or may not spread over large areas, depending on the availability of biomass fuel, physical barriers or favorable weather conditions (Bilbao et al. 2020; UNEP 2022).

It is essential to distinguish cultural burnings from other fire types, such as wildfires. Fire is frequently conflated with wildfires, even in scientific and technical contexts. Unlike cultural burnings, wildfires are uncontrolled fires that, under extreme weather conditions (high temperatures, dryness, and high fuel accumulation), could be catastrophic and have detrimental and unwished effects on landscapes and infrastructure (Fagundes 2019c; Bilbao et al. 2020; UNEP 2022; UNESCO 2024).

It has been reported that many ancestral fire-use practices by Indigenous communities in South America have been lost since colonial times (UNESCO 2024). Therefore, we focus here on the primary uses of fire by Indigenous Peoples and traditional Local Communities, considering evidence that these cultural practices are still used today or played a significant role in Indigenous Peoples' livelihoods in the past (Moura et al. 2019; Bilbao et al. 2022; UNESCO 2024). Some relevant cultural fire practices of local traditional communities were also considered.

The cultural burnings of Indigenous and local traditional communities, shown below, are organized according to their primary purposes and benefits.

3.3.1 Agriculture

3.3.1.1 Swidden Agriculture

Swidden agriculture, or shifting cultivation, formerly known as slash-and-burn practices, is a traditional polyculture farming method employed by many Indigenous Peoples and local traditional communities in South America, especially in tropical and subtropical regions (Levis et al. 2018; Bilbao et al. 2019; Edivaldo and Rosell 2020; UNESCO 2024). These practices involve sophisticated knowledge, where fire plays a crucial role. Swidden cultivation is considered a sustainable method for growing crops in the rainforest by using land rotation techniques, maintaining the balance between agriculture and forest regeneration.

Evidence shows that Indigenous communities along the Amazon River already engaged in swidden agriculture around 5000 years ago (Versteeg 2003; Piperno 2011; 2019). However, this practice is still present and quite common in Indigenous groups, and it is extensively used even by local non-Indigenous communities and populations, such as *Quilombolas* and *caboclos* (Pivello 2011; Fagundes 2019a, b, c, 2021), in tropical and subtropical forests. The role of fire and the sequence of the different phases of shifting cultivation are explained below.

This type of agriculture involves diverse tasks (Edivaldo and Rosell 2020; UNESCO 2024) (Fig. 3.1a). At first, the location by Indigenous communities, who carefully choose a plot based on the soil fertility, vegetation type, and proximity to their settlements. Often, the plots are selected within secondary forests that have regenerated from previous cultivations rather than in untouched primary forests. The location of agricultural plots is generally decided upon yearly during village meetings, based on the number of families, available areas, and crops the families want to grow. Soil conditions were considered when selecting a location for different crops (Edivaldo and Rosell 2020; UNESCO 2024). Forest patches were also inspected before cutting and burning to prevent the burning of patches with a high density of beneficial trees (e.g., fruit-bearing trees).

A second moment, preparation of the farming area concerns clear vegetation: After selecting the agricultural plot (usually 0.5–1 hectares), Indigenous people clear (slash) the land at the end of the rainy season/beginning of the dry season to create space for cultivation. Logs were directed towards the center of the patch during felling to prevent the spreading of the fire during burning.

Drying the Slashed Vegetation: The cut plants and trees are left to dry for a period, typically a few weeks to a few months, until the middle of the dry season, depending on the climate and the moisture content of the vegetation (UNESCO 2024). The drying phase is crucial as very wet vegetation does not burn well and can result in incomplete burns that do not release sufficient nutrients into the soil.

The burning phase entails different tasks. The burn generally begins at the edge of the cleared plot. Starting the fire at the edges allows for a more controlled burn, preventing the fire from escaping into surrounding forest areas. The direction of the wind is a critical factor. Farmers start the fire on the side from which the wind blows to drive the fire across the plot. This technique ensures the fire spreads evenly and the heat is distributed across the entire area, turning the biomass into ashes.

The fire must be carefully monitored and controlled. Community members or family units watch the fire to ensure it does not spread beyond the intended boundaries. They may create firebreaks around the plot by clearing lines down to the bare earth to contain the fire within the plot. The controlled use of fire helps managing weeds and unwanted vegetation. Burning away excessive plant growth can create a more favorable environment for desired crop growth. Fire can also be employed to control pests and diseases. Burning can disrupt the life cycles of certain insects and pathogens, helping to manage potential crop threats.

Finally, the post-burn phase involves charcoal and ash fertilization. The ash and burned residues left behind after a controlled burn act as natural fertilizers. They contain essential minerals that enrich the soil, increase the pH, and reduce Aluminum toxicity (especially beneficial in very acid tropical soils), promoting healthier crop growth.

The cultivation after burning involves several steps:

Once the fire has consumed all the dried vegetation and only ash remains, the land is ready for planting.

Crops are planted directly into the ash-covered soil, often without any need for ploughing or turning the soil.

The fallow phase (forest regeneration) in the shifting cultivation system involves rotating fields to allow the land to fallow and regenerate (Bilbao et al. 2017). Using fire in this context helps stimulate new growth by clearing old vegetation and allowing sunlight to reach the forest floor. Usually, the cultivation plots are utilized for 2–3 years and then left to rest, transitioning into a stage known as *barbecho* (a Spanish term defining secondary vegetation that regenerates during the resting phase). This *barbecho* period can extend over longer periods, typically between 5 and 20 years, depending on the soil characteristics and the surrounding forest vegetation (UNESCO 2024). For Carib and Arawak Indigenous Peoples in North SA, the plots in the fallow stand are not considered wholly abandoned. For example, some crops (like napi and yams - *Dioscorea* spp.) would still grow, and food could be harvested from these plots, incidentally, when other plots failed to produce (sufficient) food (UNESCO 2024).

Fire in shifting cultivation is not just a practical tool but often has cultural significance. Indigenous communities may have traditional rituals or ceremonies associated with using fire in farming, connecting their practices to broader cultural beliefs. For example, as part of the Pemón Arekuna community's swidden cultivation practices, in Kavanayén, Gran Sabana, Venezuela, they ask permission from God Wacarampü, at the moment to start a fire in the *conuco* (swidden cultivation plot in the forest) to have a good and controlled fire to have more productive crops (UNESCO 2024) (Fig. 3.2d).

Various Indigenous Peoples have historically employed swidden cultivation practices across the continent since prehistoric times. According to the different ecoregions, some of these Indigenous groups include:

Amazon Basin: Kayapo People (Brazil), Yanomami People, Pemón (Venezuela); Wapishana and Macushi (Guyana, Brazil); Ashaninka (Peru and Brazil); Munduruku Peoples (Brazil); Kichwa (Ecuador); Jivaroan Peoples (Ecuador and Peru); Tucano Peoples (Colombia).

Guiana Shield: Wayampi, Palikur, and Emerillon (French Guyana), Wayana, Lokono (Arawak) Kari'na (Carib) (Suriname, French Guyana and Brazil); Pemón and Kari'na (Carib) (Venezuela); Wapishana and Macushi (Guyana, Brazil)

Andean Systems: swidden cultivation is less commonly practiced in the Andean region than in lowland tropical areas. However, some Indigenous groups in the

Andes have historically engaged in similar agricultural practices: Quechua and Aymara**:** They inhabit the Andean highlands of Peru, Bolivia, Ecuador, and parts of Colombia and Chile, have historically practiced agricultural techniques that involve clearing and burning small plots of land for cultivation. Asháninka**:** This group inhabit both the Amazon rainforest and the eastern slopes of the Andes in Peru and practices swidden cultivation in the lowland tropical areas of their territory. It's essential to recognize that the agricultural practices in the Andean region often involve terrace farming, where stepped platforms are constructed on hillsides to maximize arable land.

Other ecosystems: While shifting cultivation has been a sustainable agricultural practice for many Indigenous communities, external pressures, including deforestation and government policies, have sometimes challenged these traditional practices, prohibiting or forcing communities to carry them out in secondary and degraded ecosystems (e.g. Carmenta et al. 2013; Bilbao et al. 2017).

3.3.1.2 Terra Preta

Terra preta, meaning "black soil" in Portuguese, is a highly fertile and intentionally created soil in the Amazon rainforest by pre-Columbian Indigenous communities (Neves et al. 2003). It features rich nutrient content, including organic matter, nitrogen, phosphorus, and calcium, a dark color from charcoal and organic matter, and longevity in maintaining fertility. *Terra preta* soils can maintain their fertility for extended periods, even over centuries, contributing to the sustained productivity of the typically nutrient-poor tropical soils over centuries (Glaser et al. 2001; Neves et al. 2003).

The creation of *terra preta* is thought to involve a combination of traditional practices by Indigenous communities, including:

Biochar addition: Indigenous Peoples likely intentionally added charcoal or biochar to the soil, which helps improve nutrient retention, water-holding capacity, and microbial activity.

Organic matter incorporation: The addition of organic materials, such as plant residues and kitchen waste, enhanced the soil's organic content and fertility.

Ceramic fragments: Some *terra preta* sites contain broken fragments of pottery, which may have been deliberately incorporated to enhance the soil's structure and nutrient retention.

The knowledge and practices associated with *terra preta* played a significant role in supporting sustainable and sophisticated agricultural practices to enhance soil fertility in the nutrient-poor tropical environment (Liang et al. 2010). They contributed to developing complex societies in the Amazon. Archaeological evidence suggests that multiple Indigenous communities in the Amazon region intentionally produced *terra preta* or anthropogenic dark earth during pre-Columbian times. Some estimates suggest these soils could be as old as 2500–3000 years old (Neves et al. 2003).

Understanding these ancient techniques is crucial for contemporary sustainable agriculture, soil management, and carbon sequestration efforts. It emphasizes the value of integrating traditional ecological knowledge into modern environmental stewardship.

3.3.1.3 Raised Beds

Around 2000 BP, a method of agriculture was developed by Indigenous coastal communities in seasonally flooded freshwater swamps. This method involved the establishment of arrays of small square raised fields ('beds'), likely by piling up material dug up from spaces between the beds (UNESCO 2024).These so-called Pre-Colombian raised beds are thought to have been established by Indigenous Peoples from the Arauquinoid cultures who progressively expanded from west to east, from Guyana to the island of Cayenne, identifying four cultures of the same tradition: Hertenrits, Kwatta, Barbakoeba and Themire (Rostain 2010; Versteeg 2003). Remnants of these raised beds can be found in Hertenrits, Buckleburg, and several sites in northeast Suriname (Versteeg 2003).

Teunissen (2020) argues that fire must have been used to establish and maintain raised beds. Not only would fire be needed to clear the area, but frequent fires were also probably used to prevent the build-up of peat.

3.3.1.4 Flowering and Fruit Production

Indigenous Peoples and local traditional communities utilize fire for various ecological, agricultural, and cultural practices, including enhancing the flowering and fruiting of particular species in certain ecosystems. This practice is particularly significant in regions where specific plant species require fire cues to bloom, fruiting or regenerate effectively (Bond and Van Wilgen 1996; Fidelis et al. 2022).

Flowering also used to be promoted by fire. Specific plant species have evolved and respond to fire by flowering in some ecosystems, particularly savannas and certain forested areas. The heat from the fire can trigger seed germination or stimulate flowering. For instance, in savanna ecosystems, many plant species bloom only after fire events, reducing the grassy competition and increasing sunlight exposure to the soil. Thus, Indigenous and Local Communities frequently use fire to manage the landscape, promoting the growth of particular plants that are valuable for their flowers and might be used for medicinal purposes, food, or craft materials. It is possible to find some examples of these flowering systems through cultural burnings in the Serra do Espinhaço en Minas Gerais. It is structured from the collection of flowers popularly called *sempre-vivas* (evergreen) flowers. The Serra do Espinhaço is a unique mountain range in Minas Gerais, Brazil. It is noted for its biodiversity and the economically valuable *sempre-vivas* flowers, which thrive there due to the region's ecological conditions. This area is predominantly inhabited by traditional communities, including many *Quilombola* groups known as the "flower harvesters of *sempre-vivas*." These communities have sustainably managed the landscape using controlled fire practices to support the growth of these flowers, clear underbrush, and control invasive species, thus maintaining the ecological balance and enhancing biodiversity. These flowers are picked for commercial purposes, and the practice is passed down through generations, embedding deep cultural values (Fagundes 2019a, b, c, 2021).

However, establishing the Sempre Vivas National Park in 2002 posed significant challenges. The traditional burning practices, crucial for managing *sempre-vivas*, faced heavy scrutiny and legal restrictions, threatening the local way of life and economic survival. In a significant development in 2020, the SAT-Sempre Vivas System was recognized by the FAO as Brazil's first Important Agricultural Heritage System (IAHS) and a World Agricultural Heritage site, legitimizing these traditional practices and promoting sustainable development. This recognition helps protect local rights and supports the communities' fire-based agricultural practices, ensuring the preservation of their cultural heritage and the Serra do Espinhaço's unique ecosystem.

This example highlights the critical role of integrating traditional ecological knowledge and fire management practices in conserving biodiversity and supporting local economies. International bodies like the FAO recognize the importance of these practices, ensuring that traditional methods continue alongside modern conservation efforts. This approach not only preserves the cultural heritage but also enhances the sustainability of these unique ecosystems.

Similarly, one example of fruit production promoted by cultural burns is the edible fruits in the Brazilian savannas, which represent one of the primary natural resources in this region and are essential for both Indigenous Peoples and native fauna. These edible fruits are found abundantly from the herbaceous to the arboreal layer. The fruit-bearing species of these fire-prone ecosystems have a close relationship with fire. The Indigenous Peoples of the Cerrado are well acquainted with the ecology of these species and use fire to produce fruits and protect native orchards. Recently, this knowledge has been accessed, systematized, and used to define the fire regimes applied by Indigenous fire brigades of the Federal Brigades Program-BRIFs (Falleiro et al. 2021) of the National Center Prevfogo, linked to the Brazilian Institute of the Environment and Renewable Natural Resources-Prevfogo/Ibama. Fire regimes varied considerably according to objectives and ethnic groups. The frequency of burning could not be precisely defined, as it heavily depended on the ecosystems, species involved, and the accumulation of combustible material. However, the timing of the burns was more stable, with a period notably aimed at stimulating fruiting occurring at the start of the dry season (Fig. 3.4).

The National Indian Foundation-FUNAI supported access to this traditional knowledge, involving more than 600 Indigenous representatives, especially elders, shamans, herbalists, and hunters, from the Indigenous groups where fire management was implemented: Alto Xingu (Aweti, Kalapalo, Kamayurá, Kuikuro, Matipu, Mehinako, Nahukuá, Trumai, Wauja, and Yawalapiti), Kadiwéu, Bakairi, Paresi, Krahô, Javaé, Karajá, Xerente, Tenharim, Makuxi, Wapichana, Taurepang, Guajajara, Kanela, Gavião, and Krikati, as well as the traditional *Quilombola*'s community Kalunga. During the interviews, which were often informal, they reported various fire uses and traditional management of the savannas. They also defined the goals, burning methods, and fire regimes (Falleiro et al. 2021, 2022). The results can be summarized in Fig. 3.4a, b, where it is shown that the prescribed burns (PB) seem to be the most effective burns to increase the fruit production of native fruit-bearing plants (Fig. 3.4a). Additionally, BDS - beginning the dry season

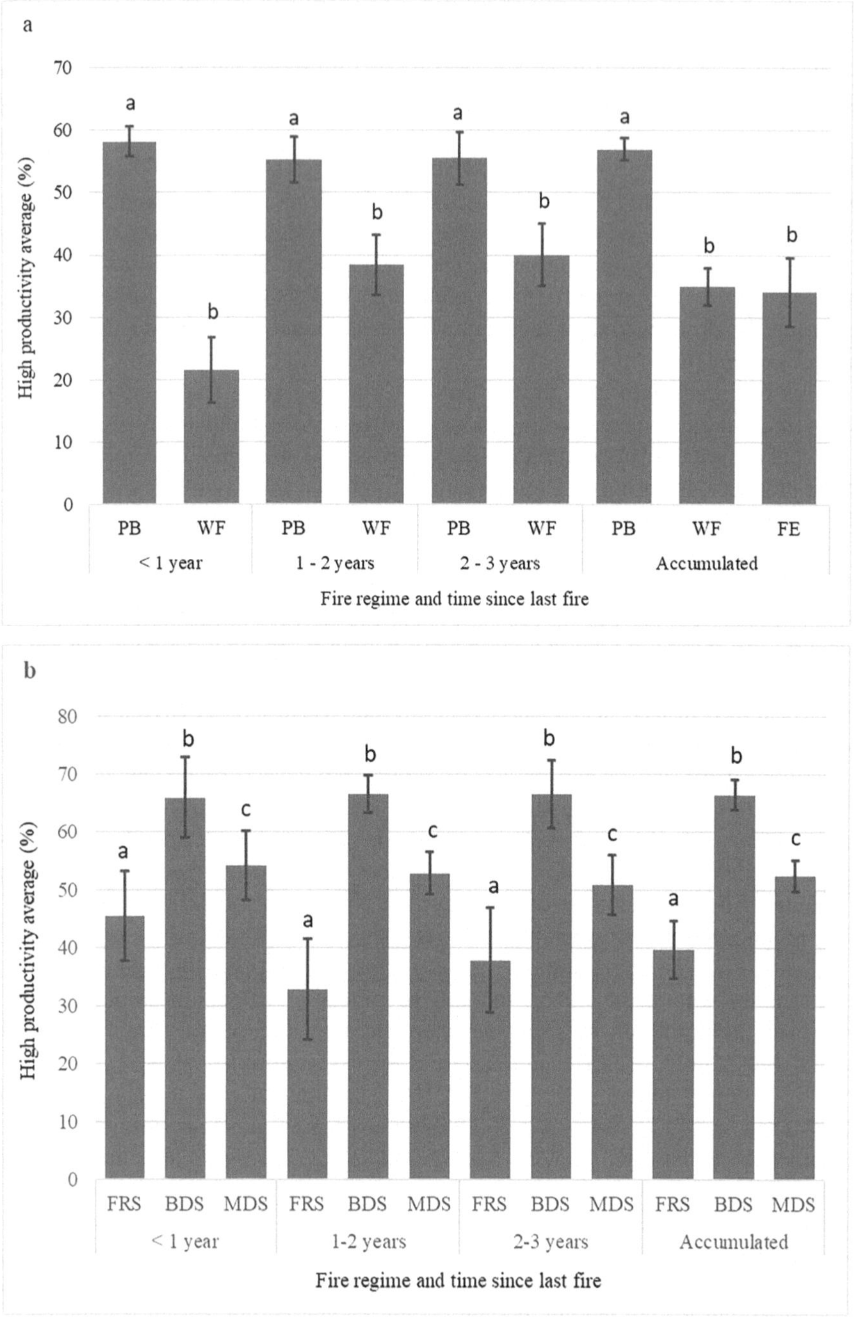

Fig. 3.4 (**a**) Effects of prescribed burns (PB), wildfires (WF), and fire exclusion (FE) on fruit production across different communities (see details in text), evaluated on number of native fruit-bearing plants; (**b**) Effects of the timing of prescribed burns (*FRS* final rainy season, *BDS* beginning dry season, *MDS* mid dry season) on fruit production across the entire community of evaluated number of native fruit-bearing plants in the Cerrado and Cerrado-Amazon transition regions (Source: R.M. Falleiro, unpublished data)

burns resulted in the most effective timing of prescribed burns on edible fruit production (Fig. 3.4b).

Similar results were found by Falleiro et al. (2024), in which *Hancornia speciosa* (Apocynaceae) (known as *mangabeira*) populations exposed to different fire histories showed higher tree mortality, canopy damage and loss of reproductive structures in areas affected by wildfires, which also had the lowest fruit production per plant compared to PB (prescribed burns) and FE (fire exclusion) areas. The authors also found higher mortality in FE areas than PB ones, probably due to plant diseases in areas with extended FE.

This study supports the Integrated Fire Management (IFM) National Policy currently implemented in federally protected areas in Brazil—focused on early dry season prescribed burns and many other preventive actions such as environmental education, monitoring, capacity building, etc—as an effective management strategy. Thus, burns conducted at the start of the dry season, as traditionally practiced by the Indigenous communities, which increase fruit production is highly valued by the communities, alongside other benefits such as biodiversity conservation and greenhouse gas emission reductions (Santos et al. 2021), food security, and cultural preservation. These outcomes have reinforced Integrated Fire Management in Brazil and prompted significant environmental policy shifts (see Sect. 3.5).

3.3.2 Pest Control

Burning is also necessary and frequently used to clear the area of deadly snakes, spiders, and mosquitoes and for general pest management in North Amazonian communities (Rodríguez et al. 2011; Bilbao et al. 2019). Local Indigenous farmers mentioned an accelerating need for pest management in farming areas associated with climate change. This supports the claim made by recent studies that invasive pest populations are rising due to climate change (IPCC 2021).

3.3.3 Hunting and Fishing

Indigenous Peoples across South America have long utilized fire as a tool in their hunting and fishing practices. These fire-based techniques are adapted to the diverse ecosystems of the continent and are deeply rooted in the communities' cultural traditions and ecological knowledge. Here are several examples illustrating how fire is integral to Indigenous methods of hunting and fishing:

3.3.3.1 Fire Use in Hunting

Some Indigenous communities practiced a technique known as "fire-stick farming" or "fire hunting". This involves controlled burning specific areas to promote new vegetation growth and attract game animals. These practices create open areas with fresh grass and shoots, attracting herbivores. This, in turn, could attract predators, making prey more accessible. An example is the Pemón burning the savanna to stimulate the production of tender grass shoots, transforming these places into feeding sites and hunting grounds for deer (Bilbao et al. 2019). Another example is the deer and rhea hunting practices in the Brazilian Cerrado (Fagundes 2019a), where hunters also burn to stimulate grass shoots. But while hunting deer, the hunter must dodge from the mammal's sense of smell, the rhea hunting plays on bird's vision inside burned areas.

Another interesting example is the use of fire to hunt in the grasslands and bush of the Chaco region in Argentina. This method, widely practiced by Indigenous Peoples, involved setting controlled fires to flush out and capture games. The charred carcasses of animals caught in the fire were immediately collected and consumed. Later, the Indigenous hunters would return to the burned area to hunt the numerous deer attracted by the salty ashes or the tender new growth of grasses (Kunst et al. 2015).

Another objective of burning in hunting is to increase visibility. One example is the Xavante, an Indigenous group living in the state of Mato Grosso, Brazil, who are highlighted for their sophisticated use of fire as a land management tool. They employ fire to clear underbrush, create pathways, improve visibility for hunting, and manage resources such as game and plants (Mistry et al. 2005). These authors' study also underscores how these practices help maintain biodiversity and ecological balance in the Cerrado.

Fire is also widely used in hunting campaigns to ambush and trap game, improving hunting efficiency during dry periods. For example, Macushi and Wapishana from Guiana Shield Indigenous territories are burning large patches that trap the tortoises and armadillos commonly found in dug holes in forested areas, reducing time and effort. This method is often used for Pemón Indigenous Peoples to hunt larger mammals, such as deer or peccaries, by creating a semi-circle of fire that slowly pushes the animals toward waiting hunters. Fires are also used during collective communal hunting, allowing for better fire and game management and sharing of wild meat with family and neighbors in Pemón, Macushi and Wapishana communities (Bilbao et al. 2019; UNESCO 2024).

The decline in game populations is a common claim by South American Indigenous Peoples and local traditional communities in their territories. However, an interesting study by Welch (2015) conducted in Pimentel Barbosa Xavante Indigenous Reserve, Brazil, indicates that while past research suggested a decline in game populations within the Reserve, Xavante elders now believe the game is more abundant than ever. Younger men attribute this change to reduced hunting by newer generations and a "refuge effect" from surrounding agribusiness deforestation and local vegetation recovery. However, decreased hunting among youth raises concerns about losing traditional hunting knowledge, especially group hunting with fire, which is integral to Xavante culture. Elders use ceremonies to motivate youth participation in group hunts, linking these practices to cultural identity and ensuring knowledge transfer across generations.

3.3.3.2 Fire Use in Fishing

Some Indigenous fire techniques involve burning small segments of the forest near riverbanks (a practice also related to shifting cultivation). This improves access to water resources and positively affects aquatic ecosystems, for example, by enhancing river nutrients, which can increase fish yields. The Pemón Peoples in Venezuela, for example, use fire to fish in this indirect way, but it also directly makes insects jump into the water and attract fish (Rodríguez 2007; Bilbao et al. 2019).

3.3.4 Livestock Ranching

After European colonizers arrived and settled, new practices, including cattle raising, were introduced. These included large-scale cattle ranching, which brought new fire regimes. Fires were used to clear dense brush and create pastures, fundamentally changing the landscape and local ecosystems' ecology and fire regimes (Moura et al. 2019; Vidal et al. 2023).

Despite the significant changes to their landscapes, Indigenous groups maintained cultural practices related to fire, albeit modified for the new economic and environmental context. Thus, Indigenous Peoples and also local traditional communities adopted cattle ranching and adapted their regular burns to maintain pasture health, prevent shrub and tree encroachment, and stimulate the growth of grasses preferred by cattle (Posey 2004; Kunst et al. 2015; Eloy et al. 2018; Vidal et al. 2023) (Fig. 3.2a, b).

Below are some practices carried out by Indigenous Peoples and local traditional communities in two of the more productive cattle ranching in South America and globally:

- Argentina: In this case, the example is related to the creole people (*criollo* in Spanish) populations residing on the islands of the Paraná River in Argentina. Creole refers to the descendants of European migrants and Indigenous communities, illustrating, in this case, a non-Indigenous group but a local traditional community that utilizes fire historically for diverse sustainable practices such as cattle ranching. This does not present significant variations with other peasant communities, *caboclos* or riverside communities of South America, so we take the case of the islanders of the Paraná River as an example of this type of population. Fire here profoundly integrates into the islands' daily life and cultural practices. It is essential for cooking, heating, and managing territorial boundaries, reflecting its role in productive and domestic spaces (Millán et al. 2022; Ferrero et al. 2023).

  One of the most widespread uses of fire for the *isleños* (islanders) is related to the productive practices of renewing pastures for cattle ranching. The burning aims to eliminate dry grasses, which lose nutritional value for cattle. These burns are carried out to sprout new, greener, and more tender pastures in more significant quantities (Fig. 3.1b). Generally, this practice is performed on lands where grazing has already occurred, although in some cases, new areas are burned to expand the productive space. These burnings take place towards the end of winter and the beginning of spring, spanning from mid-August to the first weeks of October. It is carried out by a single man and sometimes with a small group of neighbors to control the fire's progress. These fires are usually started considering the strength and direction of the winds and the presence of watercourses and ponds, which can function as physical barriers. The forecast for rain in the following days is also usually considered, looking for these to extinguish any remaining embers and favor the pastures' quick regrowth (Millán et al. 2022; Ferrero et al. 2023).

- Brazil: In the Brazilian context of Jalapão (Fagundes 2019b, 2021), the *Quilombola* communities identify five types of beneficial fires used during the dry season, namely: firebreaks (*aceiros*, in Portuguese), doorway fires (*fogos de porta*, in Portuguese), controlled burns (*queimadas*, in Portuguese), precision fires (*fogos de precisão*, in Portuguese), and slash-and-burn (*fogos de roça*, in Portuguese) fires. Most of them are related to improving and stimulating the regrowth of pastures for cattle raising.

  Firebreaks are small, preventive burns conducted at the end of the rainy season to protect facilities or preserve grasslands for summer burning. Doorway fires manage cattle by preventing them from wandering into wet, boggy areas at the onset of the dry season. Controlled burns revitalize paths by encouraging the regrowth of tough grasses in moist soils as summer begins. Precision fires, executed only in lowlands during the peak dry season, address urgent needs and utilize the area's moisture retention capabilities to facilitate regrowth (Fagundes 2019a, b, c, 2021).

  Slash-and-burn fires, typically set between late September and early October during the "heat of the earth" period, prepare fields for pasture or cassava by leveraging the warm, moist soil conditions. These small, controlled fires produce visible smoke plumes over the region, contained by strategic firebreaks to prevent wider spread (Fagundes 2019a, b, c, 2021).

  Another example is in the Serra da Canastra region of Minas Gerais, where the use of fire is central to the traditional agropastoral system that produces the renowned "Canastra cheese", linked to the area's unique micro biodiversity and protected as a state and national heritage. This system, documented since the seventeenth century by observers like Augustin Saint-Hilaire (2000), involves burning native pastures to support cattle for milk and cheese production. Despite conflicts with environmental agencies, especially post-1972 with the establishment of Serra da Canastra National Park during Brazil's military regime, the local *canastreiros* were recognized as traditional people in 2018. Efforts are now being made to integrate these communities into fire management practices to preserve their heritage and the environment (Fagundes 2019a, b, c, 2021).

- Guiana Shield countries: In the case of ranching for Makushi and Wapishana Peoples, belonging to the Carib and Arawak Linguistic Families, respectively, in the Essequibo and Roraima regions of the Guiana Shield, savanna patches used for grazing are burnt once annually by the vaquero during the middle of the dry season, which usually begins in February and March and also in October and

> November. This approach to ranching favors grass regrowth since it ensures that the cattle feed within manageable boundaries instead of moving away from the community to graze. According to ranchers, this would increase the cost (energy, time) of ranching and reinforce the limits of the territory). Other uses of fire in ranching are to remove ticks from cows and herds and mark them (Rodríguez et al. 2013; Bilbao et al. 2019; UNESCO 2024).

Frequently using fire by traditional local cattle ranchers in tropical savannas is often a significant source of conflict between conservationists and Indigenous and Local Communities (IPLCs). However, some studies revealed the conservationist role of traditional pastoral management in wildfire prevention compared with governmental Protected Areas. For example, within the Brazilian savanna, Eloy et al. (2018) compared fire regimes in community-managed areas with those in government-managed areas. The findings revealed that community-managed regions, which use controlled burns for ranching productive activities and landscape management, experience less damage from wildfires than government-managed areas, which are often affected by large, late dry-season wildfires. The study highlights the need for a deeper understanding of traditional fire management practices to improve wildfire management policies in tropical savannas.

On the contrary, associated with agro-industrial large-scale practices, since the 1960s, the cattle population in the Amazon Basin has grown from five million to over 70–80 million animals. This expansion has replaced about 15% of the Amazon forest, with approximately 80% of deforested areas—around 900,000 km^2—converted to pastureland. Cattle ranching has advanced along the new agricultural frontier known as the "Arc of Deforestation," stretching from the eastern Brazilian Amazon (Maranhão and Pará) through the southern Brazilian Amazon (Tocantins, Mato Grosso, and Rondônia, considering the Brazilian Amazon alone, it is estimated that 80% of deforested areas are occupied by pastures (MMA 2018), into the Bolivian rainforests, and further into the Andean Amazon regions of Bolivia, Peru, Ecuador, Colombia, and Venezuela (Veiga et al. 2002).

Similarly, the Amazon Assessment Report (2021) reported a similar loss of primary forests in Amazonia, adding at least 1,036,080 km2 of degraded Amazonian forests due to the agricultural expansion, mainly cattle ranching as the principal driver of deforestation and degradation in the Amazon (Berenguer et al. 2021). In addition to greenhouse gas emissions from deforestation, pastures in the Amazon further contribute to emissions through regular burning and bovine enteric fermentation (Bustamante et al. 2012), causing significant changes in soil properties, including compaction and altered nutrient concentrations (Berenguer et al. 2021). Unfortunately, although Indigenous Peoples and Local Community's (IPLCs) cattle ranching practices differ significantly from the large-scale, intensive agro-industrial cattle ranching observed in the Amazon Basin, this last publication, similar to others, does not emphasize the difference between agro-industrial activities and Indigenous and Local Communities' traditional practices. Frequently, the positive effects of cattle ranching traditional practices of traditional IPLCs are commonly underestimated and included as other threatening risk factors of large-scale deforestation.

3.3.5 Fire Risk Reduction Through Fuel Management

Indigenous Peoples have long used fire as a tool to manage landscapes and mitigate the risk of wildfires, a practice that demonstrates an intricate understanding of local ecosystems. These traditional techniques, known as cultural or prescribed burns, involve the strategic and controlled application of fire to clear out dead and accumulated organic materials that can fuel large fires. It is important to mention three key aspects of how Indigenous Peoples use fuel management to reduce fire risks. On the one hand, Indigenous uses of fire reduce fire severity and spread. It is due to controlled burns reduce the volume of flammable materials, directly decreasing the intensity and spread of subsequent wildfires (Bilbao et al. 2009, 2010; Schmidt et al. 2018; Pivello et al. 2021). But it also contributes to ecosystem health. These fires help regenerate fire-adapted plant species, control invasive species, and maintain habitat conditions for various wildlife species.

Additionally, the heterogeneity created by patches with different fire histories promotes pyrobiodiversity. Pyrobiodiversity, defined by Ponce-Calderón et al. (2022) refers to the variations in how species and ecosystems respond to fire, highlighting its role in creating diverse habitats and enhancing biodiversity. This concept is crucial in fire-dependent ecosystems where periodic fires are vital for maintaining ecological balance. Species within these ecosystems may depend on specific fire regimes for their life processes, including germination, growth, and reproduction. Therefore, managing fire regimes is essential for preserving and promoting biodiversity in such environments (Bowman et al. 2016; Kelly et al. 2020). Indigenous uses of fire also contribute to nutrient cycling by decomposing dead biomass and converting it into charred material and ashes, which enriches the soil and promotes healthier plant growth (Hoffman et al. 2021; Roshan and Biswas 2023).

There are two types of fuel load management (sometimes combined): dry season burning, namely Savanna Fire Management (SFiM), and Patch Mosaic Burning (PMB), described below.

3.3.5.1 Early Dry Season vs. Late Dry Season Burning

One of the Indigenous core burning practices in tropical savanna areas, or fire-prone ecosystems, is the implementation of controlled burns early in the dry season (EDS) or at the end of the wet season, when the vegetation is less dry, with calmer winds and fires are less intense. This helps reduce the fuel load and lowers the risk of high-intensity wildfires later in the season (LDS), which can be more destructive and more challenging to control. This Indigenous technique is also referred as Savanna Fire Management (SFiM), which is the name of the tropical north Australian savanna experience based on aboriginal traditional knowledge and practices of fire management (SFiM) (Russell-Smith et al. 2009, 2017).

3.3.5.2 Patch Mosaic Burning (PMB)

On the other hand, Indigenous patch mosaic burning refers to creating a landscape mosaic of various patches representing different fire histories. This method aims to enhance biodiversity and reduce fire hazards by introducing spatial and temporal diversity in fuel loads and species composition (Parr and Brocket 1999; Bilbao et al. 2010, 2021) (Fig. 3.2c). The concept revolves around using fire as a tool for ecosystem management, informed by deep knowledge of the land, seasons, and ecological processes. It involves "cool burns," controlled fires applied at specific times and places to manage vegetation, promote certain plant species or hunting opportunities, and reduce the risk of more extensive, uncontrolled wildfires. This practice demonstrates an intricate understanding of fire's role in landscape management and biodiversity conservation (Bilbao et al. 2010, 2022; Kimber and Friedel 2015).

3.3.6 Non-land Management Uses of Cultural Burnings

3.3.6.1 Cure of Diseases

Indigenous Peoples use fire to prepare traditional medicines, heal (smoke out evil spirits), perform ceremonial practices, chase away dangerous spirits, or sometimes call them (e.g., rain spirits). They also smoke their people to chase spirits away before meetings or essential activities (Rodríguez et al. 2011; Mistry et al. 2016; Bilbao et al. 2019; UNESCO 2024).

3.3.6.2 Ceremonial Practices

Indigenous Peoples of South America employ fire not only for land management and ecological purposes but also within a variety of ceremonial contexts. These ceremonies are deeply embedded in cultural traditions and spiritual beliefs, reflecting the profound relationship between these communities and their natural environments. In these contexts, fire symbolizes purification, renewal, and the cyclical nature of life. Here are some examples:

In the Andes region, the Inti Raymi, or Festival of the Sun, is an ancient Inca religious ceremony in honour of the god Inti (sun god), revered by the *quechua* peoples in Cuzco Peru and other cities of the ancient Inca empire. It coincides with the winter solstice and harvest time. Large bonfires are part of this celebration, and people throw in offerings to be consumed by the fire. The fire's light and heat are meant to reinvigorate the sun as the days begin to get shorter, ensuring the sun's return and thereby guaranteeing a fruitful growing season (Vega and Guzmán Palomino 2005).

In the Indigenous lands: Krikati, Governador, Porquinhos (all 3 in Maranhão state, Brazil), Maxakali (Minas Gerais state, Brazil), and Parque Indígena do Xingu (Mato Grosso state, Brazil) - it was registered that the Indigenous Peoples use fire in rituals and celebrations for swidden agriculture practices (Araujo et al. 2020).

We showed here a selection of the main uses of fire by traditional South America's Indigenous and Local Communities (IPLCs). However, the literature shows a vari ety of practices, indicating the great diversity of techniques adapted to environmental conditions, guided by deliberate actions rather than random occurrences, and responding to different purposes. This continuous use of fire reflects a complex interaction between cultural traditions, ecological management, and modern regulatory challenges, underscoring its multifaceted role in South American's social and environmental landscapes.

3.4 Benefits, Threats and Challenges of Indigenous and Local Communities (IPLCs) Traditional Fire Practices

3.4.1 Benefits

Indigenous Peoples possess valuable knowledge for managing lands with coexisting fire-sensitive and fire-prone ecosystems, conserving biodiversity, reducing deforestation, reducing GHG emissions and mitigating climate impacts. Indigenous Peoples in these landscapes have demonstrated sophisticated knowledge of fire regimes, implementing this knowledge to make fires to enhance biodiversity and resource management, as well as reduce risks of dangerous wildfires and out-of-control fire events (Mistry et al. 2016; MacLean et al. 2023; UNESCO 2024).

Thus, emerging research underscores the pivotal role of Indigenous land-use practices contributing to these benefits, particularly controlling deforestation and reducing CO_2 emissions. Analysis of satellite imagery indicates that Indigenous lands exhibit lower rates of deforestation and habitat conversion, along with reduced greenhouse gas (GHG) emissions compared to surrounding areas. For instance, Stevens et al. (2014) found that in Bolivia, from 2000 to 2010, only 0.5% of the Indigenous territories were deforested, compared to 3.2% overall in the Bolivian Amazon, signifying that rates of deforestation were six times lower on Indigenous lands than in other forests. In the Brazilian Amazon, Petersen and Stevens (2014) observed that between 2000 and 2012, forest loss was merely 0.6% inside Indigenous lands, against 7.0% outside, indicating a rate more than ten times lower. Walker et al. (2020), when analyzing net changes in carbon stock from 2003 to 2016 across the Amazon region within Indigenous territories (ITs) and protected natural areas (PNAs), compared to other lands, the data shows distinct differences. Amazon ITs, excluding areas that overlap with PNAs, underwent a 0.1% net change, corresponding to a decrease of 24 million tons of carbon (Mt C), the lowest loss among all land categories. In contrast, PNAs, not considering areas of overlap with ITs, experienced a 0.6% net loss, which equals a reduction of 96 Mt. C. In another sense, the other land category exhibited a significant and extraordinary 3.6% net loss, amounting to a decrease of 1160 Mt. C (12 and 50 times higher than PNAs and Its respectively). These findings suggest that both ITs and, to a lesser degree, PNAs, independently and collectively, were more effective than the Other Lands category in achieving a balance between carbon losses and gains, thereby preserving their overall carbon stocks.

Similarly, Carranza et al. (2014) demonstrated that in the Brazilian Cerrado, Indigenous lands experienced lower rates of habitat conversion from 2002 to 2009 than unprotected areas. Similarly, Flantua et al. (2013) noted that in the western Indigenous Pemón-inhabited sector of Canaima National Park, Venezuela, the deforestation rate between 1986 and 2006 was significantly lower at 0.037%, associated with traditional shifting cultivation farming, compared with the 10 km unprotected buffer zone at 0.17% and the average national rate at 0.63%.

These examples and figures underscore the effectiveness of ecological benefits of Indigenous fire management practices in promoting environmental health, enhancing biodiversity, reducing wildfire risks, and mitigating climate change impacts. Such practices not only maintain the ecological and cultural landscapes but also offer scalable solutions that can be integrated into national and global environmental management strategies, especially under extreme climate and land use change scenarios. The continued documentation and integration of these practices are crucial for sustainable global change mitigation and adaptation strategies development, as well as biodiversity conservation efforts.

3.4.2 Threats

Many threats have emerged in recent decades that menace Indigenous Peoples and local traditional fire practices. These include climate change, land-use change, changes in population density patterns and migrations, loss of territory (or land rights), and suppression policies that prohibit or restrict traditional uses of fire. In addition, a new fire dynamic in the region is promoting fires of great intensity and severity ('megafires') that are impossible to control, affecting territories not usually exposed to this type of fire (e.g., wetlands) (Bilbao et al. 2020; UNEP 2022, MacLean et al. 2023).

Since the conquest of America, the uses of fire made by Indigenous Peoples have been frequently prohibited, sanctioned and persecuted. The prohibition of traditional uses of fire has affected their basic productive and reproductive practices, as well as the seizure of their lands and appropriation of their natural resources. These restrictive practices have continued to the present day, where the use of fire in burning practices is still prohibited in many countries (Mistry et al. 2019; Moura et al. 2019; Bilbao et al. 2022; Meza et al. 2023). Most policies for conserving natural resources, forests, grasslands, and savannas particularly affect Indigenous and rural communities, which are the most heavily dependent on fire. Zero-fire policies not only aim to protect nature but usually result in a restriction on the livelihoods of Indigenous and rural communities. This scenario coincides with Indigenous Peoples' decreased management of these forests, including their traditional burning practices using their traditional knowledge systems (Barradas et al. 2020; Fagundes 2021; Smith 2022).

Similarly, the advent of climate change advances, characterized by larger heat waves, hotter temperatures and longer and more intense dry seasons (IPCC 2021; UNEP 2022), together with the increase in wildfire risk, poses threats to ecosystems, wildlife, people and livelihoods. Indigenous and Local Communities (IPLCs) are among the most vulnerable to the increased multidimensional challenges associated with the increasing wildfire risk in their territories. Here are some negative impacts represented by one of the main threats represented by the new dynamic of fire related to the increase of severe wildfires:

Ecosystem Services: Wildfires can disrupt essential ecosystem services of Indigenous territories, including water purification, soil fertility, and carbon sequestration. The loss of these services can have cascading effects on both natural and human systems.

Loss of Natural Resources and Biodiversity: Wildfires can lead to the destruction of Indigenous and local traditional communities' habitats, resulting in the loss of plant and animal species. This biodiversity loss affects ecosystems' resilience and undermines their capacity to adapt to changing environmental conditions.

Erosion: The intense heat generated by wildfires can contribute to soil erosion, impacting the fertility of the land. Erosion can further lead to water quality degradation and pose challenges for agriculture and other land-based activities.

Threats to Food Security: Wildfires can directly impact agricultural Indigenous practices, destroying crops and reducing yields. The resulting food shortages can threaten the food security of communities dependent on agriculture.

Territorial Impact: Indigenous Peoples and local traditional communities often have deep connections to their territories, relying on them for cultural, spiritual, and economic sustenance. Wildfires can threaten these territories, leading to displacement, loss of cultural sites, and challenges to traditional livelihoods.

Infrastructure and Settlements: The spread of wildfires can endanger human settlements and critical infrastructure. Evacuations, property damage, and the destruction of infrastructure can have long-lasting social and economic implications for affected communities.

Cultural Impact: Wildfires can destroy cultural heritage sites and artefacts, eroding cultural identity's tangible and intangible aspects. This loss is particularly significant for Indigenous Peoples, whose cultural practices are intricately tied to the natural environment.

Increase in Natural Vulnerability: The overall increase in the frequency and intensity of wildfires contributes to the natural vulnerability of ecosystems and human communities. It creates challenges in adapting to rapidly changing environmental conditions and requires innovative strategies for resilience.

Additionally, the combination of multiple and interconnected socio-environmental and economic threats is leading to a heightened vulnerability among Indigenous and Local Communities in a multifaceted and complex way with severe consequences for their livelihoods, cultures, and overall well-being putting them at risk of disappearing and causing forced displacements (IUCN 2008; ECLAC 2014). According to the ECLAC (2014), many Indigenous cultures live in conditions of extreme demographic fragility and are in danger of physical or cultural disappearance, particularly in Bolivia, Brazil, Colombia and Peru. There are also approximately 200 Indigenous groups living in voluntary isolation in conditions of extreme vulnerability in Bolivia, Brazil, Colombia, Ecuador, Paraguay, Peru and Venezuela. Among the most severe impacts we can mention (ILO 2018):

Increased Vulnerability to Diseases: Climate change can alter the distribution of vector-borne diseases like malaria and dengue fever, exposing Indigenous populations to new health risks. Limited access to healthcare facilities and resources exacerbates the impact of these diseases.

Cultural Erosion and Disappearance: The disappearance of Indigenous cultures signifies the loss of valuable knowledge and practices that have historically enabled adaptation to changing environmental conditions. This cultural erosion is a significant loss not only for Indigenous Peoples but also for humanity's collective understanding of sustainable living.

Migration to Urban Areas: Environmental pressures and the loss of traditional livelihoods often force Indigenous Peoples to migrate to urban areas in search of employment and economic opportunities. This migration exposes them to new challenges, including discrimination, exploitation, and the risk of losing their cultural identity in unfamiliar urban contexts.

Territorial circumscription, increased "sedentarization", and economic changes threaten Indigenous Peoples and local traditional communities in South America by restricting access to ancestral lands, disrupting traditional lifestyles, and causing environmental degradation. These forces undermine cultural practices, food security, and self-sufficiency while prioritizing external interests over Indigenous rights and autonomy.

The loss of Indigenous cultures erodes unique ways of life, as demonstrated in the archaeological and present research, which has an irreplaceable legacy of adaptation and coexistence with biodiversity and fire management of their territories.

3.4.3 Challenges

Addressing Indigenous Peoples' and local traditional communities' vulnerability requires holistic approaches that consider protecting Indigenous rights, traditional knowledge and cultural practices and territories. Collaborative efforts involving Indigenous communities in decision-making over their territories are crucial to developing strategies that promote resilience, respect cultural diversity, and adaptation to the effects of climate change on these vulnerable populations.

Despite the growing body of evidence underscoring the value of traditional fire management practices, significant challenges remain in policy transformation. These include respecting the original Indigenous Peoples' cultural legacies and ancient lifestyles and acknowledging the *criolla* (creole) cultures that have adapted their traditional knowledge to new socio-environmental and economic realities post-colonization.

Another key challenge is ensuring that Indigenous and Local Communities have a voice and active participation in actions and decision-making processes, particularly regarding fire management policies and strategies on their territories and the protection and preservation of their cultural heritage (Bilbao et al. 2019; UNESCO 2024). Similarly, the following section will discuss major challenges and contributions of recognizing Indigenous and traditional fire management approaches specific to each locality, emphasizing their crucial role in developing more sustainable and inclusive fire management strategies.

3.5 The Role of Indigenous and Local Communities' Cultural Heritage in Shaping New Fire Management Paradigms in South America

As evidenced by some South American experiences, traditional and Indigenous fire knowledge systems and practices have contributed to and inspired change through a new fire management paradigm, moving away from the classic and dominant suppression policies towards more effective and nature-based forms and consolidating the cultural component of Integrated Fire Management. While Indigenous and local traditional fire management has many benefits (discussed in the previous section), the pillars of change are the ability to reduce the incidence and scope of out-of-control wildfires, protect carbon stocks and reduce greenhouse gases, make ecosystems more resilient, and preserve biodiversity and ecosystem functionality.

Effective traditional fire management policies also hinge on proper legal frameworks, institutional capacities, community support, and a commitment to protecting these ways of life. Although fire management policy frameworks typically exhibit top-down control, recognizing traditional management knowledge and practices offers clear benefits for enhancing the effectiveness of fire management strategies. Thus, acknowledging locality-specific Indigenous and traditional approaches to fire management is crucial (Fig. 3.5). Some case study examples that illustrate the important contribution of Indigenous and Local Communities to new paradigms of fire management in South America are presented in Boxes 3.1, 3.2 and 3.3.

Fig. 3.5 (**a**) Assessment of traditional fire management carried out in Aldeia Ipatse in Xingu Indigenous Land (Photo: Rodrigo M. Falleiro); (**b**) Survey of traditional knowledge of fire (Xingu Indigenous Lands - Querência/Brazil - MT 2017) (Photo: Rodrigo M. Falleiro); (**c**) Chief Manoel Irantxe (Manoki ethnicity) planning traditional Cerrado management with volunteer brigadiers (Irantxe Indigenous Land—Brasnorte/Brazil—MT 2008) (Photo: Rodrigo M. Falleiro); (**d**) Indigenous, researchers, fire-fighters and governmental policy makers participants in the workshops to formulate a legitimate intercultural and participatory fire management policy in Venezuela. (Photo: Maiquel Torcatt); (**e**) Presentation of rich pictures illustrating the challenges for intercultural and participatory fire management policies in Canaima National Park (CNP, Venezuela). These insights are drawn from Pemón Indigenous leaders, academics from various universities, policymakers, and firefighters during the workshop titled "Design of Action Plans on Climate Change: Integrating Local Indigenous Perspectives with Academia and Institutions in the CNP," held at IVIC (Spanish acronym—Instituto Venezolano de Investigaciones Científicas -, for Venezuelan Institute of Scientific Research), Altos de Pipe, in January 2017 (Photo: Maiquel Torcatt); (**f**) Learning together academics, students, emblematic grandfather (*abuelo*) and other inhabitants of Kavanayén Indigenous community, Gran Sabana, Canaima National Park, Venezuela, about traditional Indigenous fire practices (Photo: Bibiana Bilbao)

Box 3.1 Case Study: Building Integrated Fire Management in Brazil with Indigenous and Local Communities ECLAC Inclusion

Over the past decade, Brazil has made significant progress in protecting against wildfires, particularly in its fire-prone tropical savannas, by shifting from fire exclusion to an Integrated Fire Management (IFM) policy. This IFM policy approach prominently incorporates Indigenous and Local Communities' traditional knowledge, overcoming earlier biases against local fire uses that prevailed in protected areas and Indigenous lands until the early 2000s (Berlinck and Batista 2020; Falleiro et al. 2021; Schmidt et al. 2018; Eloy et al. 2019; Bilbao et al. 2020).

Although the importance of fire to rural populations and Brazilian ecosystems has been recognized since the 1960s (Brasil 1965, 1979; Coutinho 1982; Miranda 2010), fire exclusion strategies dominated until the early 2000s (Moura et al. 2019). In Indigenous lands and traditional territories, there was no official program for wildfire protection despite these areas being primary targets of prejudice regarding fire use (Falleiro et al. 2021). Progress accelerated in 2013 with two key initiatives: the Cerrado-Jalapão Project and the Federal Brigades Program (BRIFs). The Cerrado-Jalapão Project, in collaboration with Germany, introduced prescribed burns in Conservation Units to protect the Cerrado biome. At the same time, the BRIFs program, under Prevfogo/Ibama, addressed longstanding conflicts over fire use in Indigenous lands, marking a pivotal shift toward modern fire management strategies in Brazil (Falleiro et al. 2016) (Fig. 3.5 a–c).

Before the BRIFs Program, Indigenous communities had already rejected fire exclusion policies proposed in 2007. Following criticism from Indigenous leaders, technicians from Prevfogo/Ibama and Funai began to understand the reasons behind the communities' systematic use of fire, uncovering a rich body of traditional fire knowledge that was incorporated into new protection strategies (Falleiro 2011). This experience highlighted the need to explore alternative fire protection measures, even if initial plans were not fully implemented due to limited resources.

Experiences from countries with similar ecosystems, such as South Africa and Australia, advances in research in Brazil (Miranda 2010), and global fire management guidelines (Myers 2006) supported the principles of traditional Indigenous management. As pressure mounted for government programs that respected Indigenous cultural practices, public environmental agencies like Prevfogo/Ibama and ICMBio made concerted efforts to improve dialogue with communities in over 113 protected areas (Barradas and Ribeiro 2021) (Fig. 3.5 a, b).

The creation of the BRIFs Program in 2013 included a component dedicated to protecting Indigenous lands, relying on local community members and valuing traditional knowledge. This enabled the implementation of prescribed burns on a large scale, supported by a survey of traditional knowledge

(continued)

among served diverse Indigenous groups (IBAMA 2018). Fire management on Indigenous lands rapidly expanded to cover six million hectares across the Cerrado, Amazon, and Pantanal biomes, with one million hectares undergoing prescribed burns annually (Falleiro et al. 2021). This approach reduced wildfire intensity and size, firefighting costs, and greenhouse gas emissions while conserving natural resources vital to Indigenous communities and protecting sociobiodiversity (Andrade et al. 2021; Santos et al. 2021; Oliveira et al. 2022).

The successful integration of traditional knowledge in Indigenous lands, *Quilombola's* territories, and protected areas strengthened IFM as an alternative to outdated zero-fire policies (Schmidt and Eloy 2020; Santos et al. 2021; Oliveira et al. 2022) (Fig. 3.5 a–c). New legislation was drafted in 2014 to align with the Brazilian Forest Code (Brasil 2012), promoting Integrated Fire Management as a fundamental principle in the National Policy of Integrated Fire Management, recently approved in the Senate (Brasil 2024).

Enhanced dialogue has improved understanding of traditional fire practices in Brazil, leading to the systematic recording of Indigenous and *Quilombola*'s fire management knowledge by Prevfogo/Ibama (Bilbao et al. 2019; Fagundes 2019b; Falleiro et al. 2021) (Fig. 3.5 a–c). This integration of traditional knowledge is vital in climate change context and offers a path for Indigenous and Local Communities (IPLCs) to be recognized and compensated for their contributions to environmental conservation (UNEP 2022; UNU 2015; Ribeiro and Pereira 2023).

Box 3.2 Case Study: Building Integrated Fire Management in Bolivia with Indigenous Inclusion

In 2011, Fundación Amigos de la Naturaleza (FAN), alongside local partners, began a pilot initiative for community fire management and climate change adaptation in the Chiquitano Forest region of Bolivia. This area extends along a corridor of protected areas in the municipalities of San José de Chiquitos and Roboré, Santa Cruz, involving 36 Indigenous and rural farming communities and 3 Mennonite colonies. In 2013, the initiative expanded to 20 communities in Riberalta, Beni, and by 2016, it included five communities in Puerto Gonzalo Moreno, Pando.

The initiative promoted best practices in fire management by creating an organizational, planning, and coordination system at the community and intercommunity levels, supported by a monitoring and early warning system for wildfire risks. This approach helped in generating and disseminating timely information to guide decision-making during various stages of wildfire

(continued)

management. Local institutional capacities were further enhanced through regular training on geographic information systems tailored for forest and wildfire monitoring, aimed at technicians from government institutions and civil society organizations.

These efforts have been vital in combating deforestation and maintaining the ecological integrity of forests that host diverse flora and fauna and are culturally significant for Indigenous communities (FAN 2019; Bilbao et al. 2020; Maillard 2023. To bolster these efforts, the Bolivian Chamber of Deputies approved the Legislative Project "Use and Management of Burns" in 2018, which led to the enactment of Law 1171 in 2019 (Bolivia 2019). This law establishes guidelines for Integrated Fire Management and introduces administrative penalties for unauthorized burning activities, promoting controlled and planned burning practices under the mandate of the Ministry of Environment and Water.

Law 1171 (Bolivia 2019) applies to both individuals and legal entities using fire in their productive operations. It sets a framework for the rational use and management of fire, particularly in agriculture and livestock, and establishes a regime of sanctions for unauthorized burns. The law allows burning authorizations for agricultural purposes for up to 3 years and up to 5 years for livestock-related activities, with a temporary period for regularizing past offenses. However, while the 2009 Political Constitution of Bolivia protects Indigenous rights, including cultural identity and traditional knowledge, these rights are not explicitly safeguarded in fire management laws like Law 1171 (Bolivia 2019). The newly enacted "Condor Law" (Law No. 1525) of November 2023 amends the Penal Code to enforce stricter penalties for damaging fires in protected areas, aiming to deter illegal burning and enhance protection under Environmental Law No. 1333.

Despite integrating the MIF (Integrated Fire Management) approach into its legislation (Meza et al. 2023), Bolivia's fire management policies remain focusing on suppression and penalties. Nevertheless, organizations like Fundación Amigos de la Naturaleza (FAN) have advocated for a more holistic, community-based strategy over the past decade to enhance ecosystem resilience and reduce risks for vulnerable communities.

Box 3.3 Case Study: Building Integrated Fire Management in Venezuela with Indigenous Inclusion

Over the past two decades, Venezuela has developed an intercultural and participative fire management approach, beginning in Canaima National Park (CNP) in the northern Amazon basin and expanding nationally. This shift integrates the knowledge of the Pemón Indigenous Peoples, scientists, and government institutions. Canaima National Park, a UNESCO World Heritage

(continued)

site since 1994, is vital for Venezuela's hydroelectric energy, providing 80% of the country's electricity via the Caroní River dam system managed by CORPOELEC-EDELCA. Tensions arose between the Pemón's traditional fire use and fire exclusion policies enforced by the hydroelectric company and park management (Gómez et al. 2000). The shift in fire management resulted from recognizing the Pemón's fire practices, incorporating scientific research, and fostering collaboration among diverse stakeholders, transforming national fire management policies. Here are the key steps in this integrative process:

- Collaborative Long-Term Fire Experiments: In 1999, researchers from Universidad Simón Bolívar, funded by the Ministry of Science and Technology, initiated a long-term fire experiment to study fire behavior and its impacts on vegetation, soil, and the atmosphere. Conducted with the Pemón communities and CORPOELEC fire brigades, these studies replicated traditional burning practices across the savanna-forest gradients of the Gran Sabana in Canaima National Park (CNP). The research revealed significant variability in fire intensity and its effects on biomass and biodiversity, showing that fire exclusion increased the risk of high-intensity fires due to dry fuel accumulation, while areas with diverse fire histories exhibited varied patterns in biomass and species composition, supporting the "Patch Mosaic Burning" (PMB) strategy (Bilbao et al. 2009, 2010). These findings challenged existing fire exclusion policies and highlighted the need to incorporate Indigenous fire knowledge and practices, leading to more inclusive and effective fire management strategies (Bilbao et al. 2021, 2022).
- Preserving Indigenous fire knowledge: The research also highlighted Pemón's sophisticated fire-use knowledge, which includes activities such as shifting cultivation and controlled burns along savanna-forest boundaries to prevent wildfires. However, this cultural heritage was considered at risk due to socio-environmental threats, lifestyle changes, transculturation, and inadequate dissemination of its values within and outside the community.
- Efforts have been made to revitalize Pemón culture by training Indigenous researchers to document and share this knowledge. Intercultural spaces have been created for exchanging fire management perspectives among diverse stakeholders and interdisciplinary teams at various levels (Bilbao et al. 2017, 2019; Mistry et al. 2019; Eloy et al. 2019; UNESCO 2024) (Fig. 3.5 d, e, and f).
- Formulating an Intercultural Fire Management Policy: A sustainable fire management approach was developed in the Gran Sabana through collaboration among academics, government officials, and Indigenous communities. In 2018, an Integrated Fire Management (IFM) program with an intercultural vision was launched, encouraging mutual training exchanges

(continued)

between Indigenous communities and fire brigades to foster cooperation (Mistry et al. 2019; Eloy et al. 2019; Bilbao et al. 2019; UNESCO 2024).
- By 2019, the national forest fire brigades expanded to 10,000 personnel, with 3400 trained in IFM principles. By 2021, an IFM working group, including various stakeholders, was established to promote this approach nationwide through workshops and webinars. These efforts were incorporated into national training programs for forest firefighters, supported by an intersectoral team involving INPARQUES, the Ministry of Ecosocialism, and academics (Bilbao et al. 2021, 2022).
- Legal Framework and Future Directions: Venezuela acknowledges Indigenous Peoples' rights to maintain their practices and protects collective intellectual property but lacks specific laws for Indigenous traditional fire management. Restrictive laws, such as the Environmental Criminal Law of 2012, limit cultural burning practices without authorization, potentially conflicting with Indigenous traditions (Rodríguez et al. 2013; Mistry et al. 2016).

Current initiatives, funded by the Ministry of Science and Technology and the LANDMARC project (EU, H2020 Program), seek to establish a formal Integrated Fire Management (IFM) legal framework that combines Indigenous, technical, and scientific knowledge to support climate change mitigation and adaptation while preserving cultural heritage (Bilbao et al. 2021, 2022).

3.6 Conclusions

South America comprises one of the world's highest diversity of Indigenous Peoples (between 500 and 600 Indigenous groups). Fire has been a fundamental tool for obtaining and producing food, domesticating plants and improving soils, reducing wildfires, and promoting the health and cultural evolution of the South American Indigenous Peoples.

The design of sustainable fire management techniques by traditional Indigenous Peoples and Local Communities (IPLCs) represents clear benefits for systems that include the protection of biodiversity, water and nutrient cycles, deforestation and forest fires (protecting carbon stocks and reducing greenhouse gases).

Remarkably, these practices were often profoundly intertwined with Indigenous and Local Communities' traditional ecological knowledge and sustainable practices. They understood the delicate balance of ecosystems and used fire as a tool to manage landscapes rather than exploit them. This contrasts with contemporary agricultural and industrial practices that can have more detrimental environmental effects.

A paradox is the great diversity of Indigenous groups and Local Communities (IPLCs) in South America, yet few laws and actions protect traditional practices from fire. Likewise, despite the socio-environmental benefits, Indigenous Peoples and Local Communities (IPLCs) have little interference in decision-making and participation in their territory.

With a growing recognition across the world that combatting landscape fires is not only ecologically but also socially and economically unviable, drawing on traditional Indigenous and Local Communities (IPLCs) fire management could be a useful lens through which to find practical fire management solutions, but also lessons on how environmental governance could be structured and implemented more widely.

Several initiatives in South America aim to incorporate this knowledge and foster the inclusion and participation of Indigenous and Local Communities (IPLCs). These initiatives represent valuable lessons and insights that we hope can be expanded throughout the region. They seek mechanisms to continue learning from and creating fire management policies that are holistic, comprehensive, equitable, and in harmony with our cultures' heritage and nature.

The authors hope this chapter addresses gaps in academic and policymaker narratives by highlighting the extraordinary cultural diversity of South American Indigenous and Local Communities (IPLCs)' fire practices and encourage inter- and transdisciplinary studies that integrate the triple focus on the ecological, human, and fire management dimensions, essential pillars of our continent's pyrobiocultural legacy.

References

Agência Brasil (2023) Censo 2022: Brasil alberga 1,69 millones de indígenas—Norte y el Nordeste concentran el 75% de la población nativa de Brasil. https://agenciabrasil.ebc.com.br/es/geral/noticia/2023-08/censo-2022-brasil-alberga-169-millones-de-indigenas. Accessed 23 Feb 2024

Amazon Assessment Report (2021) United Nations Sustainable Development Solutions Network, New York, USA. https://doi.org/10.55161/AIZJ1133. https://www.theamazonwewant.org/spa-reports/. Accessed 4 Mar 2024

Andrade ASR, Ramos RM, Sano EE, Libonati R, Santos FL et al (2021) Implementation of fire policies in Brazil: an assessment of fire dynamics in a Brazilian savanna. Sustain For 13:11532. https://doi.org/10.3390/su132011532

Araujo Nascimento A, Ono K, Horta A (2020) Shaping landscapes, remaking knowledges: practices, myths and fire management in Xingu Indigenous Territory (MT, Brazil). Biodivers. Brasil 10(1). https://doi.org/10.37002/biobrasil.v10i1.1609

Athayde S, Larrea D, Ardayac ZL, Neves E, Rosero-Peña MC (2021) Chapter 10: Definition of indigenous peoples and local communities for the science panel for the Amazon 2021. In: Nobre C, Encalada A, Anderson E, Roca Alcazar FH, Bustamante M et al (eds) Amazon assessment report 2021. United Nations Sustainable Development Solutions Network, New York, USA. https://www.theamazonwewant.org/spa-reports/. Accessed 10 Jun 2024. https://doi.org/10.55161/AIZJ1133

Barradas ACS (2017) A gestão do fogo na Estação Ecológica Serra Geral do Tocantins. Brasil, Trabalho de conclusão de curso, Escola 9pNacional de Botânica Tropical/JBRJ

Barradas ACS, Ribeiro KT (2021) Manejo Integrado do Fogo: Trajetória da Estação Ecológica Serra Geral do Tocantins (2001 a 2020). Biodivers Brasil 11:139–152. https://doi.org/10.37002/biodiversidadebrasileira.v11i2.1739

Barradas ACS, Borges MA, Costa MM, Ribeiro KT (2020) Paradigmas da gestão do fogo em áreas protegidas no mundo e o caso da Estação Ecológica Serra Geral do Tocantins. Biodiversidade Brasileira 10:71–86. https://doi.org/10.37002/biodiversidadebrasileira.v10i2.1474

Barretto Filho HT, Ramos A, Bilbao B, Barra CS, Rodríguez C et al (2021) Amazon assessment report 2021. United Nations Sustainable Development Solutions Network, New York, USA. https://www.theamazonwewant.org/spa-reports/. Accessed 4 Mar 2024. https://doi.org/10.55161/AIZJ1133

Berenguer E, Armenteras D, Lees AC, Fearnside PM, Smith CC et al (eds) (2021) Amazon assessment report 2021. United Nations Sustainable Development Solutions Network, New York, USA. https://www.theamazonwewant.org/spa-reports/. Accessed 17 Feb 2024. https://doi.org/10.55161/AIZJ1133

Berlinck CN, Batista KLE (2020) Good fire, bad fire: it depends on who burns. Flora 268:151610. https://doi.org/10.1016/j.flora.2020.151610

Bilbao B, Leal A, Méndez C, Delgado-Cartay MD (2009) The role of fire on vegetation dynamics of upland savannas of the Venezuelan Guayana. In: Cochrane MA (ed) Tropical fire ecology: climate change, land use and ecosystem dynamics. Springer-Praxis, Heidelberg, pp 451–480

Bilbao BA, Leal AV, Méndez CL (2010) Indigenous use of fire and forest loss in Canaima National Park, Venezuela—assessment of and tools for alternative strategies of fire management in Pemón Indigenous lands. Hum Ecol 38:663–673. https://doi.org/10.1007/s10745-010-9344-0

Bilbao B, Rosales J, Marín S, Millán A, Salazar-Gascón R et al (2017) Chureta ru to pomupök integration of Indigenous and ecological knowledge for the restoration of degraded environments. In: Ceccon E, Pérez D (eds) Beyond restoration ecology: social perspectives in Latin America and the Caribbean. Vázquez Mazzini Editores, Buenos Aires, pp 331–353. https://doi.org/10.1111/rec.12833

Bilbao B, Mistry J, Millán A, Berardi A (2019) Sharing multiple perspectives on burning: towards a participatory and intercultural fire management policy in Venezuela, Brazil, and Guyana. Fire 2:1–33. https://doi.org/10.3390/fire2030039

Bilbao BA, Steil L, Urbieta IR, Anderson L, Pinto C et al (2020) Wildfires. In: Moreno JM, Laguna-Defior C, Barros V et al (eds) Adaptation to climate change risks in Ibero-American Countries—RIOCCADAPT report. McGraw-Hill, Madrid, pp 435–496

Bilbao B, Millán A, Vessuri H, Mistry J, Salazar-Gascón R et al (2021) To burn or not to burn? The history behind the construction of a new paradigm of fire management in Venezuela through interculturality: local actions of national and regional impact. Biodivers Brasil 2(2021):99–127. https://doi.org/10.37002/biobrasil.v11i2.1878

Bilbao BA, Millán A, Matany Luque M, Mistry J, Gómez-Martínez R et al (2022) An intercultural vision for integrated fire management in Venezuela. Tropical Forest Issues 61:39–46. https://doi.org/10.55515/CNUU7417

Bolivia (2019) Uso y Manejo Racional de Quemas. Ley n 1171 de 25 Apr 2019

Bond WJ, Van Wilgen BW (1996) Fire and plants. population and community biology series. Springer, Dordrecht, p 263. https://doi.org/10.1007/978-94-009-1499-5

Bowman DMJS, Perry GLW, Higgins SI, Johnson CN, Fuhlendorf SD, Murphy BP (2016) Pyrodiversity is the coupling of biodiversity and fire regimes in food webs. Philos Trans R Soc B 371:20150169. https://doi.org/10.1098/rstb.2015.0169

Brasil (1965) Código florestal brasileiro. Lei n. 4.771, de 15 de setembro de 1965. Diário Oficial da União. https://www2.camara.leg.br/legin/fed/lei/1960-1969/lei-4771-15-setembro-1965-369026-publicacaooriginal-1-pl.html. Accessed 23 Nov 2023

Brasil (1979) Regulamento dos Parques Nacionais brasileiros. Decreto Federal n. 84.017, de 21 de setembro de 1979. https://www.planalto.gov.br/ccivil_03/Atos/decretos/1979/D84017.html. Accessed 23 Nov 2023

Brasil (2012) Código florestal brasileiro. Lei n. 12.651, de 25 maio de 2012. https://www.planalto.gov.br/ccivil_03/_ato2011-2014/2012/lei/l12651.htm. Accessed 23 Nov 2023

Brasil (2024) Política Nacional de Manejo Integrado do Fogo. Lei n. 14.944, de 31 de julho de 2024. Casa Civil. https://www.planalto.gov.br/ccivil_03/_ato2023-2026/2024/lei/L14944.htm. Accessed 23 Ago 2024

Bustamante MMC, Nobre CA, Smeraldi R, Aguiar APD, Barioni LG et al (2012) Estimating greenhouse gas emissions from cattle raising in Brazil. Clim Chang 115:559–577. https://doi.org/10.1007/s10584-012-0443-3

Carmenta R, Vermeylen S, Parry L, Barlow J (2013) Shifting cultivation and fire policy: insights from the Brazilian Amazon. Hum Ecol 41:603–614. https://doi.org/10.1007/s10745-013-9600-1
Carranza T, Balmford A, Kapos V, Manica A (2014) Protected area effectiveness in reducing conversion in a rapidly vanishing ecosystem: the Brazilian Cerrado. Conserv Lett 7:216–223. https://doi.org/10.1111/conl.12049
Coutinho LM (1982) Ecological effects of fire in Brazilian Cerrado. In: Huntley BJ, Walker B (eds) Ecology of tropical savanna, Ecological studies, vol 42. Springer Verlag, Berlin, pp 273–291
ECLAC—Economic Commission for Latin America and the Caribbean (2014) Guaranteeing Indigenous people's rights in Latin America: progress in the past decade and remaining challenges. https://repositorio.cepal.org/server/api/core/bitstreams/d7a89fa6-5c4e-4f57-8582-06f8cb0a94d4/content. Accessed 5 May 2024
ECLAC—Economic Commission for Latin America and the Caribbean (2022) Latin America and the Caribbean: population estimates and projections. https://www.cepal.org/en/subtopics/demographic-projections/latin-america-and-caribbean-population-estimates-and-projections. Accessed 8 Apr 2024
Edivaldo T, Rosell S (2020) Slash-and-burn agriculture in southern Brazil: characteristics, food production and prospects. Scottish Geogr J 136:176–194. https://doi.org/10.1080/14702541.2020.1776893
Eloy L, Schmidt IB, Borges SL, Ferreira MC, dos Santos T (2018) Seasonal fire management by traditional cattle ranchers prevents the spread of wildfire in the Brazilian Cerrado. Ambio 48:890–899. https://doi.org/10.1007/s13280-018-1118-8
Eloy L, Bilbao B, Mistry J, Schmidt IB (2019) From fire suppression to fire management: advances and resistances to changes in fire policy in the savannas of Brazil and Venezuela. Geogr J 185:10–22. https://doi.org/10.1111/geoj.12245
Fagundes GM (2019a) Fogos gerais: transformações tecnopolíticas na conservação do Cerrado. Universidade de Brasília
Fagundes GM (2019b) Fire normativities: environmental conservation and Quilombola forms of life in the Brazilian savanna. Vibrant (Florianópolis) 16:1–22. https://doi.org/10.1590/1809-43412019v16d501
Fagundes GM (2019c) Fazer o fogo fazer: manipulações e agenciamentos técnicos na conservação do Jalapão (TO). Equatorial—Revista do Programa de Pós-Graduação em Antropologia Social 6(10):16–49. https://doi.org/10.21680/2446-5674.2019v6n10ID15640
Fagundes GM (2021) Free to burn—autonomy and heteronomy toward fire in the Brazilian savannah. J Anthropol 164:67–87. https://doi.org/10.4000/jda.10569
Falleiro RM (2011) Resgate do manejo tradicional do cerrado com fogo para proteção das terras indígenas do oeste do Mato Grosso: um estudo de caso. Biodivers Brasil 2:86–96. https://doi.org/10.37002/biodiversidadebrasileira.v1i2.114
Falleiro RM, Santana MT, Berni CR (2016) As Contribuições do manejo integrado do fogo para o controle dos incêndios florestais nas Terras Indígenas do Brasil. Biodivers Brasil 6:88–105. https://doi.org/10.37002/biodiversidadebrasileira.v6i2.655
Falleiro RM, Steil L, Oliveira MS, Lando I, Machado LOR et al (2021) Histórico, avaliação, oportunidades e desafios do manejo integrado do fogo nas terras indígenas brasileiras. Biodivers Brasil 11:75–98. https://doi.org/10.37002/biodiversidadebrasileira.v11i2.1742
Falleiro RM, Santana MS, Xerente AJK, Krahô JJ, Zomaizokiece AA et al (2022) Efeitos de diferentes regimes de fogo nas árvores do gênero Mouriri do Cerrado brasileiro. In: Trajetórias, perspectivas e desafios da gestão ambiental pública, vol 3. IBAMA, Brasília, DF. ISSN: 2675-4762. http://www.biodiversity4all.org/taxa/185927-Mouriri. Accessed 8 Apr 2024
Falleiro RM, Moura LC, Xerente PP, Pinto CP, Santana MT et al (2024) Using a cultural keystone species in participatory monitoring of fire management in Indigenous lands in the Brazilian savanna. Fire 7:231. https://doi.org/10.3390/fire7070231
FAN—Fundación Amigos de la Naturaleza (2019) Reporte de incendios forestales a nivel nacional, Bolivia. Fundación Amigos de la Naturaleza. http://incendios.fan-bo.org/Satrifo/publicaciones/#tab-4. Accessed 17 Feb 2024

FAO—Organización de las Naciones Unidas para la Alimentación y la Agricultura, FILAC—Fondo para el Desarrollo de los Pueblos Indígenas de América Latina y el Caribe (2021) Los pueblos indígenas y tribales y la gobernanza de los bosques: una oportunidad para la acción climática en América Latina y el Caribe. FAO, Santiago. https://doi.org/10.4060/cb2953es

Ferrero B, Millán A, Bilbao B (2023) Sin fuego no hay isla: Los usos del fuego en el delta superior del río Paraná. In: Wertheimer M, Fernández Bouzo S (eds) Argentina en llamas. Voces urgentes para una ecología política del fuego, El Colectivo, Buenos Aires, pp 183–206

Fidelis A, Alvarado S, Barradas A, Pivello V (2018) The year 2017: Megafires and management in the Cerrado. Fire 1(3):49. https://doi.org/10.3390/fire1030049

Fidelis A, Zirondi HI, Rossatto DR, Zanzarini V (2022) Fire stimulates grass flowering. J Vegetation Sci 33(2). https://doi.org/10.1111/jvs.13125

Flantua S, Bilbao B, Rosales J (2013) Indigenous land. use and land cover changes in the National Park of Canaima, Venezuela: Pemón Sector II Kamarata. In: Lopes LT, Arley S, Silveira da Costa J, Baptista ER (eds) Libro Escudo Guiane^s, biodiversidade, conservacao dos recursos naturais e cultura. Georgetown, NAEA, GSF, Belém, pp 209–224

Glaser B, Haumaier L, Guggenberger G, Zech W (2001) The 'Terra Preta' phenomenon: a model for sustainable agriculture in the humid tropics. Naturwissenschaften 88:37–41. https://doi.org/10.1007/s001140000193

Gómez E, Picon G, Bilbao BA (2000) Los incendios forestales en Iberoamérica. Caso Venezuela. In: Vélez Muñoz R (ed) La Defensa Contra Incendios Forestales. Fundamentos and experiencias. McGraw-Hill, Madrid, pp 22.3–22.42. https://dialnet.unirioja.es/servlet/libro?codigo=872707. Accessed 8 Apr 2024

Guidon N, Delibrias G (1986) Carbon-14 dates point to man in the Americas 32,000 years ago. Nature 321:769–771. https://doi.org/10.1038/321769a0

Halbmayer E (2013) Contemporary Carib-Speaking Amerindians: a bibliography of social anthropological and linguistic resources. In: Reihe Curupira-workshop, vol 16. Curupira, Marburg

Hoffman KM, Davis EL, Wickham SB, Schang K, Johnson A et al (2021) Conservation of Earth's biodiversity is embedded in Indigenous fire stewardship. PNAS 118:e2105073118. https://doi.org/10.1073/pnas.2105073118

Huber O, Zent S (1995) Indigenous People and vegetation in the Venezuelan Guayana: some ecological consideration. Scientia Guaianae 5:37–64

IBAMA—Instituto Brasileiro do Meio Ambiente e dos Recursos Naturais Renováveis (2018, unpublished). Relatório da compilação dos resgates do conhecimento tradicional sobre o uso do fogo em Terras Indígenas Brasileiras. Internal archive of the Prevfogo Specialized Centre. Brasília, DF

IBGE—Instituto Brasileiro de Geografia e Estatística (2010) Censo 2010. https://censo2010.ibge.gov.br/. Accessed 11 Jan 2024

ILO—International Labor Organization (1989) Convenio Número 169 de la OIT sobre Pueblos Indígenas y Tribales en países independientes. Declaración de las Naciones Unidas sobre los Derechos de los Pueblos Indígenas. Lima, OIT/Oficina Regional para América Latina y el Caribe, pp. 134

ILO—International Labor Organization (2018) Social Protection for Indigenous Peoples. https://www.ilo.org/publications/social-protection-Indigenous-peoples. Accessed 3 Feb 2024

INE—Instituto Nacional de Estadísticas (2018) Síntesis resultados Censo 2017. INE, Santiago de Chile. https://comunidadcreativalosrios.cultura.gob.cl/wp-content/uploads/2019/11/sintesis-de-resultados-censo2017.pdf. Accessed 25 May 2024

Ingold T (2004) Two reflections on ecological knowledge. In: Ortalli G, Sanga G (eds) Nature knowledge: ethnoscience, cognition, identity. Berghahn, New York, pp 301–311

IPBES—Intergovernmental Science-Policy Platform on Biodiversity and Ecosystem Services (2016) Indigenous and Local Knowledge Systems (deliverable 1 (c)). IPBES/5/4. United Nations Intergovernmental Science-Policy Platform on Biodiversity and Ecosystem Service. https://ipbes.net/sites/default/files/downloads/pdf/ip bes-5-inf-4.pdf. Accessed 14 Mar 2024

IPBES—Intergovernmental Science-Policy Platform on Biodiversity and Ecosystem Services (2019) Summary for policymakers of the global assessment report on biodiversity and ecosystem services of the Intergovernmental Science-Policy Platform on Biodiversity and Ecosystem Services. In: Díaz S, Settele J, Brondízio ES, Ngo HT, Guèze M et al (eds) IPBES secretariat, Bonn, Germany. 56 pages. https://ipbes.net/sites/default/files/2020-02/ipbes_global_assessment_report_summary_for_policymakers_en.pdf

IPCC—Intergovernmental Panel on Climate Change (2021) The physical science basis. In: Contribution of working group i to the sixth assessment report of the intergovernmental panel on climate change. Cambridge University Press, Cambridge, and New York, NY. https://doi.org/10.1017/9781009157896

IUCN—International Union for Conservation of Nature (2008) Learning from Landscapes. Arbor Vitae Special Issue. IUCN, Gland, Switzerland. https://iucn.org/sites/default/files/import/downloads/a_avspecial_learning_from_landscapes.pdf. Accessed 2 Feb 2024

IWGIA—International Work Group for Indigenous Affairs (2019) The Indigenous World 2019. General Editor: David Nathaniel Berger. Ministry of Foreign Affairs of Denmark (DANIDA). ISSN: 1024–0217 ISBN: 978–87–92786-93-7. 680 pp. https://www.iwgia.org/en/documents-and-publications/documents/publications-pdfs/english-publications/4-the-Indigenous-world-2019/file.html. Accessed 22 Apr 2024

IWGIA—International Work Group for Indigenous Affairs (2024) The Indigenous World 2024. 8th ed. General Editor: Dwayne Mam. Ministry of Foreign Affairs of Denmark (DANIDA). ISSN: 1024–0217 ISBN: 978–87–93961-64-7. https://www.iwgia.org/en/resources/publications/5508-the-Indigenous-world-2024.html. Accessed 22 Apr 2024

Kelly L, Giljohann K, Duane A, Aquilué N, Archibald S et al (2020) Fire and biodiversity in the Anthropocene. Science 370:eabb0355. https://doi.org/10.1126/science.abb0355

Kimber RG, Friedel MH (2015) Challenging the concept of Aboriginal mosaic fire practices in the Lake Eyre Basin. Rangel J 37(6):623–630. https://doi.org/10.1071/RJ15057

Kreider MR, Higuera PE, Parks SA, Rice WL, White N et al (2024) Fire suppression makes wildfires more severe and accentuates impacts of climate change and fuel accumulation. Nat Commun 15:2412. https://doi.org/10.1038/s41467-024-46702-0

Kunst C, Bravo S, Ledesma R, Navall M, Anríquez A et al (2015) Chapter 8. Current international perspectives on wildland fires, mankind and the environment. In: Leblon B, Alexander M (eds) Ecology and management of the dry forests and savannas of the western Chaco region, Argentina. Nova Sciences Publishers, pp 211–234

Leonell M (2000) O uso do fogo: o manejo indígena e a piromania da monocultura. Estudos Avançados 14:231–250. https://doi.org/10.1590/S0103-40142000000300019

Levis C, Flores BM, Moreira PA, Luize BG, Alves RP et al (2018) How people domesticated Amazonian forests. Front Ecol Evol 17:171. https://doi.org/10.3389/fevo.2017.00171

Liang B, Lehmann J, Sohi SP, Thies JE, Bl O'N et al (2010) Black carbon affects the cycling of non-black carbon in soil. Org Geochem 41.:206–213, ISSN 0146-6380,. https://doi.org/10.1016/j.orggeochem.2009.09.007

Maclean K, Hankins DL, Christianson AC, Oliveras Menor I, Bilbao BA et al (2023) Revitalising Indigenous cultural fire practice: benefits and partnerships. Trends Ecol Evol 38:899–902. https://doi.org/10.1016/j.tree.2023.07.001

Maillard O (2023) Post-fire natural regeneration trends in Bolivia: 2001–2021. Fire 6:18. https://doi.org/10.3390/fire6010018

Mauro F, Hardison P (2000) Traditional knowledge of indigenous and local communities: international debate and policy initiatives. Ecol Appl 10:1263–1269. https://doi.org/10.1890/1051-0761(2000)010[1263:TKOIAL]2.0.CO;2

Meza MC, González T, Armenteras D (2023) De la paradoja del fuego: del contexto internacional al caso de Colombia. Universidad Nacional de Colombia, Bogotá, D.C., Colombia

Millán A, Ferrero B, Bilbao B (2022) Traditional knowledge of fire use by islanders in the Paraná Delta, Argentina. Tropical forest Issues 61: 60-65. https://doi.org/10.55515/DVRK2501

Miranda HS (2010) Efeitos do regime de fogo sobre a estrutura de comunidades de Cerrado: Projeto Fogo. Ibama, Brasília. 144p. ISBN 978-85-7300-305-5. https://www.ibama.gov.br/sophia/cnia/livros/efeitosdoregimedofogodigital.pdf. Accessed 22 Jan 2024

Mistry J, Berardi A, Andrade Vianez JAR (2005) Indigenous fire management in the Cerrado of Brazil. Hum Ecol 33:365–386. https://doi.org/10.1007/s10745-005-4143-8

Mistry J, Bilbao BA, Berardi A (2016) Community owned solutions for fire management in tropical ecosystems: case studies from Indigenous communities of South America. Philos Trans R Soc B Biol Sci 371(1696):20150174. https://doi.org/10.1098/rstb.2015.0174

Mistry J, Schmidt IB, Eloy L, Bilbao B (2019) New perspectives in fire management in South American savannas: the importance of intercultural governance. Ambio 48:172–179. https://doi.org/10.1007/s13280-018-1054-7

MMA—Ministério do Meio Ambiente (2018) Projeto Cerrado-Jalapão. http://cerradojalapaom-magovbr/biblioteca/arquivos .Acessed 23 Nov 2023

Moura LC, Scariot AO, Schmidt IB, Beatty R, Russell-Smith J (2019) The legacy of colonial fire management policies on traditional livelihoods and ecological sustainability in savannas: impacts, consequences, new directions. J Environ Manag 232:600–606. https://doi.org/10.1016/j.jenvman.2018.11.057

Myers RL (2006) Living with fire: sustaining ecosystems and livelihoods through integrated fire management. The Nature Conservancy, Global Fire Initiative, Tallahassee, USA

Neves EG, Petersen JB, Bartone RN, Augusto SC (2003) Historical and socio-cultural origins of Amazonian dark earth. In: Lehmann J, Kern DC, Glaser B, Wodos WI (eds) Amazonian dark earths. Springer, Dordrecht. https://doi.org/10.1007/1-4020-2597-1_3

Oliveira MR, Ferreira BH, Souza EB, Lopes AA, Bolzan FP et al (2022) Indigenous brigades change the spatial patterns of wildfires, and the influence of climate on fire regimes. J Appl Ecol 59:1279–1290. https://doi.org/10.1111/1365-2664.14139

Parr C, Brockett B (1999) Patch-mosaic burning: a new paradigm for savanna fire management in protected areas? Koedoe Afr Protect Area Conserv Sci 42:117–130. https://doi.org/10.4102/koedoe.v42i2.237

Pereira ACJ, Oliveira SLJ, Pereira JMC, Turkman MAA (2014) Modelling fire frequency in a Cerrado Savanna protected area. PLoS One 9:e102380. https://doi.org/10.1371/journal.pone.0102380

Petersen R, Stevens C (2014) Three maps show importance of local communities in forest conservation. World Resource Institute. https://www.wri.org/insights/3-maps-show-importance-local-communities-forest-conservation. Accessed 11 Apr 2024

Piperno DR (2011) The origins of plant cultivation and domestication in the new world tropics: patterns, process, and new developments. Curr Anthropol 52:S453–S470. https://doi.org/10.1086/659998

Piperno DR, McMichael CNH, Bush MB (2019) Finding forest management in prehistoric Amazonia. Anthropocene 26:100211. https://doi.org/10.1016/j.ancene.2019.100211

Pivello VR (2011) The use of fire in the Cerrado and Amazonian rainforests of Brazil: Past and present. Fire Ecol 7:24–39. https://doi.org/10.4996/fireecology.0701024

Pivello VR, Vieira I, Christianini AV, Ribeiro DB, da Silva Menezes L, Berlinck CN et al (2021) Understanding Brazil's catastrophic fires: causes, consequences and policy needed to prevent future tragedies. Perspect Ecol Conserv 19:233–255. https://doi.org/10.1016/j.pecon.2021.06.005

Ponce-Calderón LP, Limón-Aguirre F, Rodríguez I, Rodríguez-Trejo DA, Bilbao BA et al (2022) Fire management in pyrobiocultural landscapes, Chiapas, Mexico. In: Pasiecznik N, Goldammer JG (eds) Towards fire-smart landscapes. Tropical forest Issues 61. Tropenbos International, Ede, The Netherlands. https://doi.org/10.55515/DVRK2501

Posey DA (2004) Indigenous knowledge and ethics: a Darrell Posey reader. Routledge, New York and London

Ribeiro D, Pereira AM (2023) Solving the problem of wildfires in the Pantanal wetlands. Perspect Ecol Conserv 21:271–273. https://doi.org/10.1016/j.pecon.2023.10.004

Rodríguez I (2007) Pemon perspectives of fire management in Canaima National Park, southeastern Venezuela. Hum Ecol 35:331–343. https://doi.org/10.1007/s10745-006-9064-7

Rodríguez I, Albert P, La Rose C, Sharpe CJ (2011) A study of the use of fire by Amerindian Communities in South Rupununi, Guyana, with recommendations for sustainable land management. Technical Report, South-Central Peoples Development Association (SCPDA), Forest Peoples Programme. https://www.researchgate.net/publication/348048984_A_study_of_the_use_of_fire_and_by_Amerindian_communities_the_South_Rupununi_Guyana_with_recommendations_for_sustainable_land_management. Accessed 10 Apr 2024

Rodríguez I, Sletto B, Bilbao B, Sánchez-Rose I, Leal A (2013) Speaking of fire: reflexive governance in landscapes of social change and shifting local identities. J Environ Policy Plan 15:569–588. https://doi.org/10.1080/1523908X.2013.766579

Roosevelt AC (1994) Amazonian anthropology: strategy for a new synthesis. In: Roosevelt AC (ed) Amazonian Indians from prehistory to present: anthropological perspective. University of Arizona Press, Tucson, pp 1–29

Roshan A, Biswas A (2023) Fire-induced geochemical changes in soil: implication for the element cycling. Sci Total Environ 868:161714. https://doi.org/10.1016/j.scitotenv.2023.161714

Rostain S (2010) Pre-Columbian Earthworks in Coastal Amazonia Diversity 2:331–352. https://doi.org/10.3390/d2030331

Russell-Smith J, Whitehead P, Cooke P (2009) Culture, ecology and economy of fire management in north Australian savannas: rekindling the Wurrk tradition. CSIRO, Victoria. ISBN 978–0–6430–9402-4

Russell-Smith J, Monagle C, Jacobsohn M, Beatty R, Bilbao BA et al (2017) Can savanna burning projects deliver measurable greenhouse emissions reductions and sustainable livelihood opportunities in fire-prone settings? Clim Chang 140:47–61. https://doi.org/10.1007/s10584-013-0910-5

Saint Hilaire A (2000) Viagens pelas províncias do Rio de Janeiro e Minas Gerais. Ed. Itatiaia, Belo Horizonte

Santos FLM, Nogueira J, Souza RAF, Falleiro RM, Schmidt IB, Libonati R (2021) Prescribed burning reduces large, high intensity wildfires and emissions in the Brazilian savanna. Fire 4:56. https://doi.org/10.3390/fire4030056

Schmidt IB, Eloy L (2020) Fire regime in the Brazilian savanna: recent changes, policy and management. Flora 268:151613. https://doi.org/10.1016/j.flora.2020.151613

Schmidt IB, Moura LC, Ferreira MC, Eloy L, Sampaio AB et al (2018) Fire management in the Brazilian Savanna: first steps and the way forward. Jf Appl Ecol 55:2094–2101. https://doi.org/10.1111/1365-2664.13118

Smith T (2022) Forest management and conservation governance in relation to Indigenous Food Sovereignty with the Lílwat First Nation in British Columbia, Canada. Thesis, University of British Columbia. https://open.library.ubc.ca/soa/cIRcle/collections/ubctheses/24/items/1.0415891. Accessed 28 Jun 2024

Steffensen V (2020) Fire country: how indigenous fire management could help save Australia. Hardie Grant Travel, Sydney

Stevens C, Winterbottom R, Springer J, Reytar K (2014) Securing rights, combating climate change. How strengthening community forest rights mitigates climate change. World Resources Institute, Washington, DC. www.wri.org/securing-rights. Accessed 24 Nov 2023

Sutter RC (2021) The Pre-Columbian peopling and population dispersals of South America. J Archaeol Res 29:93–151. https://doi.org/10.1007/s10814-020-09146-w

Teunissen van Manen ML (2020) Plant wax n-alkane biomarkers in the tropical Andes (Ecuador). Thesis, Universiteit van Amsterdam. https://pure.uva.nl/ws/files/47483312/Thesis_complete_pdf. Accessed 16 Jun 2024

Trauernicht C, Brook BW, Murphy BP, Williamson GJ. Bowman DMJS (2015) Local and global pyrogeographic evidence that Indigenous fire management creates pyrodiversity. Ecol Evol 5:1908–1918. doi: https://doi.org/10.1002/ece3.1494

UNEP—United Nations Environment Programme (2022) Spreading like wildfire—the rising threat of extraordinary landscape fires. A UNEP Rapid Response Assessment. Nairobi https://www.unep.org/resources/report/spreading-wildfire-rising-threat-extraordinary-landscape-fires. Accessed 16 Mar 2024

UNEP-WCMC—United Nations Environment Programme World Conservation Monitoring Centre, IUCN—International Union for Conservation of Nature (2016) Protected Planet Report

2016. UNEP-WCMC and IUCN: Cambridge UK and Gland, Switzerland. https://resources.unep-wcmc.org/products/WCMC_RT062. Accessed 20 Feb 2024

UNESCO—United Nations Educational, Scientific and Cultural Organization (2024) Indigenous Peoples' Traditional Knowledge of Fire: Observing, Understanding, and Coping with Climate Change and the vulnerabilities of Socio-Environmental Systems - Case studies from the Guiana Shield (Guyana, Suriname, Venezuela). UNESCO, Paris

UNICEF—Fondo de las Naciones Unidas para la Infancia (2009) Atlas Sociolinguistico de pueblos Indígenas en América Latina. UNICEF and FUNPROEBID Andes

UNU—United Nations University (2015) The global potential of Indigenous fire management: findings of the regional feasibility assessment. International Savanna Fire Management Initiative-UN. https://collections.unu.edu/eserv/UNU:5605/Indigenous_fire_management.pdf

Vega JJ, Guzmán Palomino LG (2005) El Inti Raymi Inkaico. La verdadera historia de la Gran Fiesta Del Sol. Bol Mus Arqueol Antropol (Unmsm) 6:37–71. https://sisbib.unmsm.edu.pe/bibvirtual/publicaciones/antropologia/2003_n01/a05.htm. Accessed 5 May 2024

Veiga JB, Tourrand JF, Poccard-Chapuis R, Piketty MG (2002) Cattle ranching in the Amazon rainforest. Anim Prod Aust 24:253–256. http://www.livestocklibrary.com.au/bitstream/handle/1234/9232/tourrand1B.pdf?sequence=1. Accessed 24 Jun 2024

Versteeg AH (2003) Suriname before Columbus. Stichting Surinaams Museum, Paramaribo. ISBN: 9789991493718

Vidal-Riveros C, Souza-Alonso P, Bravo S, Laino R, Ngo Bieng MA (2023) A review of wildfires effects across the Gran Chaco region. For Ecol Manag 549:121432. https://doi.org/10.1016/j.foreco.2023.121432

Walker W, Gorelika SR, Baccini A, Aragon-Osejo JL, Josseb C (2020) The role of forest conversion, degradation, and disturbance in the carbon dynamics of Amazon Indigenous territories and protected areas. Proc Natl Acad Sci USA 117:3015–3025. https://doi.org/10.1073/pnas.1913321117

Welch JR (2015) Learning to hunt by tending the fire: Xavante youth, ethnoecology, and ceremony in central Brazil. J Ethnobiol 35:183–208. https://doi.org/10.2993/0278-0771-35.1.183

Welch JR, Coimbra CEA Jr (2021) Indigenous fire ecologies, restoration, and territorial sovereignty in the Brazilian Cerrado: the case of two Xavante reserves. Land Use Policy 104:104055. https://doi.org/10.1016/j.landusepol.2019.104055

Fire in Subtropical and Temperate Grasslands

4

Luis López-Mársico, Luciana R. Podgaiski, Sergio M. Zalba, Claudia Rodríguez, and Gerhard E. Overbeck

Abstract

Grasslands are open ecosystems that can be determined both by climatic patterns and disturbance occurrence. This chapter aims to present the current state of information on the incidence of fire in subtropical and temperate grasslands of South America (Brazil, Uruguay and Argentina). Fires, natural or human-induced, cause changes in grasslands by reducing plant biomass, opening up space and increasing the incidence of sunlight on the ground. Most of the existing studies in these grasslands have examined the effect of fire on plants and animals, and we provide a comprehensive compilation of existing studies on this topic. Some existing information about the impact of fire on soil and decomposition variables has also been compiled, as well as regarding the use of fire as a management tool by livestock ranchers. We also discuss how fire (in addition to grazing) has been able to maintain grasslands in a region where the climate allows for forest development. Lastly, we address the issue of the knowledge and policy gaps that exist. There is a need to continue generating knowledge on the

L. López-Mársico (✉) · C. Rodríguez
Instituto de Ecología y Ciencias Ambientales, Universidad de la República, Montevideo, Uruguay
e-mail: luislopez@fcien.edu.uy; claudia@fcien.edu.uy

L. R. Podgaiski · G. E. Overbeck
Instituto de Biociências, Universidade Federal do Rio Grande do Sul, Porto Alegre, Brazil
e-mail: gerhard.overbeck@ufrgs.br

S. M. Zalba
Departamento de Biología, Bioquímica y Farmacia, Universidad Nacional del Sur, Bahía Blanca, Argentina
e-mail: szalba@criba.edu.ar

A. Fidelis, V. R. Pivello (eds.), *Fire in the South American Ecosystems*, Ecological Studies 250, https://doi.org/10.1007/978-3-031-89372-8_4

fire effects, so that decision makers can evaluate its potential use as a conservation or management tool.

Resumen

Los pastizales son ecosistemas abiertos que pueden ser determinados por patrones climáticos y por perturbaciones. El objetivo de este capítulo es presentar el estado actual del arte sobre la incidencia del fuego en pastizales subtropicales y templados de Sudamérica (Brasil, Uruguay y Argentina). Los incendios, naturales o provocados por el ser humano, generan cambios en los pastizales al reducir la biomasa vegetal, abrir espacio y aumentar la incidencia de luz solar sobre el suelo. Existen estudios en la región que han evaluado el efecto del fuego sobre las plantas y los animales. Aquí se presenta una recopilación actualizada. También se compilaron estudios que evaluaron el impacto del fuego sobre variables del suelo y de la descomposición. Se presentan estudios en los que el fuego es empleado por productores como una herramienta de manejo. Asimismo, se plantea cómo el fuego (además del pastoreo) ha podido mantener a los pastizales en una región en la que el clima permite el desarrollo de bosques. Finalmente abordamos el tema de los vacíos de conocimiento y político que existen. Es necesario seguir generando conocimiento sobre los efectos del fuego, para que los tomadores de decisiones puedan evaluar su uso potencial como herramienta de conservación o gestión.

Resumo

Campos são ecossistemas abertos, determinados tanto por padrões climáticos quanto pela ocorrência de distúrbios. Este capítulo apresenta o estado da arte sobre a incidência do fogo em campos temperados e subtropicais da América do Sul (Brasil, Uruguai e Argentina). Queimadas, sejam elas naturais ou antropogênicas, causam mudanças nos campos pela redução da biomassa vegetal, abertura de clareiras e aumento da incidência solar sobre o solo. A maioria dos estudos existentes na região investigou o efeito do fogo em plantas e em animais e aqui apresentamos uma compilação atualizada destes estudos, assim como sobre avaliações do impacto do fogo no solo e variáveis de decomposição. Outros estudos avaliaram o uso do fogo por pecuaristas como ferramenta de manejo, cujos resultados também são também apresentados. Discutimos ainda como o fogo (além do pastejo) tem conseguido manter a fisionomia campestre em regiões que favorecem o avanço florestal. Finalmente, são abordados os problemas relacionados à falta de conhecimento e de políticas públicas para lidar com a questão do fogo nestes campos, ressaltando a necessidade de mais pesquisas, a fim de gerar conhecimento aos tomadores de decisão, para que possam avaliar o uso potencial do fogo como ferramenta de conservação e manejo.

4.1 Grasslands: Climate, Vegetation and Disturbances

Grasslands are open ecosystems that can be determined both by climatic patterns and the occurrence of disturbances (Lauenroth and Sala 1992; Noy-Meir 1995; Oesterheld et al. 1999; Pausas and Bond 2020). Grasslands, simply defined by Gibson (2009) as "a habitat dominated by grasses" occur in a wide range of mean annual precipitation, from 200 to 1500 mm, and mean annual temperature, from 0 to 26 °C (Lauenroth 1979; Sala et al. 2001; Bond 2019). A relatively continuous herbaceous stratum without or with scattered woody components, as in the case of savannas, determines their main physiognomic characteristics. The main components of this herbaceous stratum are, in the majority of grasslands, grasses (Poaceae), which do not shed their leaves as they age and accumulate them as standing dead biomass if not removed (Bond and Parr 2010).

Grasslands are represented in every continent, except Antarctica. In South America, subtropical and temperate grasslands extend from 23° to 39°S and 50° to 66°W. The extension of these grasslands involves territories of three countries (central-eastern Argentina, southern Brazil, and Uruguay; Fig. 4.1). They can be divided into the Río de la Plata grasslands and the Highland grasslands of Southern Brazil. The main economic activity of the area is cattle and sheep ranching (Nabinger et al. 2000; Oyarzabal et al. 2020; Guido et al. 2025), but the grassland area is in

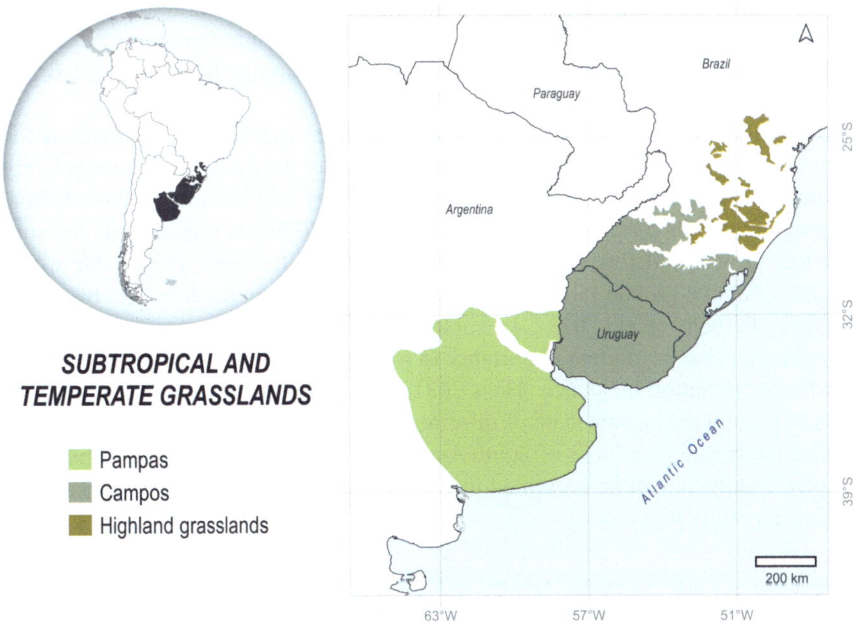

Fig. 4.1 Distribution of the subtropical and temperate grasslands of South America, with regional sub-divisions of the Río de la Plata grasslands (Pampas and Campos) and Highland grasslands of Southern Brazil. (Sources: Andrade et al. 2019; Soriano et al. 1991. Produced by Júlia Abrantes Rodrigues)

continuous decline due to the land use change to cropland and implanted forests (Vega et al. 2009). Beyond their apparent physiognomic homogeneity, they are characterized by a remarkable species richness and a continuous photosynthetic activity throughout the year (Soriano et al. 1991; Boldrini 2009. Lezama et al. 2019; Oyarzabal et al. 2020). The dominant plant families are Poaceae, Asteraceae, Fabaceae, and Cyperaceae. C3 and C4 grasses co-occur across the region, although their proportion shifts along the latitudinal gradient, with C4 pathway increasing with decreasing latitude (Soriano et al. 1991; Perelman et al. 2001; Overbeck et al. 2007; Boldrini 2009; Andrade et al. 2019; Lezama et al. 2019). Woody vegetation within the region is mostly restricted to small areas near water courses, such as the gallery forests along the large Paraná and Uruguay rivers and their tributary streams (Paruelo et al. 2007), and to mountainous regions, such as the Serra do Sudeste in Southern Brazil. Only in the South Brazilian highland region, forest, in this case *Araucaria* forest, occur in larger extensions, forming mosaics with the grasslands.

Climate in this region is humid subtropical according to the Köppen classification (Cfa Type over the larger part of the region, Cfb in the highlands in Southern Brazil). A key climatic feature of southern South America is the relatively low thermal amplitude caused by the low ratio of land mass to ocean in this portion of the world (Paruelo et al. 2007). Mean temperature presents a gradient between 13 °C in the south and 20 °C in the north, while mean annual precipitation ranges between 400 mm in the southwest and 1500 mm in the northeast, although there is no seasonal variation in precipitation (Oyarzabal et al. 2020). Depending on topography and orography, the presence of woody vegetation can acquire greater coverage in the landscape, as can be observed in the highland grasslands of southern Brazil (Overbeck et al. 2007; Andrade et al. 2019).

Grasslands are often found in climate zones where other types of ecosystems can also occur (Bond 2019). This means that climate is not the (only) determining factor for their occurrence, and disturbances like herbivory and fire are decisive factors shaping grasslands (Oesterheld et al. 1999; Gibson 2009). In regions that can support tree growth and forest development, the main disturbance agents, due to the amount of above-ground biomass they remove, are grazing by domestic livestock, and fire (Paruelo et al. 2022). In subtropical and temperate grassland regions, droughts can also be of great importance in some years, particularly in summer, due to a higher radiation incidence (Hoss 2021; Tommasino et al. 2022). This chapter aims to present the current state of information on the incidence of fire in subtropical and temperate grasslands of South America. In addition, in some sections of the chapter, examples will be presented in which fire interacts with other types of disturbance, mainly herbivory.

4.2 Fire as a Historical Disturbance

When plant biomass is not removed by herbivores, grasslands can become a fire-prone environment (Gibson 2009; Bond and Keeley 2005; Pivello et al. 2011). Fires occurring in the subtropical and temperate grasslands of South America can be natural or human-induced (Fig. 4.2); either way, they play an important role in the conservation of these ecosystems (Laterra et al. 1998; Overbeck et al. 2005; Loydi et al. 2020; Ferreira et al. 2021; López-Mársico et al. 2021). Fires cause changes in grasslands by reducing plant biomass, opening up space, and increasing the incidence of sunlight on the ground. Natural fires are caused by lightning strikes and have occurred in these grasslands over the last millions of years (Veldman et al. 2015). In most cases, however, fires in this region are human-induced. Today, two main types of events can be distinguished: 1—those caused by livestock ranchers as a management tool aimed to remove dead biomass and promote resprouting of the vegetation and 2—those caused accidentally. During a fire, the flames consume the

Fig. 4.2 (**a**) Experimental fire in Sierras del Este grasslands, Treinta y Tres, Uruguay (Photo: Federico Gallego); (**b**) Fire in Sierra de la Ventana, Buenos Aires, Argentina (Photo: Sergio Marto)

plant biomass, not much affecting the buds that are above- or belowground, allowing them to resprout (Pausas et al. 2004; Archibald et al. 2019). In grasslands dominated by tussock grasses, fuel elements are sufficiently separated to allow oxygenation, and sufficiently close together to facilitate fire spread (López-Mársico et al. 2021). Because the combustible material generally consists of thin elements (grass leaves), fires spread quickly and are of short duration (Fidelis et al. 2010; López-Mársico et al. 2019a). Recent studies have shown that patchy burning promotes heterogeneity and contributes to biodiversity conservation (Beal-Neves et al. 2020a; López-Mársico et al. 2021; Goldas et al. 2022a). Despite the historical incidence of fire in South American grasslands, scientific knowledge has increased markedly only in the last two decades.

In South America, the expansion of grass-dominated ecosystems initiated at the end of the early Oligocene (~30 Ma), about 15 Ma earlier than in North America (Jacobs et al. 1999; MacFadden 2000). Initially, vegetation was dominated by C3 grasses, whereas C4 species did not contribute significantly to plant biomass (Cerling et al. 1993; Jacobs et al. 1999; Kellogg 2000; Osborne 2008). The expansion of C4 grass-dominated ecosystems occurred between the late Miocene and Pliocene (8–4 Ma), when C4 grasslands expanded across at least 4 continents (Cerling et al. 1993; Jacobs et al. 1999; Sage 2004; Osborne 2008). Some authors have pointed out that disturbances such as fire and large herbivore activity gained importance during the expansion of C4 grasses in the late Miocene. Fire frequency and intensity increased because of the drying climate, while grazing increased with the evolutionary diversification of the mammalian megafauna (Sage 2001; Keeley and Rundel 2005; Christin et al. 2008; Edwards and Still 2008). During the Pleistocene-Holocene transition, the warmer and wetter conditions (which increased primary production) and the lack of large grazers after the megafaunal extinction (which allowed the accumulation of standing dead biomass) would have promoted the increase of fire frequency in some regions of the Río de la Plata grasslands (Behling et al. 2005; Paruelo et al. 2022). In the mid Holocene (7000–5000 BP), palaeoecological data suggest that fire was very frequent, associated with the arrival and settlement of human populations (Behling and Pillar 2007; Kaal et al. 2019). Thus, during the temporary window of around 10,000 years without grazing, until the introduction of livestock by the Europeans settlers, fire may have played the same role as grazers, maintaining the same selective pressure on certain plant attributes, such as basal meristems, underground reserves and rapid growth.

4.3 Fire Effects on Plants

Plants respond differently to fire. While fire is capable of killing some plants, most of them can resprout within a few days. Most species of grasses, and many species from other plant families, have their resprouting buds at the ground level (hemicryptophytes) or below-ground (geophytes) (Raunkiaer 1934), and are therefore able to resprout after a fire event (Overbeck and Pfadenhauer 2007). These species have been of great interest as they constitute a bud bank (Benson et al. 2004; Dalgleish and Hartnett 2009) that has been linked to the resilience of plant communities in response to different disturbances (Collins and Wallace 1990; Hartnett et al. 2006). Tussock grasses and clump-forming grasses have a "protection system", determined by a bunch of tightly compressed sheaths that protect the regrowth buds, which are exposed after the above-ground biomass is removed (Walter 1979; Prober et al. 2007; Pausas and Paula 2020; Pilon et al. 2020). These characteristics give these grasses an advantage over species without such structures, as they can resprout rapidly after removal of the standing dead biomass (Bond and van Wilgen 1996). Most of the studies evaluated in grasslands refer to the effects of fire on the above-ground portion of plants. However, a recent review of studies on the below-ground portion of plants in Río de la Plata grasslands (López-Mársico et al. 2024a) showed that there are few studies that considered fire. Fidelis et al. (2009, 2014) observed that below-ground biomass was higher in a recently burned site than in an unburned site for 6 years.

Recently, plant species have been categorized based on the presence of certain traits that respond to disturbances such as fire and herbivory (Archibald et al. 2019). Here we will only mention fire. This proposal uses grasses as an example and is based on a set of filters in which the presence of certain traits assessed before, during, or after a fire can differentiate species into avoidant, resistant, or tolerant. The authors propose that low flammability species have traits associated with avoidance. Within this group, prostrate grasses and rosette forbs are illustrative. In this sense, it could be understood that these species have converged to avoid being burned or consumed by herbivores. In contrast, plants that can maintain viable above-ground buds during a fire have traits associated with resistance. Species whose above-ground portion is completely consumed by fire but can maintain their resprouting buds at the ground level or below-ground, exhibit traits associated with tolerance at the individual level. Within this group we can mention tussock grasses and clump-forming grasses.

Most species found in subtropical and temperate grasslands in South America can resprout from below-ground systems, including bulbs and corms, rhizomes, xylopodia, and root tubers (Fidelis et al. 2009, 2014). Additionally, some species depend on germination as a mechanism to rebuild their populations after a fire event. Also, several perennial species, which resprout after disturbance, can germinate due to the release of space caused by fire (Overbeck et al. 2006; Arcamone and Jaureguiberry 2018; López-Mársico et al. 2019b). As proposed by Archibald et al. (2019), some species can be consumed by fire and cannot resprout from above- and below-ground structures although they retain viable seeds. These species have traits associated with tolerance to fire at the population level. Species that cannot resprout or germinate after a fire, but can arrive from nearby areas, are mentioned as having traits associated with tolerance at the landscape level. Finally, those species that do not tolerate any situations mentioned above are locally extinct from the community (Archibald et al. 2019). On the other hand, the response of grassland plants can vary according to fire intensity, frequency, and season (Bond and van Wilgen 1996). Therefore, although the above-ground biomass of grassland plants is prone to be consumed by fire, a large pool of species is maintained due to the combination of different species traits in a given space and time.

Studies on the impacts of fire on vegetation in subtropical and temperate grasslands in South America come mainly from the Campos Sulinos (Brazil), some regions of the Pampas (Argentina), and the Sierras del Este region (Uruguay). Grasslands in Brazil, known as Campos Sulinos, are represented in two distinct regions: the Brazilian Pampa (located in the southern half of the state of Rio Grande do Sul) and the southern part of the Atlantic Forest region (located in northern Rio Grande do Sul, Santa Catarina, and Paraná; IBGE 2004), where they occur in mosaics with *Araucaria* forest. In the north of the latter, grasslands already are in a transitional situation to savannas of the Central Brazilian Cerrado (See Chap. 5). In the Campos Sulinos, it has been observed that in recently burned areas, the cover of rosette forbs and graminoids increases (Overbeck et al. 2005; Müller et al. 2007; Overbeck and Pfadenhauer 2007; Fidelis et al. 2012; Podgaiski et al. 2013), while species richness may increase (Overbeck et al. 2005; Joner et al. 2021), remain the same (Fidelis et al. 2012), or even peaked in relation to fire (i.e. Intermediately-burnt sites may harbor higher habitat heterogeneity and plant species richness in comparison with recently or long-burnt sites (Beal-Neves et al. 2020a) (Fig. 4.3a, b). As time since last burn increases, the dominant C_4 tussock grasses (*Andropogon lateralis, Anthaenantia lanata, Elyonurus muticus*) accumulate live and standing dead biomass (Overbeck and Pfadenhauer 2007). Through competitive exclusion, the establishment of other species may be prevented (decreases species richness, Overbeck et al. 2005) or the density of some species may be reduced (decreases the size and number of reproductive structures, da Silva et al. 2020). The absence of burning or grazing for a long time promotes an increase in the cover of woody species (Overbeck and Pfadenhauer 2007; Fidelis et al. 2012; Ferreira et al. 2020).

In some regions of Uruguay, fire is an important management tool employed by livestock ranchers, and most studies have been carried out in the Sierras del Este region (Fig. 4.2a). In this region, the unpalatable C_4 tussock grass (clump-forming) *Saccharum angustifolium* develops in a grassland community. This species can be dominant in wetter parts of the landscape. Its large size and its ability to accumulate large amounts of standing dead biomass provide the ideal fuel for fire spread (López-Mársico et al. 2021; Fig. 4.3c). Asynchronous and patchy burning carried out by livestock ranchers in sites under livestock grazing temporarily reduces the cover of *Saccharum angustifolium* (López-Mársico et al. 2021). After a burning event, it resprouts rapidly from vegetative buds. The resprouting of fresh leaf blades causes an improvement in forage quality and promotes consumption by domestic livestock (Rosengurtt 1979; López-Mársico et al. 2019a, 2021). The patches, which are evident in the landscape due to recent fires, are generally less than 0.5 ha (Bruzzone 2019). Evidence suggests that fire can have immediate effects on vegetation structure by modifying species composition, increasing species richness, and decreasing the cover of dominant species (López-Mársico et al. 2021; Fig. 4.3d). While no positive germination response has been observed in relation to the main fire cues (i.e. heat shock and smoke; López-Mársico et al. 2019b), it has been shown that a large number of species, present in the local species pool, rapidly colonize

Fig. 4.3 Images showing two moments: during (or immediately after) and a few months after a fire event along the gradient of subtropical and temperate grasslands of South America. (**a**) Experimental fire at the Agronomic Experimental Station of the Universidade Federal do Rio Grande do Sul, Eldorado do Sul, Brazil (Photo: Camila S. Goldas); (**b**) Few weeks after fire on Highland Grasslands, São Francisco de Paula, Brazil (Photo: Luciana R. Podgaiski); (**c**) Experimental fire in Quebrada de los Cuervos y Sierras del Yerbal Protected Landscape, Treinta y Tres, Uruguay (Photo: Luis López-Mársico); (**d**) Two months after fire in Quebrada de los Cuervos y Sierras del Yerbal Protected Landscape, Treinta y Tres, Uruguay (Photo: Luis López-Mársico); (**e**) Fire in an area invaded by pines in Sierra de la Ventana, Buenos Aires, Argentina (Photo: Gabriela Brancatelli); (**f**) Pine seedlings some months after a fire in Sierra de la Ventana, Buenos Aires, Argentina (Photo: Sergio Zalba)

gaps of bare soil (Cuello et al. 2020). As time since last fire increases, the standing dead biomass of the dominant tussock grass accumulates, prevents the establishment of other species through competitive exclusion, and causes a decrease in species richness (López-Mársico et al. 2021). A study conducted in the eastern

Uruguayan grasslands using NDVI (Normalized Difference Vegetation Index), a spectral index closely related to primary productivity showed that after fire, NDVI experience a rapid recovery (72 ± 37 days), and then values of burned sites can even exceed those of adjacent unburned sites (Bruzzone 2019).

The effects of fire on vegetation have also been studied in different regional subdivisions of the Pampa (Argentina). In the *Flooding Pampa* (Buenos Aires, Argentina), burning of *Paspalum quadrifarium*-dominated grasslands can lead to a change in species composition, an enhancement of forage quality, and a short-term increase in above-ground productivity (Laterra et al. 1998). Regarding the dominant species, an increase in fire frequency leads to a decrease in size of the individuals (Vignolio et al. 2003). In the *Mesopotamian Pampa* (Entre Ríos, Argentina) it has been observed that shrub species typical of the region show different responses to fire (Biganzoli et al. 2009; Galíndez et al. 2009). While a burning event can be lethal for one species (*Baccharis dracunculifolia*), other species (*Baccharis medulosa*, *Acanthostyles buniifolius*) can resprout (Galíndez et al. 2009). Mortality of *Acanthostyles buniifolius* was only observed on sites with high shrub density (Biganzoli et al. 2009). In the *Southern Pampa* (Buenos Aires, Argentina; Fig. 4.2b) it has been observed that vegetation cover decreases and that it recovers within 2 years after a fire in two types of communities, a grassland and an open low shrublands (Loydi et al. 2020). The main plant groups explaining the reduction in cover were shrubs and grasses in both communities. Species richness, recorded 2 years after burning, was higher in burned sites than in unburned sites in the grassland community, while it was similar between burned and unburned sites in the shrubland community. This study shows that a fire event can generate a change in plant composition, and that it can be maintained for at least 2 years (in the grassland community).

4.4 Fire Effects on Animals

Knowledge about the effects of fire on grassland wildlife is comparatively scarce, fragmentary, and preliminary (Driscoll et al. 2021; Saunders et al. 2021), especially in the case of South American grasslands (Behling et al. 2005). The reorganization of the grassland fauna after a fire depends on the particular response of each species and has been studied in invertebrates (Swengel 2001; Engstrom 2010; Driessen and Kirkpatrick 2017; Andersen 2019; Carbone et al. 2019; Certini et al. 2021), amphibians and reptiles (Larson 2014), birds (Brawn et al. 2001; Vickery et al. 2005; Fuhlendorf et al. 2006), and mammals (Sutherland and Dickman 1999; Griffiths and Brook 2014; Nieman et al. 2022). Fire effects on animal communities can be direct or indirect. The direct fire effects are those caused by heating during biomass combustion, leading to death or body damage to those animals that do not escape fire by flying, running, or sheltering. Fire intensity and the patterns of combustion spread in grasslands will influence the survival chances of the less mobile individuals. For example, fire that spreads rapidly and heterogeneously according to the flammable biomass, wind and relief conditions, and presence of moist microhabitats may promote the formation of patches with incomplete combustion, which may in turn increase the survival chances for animals that use these patches. Plants such as *Eryngium horridum* (a rosette plant with thorny, fibrous leaves that form phytotelmata, Apiaceae) may not be entirely consumed by fire, and thus can protect the associated fauna (e.g., invertebrates) from heat injuries (Campos 2010; Podgaiski et al. 2013; de Omena et al. 2017; Fig. 4.4a). There is a generalized lack of studies on the direct effects of fire on grassland animals, and the patterns of population reductions after fire (Swengel 2001). Evidence of fire mortality of animals is implied just from local observations of the post-burned environment.

The indirect fire effects on animal communities are those related to fire-induced habitat changes, such as the post-burn vegetation structure, resource availability, shelters, microclimate and predation (Whelan 1995). Recolonization patterns will largely depend on the size of the burning area and the proximity to the unburned matrix (Swengel 2001). Restrictions to animal colonization are proportional to burned patch size, i.e., the lower the burned patch, the more the established community will reflect its habitat preferences or constraints. Experimental studies using small burned plots in grasslands of Southern Brazil (10 × 10 m, mimicking a mosaic of burnt patches) indicate an extremely fast recolonization of arthropods, such as grasshoppers (Ferrando et al. 2016), hemipterans (Goldas 2014), thrips (Podgaiski et al. 2018), spiders (Podgaiski et al. 2013), ants (Barbosa et al. 2022), and other arthropods from different guilds (Goldas 2014). These groups showed increased abundance in burned plots, or did not differ in abundance when compared to paired unmanaged plots already within the first month after fire. Another study, using larger experimental plots (22 × 22 m; Ramos 2019) also reported no differences among burned (about 2 months after fire), traditionally grazed, and burned + grazed plots regarding abundance of arthropods. Recolonization of burned grasslands by animals associated with the vegetation is usually faster than by those animals more associated with the ground stratum because the resprout of vegetation is extremely fast, while the ground microhabitats, such as the litter layer, take longer to re-establish (Goldas 2014).

Post-burn habitat and the resprouted vegetation in subtropical and temperate grasslands usually offer a high resource availability to fauna. Palatable tissues of resprouted plants present high nutrient concentrations (e.g., nitrogen) and lack of induced defensive compounds (Aide 1993), and thus can be very attractive to herbivores such as grasshoppers, hemipterans, and leaf-feeding thrips (Goldas 2014; Ferrando et al. 2016; Podgaiski et al. 2017). Flowering after fire is also common, especially by forbs that use stored energy from below-ground organs (Fidelis and Blanco 2014), and the formation of mass flowering patches after fire represent resource-rich spots for pollinators (Fig. 4.4c–e). In remnant grasslands inserted in an urban matrix from Southern Brazil, four times more flowers were found in freshly-burned grasslands (varying from two to 6 months after fire) in comparison with old-burn grasslands (about 2 years after fire) (Goldas et al. 2022a). Bee abundance and richness, and butterfly abundance were increased in freshly-burned grasslands, and directly predicted by flower abundance patterns. Pollinator species such as the bumblebee *Bombus morio* and the sweet bee *Augochlorella acarinata* were

Fig. 4.4 (**a**) Experimental plot showing partially burned *Eryngium horridum* plants in unmanaged grasslands (ungrazed/unburned) in Eldorado do Sul, Brazil (Photo: Luciana R. Podgaiski); (**b**) Spider orb web in *Eryngium horridum* (Photo: Luciana R. Podgaiski); (**c**) Freshly-burned grassland a few weeks after fire in the Parque Natural Municipal Saint' Hilaire in South Brazil (Photo: Camila S. Goldas); (**d**) *Aspilia montevidensis* and the butterfly *Vanessa braziliensis* (Nymphalidae) (Photo: Camila S. Goldas); (**e**) *Peltodon longipes* (Lamiaceae) and non-identified wasp (Vespidae) (Photo: Camila S. Goldas)

associated with these grasslands, since they also benefit from the bare soil available to excavate underground nests. In this same system, populations of *Chamaecrista repens* (Fabaceae, Caesalpinioideae) were more abundant, with larger individuals, more reproductive structures (buds, flowers, and fruits), and more active extrafloral nectaries in freshly-burned grasslands (da Silva et al. 2020). Both the great resource offer (density of extrafloral nectaries) and the habitat openness (high % bare ground around the plants) led to a higher density of extrafloral nectaries-visiting ants. Other study found higher rates of seed removal by ants in burned than in unburned grasslands, attributed to the simplification of the post-fire habitat, where it is easier for ants to locomote, forage, and find resources (Barbosa et al. 2022). Finally, freshly sprouted biomass, flowers, and nectaries are also optimal foraging spots for predators, such as spiders, that draw benefits from the post-burn environment likely because of increased prey availability (Podgaiski et al. 2013).

Grassland plant species and functional diversity usually increases in the following growing season after fire, by allowing several less competitive and functionally diverse plant species to resprout, germinate, grow, and establish (Fidelis and Blanco 2014; Overbeck et al. 2005). Animals from all trophic levels usually follow the dynamics of plant communities, increasing diversity when plants increase diversity, which has been considered a proxy of resource diversity, or habitat heterogeneity. Clear positive relationships between grassland plant and animal species diversities after fire has been shown for thrips (Podgaiski et al. 2018), flower visitors (Beal-Neves et al. 2020b), and ants (Dröse et al. 2021; Barbosa et al. 2022) in South Brazil. Spider communities occurring in more functionally diverse plant communities were more diverse regarding their body traits and web building type, probably because of higher possibility of resource exploitation (foraging sites, prey, shelters) in burned grasslands (Podgaiski et al. 2013).

While post-fire plant community assembly patterns are relatively well-known in subtropical and temperate grasslands, the role that post-burn habitats play in animal assembly is still largely unknown. Fire-induced habitat changes can trigger animal trait-based community assembly through environmental filtering processes, e.g., predict trait states favored or disfavored under post-burn conditions, which is expected to be a very dynamic process associated with habitat recovery patterns (Podgaiski et al. 2013). The post-burn grasslands can present harsh environmental conditions for some invertebrate species by creating an open and less complex environment, more xeric (e.g., high sun exposure, soil temperatures, and low moisture retention; Podgaiski et al. 2014), with darkened substrate due to the ashes and soil exposition. Studying color patterns of arthropods in highland grasslands of South Brazil, was found in recently burned sites an increased frequency of darker phenotypes of grasshoppers and beetles (Monoplatina group/Alticini) compared to unburned grassland (Rocha et al. 2015). This is likely a consequence of higher predation pressure on lighter phenotypes by visually-oriented predators. Spiders that build small sheet or spatial webs, and ground hunters may be also more limited in recently burned sites probably because they lack appropriate physical structures for web attachment and buffered shelters (Podgaiski et al. 2013). Conversely, orb weavers that can spin webs across wider spaces may increase in proportion after fire, benefiting from the higher proportions of rosette plants, especially *Eryngium horridum*, which present a vertical architectural stratification (Fig. 4.4b).

Grassland birds, that is, those species that spend a main part or their entire life cycle in grasslands (Azpiroz et al. 2012; Vickery et al. 1999), and mammals, are especially vulnerable to changes in vegetation structure (Askins et al. 2007; Cueto and Lopez de Casenave 1999; Loggins et al. 2019; Rowland-Schaefer et al. 2023). In natural grasslands these structural changes are driven, fundamentally, by herbivory and fire and their interactions. In the southern Pampa (Argentina), overgrazing by exotic herbivores increases the exposure of the birds' nests to predators, reducing their reproductive success (Cozzani and Zalba 2012; Zalba and Cozzani 2004). The plant communities that are least affected by the action of livestock, such as those dominated by tall perennial grasses C_4 like *Paspalum quadrifarium* and *Cortaderia selloana* become key environments to sustain grassland bird populations (Cozzani and Zalba 2009). These habitats usually associated with shallows and stream edges, are, however, especially vulnerable to fire due to the accumulation of biomass. A study in the southern Pampas showed that the abundance, richness and diversity of birds decreased after a fire affecting riverbanks, but increased significantly 2 years later, in association with the recovery of vegetation height and complexity (Bahía and Zalba 2019). However, differences were observed in the particular response of certain species that might be due to their habitat requirements or preferences, as it was the case for the Great Pampa Finch (*Embernagra platensis*), the Brown-and-Yellow Marshbird (*Pseudoleistes virescens*) and the Grassland Yellow-Finch (*Sicalis luteola*), typical grassland birds which were absent or only present in very low numbers in recently burned areas. This same effect of impoverishment of bird diversity in grasslands dominated by the same species after fire had already been cited previously by Comparatore et al. (1996), with similar results also for small mammal communities. As if it were the negative of this picture, another study conducted in highland grasslands of southern Brazil, point out the importance of maintaining areas dominated by low grass, such as those resulting from fires, to guarantee the persistence of populations of the Saffron-Cowled Blackbird (*Xanthopsar flavus*) (Petry and Krüger 2010). Grassland birds also showed a particular delay in resuming reproductive activity in recently burned areas with respect to other more generalist species (Bahía and Zalba 2019). Experiments with artificial nests showed a negative effect of the habitat changes associated with fire on the breeding success of birds, and this situation may have consequences on the dynamics of their populations. Fire frequency was also identified as a modulator of the vegetation structure and, consequently, of the associated bird communities in highland grasslands of southern Brazil (Beal-Neves et al. 2020a). Studying grasslands, three to 180 months after the last fire, they observed that the complexity of the vegetation was highest at intermediately burnt sites, and that the abundance and richness of bird species decreased with the time elapsed since last fire, while changes in the composition of the functional groups of birds showed more complex patterns. Based on similar results, a study conducted in grasslands and palm groves of El Palmar National Park (Argentina) recommended the use of prescribed fire as a tool to maximize the richness and diversity of small mammal communities (Marconi and Kravetz 1986). The use of prescribed fire has also been proposed by different authors as a tool to increase the forage quality of the environments used by the Pampas deer (*Ozotocerus bezoarticus celer*) in Samborombón Bay, in the Flooding Pampas of Argentina (Jackson and Giulietti 1988; Vila et al. 2008), but discouraged for the management of similar environments within the same region in the case of the conservation of the Red-capped wren-spinetail (*Spartonoica maluroides*) (Isacch et al. 2004).

A few studies in subtropical and temperate grasslands (specifically in grassland of south Brazil) have investigated fire effects on animal and plant interactions described by ecological networks. Plant–pollinator networks were found to be highly resilient, rapidly reassembling after fire and maintaining their structure across grassland post-fire recovery stages (Goldas et al. 2022b; Beal-Neves et al. 2020b). These networks are shaped by plants and pollinators highly capable of adjusting their roles and switching their interaction partners as post-fire dynamics take place. Da Silva et al. (2020) described networks of grassland ant-plant interaction mediated by extrafloral nectaries across a post-fire successional gradient. Ecological networks in freshly burned grasslands presented ant species with lower individual interaction specialization, i.e., interacting with more plants than those networks longer without burning. Moreover, the core of generalist ant species (i.e., ants that comprise most of the interactions, usually formed by competitively superior species) changed along the time-since-fire gradient, probably because the habitat structure altered the competitive ability of these ant species, shifting their roles within the nested ant-plant networks. Therefore, fire-plant-animal interaction networks must be addressed to better understand the functioning of grasslands that are prone to burning.

4.5 Fire Effects on Soil Properties and Decomposition

Fire, by reducing vegetation cover, generates a change in the incidence of solar radiation, which affects soil temperature and moisture. Burning in temperate and subtropical grasslands is generally carried out with the prevailing wind (head fires), spreads quickly and therefore soil heating is limited to the passage of surface fires (Laterra et al. 2006; Fidelis et al. 2010; López-Mársico et al. 2024b). A study conducted in the South Brazilian Campos showed that soil temperature and residence time was higher in sites excluded from fire since 6 years than in frequently burned sites (Fidelis et al. 2010). Also, the temperature reached at the soil surface was variable (between 47 and 538 °C), partly due to heterogeneity in vegetation distribution. In a burning experiment with two fuel load levels, conducted in an area dominated by *Paspalum quadrifarium* in the flooding Pampas (Argentina), it was observed that the maximum temperatures during fire decreased with increasing soil depth (i.e., 1 to 3 cm depth) (Laterra et al. 2006). At 3 cm depth, no differences were detected concerning the initial fuel load levels. Another experimental study, conducted in an area dominated by *Saccharum angustifolium* in Sierras del Este (Uruguay) recorded that the temperature at the ground surface increased by 40% (6 °C) after burning, while soil moisture was not modified by burning (López-Mársico et al. 2024b). Six months after the experiment started, it was recorded that soil temperature was approximately 40% higher in burned and grazed sites than in unburned sites, both grazed and ungrazed. In turn, soil moisture was 45% lower in the burned sites than in the unburned sites. In Southern Brazil, decreased relative air moisture at the soil surface was found in burned grasslands 1 month after fire compared to unburned grasslands (Podgaiski et al. 2014). However, 6 months after fire, no differences were detected anymore. By comparing grasslands with different time since last fire, Goldas et al. (2022a) found recent burned grasslands (2–6 months after fire) with higher air temperature (mean: 34.1 °C) and lower moisture (mean: 61.2%) at the soil surface than old-burnt grasslands (2 years or more) (mean: 28.1 °C; 76.3%). These grasslands contrasted mostly between 8 a.m. and 6 p.m.

Fire modifies nutrient dynamics through volatilization during combustion and subsequent deposition in the form of ash (Bond and Keeley 2005; Archibald et al. 2019). Contrasting results regarding effects on soil nutrients and their effect on vegetation have been observed in North American grasslands (Collins et al. 1995; Blair 1997; Kohmann et al. 2023). For the South American grassland treated here, studies on fire effects on soil are still scarce. A study conducted in highland grasslands in southern Brazil reported a short-time increase in several soil nutrients (total N, K, Ca, Mg) in the soil surface layer (Rheinheimer et al. 2003). However, clearly more studies, and at sites with different soil conditions, are necessary. On the other hand, ash deposition can impermeabilize and compact the soil (O'Dea and Guertin 2003; Vermeire et al. 2005). In a Uruguayan grassland, it was observed that soil density was approximately 10% higher in burned than in unburned sites (López-Mársico et al. 2024b). This implies that ash deposition generates compaction in the topsoil layer. In grasslands of southern Brazil, it was detected altered soil hydro-physical properties in the soil surface of recently burned grasslands (i.e., decreased air permeability and increased water repellency), which was restored to pre-burn conditions in less than 6 months (Vogelmann et al. 2012).

Only a few studies have evaluated the effects of fire on soil detritivore fauna and decomposition processes in subtropical and temperate grasslands of South America. Podgaiski et al. (2014) performed an experimental study at a grassland site in southern Brazil to investigate microsite scale fire effects on soil fauna feeding activity (i.e., detritivory process) and decomposition process. One month after the fire, grassland patches presented reduced soil fauna densities (springtails and mites). This was probably caused by mortality by peak temperatures during combustion, and/or by fire-induced habitat changes. These changes resulted in a more extreme post-burn environment that was drier, had higher solar radiation incidence, and lacked surface organic matter, thus representing a very harsh environment for soil fauna (Certini et al. 2021). At this time, feeding activity of soil fauna at the soil surface, measured by the bait-lamina assay (von Törne 1990), was indeed lower. However, feeding activity of soil fauna measured belowground (up to 100 mm deep soil) did not differ from those of unburned plots, indicating that fire effects were only superficial. About 6 months after fire, soil fauna densities and its feeding activity in all soil layers reached the values of unburned plots, indicating high resilience. This study also accessed litter decomposition patterns after fire with the litter bag method using a standard mixture of leaf-litter material (Wieder and Lang 1982). Litter decay was faster in burned plots, probably due to major contributions of abiotic processes such as photodegradation (Brandt et al. 2010), due to the absence of a canopy (Podgaiski et al. 2014). Decomposition of sun-exposed litter is usually higher in comparison with shaded litter, because UV radiation can fragment lignin molecules, turning litter more soluble and prone to leaching and decomposers.

4.6 Fire as Tool for Conservation and Management

In this chapter, we have described how fire is able to reduce the biomass of dominant plant species and create heterogeneity in the landscape. Fire maintains grassland in some areas in a region where climate allows for forest development. Heterogeneity favors diversity, of both plants and animals. By favoring or disfavoring species based on their traits related to habitat preferences or limitations, fire-induced habitat changes may trigger modifications in species community composition. Sharp species turnover across chronosequences of grassland patches under different times since last fire has been reported for plant and bird communities (Laterra et al. 1998; Overbeck et al. 2005; Beal-Neves et al. 2020a; López-Màrsico et al. 2021), and flower visitors (Goldas et al. 2022a), with indicator species found for all post-fire stages. For each stage of grassland recovery after fire (or each time since last fire) one can find a different composition of species. These outcomes contribute to the emergent idea that maintaining grassland patches under different fire histories (i.e., with high pyrodiversity; Parr and Brockett 1999; Ponisio et al. 2016), which results in a mosaic of diverse grassland habitat structures, could be the optimal conservation strategy at the landscape level because it would conserve more species. The relationship between fires and grassland biodiversity is not as direct: fire effects depend on the frequency, intensity, and seasonality of fires, as well as the area burned and landscape context, along with concurrent factors such as herbivory. From a conservation perspective, it thus seems advisable to avoid uniform fire regimes across large areas. Instead, mosaics of sites under different fire regimes should increase habitat heterogeneity and, consequently, biodiversity. Regardless, it is important to monitor target species of conservation to evaluate if fire regimes area is adequate.

A case of particular interest in the conservation and management of grasslands is the role of fire as a modulator of the establishment and spread of alien plants, mainly woody species. Changes in fire dynamics are resulting in a global pattern of replacement of open ecosystems by woody-dominated communities known as *woody encroachment* (Wilcox et al. 2018). On the other hand, and although the general pattern would seem to indicate that fire contains the advance of woody plants on the grasslands, many of the alien trees and shrubs invading South American grasslands have evolved in fire-prone ecosystems (e.g., Mediterranean ecosystems), and their life cycles are adjusted to the occurrence of fires. Thus, fires have been associated with significant expansions of the area covered by invasive trees and shrubs, such as pines (*Pinus radiata*, *Pinus halepensis*), and shrub legumes (*Spartium junceum* and

Genista monspessulana) in mountain grasslands (Zalba and Villamil 2002; Zalba et al. 2008; Sanhueza and Zalba 2012; Ariza 2018; Brancatelli et al. 2020, 2021; Ramírez 2020), and *Pinus pinaster* and *Acacia longifolia* (Fabaceae) in psammophilous coastal grasslands (Lecanda et al. 2013; Yezzi et al. 2018, 2021) of the Southern Pampa (Fig. 4.3e, f), and similarly promotes the advancement of exotic woody plants throughout the region (Guadagnin et al. 2009; Simberloff et al. 2010). Fires lead to an intensification of the fragmentation associated with pine plantations and with the spontaneous advance of pines and acacias, and the combination of both forces affects the successional trajectory of the grasslands (Yezzi et al. 2019, 2020). However, as expected, the effects of fire on the dynamics of the invasion will depend, among other things, on the time elapsed between successive burnings (Pauchard et al. 2016). Thus, Sanhueza and Zalba (2009, 2014) demonstrated that fires deplete the seed bank of *Spartium junceum*, exposing the entire invasive population to eventual management actions, and creating a window of opportunity for control before new recruits reach reproductive age. Similarly, demographic models developed for *Pinus halepensis* predict that burning carried out at intervals of less than 7 years will result in the population retraction of the invader and, eventually, in its local eradication (Brancatelli et al. 2022, 2023). In the South Brazilian coastal region, an experimental study has shown the positive role of fire at the beginning of restoration of a former grassland site degraded by an exotic pine plantation: fire was effective in removing the accumulated pine litter layer and in killing pine seeds in the system (Porto et al. 2023).

In Uruguayan grasslands, there are very few antecedents, and they date back to a couple of trials for the control of the woody legume *Ulex europaeus* (Quiñones et al. 2015; Bresciano et al. 2022) and the grass *Eragrostis plana* (Ashfield 2015). Quiñones et al. (2015) evaluated the effect of autumn and spring burning on regrowth, viability, seed bank and recruitment of *Ulex europaeus* under grazing conditions. Resprout density was higher in autumn burning than in spring burning. The proportion of viable seeds did not differ between treatments and was higher than 80% in all cases. The seed bank was similar among burned treatments but was lower than in unburned sites. The burning season affected seedling density, being lower in both burned treatments than in unburned sites. Bresciano et al. (2022) carried out an experiment in which they combined fire with other treatments, such as mechanical mowing, grazing by domestic livestock and application of commercial herbicide. Their results show that fire is not sufficient as the only control mechanism and that a larger set of control treatments is necessary to slow down the advance of *Ulex europaeus* in Uruguayan Sierras del Este grasslands. On the other hand, there is a species that has a great invasive potential: *Eragrostis plana* (Poacaeae). This species is a warm-season C_4 grass, originally from southern Africa, which was accidentally introduced in southern Brazil as a forage species. Its presence is causing negative impacts in the grasslands of southern Brazil and Uruguay on diversity and livestock production due to its low palatability (Medeiros et al. 2009; Guido and Pillar 2017; Dresseno et al. 2018; Guido et al. 2021). Although it does not generate a large accumulation of standing dead biomass a study has shown that burning resulted in an increase in basal area and in the number of panicles compared to an unburned site (Ashfield 2015).

Grazing is the most influential disturbance for biomass consumption in subtropical and temperate grasslands of South America (Paruelo et al. 2022) and grassland burning has a lesser incidence (Bernardi et al. 2016, 2019). However, in some regions, livestock ranchers use fire as a management tool. The objective is to achieve a reduction of accumulated non-palatable biomass of the dominant species: woody biomass in the case of shrubs and standing dead biomass in the case of tussock grasses (Royo Pallarés et al. 2005). Fire can kill shrub species (Galíndez et al. 2009) but is unlikely to kill adult perennial grasses (Overbeck and Pfadenhauer 2007). Therefore, another aim of the livestock ranchers is to stimulate the resprouting of fresh biomass (Boldrini 1997; Laterra et al. 1998; Overbeck and Pfadenhauer 2007; López-Mársico et al. 2021). This allows for improved forage quality and increased consumption by livestock. In several grassland regions of the world, it has been observed that after a fire event, herbivores are attracted to burned areas (Laterra et al. 2003; Fuhlendorf and Engle 2004). In the region, grasslands are usually burned in late winter or early spring, approximately with a frequency of 2 to 5 years depending on biomass accumulation (Vincent 1935; Laterra et al. 1998; Overbeck et al. 2007; Ferreira et al. 2021; López-Mársico et al. 2021).

In the highland grasslands of southern Brazil, it has been suggested that fire is one of the factors responsible, together with grazing, for preventing forest expansion, which would be expected under the actual climatic conditions (Oliveira and Pillar 2004). In these grasslands, the low carrying capacity during winter determines the year-round rather low stocking rate and leads to the accumulation of standing dead biomass (Andrade et al. 2019). In other regions (e.g. Northern Campos, Flooding Pampa), a large part of the biomass produced by highly productive C_4 grasses in summer is not consumed by livestock (Laterra et al. 2003; López-Mársico et al. 2021). Some tussock grass-dominated grasslands need the joint action of fire and grazing to be conserved in space and time (Fuhlendorf et al. 2009; Pausas and Bond 2020). A study in Sierras del Este (Uruguay) showed that grazing is not sufficient to control the growth rates of the tussock grass *Saccharum angustifolium*, which is dominant in the landscape. A patch scale cycle based on the Pyric Herbivory model by Fuhlendorf et al. (2009) was proposed (López-Mársico et al. 2021). The patch scale cycle starts when the biomass and standing dead biomass of *Saccharum angustifolium* is burned and the grazing pressure increases. The cycle ends when *Saccharum angustifolium* becomes dominant again (3 to 5 years after), and a large amount of standing dead biomass is accumulated.

So far, we have described how fire has been used as a management tool, mainly to prevent the dominance of certain species and/or to promote grazing by domestic livestock. However, can fire be used as a tool for the conservation of subtropical and temperate grassland diversity in South America? Although there are data that may support an affirmative answer to this question, it might be more accurate to say, it depends. As mentioned earlier, it would be advisable to avoid burning large areas and instead generate patch burns to promote the presence of areas with different vegetation structures. Concerning invasive exotic species, several originate from fire-dependent areas and are therefore evolutionarily associated with fire. Attempting to control them with burns could produce results contrary to those intended. Another important aspect to consider is evaluating the possibility of using fire as a tool for the restoration of degraded areas. Fire, not only indirectly through heat, but also through chemical components found in the smoke, could have a beneficial effect for some pioneer species in the recovery of areas with sparse vegetation cover.

4.7 Policy and Knowledge Gaps

Fire in grassland has only recently become a topic of research in Southeastern South America. In Brazil, where focus of conservation and also conservation science traditionally had been on forest ecosystems, fire was, for a long time, considered as a negative process to be avoided and thus was not used as a management tool in conservation and was illegal on private properties in Southern Brazil. Today, this has changed somewhat, and fire can be used as a management tool on private land, when the local characteristics justify its use and with a license from state environmental agencies (Law 12.651). These burns usually are conducted in winter, with the objective to remove accumulated grass biomass (Brunel et al. 2021). In protected areas, use of controlled burned has increases in the last years, also in southern Brazil (Pivello et al. 2021). In Uruguay, it is established that from 1 November of each year until the second half of April of the following year, fires and open burning of any kind are prohibited throughout the territory, both in public and private areas (Decree N° 436/007). Regarding protected areas, in at least one (Quebrada de los Cuervos y Sierras del Yerbal) burning of vegetation dominated by tussock grasses is allowed during the late autumn, winter and early spring to avoid the breeding season of wildlife species (Bianco and Lapetina 2013). However, burning by livestock producers must be communicated in advance to the area's management to obtain authorization. In Argentina, national law 26562 establishes the prohibition of burning without the relevant authorization. Law 26815, in turn, establishes that in the case of voluntary or accidental fires affecting grasslands, it is prohibited, for a period of 30 years from the fire, to carry out real estate projects and change the agricultural use prior to the one that the area had before the fire. This regulation seeks to prevent illegal fires based on productive or real estate speculation, and "to guarantee the conditions for the burned surfaces to be restored". Provincial jurisdictions establish absolute ban periods for burning rural properties particularly during the summer months. Within national parks, those exposed to fires according to their ecological characteristics, are obliged to develop fire management plans, including the use of prescribed burning, although the main motivation continues to be to avoid the effects considered negative, especially in forest ecosystems.

While use of fire thus is slowly becoming a bit more common, it is clear that overall knowledge on the effects of fire and thus its potential use as a conservation or management tool is still scanty in the region. Considering the rather low number of studies, many rather local in their scope, it is still hard to make science-based recommendations for fire-management in the region as a whole. The timing of burning can determine the temporary appearance or disappearance of some species. Clearly, we need more studies, especially conducting long-term studies will enable us to draw better conclusions about the relationship with extreme climatic events (e.g., drought, floods). Ideally, they go beyond the plot-scale commonly used in scientific experiments but are conducted in the context of land management, be it on private properties (i.e., in a grazing context), or in protected areas. There still are few scientific studies on the use of fire that provide clear data when—i.e., under which conditions—and how—i.e., in which concrete way—fire can and should be used for conservation or management purposes. Also, if fire can also be a tool at subsequent phases of restoration management remains to be seen. The use of fire often still is a taboo subject for land managers and decision makers. However, managers, decision makers and policy makers must be open for the use of fire and see the need for experimentation, which often still is not the case. Grasslands suffer not only from 'biome awareness disparity' (Silveira et al. 2022), but the discussion of the role of fire is especially problematic and often polarized, despite the existing evidence of its positive effects in the region and beyond. More studies on fire also seem necessary concerning climate change: changing environmental conditions may also lead to increased risk of fire and changed fire regimes (Pivello et al. 2021), emphasizing the need for awareness and monitoring, but also for Integrated Fire Management strategies.

References

Aide TM (1993) Patterns of leaf development and herbivory in a tropical understory community. Ecology 74:455–466. https://doi.org/10.2307/1939307

Andersen AN (2019) Responses of ant communities to disturbance: five principles for understanding the disturbance dynamics of a globally dominant faunal group. J Anim Ecol 88:350–362. https://doi.org/10.1111/1365-2656.12907

Andrade BO, Bonilha CL, Overbeck GE, Vélez-Martin E, Rolim RG et al (2019) Classification of South Brazilian grasslands: implications for conservation. Appl Veg Sci 22:168–184. https://doi.org/10.1111/avsc.12413

Arcamone JR, Jaureguiberry P (2018) Germination response of common annual and perennial forbs to heat shock and smoke treatments in the Chaco Serrano, Central Argentina. Austral Ecol 43:567–577. https://doi.org/10.1111/aec.12593

Archibald S, Hempson GP, Lehmann C (2019) A unified framework for plant life-history strategies shaped by fire and herbivory. New Phytol 224:1490–1503. https://doi.org/10.1111/nph.15986

Ariza A (2018) Estudio y comparación de dos especies de plantas exóticas invasoras en el Parque Provincial Ernesto Tornquist, Buenos Aires. Tesis de Licenciatura en Ciencias Biológicas, Universidad Nacional del Sur, Bahía Blanca

Ashfield L (2015) Ecología de *Eragrostis plana* Ness y respuesta a alternativas culturales de manejo. Tesis de Ingeniería Agronómica, Facultad de Agronomía, Montevideo

Askins RA, Chávez-Ramírez F, Dale BC, Haas CA, Herkert JR, Knopf FL, Vickery PD (2007) Conservation of grassland birds in North America: understanding ecological processes in different regions. Ornithol Monogr 64:1–46. https://doi.org/10.2307/40166905

Azpiroz AB, Isacch JP, Dias RA, Di Giacomo SA, Fontana CS, Palarea CM (2012) Ecology and conservation of grassland birds in southeastern South America: a review. J Field Ornithol 83:217–246. https://doi.org/10.1111/j.1557-9263.2012.00372.x

Bahía R, Zalba S (2019) Changes in grassland bird communities and breeding success after a fire in the Argentinian Pampas. Biodivers Conserv 28:3767–3786. https://doi.org/10.1007/s10531-019-01850-4
Barbosa GG, Ferrando CPR, Mendonça MS Jr, Podgaiski LR (2022) How do patch burnings affect ant communities and seed removal in a subtropical grassland? Perspect Ecol Conserv 20(4):322–329. https://doi.org/10.1016/j.pecon.2022.07.002
Beal-Neves M, Chiarani E, Ferreira PMA, Fontana CS (2020a) The role of fire disturbance on habitat structure and bird communities in South Brazilian Highland Grasslands. Sci Rep 10:19708. https://doi.org/10.1038/s41598-020-76758-z
Beal-Neves M, Ely CV, Esteves MW et al (2020b) The influence of urbanization and fire disturbance on plant-floral visitor mutualistic networks. Diversity 12:141. https://doi.org/10.3390/d12040141
Behling H, Pillar VP (2007) Late quaternary vegetation, biodiversity and fire dynamics on the southern Brazilian highland and their implication for conservation and management of modern Araucaria forest and grassland ecosystems. Philos Trans R Soc B 362:243–251. https://doi.org/10.1098/rstb.2006.1984
Behling H, Pillar VD, Bauermann SG (2005) Late Quaternary grassland (Campos), gallery forest, fire and climate dynamics, studied by pollen, charcoal and multivariate analysis of the São Francisco de Assis core in western Rio Grande do Sul (southern Brazil). Rev Palaeobot Palynol 133:235–248. https://doi.org/10.1016/j.revpalbo.2004.10.004
Benson EJ, Hartnett DC, Mann KH (2004) Belowground bud banks and meristem limitation in tallgrass prairie plant populations. Am J Bot 91:416–421. https://doi.org/10.3732/ajb.91.3.416
Bernardi RE, Holmgren M, Arim SM (2016) Why are forests so scarce in subtropical South America? The shaping roles of climate, fire and livestock. For Ecol Manag 363:212–217. https://doi.org/10.1016/j.foreco.2015.12.032
Bernardi RE, Staal A, Xu C, Scheffer M, Holmgren M (2019) Livestock herbivory shapes fire regimes and vegetation structure across the global tropics. Ecosystems 22:1457–1465. https://doi.org/10.1007/s10021-019-00349-x
Bianco A, Lapetina J (2013) Prácticas ganaderas recomendadas para el manejo y conservación de pastizales naturales: Paisaje Protegido Quebrada de los Cuervos Programa de Manejo de Pastizales. Ministerio de Vivienda, Ordenamiento Territorial y Medio Ambiente, Montevideo
Biganzoli F, Wiegand T, Batista WB (2009) Fire-mediated interactions between shrubs in a South American temperate savannah. Oikos 118:1383–1395. https://doi.org/10.1111/j.1600-0706.2009.17349.x
Blair JM (1997) Fire, N availability, and plant response in grasslands: a test of the transient maxima hypothesis. Ecology 78:2359–2368. https://doi.org/10.1890/0012-9658(1997)078[2359:FNAAPR]2.0.CO;2
Boldrini II (1997) Campos do Rio Grande do Sul: Caracterização fisionômica e problemática ocupacional. Revista Brasileira Biociências 56:1–39
Boldrini II (2009) A flora dos Campos do Rio Grande do Sul. In: Pillar VD, Müller SC, de Souza ZM et al (eds) Campos sulinos: Conservação e uso sustentável da biodiversidade, 1st edn. MMA, Brasília, pp 63–77
Bond WJ (2019) Open ecosystems: ecology and evolution beyond the forest edge. Oxford University Press, New York
Bond WJ, Keeley JE (2005) Fire as a global "herbivore": the ecology and evolution of flammable ecosystems. Trends Ecol Evol 20:387–394. https://doi.org/10.1016/j.tree.2005.04.025
Bond WJ, Parr CL (2010) Beyond the forest edge: ecology, diversity and conservation of the grassy biomes. Biol Conserv 143:2395–2404. https://doi.org/10.1016/j.biocon.2009.12.012
Bond WJ, Van Wilgen BW (1996) Fire and plants. Chapman & Hall Ltd, London
Brancatelli G, Amodeo M, Cuevas YA, Zalba SM (2020) Invasive pines in Argentinian grasslands: lessons from control operations. Biol Invasions 22:473–484. https://doi.org/10.1007/s10530-019-02103-9

Brancatelli G, Amodeo M, Cuevas YA, Zalba SM (2021) Seedling emergence and survival of invasive pines in Argentinian mountain grasslands. Biol Invasions 23:903–918. https://link.springer.com/article/10.1007/s10530-020-02410-6

Brancatelli G, Amodeo M, Cuevas YA, Zalba SM (2022) Demographic model for Aleppo pine invading Argentinean grasslands. Ecol Model 473:110143. https://doi.org/10.1016/j.ecolmodel.2022.110143

Brancatelli G, Amodeo M, Cuevas YA, Zalba SM (2023) Modelling population dynamics of invasive pines to optimize their control in native grasslands of Argentina. J Environ Manag. https://doi.org/10.2139/ssrn.4358499

Brandt LA, King JY, Hobbie SE, Milchunas DG, Sinsabaugh RL (2010) The role of photodegradation in surface litter decomposition across a grassland ecosystem precipitation gradient. Ecosystems 13:765–781. https://doi.org/10.1007/s10021-010-9353-2

Brawn JD, Robinson SK, Thompson FR III (2001) The role of disturbance in the ecology and conservation of birds. Annu Rev Ecol Syst 32:251–276. https://doi.org/10.1146/annurev.ecolsys.32.081501.114031

Bresciano D, Olano R, De Santiago F, Borges A, Pérez L, Salvo L, Munka C (2022) Evaluación de métodos de control de invasión de *Ulex europaeus* L.: Estudio de caso en predio ganadero de la región este de Uruguay. Tercer Seminario Internacional Restauración en el bioma Pampa: Argentina, Brasil y Uruguay

Brunel M, Rammig A, Furquim F, Overbeck G, Barbosa HM et al (2021) When do farmers burn pasture in Brazil: a model-based approach to determine burning date. Rangeland Ecol Manag 79:110–125. https://doi.org/10.1016/j.rama.2021.08.003

Bruzzone L (2019) Evento de quemas en pastizales de Sierras del Este: una caracterización basada en sensores. Tesis de Licenciatura en Ciencias Biológicas, Facultad de Ciencias, Universidad de la República, Montevideo

Campos RE (2010) *Eryngium* (Apiaceae) phytotelmata and their macro-invertebrate communities, including a review and bibliography. Hydrobiologia 652:311–328

Carbone LM, Tavella J, Pausas JG, Aguilar R (2019) A global synthesis of fire effects on pollinators. Glob Ecol Biogeogr 28:1487–1498. https://doi.org/10.1111/geb.12939

Cerling TE, Wang Y, Quade J (1993) Expansion of C4 ecosystems as an indicator of global ecological change in the late Miocene. Nature 361:344–345. https://doi.org/10.1038/361344a0

Certini G, Moya D, Lucas-Borja ME, Mastrolonardo G (2021) The impact of fire on soil-dwelling biota: a review. For Ecol Manag 488:118989. https://doi.org/10.1016/j.foreco.2021.118989

Christin PA, Besnard G, Samaritani E, Duvall MR, Hodkinson TR (2008) Oligocene CO2 decline promoted C4 photosynthesis in grasses. Curr Biol 18:37–43. https://doi.org/10.1016/j.cub.2007.11.058

Collins SL, Wallace LL (1990) Fire in North American tallgrass prairies. University of Oklahoma Press, London

Collins SL, Glenn SM, Gibson DJ (1995) Experimental analysis of intermediate disturbance and initial floristic composition: decoupling cause and effect. Ecology 76:486–492. https://doi.org/10.2307/1941207

Comparatore VM, Martinez MM, Vasallo AI, Barg M, Isacch JP (1996) Abundancia y relaciones con el hábitat de aves y mamíferos en pastizales de *Paspalum quadrifarium* (paja colorada) manejados con fuego (Provincia de Buenos Aires, Argentina). Interciencia 21:228–237

Cozzani N, Zalba SM (2009) Estructura de la vegetación y selección de hábitats reproductivos en aves del pastizal pampeano. Ecol Austral 19:35–44

Cozzani N, Zalba SM (2012) Depredadores de nidos en pastizales del Parque Provincial Ernesto Tornquist (Provincia de Buenos Aires, Argentina), su importancia relativa bajo distintas inten sidades de pastoreo. El Hornero 27:137–148

Cuello N, López-Mársico L, Rodríguez C (2020) Field burn versus fire-related cues: germination from the soil seed bank of a South American temperate grassland. Seed Sci Res 30:206–214. https://doi.org/10.1017/S0960258520000288

Cueto VR, Lopez de Casenave J (1999) Determinants of bird species richness: role of climate and vegetation structure at a regional scale. J Biogeogr 26:487–492. https://doi.org/10.1046/j.1365-2699.1999.00299.x

da Silva CVC, Goldas CS, Dattilo W, Dröse W, de Souza Mendonça M Jr, Podgaiski LR (2020) Effects of time-since-fire on ant-plant interactions in southern Brazilian grasslands. Ecol Indic 112:106094. https://doi.org/10.1016/j.ecolind.2020.106094

Dalgleish HJ, Hartnett DC (2009) The effects of fire frequency and grazing on tallgrass prairie productivity and plant composition are mediated through bud bank demography. Plant Ecol 201:411–420. https://doi.org/10.1007/978-90-481-2798-6_4

de Omena PM, Kersch-Becker MF, Antiqueira PAP, Bernabé TN, Benavides-Gordillo S et al (2017) Bromeliads provide shelter against fire to mutualistic spiders in a fire-prone landscape. Ecol Entomol 43:389–393. https://doi.org/10.1111/een.12497

Dresseno AL, Guido A, Balogianni V, Overbeck GE (2018) Negative effects of an invasive grass, but not of native grasses, on plant species richness along a cover gradient. Austral Ecol 43:949–954. https://doi.org/10.1111/aec.12644

Driessen MM, Kirkpatrick JB (2017) The implications of succession after fire for the conservation management of moorland invertebrate assemblages. J Insect Conserv 21:15–37. https://doi.org/10.1007/s10841-016-9948-9

Driscoll DA, Armenteras D, Bennett AF, Brotons L, Clarke MF, Doherty TS et al (2021) How fire interacts with habitat loss and fragmentation. Biol Rev 96:976–998. https://doi.org/10.1111/brv.12687

Dröse W, Podgaiski LR, Gossner MM, Meyer ST, Hermann J et al (2021) Passive restoration of subtropical grasslands leads to incomplete recovery of ant communities in early successional stages. Biol Conserv 264:109387. https://doi.org/10.1016/j.biocon.2021.109387

Edwards EJ, Still CJ (2008) Climate, phylogeny and the ecological distribution of C4 grasses. Ecol Lett 11:266–276. https://doi.org/10.1111/j.1461-0248.2007.01144.x

Engstrom RT (2010) First-order fire effects on animals: review and recommendations. Fire Ecol 6:115–130. https://doi.org/10.4996/fireecology.0601115

Ferrando CPR, Podgaiski LR, Costa MKM, Mendonça MS Jr (2016) Taxonomic and functional resilience of grasshoppers (Orthoptera, Caelifera) to fire in South Brazilian grasslands. Neotrop Entomol 45:374–381. https://doi.org/10.1007/s13744-016-0380-3

Ferreira PMA, Andrade BO, Podgaiski LR, Dias AC, Pillar VD et al (2020) Long-term ecological research in southern Brazil grasslands: effects of grazing exclusion and deferred grazing on plant and arthropod communities. PLoS One 15:e0227706. https://doi.org/10.1371/journal.pone.0227706

Ferreira PMA, Ely CV, Beal-Neves M (2021) Different post-fire stages encompass different plant community compositions in fire-prone grasslands from Southern Brazil. Flora 285:151937. https://doi.org/10.1016/j.flora.2021.151937

Fidelis A, Blanco C (2014) Does fire induce flowering in Brazilian subtropical grasslands? Appl Veg Sci 17:690–699. https://doi.org/10.1111/avsc.12098

Fidelis A, Appezzato-da-Glória B, Pfadenhauer J (2009) A importância da biomassa e das estruturas subterrâneas nos Campos Sulinos. In: Pillar VD, Müller SC, de Souza ZM, Ávila Jacques AV (eds) Campos sulinos: Conservação e uso sustentável da biodiversidade, 1st edn. MMA, Brasília, pp 88–100

Fidelis A, Delgado-Cartay MD, Blanco CC, Müller SC, Pillar VD, Pfadenhauer J (2010) Fire intensity and severity in Brazilian Campos grasslands. Interciencia 35:739–745

Fidelis A, Blanco CC, Müller SC, Pillar VD, Pfadenhauer J (2012) Short-term changes caused by fire and mowing in Brazilian Campos grasslands with different long-term fire histories. J Veg Sci 23:552–562. https://doi.org/10.1111/j.1654-1103.2011.01364.x

Fidelis A, Appezzato-da-Glória B, Pillar VD, Pfadenhauer J (2014) Does disturbance affect bud bank size and belowground structures diversity in Brazilian subtropical grasslands? Flora 209:110–116. https://doi.org/10.1016/j.flora.2013.12.003

Fuhlendorf SD, Engle DM (2004) Application of the-grazing interaction to restore a sifting mosaic on tallgrass prarie. J Appl Ecol 41:604–614. https://doi.org/10.1111/j.0021-8901.2004.00937.x

Fuhlendorf SD, Harrell WC, Engle DM, Hamilton RG, Davis CA, Leslie DM Jr (2006) Should heterogeneity be the basis for conservation? Grassland bird response to fire and grazing. Ecol Appl 16:1706–1716. https://doi.org/10.1890/1051-0761(2006)016[1706:SHBTBF]2.0.CO;2

Fuhlendorf SD, Engle DM, Kerby JA, Hamilton R (2009) Pyric herbivory: rewilding landscapes through the recoupling of fire and grazing. Conser Biol 23:588–598. https://doi.org/10.1111/j.1523-1739.2008.01139.x

Galíndez G, Biganzoli F, Ortega-Baes P, Scopel AL (2009) Fire responses of three co-occurring Asteraceae shrubs in a temperate savanna in South America. Plant Ecol 202:149–158. https://doi.org/10.1007/s11258-008-9537-4

Gibson DJ (2009) Grasses and grassland ecology. Oxford University Press, New York

Goldas CS (2014) Distúrbio por fogo nos Campos Sulinos: Artrópodes e Hemiptera como bioindicadores. Master Dissertation in Ecology, Universidade Federal do Rio Grande do Sul, Porto Alegre, Brazil

Goldas CS, Podgaiski LR, da Silva CVC, Mendonça MS Jr (2022a) Burning for grassland pollination: recently burned patches promote plant flowering and insect pollinators. Austral Ecol 47:491–506. https://doi.org/10.1111/aec.13108

Goldas CS, Podgaiski LR, da Silva CVC, Ferreira PMA, Vizentin-Bugoni J et al (2022b) Structural resilience and high interaction dissimilarity of plant–pollinator interaction networks in fire-prone grasslands. Oecologia 198:179–192. https://doi.org/10.1007/s00442-021-05071-x

Griffiths AD, Brook BW (2014) Effect of fire on small mammals: a systematic review. Int J Wildl Fire 23:1034–1043. https://doi.org/10.1071/WF14026

Guadagnin DL, Zalba SM, Costa Gorriz B, Fonseca CR, Nebbia AJ et al (2009) Árvores e arbustos exóticos invasores no Bioma Pampa—questões ecológicas, culturais e socioeconômicas de um desafio crescente. In: Pillar VD, Müller SC, de Souza ZM, Ávila Jacques AV (eds) Campos sulinos: Conservação e uso sustentável da biodiversidade, 1st edn. MMA, Brasília, pp 300–316

Guido A, Pillar VD (2017) Invasive plant removal: assessing community impact and recovery from invasion. J Appl Ecol 54:1230–1237. https://doi.org/10.1111/1365-2664.12848

Guido A, Blanco C, Pillar VD (2021) Disentangling by additive partitioning the effects of invasive species on the functional structure of communities. J Veg Sci 32:e13004. https://doi.org/10.1111/jvs.13004

Guido A, López-Mársico L, Rodríguez C, Gallego F, Lezama F et al (2025) Uruguayan Grasslands: A threatened natural asset.Austral Ecol 50:e70031. https://doi.org/10.1111/aec.70031

Hartnett DC, Setshogo MP, Dalgleish HJ (2006) Bud banks of perennial savanna grasses in Botswana. Afr J Ecol 44:256–263. https://doi.org/10.1111/j.1365-2028.2006.00646.x

Hoss D (2021) Stabilizing mechanisms of grassland communities under disturbance. Tese de Doutorado, Programa de Pós-Graduação em Ecología, Universidade Federal do Rio Grande do Sul, Porto Alegre

IBGE—Instituto Brasileiro de Geografia e Estatística (2004) Mapa da vegetação do Brasil e Mapa de Biomas do Brasil

Isacch JP, Holz S, Ricci L, Martínez M (2004) Post-fire vegetation change and bird use of a salt marsh in coastal Argentina. Wetlands 24:235–243. https://doi.org/10.1672/0277-5212(2004)024[0235:PVCABU]2.0.CO;2

Jackson JE, Giulietti JD (1988) The food habits of pampas deer *Ozotoceros bezoarticus celer* in relation to its conservation in a relict natural grassland in Argentina. Biol Conserv 45:1–10. https://doi.org/10.1016/0006-3207(88)90048-1

Jacobs BF, Kingston JD, Jacobs LL (1999) The origin of grass-dominated ecosystems. Ann Mo Bot Gard 86:590–643. https://doi.org/10.2307/2666186

Joner F, Giehl ELH, Pillar VD (2021) Functional and taxonomic alpha and beta diversity responses to burning grasslands in southern Brazil. J Veg Sci 32:e13060. https://doi.org/10.1111/jvs.13060

Kaal J, Gianotti C, del Puerto L, Criado-Boado F, Rivas M (2019) Molecular features of organic matter in anthropogenic earthen mounds, canals and lagoons in the Pago Lindo archaeological complex (Tacuarembó, Uruguayan lowlands) are controlled by pedogenetic processes and fire practices. J Archaeolog Sci Rep 26:101900. https://doi.org/10.1016/j.jasrep.2019.101900

Keeley JE, Rundel PW (2005) Fire and the Miocene expansion of C4 grasslands. Ecol Lett 8:683–690. https://doi.org/10.1111/j.1461-0248.2005.00767.x
Kellogg EA (2000) The grasses: a case study in macroevolution. Annu Rev Ecol Syst 31:217–238. https://doi.org/10.1146/annurev.ecolsys.31.1.217
Kohmann MM, Silveira ML, da Silva CA, Bracho R (2023) Short-term impacts of prescribed fire on C, N, and P dynamics in a subtropical rangeland. Plant Soil. https://doi.org/10.1007/s11104-023-06063-x
Larson DM (2014) Grassland fire and cattle grazing regulate reptile and amphibian assembly among patches. Environ Manag 54:1434–1444. https://doi.org/10.1007/s00267-014-0355-2
Laterra P, Vignolio OR, Hidalgo L, Fernández ON, Cauhépé MA, Maceira N (1998) Dinámica de pajonales de paja colorada (Paspalum spp.) manejados con fuego y pastoreo en la Pampa deprimida Argentina. Ecotropicos 11:141–149
Laterra P, Vignolio OR, Linares MP, Giaquinta A, Maceira N (2003) Cumulative effects of fire on a tussock pampa grassland. J Veg Sci 14:43–54. https://doi.org/10.1111/j.1654-1103.2003.tb02126.x
Laterra P, Ortega EZ, Del Carmen OM, Vignolio OR, Fernandez ON (2006) Interactive influences of fire intensity and vertical distribution of seed banks on post-fire recolonization of a tall-tussock grassland in Argentina. Austral Ecol 31:608–622. https://doi.org/10.1111/j.1442-9993.2006.01593.x
Lauenroth WK (1979) Grassland primary production: North American grasslands in perspective. In: Frecn NR (ed) Perspectives in grassland ecology, 1st edn. Springer, New York, pp 3–24
Lauenroth WK, Sala OE (1992) Long-term forage production of North American shortgrass steppe. Ecol Appl 2:397–403. https://doi.org/10.2307/1941874
Lecanda J, Cuevas YA, Zalba SM (2013) Aspectos biológicos de *Acacia longifolia*, una especie exótica potencialmente invasora de pastizales psamófilos. Revista de la Facultad de Agronomía de la Universidad Nacional de La Pampa 22:99–104
Lezama F, Pereira M, Altesor A, Paruelo JM (2019) Grasslands of Uruguay: classification based on vegetation plots. Phytocoenolgia 49:211–229. https://doi.org/10.1127/phyto/2019/0215
Loggins AA, Monadjem A, Kruger LM, Reichert BE, McCleery RA (2019) Vegetation structure shapes small mammal communities in African savannas. J Mammal 100:1243–1252. https://doi.org/10.1093/jmammal/gyz100
López-Mársico L, Lezama F, Altesor A (2019a) ≐Qué sabemos sobre los efectos del fuego en pastizales? In: Altesor A, López-Mársico L, Paruelo JM (eds) Bases ecológicas y tecnológicas para el manejo de los pastizales II. INIA-FPTA, Montevideo, pp 97–107
López-Mársico L, Farías-Moreira L, Lezama F, Altesor A, Rodríguez C (2019b) Light intensity triggers different germination responses to fire-related cues in temperate grassland species. Folia Geobot 54:53–63. https://doi.org/10.1007/s12224-019-09336-5
López-Mársico L, Lezama F, Altesor A (2021) Heterogeneity decreases as time since fire increases in a South American grassland. Appl Veg Sci 24:e12521. https://doi.org/10.1111/avsc.12521
López-Mársico L, Oyarzabal M, Altesor A, Paruelo JM (2024a) Grazing exclusion reduces below-ground biomass of temperate subhumid grasslands of South America: a meta-analysis and a database. Austral Ecol 49:e13304. https://doi.org/10.1111/aec.13304
López-Mársico L, Lezama F, Altesor A (2024b) Fire and grazing interaction in a subhumid grassland: effects on a dominant tussock grass. Appl Veg Sci 27:e12795. https://doi.org/10.1111/avsc.12795
Loydi A, Funk FA, García A (2020) Vegetation recovery after fire in mountain grasslands of Argentina. J Mt Sci 17:373–383. https://doi.org/10.1007/s11629-019-5669-3
MacFadden BJ (2000) Cenozoic mammalian herbivores from the Americas: reconstructing ancient diets and terrestrial communities. Annu Rev Ecol Syst 31:33–59. https://doi.org/10.1146/annurev.ecolsys.31.1.33
Marconi PN, Kravetz FO (1986) Comunidades de roedores del Parque Nacional El Palmar (Entre Ríos, Argentina) según la historia del fuego. Rev Chil Hist Nat 59:47–57
Medeiros RB, Saibro JC, Fotch T (2009) Invasão do capim-annoni (*Eragrostis plana* Nees) no bioma Pampa do Rio Grande do Sul. In: Pillar VD, Mülller SC, de Souza ZM, Ávila Jacques

AV (eds) Campos sulinos: Conservação e uso sustentável da biodiversidade, 1st edn. MMA, Brasília, pp 317–330
Müller SC, Overbeck GE, Pfadenhauer J, Pillar VD (2007) Plant functional types of woody species related to fire disturbance in forest-grassland ecotones. Plant Ecol 189:1–14. https://doi.org/10.1007/s11258-006-9162-z
Nabinger C, Moraes A, Maraschin GE (2000) Campos in Southern Brazil. In: Lemaire G, Hodgson J, Moraes A, Nabinger C, Carvalho PCF (eds) Grassland ecophysiology and grazing ecology, 1st edn. CAB International, London, pp 355–376
Nieman WA, van Wilgen BW, Radloff FG, Leslie AJ (2022) A review of the responses of medium- to large-sized African mammals to fire. Afr J Range Forage Sci 39:249–263. https://doi.org/10.2989/10220119.2021.1918765
Noy-Meir I (1995) Interactive effects of fire and grazing on structure and diversity of Mediterranean grasslands. J Veg Sci 6:701–710. https://doi.org/10.2307/3236441
O'Dea ME, Guertin DP (2003) Prescribed fire effects on erosion parameters in a perennial grassland. J Range Manag 56:27–32. https://doi.org/10.2307/4003877
Oesterheld MJ, Loreti M, Semmartin M, Paruelo JM (1999) Grazing, fire and climate effects on primary productivity of grasslands and savannas. In: Walker LR (ed) Ecosystems of disturbed ground, 1st edn. Elsevier, Amsterdam, pp 287–306
Oliveira JM, Pillar VD (2004) Vegetation dynamics on mosaics of Campos and Araucaria forest between 1974 and 1999 in Southern Brazil. Community Ecol 5:197–202. https://doi.org/10.1556/ComEc.5.2004.2.8
Osborne CP (2008) Atmosphere, ecology and evolution: what drove the Miocene expansion of C4 grasslands? J Ecol 96:35–45. https://doi.org/10.1111/j.1365-2745.2007.01323.x
Overbeck GE, Pfadenhauer J (2007) Adaptive strategies in burned subtropical grassland in southern Brazil. Flora 202:27–49. https://doi.org/10.1016/j.flora.2005.11.004
Overbeck GE, Müller SC, Pillar VD, Pfadenhauer J (2005) Fine-scale post-fire dynamics in southern Brazilian subtropical grassland. J Veg Sci 16:655–664. https://doi.org/10.1111/j.1654-1103.2005.tb02408.x
Overbeck GE, Müller SC, Pillar VD, Pfadenhauer J (2006) No heat-stimulated germination found in herbaceous species from burned subtropical grassland. Plant Ecol 184:237–243. https://doi.org/10.1007/s11258-005-9068-1
Overbeck GE, Müller SC, Fidelis A, Pfadenhauer J, Pillar VD et al (2007) Brazil's neglected biome: the South Brazilian Campos. Perspect Plant Ecol Evol System 9:101–116. https://doi.org/10.1016/j.ppees.2007.07.005
Oyarzabal M, Andrade B, Pillar VD, Paruelo JM (2020) Temperate subhumid grasslands of Southern South America. Encyclopedia of the World's Biomes:577–593. https://doi.org/10.1016/b978-0-12-409548-9.12132-3
Parr CL, Brockett BH (1999) Patch-mosaic burning: a new paradigm for savanna management in protected areas? Koedoe 42:117–130
Paruelo JM, Jobbágy EG, Oesterheld M, Golluscio RA, Aguiar MR (2007) The grasslands and steppes of Patagonia and the Río de la Plata plains. In: Veblen TT, Young KR, Orme AR (eds) The physical geography of South America, 1st edn. Oxford University Press, New York, pp 232–248
Paruelo JM, Oesterheld M, Altesor A, Piñeiro G, Rodríguez C et al (2022) Grazers and fires: their role in shaping the structure and functioning of the Río de la Plata Grasslands. Ecol Austral 32:784–805. https://doi.org/10.25260/EA.22.32.2.1.1880
Pauchard A, García R, Zalba SM, Sarasola M, Zenni RD, Ziller SR, Nunez MA (2016) Pine invasions in South America: reducing their ecological impacts trough active management. In: Canning-Clode J (ed) Biological invasions in changing ecosystems, 1st edn. De Gruyter Open Ltd, Berlin, pp 318–342
Pausas JG, Bond WJ (2020) Alternative biome states in terrestrial ecosystems. Trends Plant Sci 25:250–263. https://doi.org/10.1016/j.tplants.2019.11.003

Pausas JG, Paula S (2020) Grasses and fire: the importance of hiding buds. A response to Moore et al. 'Effects of drought and fire on resprouting capacity of 52 temperate Australian perennial native grasses'. New Phytol 226:957–959. https://doi.org/10.1111/nph.15964
Pausas JG, Bradstock RA, Keith DA, Keeley JE, Hoffman W et al (2004) Plant functional traits in relation to fire in crown-fire ecosystems. Ecology 85:1085–1100. https://doi.org/10.1890/02-4094
Perelman S, León RJC, Oesterheld M (2001) Cross-scale vegetation patterns of Flooding Pampa grasslands. J Ecol 89:562–577. https://doi.org/10.1046/j.0022-0477.2001.00579.x
Petry V, Krüger L (2010) Frequent use of burned grasslands by the vulnerable Saffron-Cowled Blackbird *Xanthopsar flavus*: implications for the conservation of the species. J Ornithol 151:599–605. https://doi.org/10.1007/s10336-009-0489-9
Pilon NA, Cava MG, Hoffmann WA, Abreu RC, Fidelis A, Durigan G (2020) The diversity of post-fire regeneration strategies in the cerrado ground layer. J Ecol 109:154–166. https://doi.org/10.1111/1365-2745.13456
Pivello VR (2011) The use of fire in the Cerrado and Amazonian rainforest of Brazil: Past and present. Fire Ecol 7:24-39. https://doi.org/10.4996/fireecology.0701024
Pivello VR, Vieira I, Christianini AV, Ribeiro DB, da Silva ML et al (2021) Understanding Brazil's catastrophic fires: causes, consequences and policy needed to prevent future tragedies. Perspect Ecol Conserv 19:233–255. https://doi.org/10.1016/j.pecon.2021.06.005
Podgaiski LR, Joner F, Lavorel S, Moretti M, Ibanez S, Mendonça MDS Jr, Pillar VD (2013) Spider Trait assembly patterns and resilience under fire-induced vegetation change in South Brazilian grasslands. PLoS One 8:e60207. https://doi.org/10.1371/journal.pone.0060207
Podgaiski LR, Goldas CS, Ferrando CP, Silveira FS, Joner F et al (2014) Burning effects on detritivory and litter decay in Campos grasslands. Austral Ecol 39:686–695. https://doi.org/10.1111/aec.12132
Podgaiski LR, Cavalleri A, Ferrando CPR, Pillar VD, Mendonça MS Jr (2018) Prescribed patch burnings increase thrips species richness and body size in grassland communities. Insect Conserv Diver 11:204–212. https://doi.org/10.1111/icad.12269
Ponisio LC, Wilkin K, M'Gonigle LK, Kulhanek K, Cook L et al (2016) Pyrodiversity begets plant-pollinator community diversity. Glob Chang Biol 22:1794–1808. https://doi.org/10.1111/gcb.13236
Porto AB, do Prado MA, LDS R, Overbeck GE (2023) Restoration of subtropical grasslands degraded by non-native pine plantations: effects of litter removal and hay transfer. Restor Ecol 31:e13773. https://doi.org/10.1111/rec.13773
Prober SM, Thiele KR, Lunt ID (2007) Fire frequency regulates tussock grass composition, structure and resilience in endangered temperate woodlands. Austral Ecol 32:808–824. https://doi.org/10.1111/j.1442-9993.2007.01762.x
Quiñones A, Lezama F, García A, Ayala W (2015) Reclutamiento y rebrote de Ulex europaeus L. luego dequemas controladas en campos de Uruguay. XXII Congreso de la ALAM, I Congreso de la ASACIM.
Ramírez F (2020) Evaluación de diferentes métodos de control de dos leguminosas invasoras en el sudoeste bonaerense. Tesis de Licenciatura en Ciencias Biológicas, Universidad Nacional del Sur, Bahía Blanca
Ramos K (2019) *Eryngium horridum* como facilitador da comunidade de plantas e artrópodes em campos sob diferentes regimes de distúrbio com fogo e pastejo. Master Dissertation in Ecology and Evolution of Biodiversity, Pontifícia Universidade Católica do Rio Grande do Sul, Porto Alegre, Brazil
Raunkiaer C (1934) The life forms of plants and statistical plant geography. Clarendon, Oxford
Rheinheimer DS, Santos JCP, Fernandes VBB, Mafra AL, Almeida JA (2003) Modificações nos atributos químicos de solo sob campo nativo submetido à queima. Ciência Rural 33:49–55
Rocha EA, Luz FA, Rossato D, Millan C (2015) Does fire disturbance change insect composition and the frequency of darker phenotypes in a grassland community? Revista Brasileira de Biociências 13:4

Rosengurtt B (1979) Tablas de comportamiento de las especies de plantas de campos naturales en el Uruguay. División Publicaciones y Ediciones de la Universidad de la República, Montevideo
Rowland-Schaefer EG, Koehn O, Jones HP (2023) Small mammal associations with habitat composition, configuration, and management in tallgrass prairies: a review. Mamm Rev. https://doi.org/10.1111/mam.12336
Royo Pallarés O, Berretta EJ, Maraschin GE (2005) The South American Campos ecosystem. In: Suttie JM, Reynolds SG, Batello C (eds) Grasslands of the world. FAO, Rome, pp 171–219
Sage RF (2001) Environmental and evolutionary preconditions for the origin and diversification of the C4 photosynthetic syndrome. Plant Biol 3:202–213. https://doi.org/10.1055/s-2001-15206
Sage RF (2004) The evolution of C4 photosynthesis. New Phytol 161:341–370. https://doi.org/10.1111/j.1469-8137.2004.00974.x
Sala OE, Austin AT, Vivanco L (2001) Temperate grassland and shrubland ecosystems. In: Levin SA (ed) Encyclopedia of biodiversity, 1st edn. Academic, pp 627–635
Sanhueza C, Zalba SM (2009) Invasión de retama (*Spartium junceum*) en las Sierras Australes Bonaerenses: investigación, acción y manejo. In: Cazzaniga NJ, Arelovich HM (eds) Ambientes y recursos naturales del sudoeste bonaerense: Producción, contaminación y conservación. Ediuns, Bahía Blanca, pp 303–314
Sanhueza CC, Zalba SM (2012) Experimental control of Spanish Broom invading natural grasslands. Manage Biol Invasions 3:97–104. https://doi.org/10.3391/mbi.2012.3.2.04
Sanhueza CC, Zalba SM (2014) Banco de semillas, germinación y longevidad de semillas de retama (*Spartium junceum*, Fabaceae): implicancias para su control. Boletín de la Sociedad Argentina de Botánica 49:67–76
Saunders ME, Barton PS, Bickerstaff JRM, Frost L, Latty T et al (2021) Limited understanding of bushfire impacts on Australian invertebrates. Insect Conserv Divers 14:285–293. https://doi.org/10.1111/icad.12493
Silveira FAO, Ordóñez-Parra CA, Moura LC, Schmidt IB, Andersen AN et al (2022) Biome Awareness Disparity is BAD for tropical ecosystem conservation and restoration. J Appl Ecol 59:1967–1975. https://doi.org/10.1111/1365-2664.14060
Simberloff D, Nuñez MA, Ledgard NJ, Pauchard A, Richardson DM et al (2010) Spread and impact of introduced conifers in South America: lessons from other southern hemisphere regions. Austral Ecol 35:489–504. https://doi.org/10.1111/j.1442-9993.2009.02058.x
Soriano A, León RJC, Sala OE, Lavado RS, Deregibus VA et al (1991) Río de la Plata grasslands. In: Coupland RT (ed) Natural grasslands. Introduction and western hemisphere, 1st edn. Elsevier, Amsterdam, pp 367–407
Sutherland EF, Dickman CR (1999) Mechanisms of recovery after fire by rodents in the Australian environment: a review. Wildl Res 26:405–419. https://doi.org/10.1071/WR97045
Swengel AB (2001) A literature review of insect responses to fire, compared to other conservation managements of open habitat. Biodivers Conserv 10:1141–1169. https://doi.org/10.1023/A:1016683807033
Tommasino A, Lezama F, Gallego F, Camba Sans G, Paruelo JM (2022) Rangeland resilience to droughts: changes across an intensification gradient. Appl Veg Sci 26:e12722. https://doi.org/10.1111/avsc.12722
Vega E, Baldi G, Jobbágy EG, Paruelo JM (2009) Land use change patterns in the Río de la Plata grasslands: the influence of phytogeographic and political boundaries. Agric Ecosyst Environ 134:287–292. https://doi.org/10.1016/j.agee.2009.07.011
Veldman JW, Buisson E, Durigan G, Fernandes GW, Le Stradic S et al (2015) Toward an old-growth concept for grasslands, savannas, and woodlands. Front Ecol Environ 13:154–162. https://doi.org/10.1890/140270
Vermeire LT, Wester DB, Mitchell RB, Fuhlendorf SD (2005) Fire and grazing effects on wind erosion, soil water content, and soil temperature. J Environ Qual 34:1559–1565. https://doi.org/10.2134/jeq2005.0006
Vickery PD, Tubaro PL, Cardoso Da Silva JM, Peterjohn BG, Herkert JR, Cavalcanti RB (1999) Conservation of grassland birds in the Western Hemisphere. Stud Avian Biol 19:2–26

Vickery PD, Zuckerberg B, Jones AL, Shriver WG, Weik AP (2005) Influence of fire and other anthropogenic practices on grassland and shrubland birds in New England. USDA Forest Service Gen. Tech. Rep. PSW-GTR-191

Vignolio OR, Laterra P, Fernandez ON, Linares MP, Maceira NO, Giaquinta A (2003) Effects of fire frequency on survival, growth and fecundity of *Paspalum quadrifarium* (Lam.) in a grassland of the Flooding Pampa (Argentina). Austral Ecol 28:263–270. https://doi.org/10.1046/j.1442-9993.2003.01286.x

Vila AR, Beade MS, Barrios Lamunière D (2008) Home range and habitat selection of pampas deer. J Zool 276:95–102. https://doi.org/10.1111/j.1469-7998.2008.00468.x

Vincent C (1935) A queima dos Campos. Rev Ind Anim 3:286–299

Vogelmann ES, Reichert JM, Prevedello J, de Barros CAP, de Quadros FLF, Mataix-Solera J (2012) Soil hydro-physical changes in natural grassland of southern Brazil subjected to burning management. Soil Res 50:465–472. https://doi.org/10.1071/SR12106

von Törne E (1990) Assessing feeding activities of soil-living animals. I. Bait-lamina-test. Pedobiologia:89–101

Walter H (1979) Vegetation of the earth and ecological systems of the geo-biosphere. Springer, New York

Whelan RJ (1995) The ecology of fire. Cambridge studies in ecology. Cambridge University Press, Cambridge

Wieder RK, Lang GE (1982) A critique of the analytical methods used in examining decomposition data obtained from litter bags. Ecology 63:1636–1642. https://doi.org/10.2307/1940104

Wilcox BP, Birt A, Fuhlendorf SD, Archer SR (2018) Emerging frameworks for understanding and mitigating woody plant encroachment in grassy biomes. Curr Opin Environ Sustain 32:46–52. https://doi.org/10.1016/j.cosust.2018.04.005

Yezzi A, Nebbia AJ, Zalba SM (2018) Fragmentación de pastizales psamófilos por plantaciones de pinos: efectos sobre la riqueza y la composición vegetal. Ecol Austral 28:133–144. https://doi.org/10.25260/EA.18.28.1.0.640

Yezzi A, Nebbia AJ, Zalba SM (2019) Interaction between fire and fragmentation in the successional stages of coastal dune grasslands of the southern Pampas, Argentina. Sci Rep 9:15109. https://www.nature.com/articles/s41598-019-51595-x

Yezzi A, Nebbia AJ, Zalba SM (2020) Synergistic interactions of grassland fragmentation, fire and invasive woody plants in the Argentine pampas. In: Daniels JA (ed) Advances in environmental research. Nova Science Publishers Inc, New York, pp 43–70

Yezzi A, Nebbia AJ, Zalba SM (2021) Fragmentation of coastal grasslands by plantations and spontaneous spread of invasive pines in the southern Pampa. Diversity 13:637. https://doi.org/10.3390/d13120637

Zalba SM, Cozzani N (2004) The impact of feral horses on grassland bird communities. Anim Conserv 7:35–44. https://doi.org/10.1017/S1367943003001094

Zalba SM, Villamil CB (2002) Invasion of woody plants in relictual native grasslands. Biol Invasions 4:55–72. https://doi.org/10.1023/A:1020532609792

Zalba SM, Cuevas YA, Boó R (2008) Aleppo pine invasion after a wildfire in an Argentinean grassland nature reserve. J Environ Manag 88:539–546. https://doi.org/10.1016/j.jenvman.2007.03.018

Fire in the Cerrado

5

Vânia R. Pivello, Márcio Martins,
Alexander V. Christianini, Ana Paula Carmignotto,
Bruna França Gomes, and Alessandra Fidelis

Abstract

The Cerrado comprises a complex of tropical grasslands and savannas that dominate the central part of Brazil, with smaller patches in the Amazonian region, Paraguay, and Bolivia. Its diverse array of landscapes and habitats renders it the world's most diverse savanna ecosystem, where plants and animals have evolved under the influence of fire. This chapter outlines various adaptations of flora and fauna to cope with the frequent, rapid, and low-intensity fires characteristic of the Cerrado: rapid regrowth and intense flowering post-fire, protected buds, well-developed roots and subterranean organs, thick and insulating barks of trees are examples for plants. Animals exhibit behaviors such as seeking refuge in burrows and capitalizing on increased resources following fires. Fire also plays a crucial role in nutrient cycling and ecosystem regeneration across the Cerrado.

V. R. Pivello (✉) · M. Martins · B. F. Gomes
Departamento de Ecologia, Instituto de Biociências, Universidade de São Paulo, São Paulo, Brazil
e-mail: vrpivel@usp.br; martinsmrc@usp.br

A. V. Christianini
Departamento de Ciências Ambientais, Universidade Federal de São Carlos, Sorocaba, Brazil
e-mail: avchrist@ufscar.br

A. P. Carmignotto
Departamento de Biologia, Universidade Federal de São Carlos, Sorocaba, Brazil
e-mail: apcarmig@ufscar.br

A. Fidelis
Departamento de Biodiversidade, Instituto de Biociências, Universidade Estadual Paulista, Rio Claro, Brazil
e-mail: alessandra.fidelis@unesp.br

A. Fidelis, V. R. Pivello (eds.), *Fire in the South American Ecosystems*, Ecological Studies 250, https://doi.org/10.1007/978-3-031-89372-8_5

Indigenous Cerrado inhabitants have learned to use fire as a management tool, potentially altering the natural fire regime towards increased frequency. Presently, anthropogenic fire regimes range from overly frequent to absent, both of which pose threats to Cerrado ecosystems. Scientific research seeks to identify suitable fire management strategies tailored to different conditions, supported by legislative frameworks. However, significant policy and knowledge gaps remain to be addressed.

Resumen

El Cerrado es un complejo de praderas y sabanas tropicales que domina el centro de Brasil, con pequeñas partes en la región Amazónica, Paraguay y Bolivia. Con su gran variedad de paisajes y hábitats, es la sabana más diversa del mundo, donde plantas y animales han evolucionado bajo la influencia del fuego. Este capítulo describe diversas adaptaciones de la flora y fauna a los incendios frecuentes, rápidos y de baja intensidad característicos del Cerrado: rápido rebrote y intensa floración post-fuego, yemas protegidas, raíces y órganos subterráneos bien desarrollados, corteza gruesa y aislante de los árboles son ejemplos en el caso de las plantas; muchos animales se refugian en huecos y se benefician de mayores recursos tras los incendios. El fuego también desempeña un papel crucial en el ciclo de los nutrientes y regeneración del ecosistema en su conjunto. Los indígenas habitantes del Cerrado utilizaban el fuego como herramienta de gestión, probablemente aumentando la frecuencia natural de las quemas. Actualmente, los regímenes antropogénicos de fuego varían entre excesivamente frecuentes y ausentes, ambos potencialmente dañinos para la sabana. La investigación científica actual busca abordar lagunas importantes sobre las estrategias y políticas adecuadas para el uso del fuego en diversas condiciones del Cerrado.

Resumo

O Cerrado compreende um complexo de campos e savanas tropicais que dominam a parte central do Brasil, com manchas menores na região Amazônica, Paraguai e Bolívia. Com grande variedade de paisagens e habitats é a savana mais diversificada do mundo, onde plantas e animais evoluíram sob a influência do fogo. Este capítulo descreve várias adaptações da flora e da fauna para lidar com os incêndios frequentes, rápidos e de baixa intensidade, característicos do Cerrado: rebrota rápida e floração intensa após o fogo, gemas protegidas, raízes e órgãos subterrâneos bem desenvolvidos, cascas grossas e isolantes das árvores são exemplos para as plantas; muitos animais se refugiam em tocas e se beneficiam de maiores recursos após as queimadas. O fogo ainda desempenha um papel crucial na ciclagem de nutrientes e na regeneração do ecossistema como um todo. Os indígenas que habitavam o Cerrado aprenderam a usar o fogo como uma ferramenta de manejo, provavelmente aumentando a frequência natural das queimadas. Hoje, os regimes de fogo antropogênico variam de excessivamente frequentes a ausentes, ambos potencialmente nocivos aos ecossistemas do Cerrado. Pesquisas científicas atuais buscam preencher importantes lacunas quanto a estratégias e políticas adequadas para o uso do fogo nas diferentes condições do Cerrado.

5.1 Cerrado Distribution and Characteristics

The Cerrado, a vast biome dominated by savannas and grasslands, lies in central-eastern South America. Under the influence of a continental tropical climate, the Cerrado extends as a continuous core area over the Brazilian Central Plateau, being almost entirely within the Brazilian territory, where it covers around 2.05 million km^2 (Sano et al. 2010). Because of that, the Cerrado is known as the "Brazilian savanna", although it also occupies small and discontinuous portions in eastern Paraguay (about 8150 km^2; Mereles 2007; Paraguay 2010) and Bolivia (about 30,000 km^2; Villarroel et al. 2016). In Brazil, the Cerrado distribution also goes beyond the large continuous core area in the Central Plateau, extending as patches towards neighboring ecoregions (Fig. 5.1). The Amazonian cerrados (around 150,000 km^2; Braga 1979) are believed to be relicts from the Quaternary climatic fluctuations when, in drier periods, savannas extended over the rainforest (more

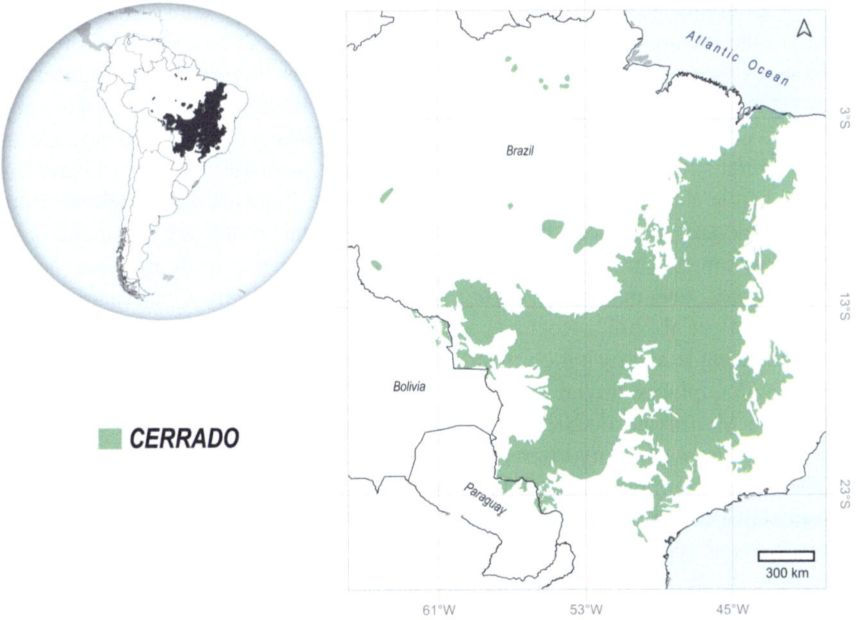

Fig. 5.1 Distribution of the Cerrado. (Source: Olson et al. 2001, modified according to Veloso 1966 and IBGE 2019. Produced by Júlia Abrantes Rodrigues)

details below) (Prance 1996; Pessenda et al. 1998; Ab'Saber 2002; Gottsberger and Silberbauer-Gottsberger 2006; Werneck 2011).

With such a wide geographic distribution that covers an extensive latitudinal/longitudinal and topographic variation, the Cerrado does not have a homogeneous climate, being subjected to different conditions of air masses and orographic influences. However, the presence of two well-defined seasons, a dry winter (May to September) and a wet summer (October to April), is the most striking feature of the Cerrado climate, and the prime determinant of this savanna-dominant vegetation at the regional scale. Then, a seasonal tropical climate—the Koeppen's Aw, or savanna climate—prevails in the Cerrado, with the Cwa subtropical seasonal climate occurring in the southern part. The average annual maximum and minimum temperatures in the Cerrado range between 28 and 18 °C, increasing from south to north along latitudinal zones. Elevated temperatures (average values >24 ° C) characterize the period from October to March, while lower temperatures (average ≤15 °C) prevail in June and July, accompanied by substantial daily temperature variations (Silva et al. 2008; Nascimento and Novais 2020; Climate Change Knowledge Portal 2023). While temperature varies mainly according to latitudinal ranges, precipitation changes primarily in response to the relief conditions and atmospheric systems that transport moisture. Nearly 70% of the Cerrado receives 1200 to 1500 mm of rain annually, heavily concentrated in summer. In the driest months (June to August) rainfall is usually less than 40 mm and the air relative humidity is lower than 30% (Silva et al. 2008; Nascimento and Novais 2020; Climate Change Knowledge Portal 2023).

The geological foundation of the Cerrado primarily consists of ancient Precambrian cratons (Lira-Martins et al. 2022; Jaramillo 2023). Over hundreds of millions of years, weathering on these aged rocks has shaped diverse landforms, observable at both regional and local spatial scales. Regionally, the Mesozoic and Cenozoic Cerrado relief is represented by large plateaus exceeding 800 m in elevation, with flat summits alongside tablelands ranging from 400 to 800 m, characterized by gentle slopes and hills, known as *chapadas*. Plateaus may originate from volcanic, metamorphic or sedimentary rocks, while tablelands are sedimentary in origin. Erosion on these elevated landforms has given rise to depressions below 400 m, ridges, and inselbergs during the Late Tertiary/Early Quaternary period (Lira-Martins et al. 2022; Rocha et al. 2022). At the local scale, Quaternary elements resulting from erosion on older landforms and material deposition are represented by floodplains, fluvial terraces and talus deposits (Ab'Saber 2007; Werneck 2011; Lira-Martins et al. 2022). The Cerrado drainage is perennial, housing the springs of several major hydrographic systems (Lima and Silva 2008).

The heavy summer rains on the complex and varied geomorphology of the Cerrado have facilitated severe weathering of the parent rocks, resulting in a wide variety of soil types and conditions (Salgado et al. 2015). Consequently, most Cerrado soils are aged, deep, worn and acidic (pH ranging from 4.0 to 5.5). These mineral soils exhibit chemical and physical properties largely influenced by the parent material, which predominantly consists of rocks of poor origin such as granites, gneisses, quartzite, and sandstones. As a result, Cerrado soils are rich in iron and aluminum ions, deficient in calcium, magnesium, sulfur, and zinc, extremely poor in potassium, rendering them infertile and aluminum-toxic for crop plants. Despite these limitations, Cerrado soils typically have a balanced texture and efficient drainage. Combined with a rich hydrography, these factors have facilitated the mechanized agricultural exploitation of the Cerrado through practices like liming and fertilization, especially over the last 50 years. About 40% of the Cerrado soils are Ferralsols, with other significant soil groups including Arenosols (around 23%), Acrisols (12%), Plinthosols (10%), and Cambisols (10%). Gleysols occur in association with flooded environments (Queiroz-Neto 1982; Montgomery and Askew 1983; Goedert 1987; Furley and Ratter 1988; Haase 1992; Haridasan 2008; Reatto et al. 2008; Lira-Martins et al. 2022).

The variety of soils, landforms and local climate has created a multiplicity of habitats in the Cerrado ecoregion. Together with fire, these conditions allowed the evolution of a great number of life forms, vegetation types and endemic species (see Sect. 5.2). About 12,700 plant species, at least one third of which endemic to the Cerrado, make this biome the most diverse savanna in the world (Klink and Machado 2005; Forzza et al. 2012). At the landscape scale, the Cerrado is a complex of several types of savannas and grasslands arranged in mosaics. Linear elements along water bodies, the riparian forests complement this landscape.

Savannas typically comprise a continuous herbaceous layer, mainly composed of graminoids (Poaceae and Cyperaceae), forbs and woody elements (trees, treelets, shrubs) at different proportions, in a gradient of increasing density of woody elements: *campo sujo*, *campo cerrado*, *cerrado sensu stricto* (following Coutinho 1978, 2006, 2006) and *cerrado denso* (Ribeiro and Walter 2008). Clusters of palm trees (especially *buriti—Mauritia flexuosa*) stand out over the grassy layer where drainage is shallow and poorly defined, or near springs (on hydromorphic soils), forming the *veredas*; on litholic soils occurs the *cerrado rupestre* (Ribeiro and Walter 2008). Moreover, dry and wet grasslands—respectively *campo limpo* and *campo úmido*—may occur in the mosaic landscape; under the influence of seasonal flooding, the *campo de murundus*—a type of wet grassland with a micro-relief of small mounds associated with termites—can be found (Oliveira-Filho 1992a, b). In the other extreme of the woody-herbaceous gradient lies the cerradão (Coutinho 1978, 2006), a forest physiognomy, where the herbaceous layer, mainly composed of tree seedlings and young, is reduced by limited sunlight (Fig. 5.2). This combination of different vegetation types determines the Cerrado. Other than cerradão, some other forest types (from the Tropical Forests ecoregion) may appear associated with particular types of soil or relief (Coutinho 2006). Cerrado patches in the Amazon region likely represent relictual remnants from the Quaternary period, still maintained due to soil constraints, as the presence of lateritic hardpans, or local lower rainfall (Eiten 1978; Desjardins et al. 1996; Sanaiotti et al. 2002; Brugger et al. 2016; Adeney et al. 2016). While the *campos rupestres* fall within the geographic scope of the Cerrado, this specific vegetation is included in the ecoregion of montane grasslands (more details in Chap. 7).

In the intricate mosaic of the Cerrado vegetation, various vegetation types coexist, prompting investigations into their associations with conditioning factors since

Fig. 5.2 Main physiognomies of the Cerrado. From left to right: a decreasing gradient of woody elements is shown and, in contrast, the increasing gradient for herbaceous elements. (Modified from Coutinho 1978, and Ribeiro and Walter 2008). (Photos: A Fidelis; Drawing: MA Chiminazzo)

the late nineteenth century (Warming 1892). Locally, these vegetation types are linked to specific landforms (Cole 1982), soil conditions encompassing fertility, moisture, texture, and depth (Furley and Ratter 1988; Goodland and Pollard 1973), as well as human activities like fire and cutting (Rawitscher 1948; Coutinho 1980, 2002; Rizzini 1997). The determinants of the presence of specific vegetation types in the Cerrado ecoregion are likely multifaceted, involving microclimate, vegetation dynamics, and evolutionary considerations (Cole 1960, 1986; Oliveira-Filho and Ratter 2002; Ruggiero et al. 2006; Souza et al. 2023b).

5.2 The Presence of Fire Along the Cerrado History

5.2.1 The Cerrado Origin

The origin of the Cerrado is incompletely understood, but climatic fluctuations and fire are recognized as fundamental for creating and maintaining this vegetation (Simon et al. 2009; Jaramillo 2023). In the Middle Miocene (around 10 Myr BP) climate drying caused the retreat of tropical forests, giving rise to South American savannas. Under warm and dry climate with low levels of atmospheric CO_2, the C3 grasses—the main component of earlier savannas—evolved to C4 grasses (Christin et al. 2008; Edwards et al. 2010; Jaramillo 2023), leading to a large diversification and expansion of C_4 grasses (more details in Chap. 2). By 7.5 Myr BP the C4 grass savannas spread in South America, covering large areas (Gottsberger and

Silberbauer-Gottsberger 2006; Spriggs et al. 2014; Jaramillo 2023). In a seasonal climate, C4 grasses produce large amounts of flammable biomass during the wet season, which dries out in the dry season and, in the presence of an ignition source such as lightning, can easily catch fire. Then, by Late Miocene fire was frequent in the Cerrado (Gottsberger and Silberbauer-Gottsberger 2006; Werneck 2011), acting as a decisive driver in the evolution of its endemic flora. Simon et al. (2009) argued for in situ fire-induced diversification of the Cerrado flora and provide evidence that several endemic plant lineages exhibiting adaptations to fire (xylopodium, dwarfing, thick cork) have diversified over the last 4 Myr BP. The Cerrado, therefore, evolved as a fire-adapted and self-maintained vegetation.

The Amazonian cerrados are more recent in origin; they emerged in the Quaternary period (600 to 10 Kyr BP) when a sequence of wet/warm and dry/cold climate periods caused the expansion and retreat of humid forests and savannas (van der Hammen 1991; van der Hammen and Hooghiemstra 2000; Ledru 2002; Salgado-Labouriau 2005; Gottsberger and Silberbauer-Gottsberger 2006). Over the last 7 Kyr, anthropogenic fires have played a significant role in shaping savanna environments amidst humid and warm climatic conditions (Pinter et al. 2011).

5.2.2 The Flammable Environment in the Cerrado

Within the mosaic of Cerrado vegetation types, from grasslands to forests, environmental flammability varies significantly, primarily determined by the herbaceous layer's quantity, composition, and continuity (Newberry et al. 2020). Grasslands and open savannas, characterized by a great proportion of grasses, especially C4 species, produce large amounts of thin and flammable leaves, being characterized as fast flammable (Zanzarini et al. 2022; Simpson et al. 2016). When desiccated, these leaves form a continuous and flammable layer (Fig. 5.3). In South American grasslands and savannas, C_4 grasses can contribute to over 90% of the total fuel load (Kauffman et al. 1994; Castro and Kauffman 1998), being thus major contributors to the high flammability of these ecosystems (Wragg et al. 2018; Zanzarini et al.

Fig. 5.3 Fire in (**a**) *campo sujo* (Photo: A. Fidelis) and (**b**) *cerrado denso* (Photo: V.R. Pivello)

2022; Simpson et al. 2016). C4 grasses are not only fire-stimulated but also fire-benefitted, playing a pivotal role in establishing recurrent fire systems (Newberry et al. 2020).

On the other side, when the woody cover is higher and, as a consequence, there is a reduction in the sun-lover C4 grasses, the system becomes less flammable. However, in dry conditions, fire can reach tree canopies (Fig. 5.3), and the impact of fire on woody species varies, causing topkill or damage, based on its intensity and severity (Miranda et al. 2009; Hoffmann et al. 2012) and the protective capacity of internal tissues and buds (Chiminazzo et al. 2021; Chiminazzo et al. 2023b). Some trees and shrubs may become "fire trapped" (Bond and van Wilgen 1996), unable to reestablish or grow despite constant resprouting from basal and belowground buds (Hoffmann 1999; Chiminazzo et al. 2021). Conversely, in the absence of a woody layer, grasses feed fires and keep an open stable state in a feedback fire-vegetation dynamic (Lehmann et al. 2014; Newberry et al. 2020; Pausas and Bond 2020). With fire exclusion, and whenever ground conditions allow, woody cover can escape the "fire trap", allowing trees to establish and grow, transforming the system into a stable, non-flammable forest state (Pausas and Bond 2020) in a process named woody encroachment (Abreu et al. 2017; Stevens et al. 2017). Therefore, fire influences vegetation and the vegetation structure and composition affect fire by modeling its behavior according to fuel characteristics (Whelan 1995), since the fuel load composition has a direct effect on the system flammability (Wragg et al. 2018).

The herbaceous layer in South American tropical grasslands and savannas harbor a very high diversity of grasses, forbs, small shrubs and belowground tree species (Coutinho 1990; Overbeck et al. 2022), creating a heterogeneous fuel bed of plants with different flammability traits (Zanzarini et al. 2022) that influence the burn characteristics and fire spread. Still, some plants may have resins that enhance their flammability (e.g. *Vellozia variabilis*, Martins and Paiva 2016). Even so, grasses are fundamental for igniting, sustaining and spreading fires (Zanzarini et al. 2022; Wragg et al. 2018).

Fuel moisture is another key driver of savannas flammability and fire intensity (Rissi et al. 2017; Rodrigues et al. 2021). Unlike other tropical savannas, where grasses completely cure during the dry season, Cerrado grasses retain some biomass as green and active, halving their moisture content from the rainy season and keeping most of dead biomass attached to the tussocks (Zanzarini et al. 2022). Moisture in forbs and shrubs is two to three-fold higher than in grasses in the dry season (Zanzarini et al. 2022; Kauffman et al. 1994). Consequently, tropical savannas flammability varies along the year, mostly driven by the amount of dead biomass and moisture content in grasses (Rissi et al. 2017; Zanzarini et al. 2022).

The presence of specific native or invasive species significantly influences fire behavior in Cerrado ecosystems. For instance, the tussock-forming grass *Tristachya leiostachya* can reach 3 m, generating substantial flammable biomass that facilitates biennial fire recurrence (Ramos-Neto and Pivello 2000). Invasive African grasses, such as *Melinis minutiflora* and *Urochloa* spp., very common in the Cerrado (Pivello et al. 1999), also contribute to increased dead biomass production, leading to more intense and severe fires with elevated temperature and flame height

(Gorgone-Barbosa et al. 2015). Invasive trees, such as *Pinus* spp., alter the local fuel load, which becomes mostly composed of flammable trees and litter, affecting fire spread. In all cases, the occurrence of more severe fires can disrupt the ecosystem's dynamics.

5.2.3 Natural and Anthropogenic Fires

Paleo-charcoal particles dating back to over 43,000 Kyr BP old have been recorded for the Cerrado region (Follador et al. 2023). While the timeframe of human arrival in Brazil is controversial, prevailing evidence suggests a range of 11 to 13 Kyr BP (Da-Gloria et al. 2018), even though estimates extend to 32 Kyr BP (Guidon and Delibrias 1986; Prous 1991; Lahaye et al. 2013). The charcoal particles from the Central Plateau likely originated from natural fires. In Amazonian cerrados, charcoal records are more recent, dating approximately 6–8 Kyr BP (Van der Hammen 1991; Cordeiro et al. 2014), possibly resulting from both natural and anthropogenic fires.

Natural fires in the Cerrado, typical of most savannas, are generally surface fires with high frequency and low intensity. Sustained by the continuous C4 grass fuel bed, these fires primarily occur during the wet season and transitional months between wet and dry seasons, often ignited by lightning during thunderstorms (Ramos-Neto and Pivello 2000). Brief dry spells within the rainy season (veranico) may allow grasses to cure sufficiently for fire sustenance (Miranda et al. 2009). These natural fires burn small areas, both because they are rapidly extinguished by rain, and because of plants high moisture. Over time, repeated fires form a mosaic of burned and unburned areas in the landscape, which has a number of advantages: recently burned patches act as firebreaks for subsequent fires; unburned areas may serve as refuge for large or flying animals during the burnings, and for less fire-prone plants; and patches with varied biomass may sustain higher biodiversity (Ramos-Neto and Pivello 2000).

Throughout the Cerrado history, natural fires have gradually been supplanted by anthropogenic fires. Over the past 11,000 years, the Cerrado has been inhabited by hunter-gatherer civilizations and agricultural societies actively using fire (Prous 1991; Barbosa and Schmitz 2008). Indigenous Peoples (IP) employed fire for various purposes, such as cultivation, hunting, food preparation, and ceremonial practices (see Chap. 3). Mosaic burnings, guided by ecological indicators, were carefully planned for specific goals, adjusting fire regimes based on season, frequency, time, and area burned (Gross et al. 1979; Anderson and Posey 1985; Mistry et al. 2005; Pivello 2011; Welch et al. 2013; Welch 2014). European settlers introduced cattle to Brazil in the late sixteenth century, with non-indigenous populations effectively occupying the region in the nineteenth century (Klink and Moreira 2002). These populations used fire for pasture management, particularly to stimulate the regrowth of natural grasses (Pivello 2011). Traditional and Local Communities (LC)—including extractivists, quilombolas, geraizeiros, flower collectors, and veredeiros—have continued the use of frequent burns for various purposes such as cultivating food and medicinal plants, raising animals, extracting native fruits, and producing raw materials for handicrafts (Rede Cerrado 2024).

With increased human density and altered land uses, most fires in the Cerrado became anthropogenic, deviating from natural regimes (Pivello 2011; Molinas-González and Florentín 2021). Anthropogenic fires are very frequent, every 2–3 years, distributed along the dry season from June to October (Paraguay 2010; Pivello 2011, 2021; Schmidt and Eloy 2020; Maillard et al. 2022). Dry-season fires occur under low air humidity and high daily temperatures, spreading to large areas that may reach as much as 100 km^2 (França et al. 2007), since the fuel load is very dry and there is no rain to extinguish them.

From the 1960s, mechanized grain agriculture of mainly soybean and corn, as well as intensive cattle ranching based on perennial pastures of African grasses (Klink and Machado 2005), have transformed the Cerrado in one of the world's main providers of food, which has caused the loss of almost 50% of its native vegetation (Klink et al. 2020; https://mapbiomas.org). Large-scale intensive agriculture has gradually reduced the traditional land management that remains in the more remote places. However, fire is still largely applied for opening new arable lands—especially in the Brazilian northern part of the Cerrado (MATOPIBA region), where most of the native vegetation is still found (Silva et al. 2021)—, for extensive cattle ranching on native pastures and in the management of crop residues. Nowadays, man-made fires account for more than 90% of the total annual burned area in the Cerrado (Silva et al. 2021), with some resulting from wildfires caused by mismanagement, becoming more frequent and intense due to global warming effects (Fidelis et al. 2018; Hofmann et al. 2021, see Sect. 5.6). Similar agricultural fire use and wildfires caused by inadequate management are observed in Paraguay (Paraguay 2010; Molinas-González and Florentín 2021) and Bolivia (Maillard et al. 2018, 2022; FAN WCS 2021). Wildfires are the main sources of greenhouse gases emissions in Brazil (Santos et al. 2021).

A different anthropogenic fire regime has been established in protected areas, based on zero-fire policy. Historically, fire was perceived as universally detrimental to the environment, and huge effort was dedicated to banishing fire even in parks and reserves aimed to protect fire-adapted ecosystems (see Pivello et al. 2021). However, fire exclusion in the Cerrado results in larger fuel accumulation (Zupo et al. 2022), which increases the risk of wildfires and extensive burning areas (Fidelis et al. 2018). Therefore, the zero-fire policy has been replaced by prescribed burns and Integrated Fire Management (IFM) policies, as detailed in Sect. 5.6.1.

5.3 Fire Parameters in the Cerrado

Although analyses at large scale show that climate instead of fuel load is the major driver of fires in the Cerrado (Alvarado et al. 2017; Santos et al. 2021), at the local scale fuel load plays an important role on fire behavior. Fires in the Cerrado are typically fast, of low or moderate intensity and severity, with a high combustion efficiency (Miranda et al. 2009; Gomes et al. 2018), but fuel load characteristics play a major role in influencing fire behavior, severity and intensity (Rissi et al. 2017; Rodrigues et al. 2021). Aboveground temperatures vary based on biomass type, amount, and vertical distribution, as well as on meteorological conditions and time since the last fire (Pivello and Norton 1996; Miranda et al. 2009; Rodrigues et al. 2021; Zupo et al. 2022). Belowground temperatures range from 25 to 55 °C at 1 cm depth (Coutinho 1990; Miranda et al. 1993), being influenced by the aboveground fuel load (Zupo et al. 2022), with no impact observed at 5 cm belowground (Miranda et al. 2009). The residence time of fire is typically short, and high temperatures (>200 °C) rarely exceed 60 s (Miranda et al. 1993; Rodrigues et al. 2021). Flame height also varies depending on fuel characteristics (composition, height, amount), fire season and time since the last fire, ranging from 1.3 m in annually burned open savannas to 4 m (e.g., savanna invaded by *Urochloa brizantha*, Table 5.1). Combustion efficiency is usually high, burning up to 95% of the vegetation (Miranda et al. 2009; Rodrigues et al. 2021). The drier and finer the fuel load, the higher the combustion efficiency (Miranda et al. 2009; Rissi et al. 2017). Grasses and small plants make up the majority of biomass consumed during grassland and savanna fires, while shrubs and trees are less affected (Miranda et al. 2009). Leaves from woody species, if not consumed by flames, are usually damaged by heat, falling to the ground soon after fires. Consequently, post-fire residue mainly consists of damaged leaves and burnt stems of shrubs and trees.

Because the proportion between herbaceous and woody layer varies a lot in the Cerrado, fuel load amount and consequently, fire intensity will vary from 503 up to 16,000 $kW.m^{-1}$ (Miranda et al. 2009; Table 1). Fires set at the beginning of the dry season (May) are of lower intensity than late-dry season fires (September–October; Rissi et al. 2017; Santos et al. 2021). In a 4-year fire exclusion savanna, fuel load doubled and thus, fire intensity was three-fold higher than in annually burned areas (Zupo et al. 2022). Rodrigues et al. (2021) found that, after five consecutive annual burns, fire decreased fuel load and discontinued the fuel bed, resulting in fires of lower intensity compared to biennial fires (Table 5.1). Still, invasive grasses may alter fire intensity, as well as the vegetation burning pattern (Table 5.1; Gorgone-Barbosa et al. 2015; Rossi et al. 2014). Therefore, the characterization of the fuel load—fuel type, continuity, moisture—are major factors affecting fire behavior in the Cerrado vegetation types.

5.4 How Do Cerrado Plants and Animals Cope with Fire?

5.4.1 Plant Strategies

In fire-prone ecosystems plants are highly adapted to the frequent removal of the aboveground biomass (Simon and Pennington 2012; Bond 2019), high temperature fluctuation (Fidelis and Blanco 2014; Daibes et al. 2017), smoke (Keeley et al. 2011) and nutrient input (Pivello and Coutinho 1992; Pellegrini et al. 2018). Traits such as bud protection, thick bark (Charles-Dominique et al. 2015; Chiminazzo et al. 2021, 2023a), and belowground bud-bearing organs with reserves (Pausas

Table 5.1 Main fire parameters according to different fire regimes in a variety of Cerrado types

Region	Vegetation type	Fire regime	Fuel load (kg. m^{-2})	Flame height (cm)	Max. Temperature (°C)	Fire intensity (kW. m^{-1})	Reference
North	Wet tropical grasslands	Early- and late-dry season; biennial	0.43–1.32 (E); 0.51–1.04 (L)	–	57–294 (E); 153–331 (L)	240.1–674.6 (E); 559.4–1083.3 (L)	Schmidt et al. (2017)
Central	Open savannas	Early-, mid- and late-dry season; biennial	0.73 (E); 0.75 (M); 0.65 (L)	250 (E); 260 (M); 320 (L)	–	503 (E);2707 (M); 3007 (L)	Rissi et al. (2017)
Central	Open savannas	Mid-dry season fires: annual and biennial	0.31 (A); 0.53 (B)	130 (A); 230 (B)	374.5 (A); 497 (B)	669.3 (A)—1850.2 (B)	Rodrigues et al. (2021)
Central	Open savannas	Late-dry season fires, biennial	0.72	–	–	10,700	Santos et al. (2003)
North	Open savannas	Mid- and late-dry season; biennial	0.42 (M); 0.4 (L)	190 (M); 250 (L)	–	898 (M); 3508 (L)	Santos et al. (2021)
Central	Open savannas	Annual, biennial, quadrennial	0.54 (A); 0.86 (B); 1.06 (Q)	305 (A); 315 (B); 560 (Q)	289 (A); 385 (B); 395 (Q)	1090.8 (A); 2781.7 (B); 3332 (Q)	Zupo et al. (2022)
Central	Open savannas	Mid-dry season fires, annual	0.37	–	–	1621	Medeiros (2002)
Central	Open savannas	Mid-dry and late-dry season, fire exclusion for 18 and 2 years	0.76 (E18); 0.99 (E2); 0.67 (L18); 0.89 (L2)	–	672 (E18); 578 (E2); 626 (L18); 598 (L2)	1390 (E18); 5889 (E2); 1256 (L18); 8134 (L2)	Miranda et al. (1996)
Central	Open savannas	Early-; mid-; late-dry season fires, biennial; quadrennial	0.7 (E); 0.86 (M); 0.93 (L); 0.8 (Q)	–	359 (E); 596 (M); 562 (L); 518 (Q)	1597 (E); 2187 (M); 12,404 (L); 3466 (Q)	Pivello et al. (2010)
Central	Grasslands, open savannas, woody savannas	Mid/late-dry season	0.71 (G); 0.73 (OS); 0.86 (WS)	–	–	>16,000 (G, OS); <4000 (WS)	Ward et al. (1992)

Region	Vegetation type	Fire regime	Fuel load (kg. m^{-2})	Flame height (cm)	Max. Temperature (°C)	Fire intensity (kW. m^{-1})	Reference
Central	Open savannas, woody savannas	Late-dry season	0.73 (OS); 0.73 (S); 1.0 (WS)	400 (OS); 200 (S); 200 (WS)	–	16,394 (OS); 2842 (S); 3455 (WS)	Kauffman et al. (1994)
Central	Invaded open savannas	Invaded savannas by *Urochloa brizantha*, late-dry season fire	0.62	404	–	788	Gorgone-Barbosa et al. (2015)
Central	Invaded and open savannas	Native and invaded savannas by *Melinis minutiflora*, mid-dry season, quadrennial	0.87 (N); 1.21 (I)	–	518 (N); 642 (I)	5722 (N); 8857 (I)	Castro-Neves (2000)
Southeastern	Invaded open savannas	Invaded savannas by *Melinis minutiflora* and *Urochloa brizantha*, early-. mid-, late-dry season fires	0.46 (ME); 0.31 (MM); 0.55 (ML); 0.63 (UE); 0.82 (UM); 0.65 (UL)	159 (ME); 224 (MM); 203 (UE); 244 (UM)	–	160.04 (ME); 465.27 (MM); 702.26 (ML); 210.21 (UE); 546.94 (UM); 1063.39 (UL)	Damasceno and Fidelis (2020), Damasceno unpublished data

A Annual fires, *E* Early-dry season fires, *M* Mid-dry season fires, *L* Late-dry season fires, *B* biennial fires, *Q* Quadrennial fires, *E2* Early-dry season fires in a 2-year fire excluded area, *L2* Late-dry season fires in a 2-year fire excluded area, *E18* Early-dry season fires in a 18-year fire excluded area, *L18* Late-dry season fires in a 18-year fire excluded area, *OS* open savanna, high cover of the herbaceous layer, *G* grasslands, *S* savanna, medium cover of the herbaceous layer, *WL* woody savannas, low cover of the herbaceous layer, *N* Native savannas, *I* Invaded savannas, *ME* Early-Dry season fires with *Melinis minutiflora*, *MM* Mid-Dry season fires with *Melinis minutiflora*, *ML* Late-Dry season fires with *Melinis minutiflora*, *UE* Early-Dry season fires with *Urochloa decumbens*, *UM* Mid-Dry season fires with *Urochloa decumbens*, *UL* Late-Dry season fires with *Urochloa decumbens* et al. 2018; Bombo et al. 2022), allow their persistence and survival after fire. Several species also show fire stimulation for resprouting (Clarke et al. 2013; Pilon et al. 2021; Zupo et al. 2021), flowering (Coutinho 1976; Fidelis et al. 2019; Fidelis and Zirondi 2021; Zirondi et al. 2021), seed dispersal (Coutinho 1990), or germination (e.g. Moreira et al. 2010; Zirondi et al. 2019; Dairel and Fidelis 2020). Most of the biomass is located belowground (Coutinho 1990; Castro and Kauffman 1998; Kauffman et al. 1994; Delitti et al. 2001) and thus, protected from being constantly removed by fire and allowing buds to resprout (Pausas et al. 2018; Rachid-Edwards 1956; Coutinho 1982; Bombo et al. 2022). Values for root:shoot ratio higher than 2 were found in cerrado-sensu-strico and cerrado-denso, and as higher as 5.6 and 7.7 in savannas and grasslands (respectively *campo limpo* and *campo sujo*) (Castro and Kauffman 1998).

Probably the most impressive post-fire effect on the Cerrado herbaceous plants is the intense and extremely rapid flowering, that may start just one day after fire, as observed for three sedge species: *Bulbostylis paradoxa* (Fidelis et al. 2019), *Rhynchospora confusa* and *Rhynchospora terminalis* (Pilon et al. 2023). European naturalists, Auguste de Saint-Hillaire and Eugene Warming, were the first to report rapid and synchronous post-fire flowering in the Cerrado, in the nineteenth century (Saint-Hilaire 1937; Warming 1892). Leopoldo Coutinho scientifically evaluated and documented this phenomenon for over 100 Cerrado species in the twentieth century (Coutinho 1976). A recent review of post-fire flowering strategies in more than 400 species of savannas of the Cerrado reveals that 85% of these species exhibit a response to fire, with 20% being fire-dependent (flower mostly only after fire), particularly those from the Poaceae family (Fidelis and Zirondi 2021; Pilon et al. 2018). However, some grasses are fire-inhibited (e.g., *Mesosetum loliiforme, Trachypogon spicatus*) (Fidelis et al. 2022). Shrubs usually invest in flowers in the first post-fire year (Zirondi et al. 2021; Hoffmann 1998).

Most Cerrado plants resprout after fire (Rachid-Edwards 1956; Zupo et al. 2021; Pilon et al. 2021) (Fig. 5.4), depending on the protection of their buds. Grasses usually have buds close to the soil surface, protected by the base of the leaf sheaths (Pilon et al. 2021; Rachid-Edwards 1956). Forbs resprout from well-protected buds located in various belowground bud-bearing storage organs, being xylopodium the most common (Pausas et al. 2018; Chiminazzo et al. 2023b). Shrubs and trees may also have buds belowground, from which they resprout after fire topkilling; they may also rely on aboveground buds protected by thick bark, trichomes, cataphils or a combination of these strategies (Santos et al. 2021; Chiminazzo et al. 2021; Chiminazzo et al. 2023b). However, the efficiency of fire protection strategies and the plants ability to resprout depend on fire season, frequency and severity (Rissi et al. 2017; Sato et al. 2010).

Post-fire germination in the Cerrado is infrequent, with few species relying solely on this process for persistence (Pilon et al. 2021). Some species experience seed dormancy breakage due to fire, while many exhibit seed resilience to temperatures up to 200 °C (Daibes et al. 2017, 2019; Zirondi et al. 2019; Fichino et al. 2016; Zupo et al. 2021). Smoke, typically recognized as a crucial cue for germination in other fire-prone ecosystems, does not significantly stimulate germination in Cerrado

Fig. 5.4 (**a–d**) Representation of a fire event and post-fire effects in herbaceous and woody elements of a Cerrado vegetation. (**b, e**) Fire quickly spreads through the herbaceous layer, scorching the tops of the trees. (**c, f**) Ash resulting from the burned biomass covers the ground as soon as the fire passes through. (**d, g–j**) A few days after the fire, plants begin to flower and sprout (**g–j**, respectively: *Lippia horridula, Bulbostylis paradoxa, Jacaranda decurrens* and *Kielmeyera* sp.). (Photos: **e–f** M. Martins, **g, h** A. Fidelis, **i, j** J.C. Motta-Junior. Drawing: MA Chiminazzo)

species (Ramos et al. 2016; Motta et al. 2024). However, a combination of heat shock and smoke has been found to enhance germination in some Fabaceae species, indicating that fire-related cues for Cerrado species may involve both direct (smoke, high temperatures) and indirect (higher daily temperature fluctuation, removal of litter) drivers (Zirondi et al. 2019; Daibes et al. 2017; Dairel and Fidelis 2020).

Fire may have different effects on the Cerrado trees, negatively affecting seedling establishment (Hoffmann 1996), survival of small individuals (Sato et al. 2010), flowering, fruiting and seed production (Hoffmann 1998) for several species, although Coutinho (1990) report beneficial effect on seed dispersion. Recurrent frequent fires may reduce woody species up to 37% in *cerrado-sensu-stricto*, with 10% of mortality of individuals with diameter below 5 cm (Medeiros and Miranda 2005). A thick corky bark provides effective protection of trunk and branches inner tissues (Coutinho 1990; Chiminazzo et al. 2023a) (Fig. 5.4j), and since young trees lack this protection, they are more susceptible to fire. Frequent fires may also topkill individuals (Gomes et al. 2020; Durigan et al. 2020) and regeneration, when it occurs, starts from the base of the plant, requiring an extended period to reach pre-fire height and cover (Hoffmann et al. 2012; Chiminazzo et al. 2021). In fire-sensitive habitats like *veredas* and gallery forests, wildfires may cause long-lasting impacts, resulting in the death of non-fire-adapted trees and canopy openings that create opportunities for the proliferation of light-demanding grasses, often of exotic origin (Flores et al. 2020; González et al. 2021).

Due to their adaptations to fire, grasses, forbs and small shrubs prevail under frequently burned areas (Rodrigues and Fidelis 2022), at expenses of the woody layer, which on the other hand, proliferate with fire exclusion (Pinheiro and Durigan 2009; Abreu et al. 2017; De Souza et al. 2022), in a fire-feedback dynamic. Even minimal woody encroachment caused by fire exclusion can result in herbaceous diversity loss (Wieczorkowski and Lehmann 2022) and affect the bud bank (Bombo et al. 2022). As a consequence, a shift from fire-tolerant species to fire-sensitive ones may occur, which may also affect the woody layer composition (Roitman et al. 2008; Lima et al. 2020; Machida et al. 2021; Geiger et al. 2011). De Souza et al. (2022) verified in wet grasslands under anthropogenic pressures and excluded from fire the dominance of the shrub *Trembleya parviflora*, which caused a 50% loss of native species. Even native grasses such as *Tristachya leiostachya* and *Axonopus siccus*, may dominate fire-excluded areas (De Souza et al. 2022), leading to homogenization in species composition.

5.4.2 Animals Strategies

Fire can have negative, neutral or positive effects on animals at different temporal and spatial scales in the Cerrado, which are related to fire regime (frequency, intensity, spatial and temporal scales, wildfire versus prescribed burning), environmental conditions (vegetation type, season, time of day, landscape attributes), and species/individual characteristics (life-stage, sex, breeding status, vagility, habitat and diet preferences, microhabitat use, competitive ability, niche requirements) (Moretti et al. 2004; Woinarski and Legge 2013; Giorgis et al. 2021; González et al. 2022, Santos et al. 2022). These characteristics interact in complex ways and influence how different taxa (or similar taxa in distinct areas) respond to fires in the Cerrado. Furthermore, the size of the area and method used to evaluate the responses (e.g., variation in species/individual detectability, selectivity of sampling methods) may also affect the results and the potential to explain general patterns.

Burnings can have both direct/immediate and indirect/long-term effects on animals. Fire and smoke may directly impact individuals by causing injury or death (Engstrom 2010; Jolly et al. 2022). Species with terrestrial and slow-moving habits (as the giant anteater *Myrmecophaga trydactyla*) (Silveira et al. 1999; Tomas et al. 2021) or certain invertebrate life stages (insect larvae, Marini-Filho 2000;

Nascimento et al. 2020) seem more susceptible to wildfires than flying taxa such as birds (Cavalcanti and Alves 1997) or adult Lepidoptera (Nascimento et al. 2020).

Differential responses between wild and prescribed fires in the Cerrado may be attributed to several factors that are challenging to separate, including variations in the burning season (Nascimento et al. 2020), fire intensity (Pinho et al. 2017; Arruda et al. 2020; Rosa et al. 2021; Nunes et al. 2023), and the spatial scale of burning (Camargo et al. 2018; Souza et al. 2023b). However, limited quantification hinders additional generalizations. For instance, a comprehensive study on the direct mortality caused by a megafire (1994) and a subsequent wildfire (1995) at Emas National Park (Brazil) revealed distinct impacts on large mammals (Silveira et al. 1999). The 1994 megafire, covering the entire park (131,868 ha), resulted in the discovery of 19 dead large mammals, including 16 giant anteaters, two giant armadillos (*Priodontes maximus*), and one tapir (*Tapirus terrestris*) in a 2000 ha survey area. The findings underscored the vulnerability of species with long, coarse fur to flames (see Batista et al. 2023), along with the demise of several snakes, lizards, and frogs. Conversely, the 1995 wildfire, affecting previously burned areas from the 1994 megafire, revealed no dead large mammals, amphibians, or reptiles. Among 45 individuals of the snake *Bothrops pauloensis* found at Emas National Park by Valdujo et al. (2002) just after prescribed fires in a savanna, only one was dead. On the other hand, several dead and/or burned lizards and snakes were found (Fig. 5.5) in a recent high effort to find dead animals after prescribed fires in a Cerrado grassland (central Brazil), particularly in areas where the time since the last fire was relatively long (4–5 years), likely resulting in more intense fires compared to areas with shorter intervals between fires (BF Gomes and M Martins, unpublished data). However, no dead frog, only one dead rodent (*Thalpomys cerradensis*) and one dead bird (*Melanopareia torquata*) were found. These findings suggest that although generally resilient to fire, certain conditions, such as high fuel loads, may impede animals from escaping or finding shelter before lethal temperatures are reached (discussed further in thermal tolerances below). Direct fire effects may also impact several birds, small mammals, lizards, and frogs nesting in lower vegetation layers by killing eggs, nestlings and juveniles (e.g., Friend 1993; Kirkland et al. 1996), or that build nests susceptible to flames, such as cartoon nests of paper wasps, bees, carpenter ants and wood-nesting ants and bees (Morais and Benson 1988; Frizzo et al. 2012; Santos 2013; Vasconcelos et al. 2017; Rosa et al. 2021). In Bolivia, especially in Beni and Santa Cruz, Maillard et al. (2022) report impact on the distribution of 54 threatened bird species, 15 endemics, due to wildfires, relating biodiversity loss to high frequency, intensity, large-scale, or absence of fires. Similarly, the home ranges of large mammals, including jaguars (*Panthera onca*), deer (*Blastocerus dichotomus*, *Ozotoceros bezoarticus*), and bush dogs (*Speothos venaticus*), are negatively impacted by wildfires in these Bolivian regions (FAN WCS 2021)

Scavengers and birds of prey (e.g. *Caracara plancus*, *Geranoaetus albicaudatus*) are attracted to burning spots, usually reaching these areas shortly after the fire starts (BF Gomes, pers. obs.), where they seek carcasses or live animals fleeing from flames and smoke (Reis et al. 2016) (Fig. 5.5). Terrestrial ants also increase foraging effort right after fires (Gibb and Parr 2010; Alcolea et al. 2022), and they

Fig. 5.5 (**a**) Representation of a fire event in the Cerrado (before fire in yellow, post-fire in grey). (**a**, **b**) Scavengers and birds of prey seek for carcasses (**c**) or live animals fleeing from flames and smoke; (**a**) terrestrial ants increase foraging activity immediately after fire; (**d, e**) small animals seek refuge using pre-existing burrows or by burying themselves in sandy soils. Species with high dispersal ability seek shelter in patches of unburned vegetation nearby (**f**) and gallery forests. (**a, g**) Within a few days after fire, newly regrown and more palatable vegetation in the burned sites attract herbivores. (Drawing: B.F. Gomes; Photos: **b**, **c**, **f–g**: B.F. Gomes; **d**: A.P. Carmignotto; **e**: G. Durigan)

are able to consume carcasses of small animals in a short time (AV Christianini, pers. obs.) (Fig. 5.5). Therefore, evaluations of fire-caused direct mortality are preferable right after fire, decreasing the chances of carcass removal by scavengers.

Although the potential evolutionary force of fire in shaping animal traits is under debate (Pausas and Parr 2018; Pausas 2019), fire may impact animal populations, species and communities in different ways at ecological time scales, and Cerrado animals show a diversity of strategies to deal with fire, usually by running away or seeking shelter in refuges underground. Species with high dispersal ability find refuge in nearby non-burned patches and gallery forests, allowing recolonization of burned patches in case of local extinction (Marini-Filho 2000; Cardoso da Silva 1995; Prada and Marinho-Filho 2004). Species with underground habits or nests, such as terrestrial termites (De Souza et al. 2003) and ants (Frizzo et al. 2012; Vasconcelos et al. 2017) demonstrate lower susceptibility to fire-induced mortality due to the insulating properties of the soil. Numerous Cerrado frogs, lizards, snakes, and small mammals commonly bury themselves in sandy soils or utilize cavities within termite mounds and pre-existing burrows constructed by armadillos and spiny rats (*Clyomys laticeps*) as refuge sites (Fig. 5.5). Armadillo shelters reach a density of 30.8 ± 23.6 per hectare (mean ± SD; data from Serra do Tombador) and become apparent when herbaceous vegetation is cleared by fire (Vitt and Caldwell 1993; Vieira 1999; Colli et al. 2002; Moreira et al. 2009; Costa et al. 2008, 2013; Duleba and Ferreira 2014; Díaz-Ricaurte et al. 2020; Fiorillo et al. 2020, 2021; Anjos et al. 2021) (Fig. 5.5). Given that the temperature belowground experiences only a marginal increase during fires (see Sect. 5.3), these subterranean retreats also tend to experience minimal heating. However, lizards concealed within shallow burrows may occasionally succumb to fire-induced mortality (BF Gomes, pers. obs.).

The voluntary thermal maximum (VTMax)—maximum temperature tolerated by individuals before actively moving to a colder place—respectively for Cerrado lizards, snakes and frogs lie between 37–41 °C, 32–39 °C and 30–34 °C (Díaz-Ricaurte et al. 2022). Cerrado frogs are typically nocturnal and usually stay inside burrows during the day, when most fires occur, but lizards and snakes may be diurnal or nocturnal. Thus, diurnal lizards and snakes start to seek refuge when temperatures get closer to their VTMax. However, their critical thermal maximum (CTMax)—temperature of losing locomotor function (Ørsted et al. 2022)—tend to be considerably higher than the VTMax, indicating a safe time interval between seeking a refuge and reaching a lethal body temperature. Perhaps this may explain the fire-resilience of the Cerrado amphibians and reptiles (Santos et al. 2022), in contrast to lizards from fire-sensitive ecosystems (Infante et al. 2021; Tomas et al. 2021).

Over time, responses of animal populations are also influenced by post-fire processes (Driscoll et al. 2010). Indeed, indirect effects such as fire-induced changes in habitat (vegetation structure) and resource availability (food, shelter) and competition or predation risk may be the most important and long-lasting effects on animals (Vasconcelos et al. 2009; Frizzo et al. 2011; Maravalhas and Vasconcelos 2014; Andersen 2018; Costa et al. 2020; Santos et al. 2022; Bergstrom et al. 2023), forcing individuals that survive fires to change behavior or relocate, thus causing local extinction (which can be reversed once suitable conditions return) (Cardoso da Silva 1995; Cavalcanti and Alves 1997; Vieira and Marinho-Filho 1998; Vasconcelos et al. 2009; Frizzo et al. 2011; Oliveira and Aguiar 2015; Santos et al. 2022). Fire-promoted changes in vegetation structure (e.g., decrease in plant cover and biomass) may increase predation susceptibility for arboreal small mammals (e.g., *Rhipidomys macrurus*) or cursorial species that lack burrows for shelter or are essentially diurnal (e.g., the rodent *Necromys lasiurus*, Vieira et al. 2005; Ghizoni et al. 2005; Camargo et al. 2018; and terrestrial tinamid birds, Pacheco et al. 1994; Falleiro et al. 2021). Simultaneously, these changes may benefit species that rely on more open environments and/or by reduced competition and favorable environmental conditions—as the burrowing-owl, *Athene cunicularia* (Tubelis and Delitti 2010), and the red-legged-seriema, *Cariama cristata* (Tubelis 2019), that select burned spots for breeding, or the greater-rhea (*Rhea americana*) that relies on open savannas (Falleiro et al. 2021).

Fire events often benefit taxa typical of grasslands and savannas—as terrestrial ants (Abreu et al. 2017), lizards (Costa et al. 2020), birds (Sanaiotti and Magnusson 1995; Coelho et al. 2023) and mammals (Henriques et al. 2006; Mendonça et al. 2015; Souza et al. 2023a)—at the expense of arboreal ones. Too frequent fires that promote the mortality of fire-sensitive vegetation types like *cerradão* or *vereda*, or the environmental homogenization by invasive species at the landscape scale benefit generalist, open-area dwellers, who colonize the modified habitat and probably disrupt the original fauna, as observed for small mammals (Mendonça et al. 2015; Camargo et al. 2018; González et al. 2021). However, it is worth noting that, as well as fire-adapted herbaceous plants, animals typical of grasslands and open savannas (including many endemics) are negatively affected by woody encroachment resulting from fire-exclusion (see Sects. 5.2.2 and 5.4.1). Therefore, the indirect effects of fire on Cerrado animals depend on shifts in habitat, the taxon involved, and the fire regime.

Community responses of terrestrial ants (Abreu et al. 2017), birds (Reis et al. 2016), lizards (Araujo et al. 1996; Costa et al. 2020), frogs (Durigan et al. 2020), and mammals (Vieira 1999; Briani et al. 2004; Souza et al. 2023a) in the Cerrado show limited or no effects of fire, often returning to pre-fire levels within a few months to 2–3 years (Durigan et al. 2020). This is due to certain traits that confer them fire protection (e.g., flight ability, use of burrows, Batista et al. 2023) and high resilience of the populations, as well as the fast recovery of the open cerrados after fire. Recovery of arthropod diversity seems to be influenced by the vigorous resprouting of vegetation, leaf litter deposition on the ground surface (Vasconcelos et al. 2009, but see Anjos et al. 2016), and proximity to unburned patches acting as sources for recolonization (Marini-Filho 2000). Other arthropod groups may show higher abundances or activity a few weeks after fire compared to non-burned spots, exploiting flushes of more palatable leaves, flowers, fruits and seeds produced by many plants after fire (see Sect. 5.4.1) or benefiting from a post-fire less complex vegetation structure for foraging (Prada et al. 1995; Vieira et al. 1996; Marini-Filho 2000; Uehara-Prado et al. 2010; Lopes and Vasconcelos 2011; Alcolea et al. 2022). While more palatable leaves in recently burned sites attract various herbivore groups (as pampas deer and tapir, Prada 2001; Nunes et al. 2023), the increased populations of certain arthropods can be attractive to a range of insectivores, including predatory arthropods, lizards, frogs, birds, and mammals (Uehara-Prado et al. 2010; Owen 2013; Vieira and Briani 2013; Sousa et al. 2015; Reis et al. 2016; Marques et al. 2022; Arrua et al. 2023), potentially enhancing foraging activity. Some of these species are considered "fire specialists," displaying regional-scale movement among burned areas, as the birds *Charitospiza eucosma* and *Geositta poeciloptera* (Cavalcanti and Alves 1997; Parker and Willis 1997; Lopes et al. 2023). Mammalian predators, including pumas, maned-wolves, skunks and anteaters (but see Falleiro et al. 2021 for anteaters) become more frequent in burned areas 1–2 post-fire years, likely drawn by increased prey abundance (e.g. insects and small mammals), the avoidance of mesopredators (e.g. ocelots), or simply due to their preference for habitats with more open vegetation (Silveira et al. 1999; Falleiro et al. 2021; Nunes et al. 2023). However, there is evidence suggesting that highly frequent fires, occurring annually, can be detrimental to grassland bird breeding (Giacomo et al. 2011) and may reduce population sizes, even for species adapted to fire (Woinarski and Legge 2013; Sousa et al. 2015). In Amazonian cerrados a single fire event can alter the community composition of birds, small mammals and lizards (Cintra and Sanaiotti 2005; Coelho et al. 2023; Souza et al. 2021), indicating potential regional variations in animal community responses to fire.

Given the extensive spatial heterogeneity of the Cerrado landscape and its profound impact on vegetation structure and resource availability, the indirect effects of fire on animal communities are likely more discernible at larger spatial scales. At the landscape level, the mosaic of patches of different fire regimes (pyrodiversity) may retain high biodiversity of ants (Maravalhas and Vasconcelos 2014), lizards (Costa et al. 2020) and mammals (Souza et al. 2023a). Thus, fire exclusion in the Cerrado reduces habitat quality and microhabitat availability for animals typical of grasslands and open-savannas, including several endemic species of ants (Abreu et al. 2017), frogs (Thomé et al. 2021), lizards (Costa et al. 2020), snakes (Fiorillo et al. 2021), birds (Parker and Willis 1997; Motta-Junior et al. 2008; Kanegae et al. 2012; Lucindo et al. 2015; Coelho et al. 2023; Lopes et al. 2010, 2023), and mammals (Furtado et al. 2021). The process of woody encroachment is causing a silent local extinction of many animal taxa typical of open environments, which fail to find suitable habitats.

The indirect effect of fire on aquatic fauna may be substantial, especially due to runoff of ashes/sediments carried into water bodies after fires, although this effect is often poorly understood in the Cerrado (Gomez Isaza et al. 2022). Bioassay studies have identified potential toxic effects of ashes on a microcrustacean (*Ceriodaphnia dubia*), earthworm reproduction (Enchytraeidae), and an aquatic mollusk (*Biomphalaria glabrata*), but no negative effects were observed on zebrafish (*Danio rerio*) (Brito et al. 2017; Oliveira-Filho et al. 2018). However, these species commonly used in toxicological studies are mostly non-native to the Cerrado, limiting the generalization of these findings to other animals. Despite evidence that Cerrado tadpoles are tolerant to water temperature increases (Ribeiro et al. 2020), many flying insects have aquatic larvae that could be affected, and any impact on aquatic fauna may cascade through terrestrial food webs (Recalde et al. 2021). Therefore, further research investigating the effect of fires on Cerrado aquatic fauna would be valuable.

5.5 The Environmental Services and Disservices of Fire/Wildfire

Peasant communities often utilize fire for land management (see Sect. 5.2.3) and livestock care. Cattle ranchers rely on the post-fire regrowth of the herbaceous layer—especially grasses—to sustain their animals. In addition to benefiting fire-adapted Cerrado plants (as presented in Sect. 5.4.1), the recycling of nutrients promoted by fire also benefits this vegetation layer. In the dry season, the decomposition of the Cerrado plants' dead material is very low, and dry leaves, especially on grass tussocks, tend to accumulate from year to year. Fire then rapidly mineralizes the organic matter, making nutrients available to plants. Grasses, with their extensive network of superficial roots, quickly absorb the nutrients from the ashes on the soil (Fig. 5.4), making them highly responsive to fire. Although a large amount of nutrients is lost with the burning of vegetation—especially the more volatile ones, such as nitrogen and sulphur—the return of macronutrients with rainfall occurs in 2–4 years (Pivello and Coutinho 1992).

Fire also stimulates flowering, fruiting and seed production of a number of Cerrado plants that have been used for centuries by Indigenous Peoples and Traditional and Local Communities as food, medicines and for crafting (see Chap. 3, Pilon et al. 2018; Fidelis and Zirondi 2021). Pequi (*Caryocar brasiliense*), mangaba (*Hancornia speciosa*), guabiroba (*Campomanesia xanthocarpa*), murici (*Byrsonima crassifolia*), araticum (*Annona montana*), among dozens of other fruits appreciated for their flavor and nutritional value, have been increasingly exploited for consumption *in natura*, in cooking and the production of sweets and ice cream, as well as for the extraction of chemical substances for cosmetic and pharmacological uses (Arruda et al. 2022; Reis and Schmiele 2019). Fire-stimulated flowers, or their components, find use in handicrafts and trade, with various species from Poaceae, Cyperaceae, Asteraceae, Eriocaulaceae, Xyrydaceae, Melastomataceae, among others (Oliveira et al. 2015; Schmidt et al. 2011). Additionally, seed production triggered by fire offers a valuable resource for plant nurseries and direct seeding in restoration projects. Indeed, there are increasing opportunities for seed harvesting to support restoration plantings in Cerrado, providing a complement of income of ca. US$ 534/person/year (Rede de Sementes do Cerrado 2002).

Occasional low-intensity fires can enhance local pollinator diversity (excluding Lepidoptera), drawn to post-fire bursts of flower resources (Carbone et al. 2019; Baronio et al. 2021). Native bees and other insects, crucial for commercial crop pollination, spill over into surrounding Cerrado patches, positively impacting crop yield (Fuzaro et al. 2018; Ramos et al. 2018; Almeida et al. 2020). Soil-nesting bees, less susceptible to fire, may even benefit from it (Lazarina et al. 2016).

However, repeated and intense burnings may reduce populations of certain native bees, especially those nesting lower in vegetation or using dead wood (Lazarina et al. 2016; Santos 2013), leading to diminished crop yield. Nevertheless, the plant-bee interaction network appears quite resilient, recovering its pattern a few years after fire (Peralta et al. 2017; Baronio et al. 2021).

The cascading impact of fire on fruit and seed production on frugivorous animals and their role in plant regeneration remains unclear. Phyllostomid bats and small rodents, vital for seed dispersal and flower pollination, experience changes in richness and abundance due to fire-induced alterations in vegetation structure, potentially affecting plant community diversity (Loeb and Blakey 2021; Cazetta and Vieira 2021). The recovery of these mammals after fire is crucial for plant recruitment and vegetation recovery, especially in woodland and gallery forest regeneration (Oliveira and Aguiar 2015; González et al. 2021).

Certain Cerrado animals are voracious predators of arthropods and may benefit crop plantations by pest control. For instance, the marsupial *Gracilinanus agilis* feeds on stink bugs (e.g., *Euschistus heros*, Hemiptera), major soybean pests (Camargo et al. 2022). This ecosystem service is estimated to be worth up to US$ 31.08 ha^{-1} of native Cerrado woodlands per year (Camargo et al. 2022). However, fire in Cerrado typically reduces the abundance of marsupials like *Gracilinanus agilis* (Mendonça et al. 2015; Camargo et al. 2018), potentially affecting the delivery of this ecosystem service. Considering the effects of fire on animals (see Sect. 5.4.2), fire impacts on animal-delivered ecosystem services in open savannas (including pollination, seed removal, and pest-control) are probably transient and taxon-specific, but can be more pronounced and long-lasting for gallery forest taxa.

5.6 Fire Management, Knowledge and Policy Gaps

5.6.1 A Paradigm Shift in the Cerrado Fire Management

Anthropogenic altered fire regimes (see Sect. 5.2.3), land use change and climate change have considerably increased the risk and frequency of wildfires across the entire Cerrado (Fidelis et al. 2018; Maillard et al. 2018, 2022; FAN WCS 2021; Hofmann et al. 2021; Molinas-González and Florentín 2021). These three factors are intricately interconnected: fire is employed for agricultural management and the creation of new agricultural zones under modified regimes. Extensive fires, especially for expanding agricultural fronts, amplify greenhouse gas (GHG) emissions, exacerbating climate change by fostering drier and hotter environmental conditions. In turn, climate change elevates the wildfire risk. The El Niño and La Niña phenomena further enhance conditions conducive to wildfires (Araújo et al. 2012). Current changes in land use and global climate are profound and irreversible: the IPCC (Intergovernmental Panel on Climate Change) climate change scenarios project an annual increase in burned area in the Cerrado, associated with drought (Silva et al. 2019; Maillard et al. 2022). Under these recent conditions, the zero-fire policy that had been adopted by South American governments for five centuries since the arrival of European colonizers is increasingly perceived as not only ineffective for the Cerrado, but also dangerous: without periodic burning, biomass accumulates and stimulates wildfires of greater intensity and more difficult to control, leading to loss of material goods and even lives, incurring substantial expenses for fire suppression (Durigan and Ratter 2016; Fidelis et al. 2018; Pivello et al. 2021).

By the end of the twentieth century, Brazil, Bolivia, and Paraguay implemented more robust fire legislation (e.g., Brazil: Decree 2661/1998, Paraguay: Law 4014/2010), primarily focused on fire prevention and control. Paraguay has also recognized the historical component of fire management by pre-Hispanic societies (Molinas-González and Florentín 2021) and this is a promising direction for other countries to follow. However, it was in the early 2000s that more adequate policies emerged to address the complex issue of fires. This shift was influenced by advancements in scientific understanding of Cerrado dynamics, heightened awareness of biodiversity value under the Convention on Biological Diversity (signed by 168 countries in 1993), and severe fires in protected areas (Paraguay 2010; Rodriguez-Montellano 2014; Durigan and Ratter 2016; FAN WCS 2021; Pivello et al. 2021). In these three countries, legislation came to accept controlled fires as a suitable management tool for traditional agriculture and native vegetation conservation (e.g., Brazil: Law 12651/2012; Paraguay: Law 6818/2021; Bolivia: Law 1171/2019) and the development of Integrated Fire Management strategies are in progress (more details in Chap. 14).

Effective fire management demands well-defined objectives and a thorough understanding of the targeted area, encompassing habitat diversity and the adaptability of species to fire. Prescribed or controlled burns should be strategically planned at a landscape scale, considering the varied vegetation types in the Cerrado, each with its unique fire tolerance levels (e.g., cerradão and wetlands), necessitating specific fire regimes or protection measures. Local nuances, such as the presence of invasive plants like highly combustible African grasses or *Pinus* spp., which increase wildfire risk, must be considered. Additional factors, including habitat fragmentation, the fire risk in the surrounding matrix, prevailing weather conditions, accumulated rainfall, woody encroachment, and the potential impact on animal sensitivity to fire, contribute to the complexity (Pivello and Norton 1996; Batista et al. 2023). Mosaic burning is generally preferable, as it mimics the natural fires. This method involves burning small areas sequentially, maintaining a mix of recently burned and long-term protected patches. The recently burned areas act as firebreaks due to reduced combustible material, creating varied conditions that provide refuges for sensitive flora and fauna, fostering pyrodiversity and maintaining a diverse landscape (Pivello and Norton 1996; Ramos-Neto and Pivello 2000; Motta-Junior et al. 2008; Pivello 2011; Maravalhas and Vasconcelos 2014; Souza et al. 2021, 2023a, b). This strategy has long been used by Indigenous Peoples and Traditional and Local Communities in the Cerrado (e.g., Mistry et al. 2005; Eloy et al. 2019), as well as in other savannas around the world (Laris 2002), and has shown to be the most sustainable. Post-fire monitoring is also essential to assess fire impacts on different taxa and population demography, and the effects of diverse fire regimes, in conjunction with other management practices like grazing (e.g., grazing, Durigan et al. 2022).

Adopting an adaptive management approach allows for adjustments based on regional specificities and targeted conservation goals (Berlinck et al. 2021; Pivello et al. 2021; Batista et al. 2023).

5.6.2 Knowledge Gaps

Despite substantial advancements in scientific understanding of fire ecology since the 1990s (mainly related to the effects of fire on abiotic and biotic components, biotic adaptations and relationships with fire) critical gaps persist, hindering the proper management of the diverse ecosystems in the Cerrado. Notably, there is a noticeable deficiency in knowledge regarding fire dynamics and effects in Cerrado wetlands, such as *vereda* and wet grassland ecosystems (Durigan et al. 2022). A few studies on the theme show that "cold fires"—in the early dry season, under mild wind and high air humidity—at low frequency (10–15 years) and in small areas (<0.5 ha) such as practiced by Traditional and Local Communities (Borges et al. 2016) are beneficial to "renew" the vegetation and decrease the invasion of the fern *Dicranopteris flexuosa* (Nascente et al. 2022). On the other side, frequent (return time 3 years or less) or late dry season fires, as well as high-intensity wildfires can be very harmful to typical *vereda* species, especially young individuals of the palm tree species *Mauritia flexuosa* and *Mauritiella* spp. (Schmidt et al. 2011; Borges et al. 2016), and also can stimulate the invasion by the African grass *Melinis minutiflora*, as observed in two protected areas in central Brazil, Serra do Tombador (V.R. Pivello, pers. comm.) and Chapada dos Veadeiros (Flores et al. 2020). Proper fire management in these wetlands is crucial to prevent poorly understood phenomena like ground fires, smoldering slow fires that occur in organic soils of Cerrado wet grasslands and *veredas*. These fires consume roots and the organic soil matter for several days or even months, and also degrade the physical quality of soils (Pivello et al. 2021; Fonseca et al. 2022).

Another important knowledge gap in the Cerrado relates to belowground dynamics. The amount and distribution of roots and belowground organs in different Cerrado vegetation types count on a few studies (Delitti et al. 2001; Lilienfein et al. 2001; Fidelis et al. 2013; Le Stradic et al. 2021) and even fewer related to depths greater than 50 cm (Castro and Kauffman 1998; Le Stradic et al. 2021), probably due to the difficulty of collecting and replicating samples at great depths (Mokany et al. 2006). Additionally, knowledge about the Cerrado microbiota is lacking, further obscuring the understanding of fire effects on the underground system. This comprehensive information is essential for a more complete understanding of the consequences of burnings on the carbon balance in the Cerrado ecosystems.

Our understanding of the impacts of fire on Cerrado animals is notably limited compared to plants. Existing studies primarily focus on invertebrates, particularly ants, and ectothermic vertebrates, predominantly lizards (Frizzo et al. 2011; Arruda et al. 2018), with a significant concentration in the core Cerrado of Central Brazil (Araujo et al. 1996; Arruda et al. 2018; Costa et al. 2013, 2020). However, less attention has been given to other taxa like bats, nocturnal birds (e.g., owls and nightjars), snakes, terrestrial mollusks, annelids, and aquatic animals. Moreover, many investigations are based on short-term (1–3 years) community responses to unplanned, intense, and large wildfires (e.g., Silveira et al. 1999; Faria et al. 2004; Vasconcelos et al. 2009; Morais et al. 2011; Berlinck et al. 2021). Considering the high beta-diversity of the Cerrado at both local and regional scales, long-term studies encompassing diverse taxa and regions are crucial for a comprehensive understanding of fire impacts on Cerrado animals and the development of effective management practices for biodiversity conservation at various scales (Schmidt et al. 2018; Pivello et al. 2021; Santos et al. 2022).

Recolonization by individuals from nearby unburned areas is likely to be higher in small, burned patches or patches close to unburned spots (Marini-Filho 2000). Given that the dispersal capacity of the vast majority of animal taxa is poorly known, it would be especially relevant to study species dispersal and long-term population persistence according to the species characteristics, the size of burned patches and the surrounding landscape features. Taken together, characteristics of the environment, fire regime and animal species may help to build a sensitivity index that may guide management practices (according to Batista et al. 2023). Regarding experimental prescribed fires, the majority of studies are derived from grasslands and savannas (Araujo et al. 1996; Vieira 1999; Briani et al. 2004; Maravalhas and Vasconcelos 2014; Durigan et al. 2020), while investigations on wildfires span various Cerrado vegetation types (Cavalcanti and Alves 1997; Morais et al. 2011; Oliveira and Aguiar 2015; Mendonça et al. 2015).

Planned studies based on prescribed burnings and considering specific ecosystem services would provide some answers we need to improve fire management aiming at animal conservation in the Cerrado. Still, detailed and accurate information at different spatial and temporal scales would permit the building of near-real-time models to predict and control wildfires.

5.6.3 Policy Gaps

Fire can also be a very useful management tool for Cerrado farmers, managers of protected areas and Traditional and Local Communities, as it is very cheap and promotes biodiversity. However, its usage requires regulated procedures aligned with evolving scientific best practices. Legislation sometimes takes time to align with scientific advancements, as seen with the Brazilian National Policy for Integrated Fire Management (PNMIF), which was submitted to the National Congress in 2018. After a period of review, it was successfully enacted in July 2024 (more details in Chap. 14). Bolivia and Paraguay lack similar legislation, though there's growing recognition of its importance in Paraguay (Molinas-González and Florentín 2021).

The Brazilian legislation Native Vegetation Protection Law (LPVN, Law No. 12.651, of May 25, 2012) mandates rural properties to allocate 20% of their area as Legal Reserve (LR), intended for natural conservation and sustainable economic uses. However, the legislation forest-centric focus inadvertently excludes non-forest vegetation types, hindering the use of prescribed burning for grassy ecosystems. Legislative adjustments are imperative to align with the reality of non-forest ecosystems (Tomas et al. 2024).

The development of regulations or guidelines for specific uses of fire, such as the collection of natural products like flowers and fruits, can be an important tool for the sustainable use of these resources. An example is the regulation established by Tocantins state (Brazil) for the collection and sustainable management of golden grass (*Syngonanthus nitens*) and buriti (*Mauritia flexuosa*), which, based on technical-scientific and traditional knowledge, defines responsibilities for collecting and commercializing these products, as well as the proper use of fire for management. Similar regulations/guidelines are necessary for fire management in the extraction of other native species, such as the *Comanthera* flowers ("sempre-vivas"). Still, policies that focus on the proper use of fire directed to ecosystem services, such as the maintenance of the carbon balance and of water provision—the Cerrado is source of the main South American river basins (see Klink et al. 2020)—and the provision of products, especially food and pharmaceutical supplies, would help conserving the Cerrado biodiversity, improve the financial income of landowners and increase national self-sufficiency.

While legislative improvements are crucial, enforcement of existing laws and penalties for illegal fires remains insufficient. Policies incentivizing small farmers with financial subsidies to adopt alternative techniques in harmful fire situations, accompanied by training and technical assistance, are vital. Collaborative policy-making involving land users, agricultural agencies, park managers, scientists, fire management agencies, and decision-makers, making use of important existing tools, such as Mapbiomas (https://mapbiomas.org)—a major program that maps land use changes and fires across South America, in order to subsidize the fight against climate change—or the FIP system—a forest fire monitoring and prevention system to predict fire risk and spread in Brazil (https://csr.ufmg.br/fipcerrado/en/; Oliveira et al. 2023) enhances the likelihood of successful fire management policies in the Cerrado.

References

Ab'Saber AN (2002) Bases para o estudo dos ecossistemas da Amazônia brasileira. Estud Av USP 16:45. https://doi.org/10.1590/s0103-40142002000200002

Ab'Saber AN (2007) Os domínios de natureza no Brasil: potencialidades paisagísticas. Ateliê Editorial

Abreu RCR, Hoffmann WA, Vasconcelos HL, Pilon N, Rossato D, Durigan G (2017) The biodiversity cost of carbon sequestration in tropical savanna. Sci Adv 3:e1701284. https://doi.org/10.1126/Sciadv.1701284

Adeney JM, Christensen NL, Vicentini A, Cohn-Haft M (2016) White-sand ecosystems in Amazonia. Biotropica 48:7–23. https://doi.org/10.1111/btp.12293

Alcolea M, Durigan G, Christianini AV (2022) Prescribed fire enhances seed removal by ants in a Neotropical savanna. Biotropica. https://doi.org/10.1111/Btp.13036i

Almeida MLS, Carvalho GS, Novais JR, Storck-Tonon D, Oliveira ML et al (2020) Contribution of the Cerrado as habitat for sunflower pollinating bees. Sociobiology 67:281–291. https://doi.org/10.13102/sociobiology.v67i2.4865

Alvarado ST, Fornazari T, Cóstola A, Morellato LPC, Silva TSF (2017) Drivers of fire occurrence in a mountainous Brazilian cerrado savanna: tracking long-term fire regimes using remote sensing. Ecol Indic 78:270–281. https://doi.org/10.1016/j.ecolind.2017.02.037

Andersen AN (2018) Responses of ant communities to disturbance: five principles for understanding the disturbance dynamics of a globally dominant faunal group. J Anim Ecol 88:350–362. https://doi.org/10.1111/1365-2656.12907

Anderson AB, Posey DA (1985) Manejo de cerrado pelos índios Kayapó. Bol Mus Para Emílio Goeldi Bot 2:77–98

Anjos D, Alves-Silva E, Ribeiro SP (2016) Do fire and seasonality affect the establishment and colonization of litter arthropods? J Insect Conserv 20:653–661. https://doi.org/10.1007/s10841-016-9896-4

Anjos AG, Solé M, Benchimol M (2021) Fire effects on anurans: what we know so far? For Ecol Manag 495:119338. https://doi.org/10.1016/j.foreco.2021.119338

Araujo AFB, Costa EMM, Oliveira RF, Ferrari RL, Simon MF et al (1996) Efeitos de queimadas na fauna de lagartos do Distrito Federal. In: Anais Do Simpósio Impacto Das Queimadas Sobre Os Ecossistemas e Mudanças Globais. 3° Congresso de Ecologia do Brasil. Brasília, DF, 6–11 October 1996, pp 148–160

Araújo FMD, Ferreira LG, Arantes AE (2012) Distribution patterns of burned areas in the Brazilian biomes: an analysis based on satellite data for the 2002–2010 period. Rem Sens 4:1929–1946. https://doi.org/10.3390/Rs4071929

Arrua BA, Carvalho LS, Teles TS, Oliveira MR, Ribeiro DB (2023) Fire has a positive effect on the abundance of sun spiders (Arachnida: Solifugae) in the Cerrado-Pantanal ecotone. Fire 6:69. https://doi.org/10.3390/fire6020069

Arruda FVD, Souza DGD, Teresa FB, Prado VHM, Cunha HF, Izzo TJ (2018) Trends and gaps in the scientific literature about the effects of fire on the Brazilian Cerrado. Biota Neotrop 18:e20170426. https://doi.org/10.1590/1676-0611-BN-2017-0426

Arruda FV, Izzo TJ, Teresa FB, Camarota F (2020) Different burning intensities affect cavity utilization patterns by arboreal ants in a tropical savanna canopy. Ecol Indic 116:106493. https://doi.org/10.1016/j.Ecolind.2020.106493

Arruda HS, Araújo MVL, Marostica Junior M (2022) Underexploited Brazilian Cerrado fruits as sources of phenolic compounds for diseases management: a review. Food Chem Mol Sci 5:100148. https://doi.org/10.1016/j.Fochms.2022.100148

Barbosa AS, Schmitz PI (2008) Ocupação indígena do Cerrado—Esboço de uma história. In: Sano SM, Almeida SP, Ribeiro JF (eds) Cerrado: ecologia e flora, vol 2. Embrapa, Brasília, pp 47–68

Baronio GJ, Souza CS, Maruyama PK, Raizer J, Sigrist MR, Aoki C (2021) Natural fire does not affect the structure and beta diversity of plant-pollinator networks, but diminishes floral-visitor specialization in Cerrado. Flora 281:151869. https://doi.org/10.1016/j.flora.2021.151869

Batista EHL, Figueira JEC, Solar RRC, Azevedo CS, Beirão MV et al (2023) In case of fire, escape or die: a trait-based approach for identifying animal species threatened by fire. Fire 6:242. https://doi.org/10.3390/Fire6060242

Bergstrom BJ, Scruggs SB, Vieira EM (2023) Tropical savanna small mammals respond to loss of cover following disturbance: a global review of field studies. Front Ecol Evol 11:1017361. https://doi.org/10.3389/fevo.2023.1017361

Berlinck CN, Lima LHA, De Carvalho Junior EAR (2021) Historical survey of research related to fire management and fauna conservation in the world and in Brazil. Biota Neotrop 21(3):e20201144. https://doi.org/10.1590/1676-0611-Bn-2020-1144

Bombo AB, Appezzato-da-Glória B, Fidelis A (2022) Fire exclusion changes belowground bud bank and bud-bearing organ composition jeopardizing open savanna resilience. Oecologia 199:153–164. https://doi.org/10.1007/s00442-022-05172-1

Bond WJ (2019) Open ecosystems: ecology and evolution beyond the forest edge. Oxford University Press, Oxford

Bond WJ, van Wilgen BW (1996) Fire and plants, Population and community biology, vol 14. Springer, Dordrecht. https://doi.org/10.1007/978-94-009-1499-5_2
Borges SL, Eloy L, Schmidt IB, Barradas ACS, dos Santos IA (2016) Fire management in veredas (palm swamps): new perspectives on traditional farming systems in Jalapão, Brazil. Ambiente Soc 19:269–294. https://doi.org/10.1590/1809-4422Asoc20150020R1V1932016
Braga PIS (1979) Subdivisão fitogeográfica, tipos de vegetação, conservação e inventário florístico da floresta Amazônica. Acta Amazon 9:53–80. https://doi.org/10.1590/1809-43921979094s053
Briani DC, Palma ART, Vieira EM, Henriques RPB (2004) Post-fire succession of small mammals in the Cerrado of Central Brazil. Biodivers Conserv 13:1023–1037. https://doi.org/10.1023/B:Bioc.0000014467.27138.0b
Brito DQ, Passos CJS, Muniz DHF, Oliveira-Filho EC (2017) Aquatic ecotoxicity of ashes from Brazilian savanna wildfires. Environ Sci Pollut R. https://doi.org/10.1007/s11356-017-9578-0
Brugger SO, Gobet E, Van Leeuwen JFN, Ledru MP, Corombaroli D et al (2016) Long-term man-environment interactions in the Bolivian Amazon: 8000 years of vegetation dynamics. Quat Sci Rev 132:114–128. https://doi.org/10.1016/j.Quascirev.2015.11.001
Camargo ACL, Barrio ROL, De Camargo ND, Mendonça AF, Ribeiro JF et al (2018) Fire affects the occurrence of small mammals at distinct spatial scales in a neotropical savanna. Eur J Wildl Res. https://doi.org/10.1007/s10344-018-1224-8
Camargo NF, Reis GG, Mendonça AF, Laumann RA, Nardoto GB et al (2022) Native marsupial acts as an in situ biological control agent of the main soybean pest (*Euschistus heros*) in the Neotropics. Eur J Wildl Res 68:62. https://doi.org/10.1007/s10344-022-01609-3
Carbone LM, Tavella J, Pausas JG, Aguilar R (2019) A global synthesis of fire effects on pollinators. Glob Ecol Biogeogr 28:1487–1498. https://doi.org/10.1111/Geb.12939
Cardoso da Silva JM (1995) Birds of the Cerrado region, South America. Steenstrupia 21:69–92
Castro EA, Kauffman JB (1998) Ecosystem structure in the Brazilian Cerrado: a vegetation gradient of aboveground biomass, root mass and consumption by fire. J Trop Ecol 14:263–283. https://doi.org/10.1017/S0266467498000212
Castro-Neves, BM (2000) Comportamento de queimadas, temperaturas do solo e recuperação da biomassa aérea em campo sujo nativo e em capim gordura (Melinis minutiflora). Master Thesis. Universidade de Brasília. 84 p
Cavalcanti R, Alves M (1997) Effects of fire on savanna birds in Central Brazil. Ornitol Neotrop 8:85–87
Cazetta TC, Vieira EM (2021) Fire occurrence mediates small-mammal seed removal of native tree species in a Neotropical savanna. Front Ecol Evol 9:793947. https://doi.org/10.3389/Fevo.2021.793947
Charles-Dominique T, Staver AC, Midgley GF, Bond WJ (2015) Functional differentiation of biomes in an African savanna/forest mosaic. S Afr J Bot 101:82–90. https://doi.org/10.1016/j.sajb.2015.05.005
Chiminazzo MA, Bombo AB, Charles-Dominique T, Fidelis A (2021) Your best buds are worth protecting: variation in bud protection in a fire-prone cerrado system. Funct Ecol 35:2424–2434. https://doi.org/10.1111/1365-2435.13907
Chiminazzo MA, Bombo AB, Charles-Dominique T, Fidelis A (2023a) Bark production of generalist and specialist species across savannas and forests in the Cerrado. Ann Bot 131:613–621. https://doi.org/10.1093/aob/mcad014
Chiminazzo MA, Bombo AB, Charles-Dominique T, Fidelis A (2023b) To protect or to hide: why not both? An investigation of fire-related strategies in Cerrado woody species. Flora 306:152350. https://doi.org/10.1016/j.flora.2023.152350
Christin PA, Besnard G, Samaritani E, Duvall MR, Hodkinson TR et al (2008) Oligocene CO2 decline promoted C4 photosynthesis in grasses. Curr Biol 18:37–43. https://doi.org/10.1016/j.Cub.2007.11.058
Cintra R, Sanaiotti TM (2005) Fire effects on the composition of a bird community in an Amazonian savanna (Brazil). Braz J Biol 65:683–695. https://doi.org/10.1590/s1519-69842005000400016

Clarke PJ, Lawes MJ, Midgley JJ, Lamont BB, Ojeda F et al (2013) Resprouting as a key functional trait: how buds, protection and resources drive persistence after fire. New Phytol 197:19–35. https://doi.org/10.1111/nph.12001
Climate Change Knowledge Portal for Development Practitioners and Policy Makers (2023) Available in: https://climateknowledgeportal.worldbank.org/. Accessed 11 June 2023
Coelho LA, Duarte Ritter C, Lima A, Cintra R, Magnusson WE, Sanaiotti TM (2023) Effects of fire regime on birds' community in an Amazonian savanna. Biodivers Conserv 32:3893–3913. https://doi.org/10.1007/s10531-023-02665-0
Cole MM (1960) Cerrado, Caatinga and Pantanal: the distribution and origin of the savanna vegetation of Brazil. Geogr J 136:168–179. https://doi.org/10.2307/1793957
Cole MM (1982) The influence of soils, geomorphology and geology on the distribution of plant communities in savanna ecosystems. In: Huntley BJ, Walker BH (eds) Ecology of tropical savannas, Ecological studies, vol 42. Springer, Berlin, pp 145–174
Cole MM (1986) The savannas: biogeography and geobotany. Academic, London. https://doi.org/10.2307/635398
Colli GR, Bastos RP, Araujo AF (2002) 12 The character and dynamics of the Cerrado herpetofauna. In: Oliveira PS, Marquis RJ (eds) The Cerrados of Brazil: ecology and natural history of a neotropical savanna. Columbia University Press, New York, pp 223–241
Cordeiro RC, Turcq B, Moreira LS, Rodrigues RAR, Simões-Filho FFL et al (2014) Palaeo fires in Amazon: interplay between land use change and palaeoclimatic events. Palaeogeogr Palaeoclimatol Palaeoecol 415:137–151. https://doi.org/10.1016/j.Palaeo.2014.07.020
Costa GC, Colli GR, Constantino R (2008) Can lizard richness be driven by termite diversity? Insights from the Brazilian Cerrado. Can J Zool 86:1–9. https://doi.org/10.1139/Z07-107
Costa BM, Pantoja DL, Vianna MCM, Colli GR (2013) Direct and short-term effects of fire on lizard assemblages from a Neotropical savanna hotspot. J Herpetol 47:502–510. https://doi.org/10.1670/12-043
Costa BM, Pantoja DL, Sousa HC, Queiroz TA, Colli GR (2020) Long-term, fire-induced changes in habitat structure and microclimate affect Cerrado lizard communities. Biodivers Conserv 29:1659–1681. https://doi.org/10.1007/s10531-019-01892-8
Coutinho LM (1976) Contribuição ao conhecimento do papel ecológico das queimadas na floração de espécies do Cerrado. Universidade de São Paulo, Brazil. https://doi.org/10.11606/T.41.2014.tde-04092014-095647
Coutinho LM (1978) O conceito de cerrado. Rev Bras Bot 1:17–23
Coutinho LM (1980) As queimadas e seu papel ecológico. Bras Flor 10:7–23
Coutinho LM (1982) Ecological effects of fire in Brazilian cerrado. In: Huntley BJ, Walker BH (eds) Ecology of tropical savannas. Springer, Berlin, pp 273–291
Coutinho LM (1990) Fire in the ecology of Brazilian cerrado. In: Goldamer JG (ed) Fire in the tropical biota ecosystem: processes and global challenges, Ecological studies, vol 84. Springer, Berlin, pp 81–105
Coutinho LM (2002) O bioma do cerrado. In: Eugen Warming e o Cerrado Brasileiro: um século depois. Ed. UNESP, São Paulo, Brazil
Coutinho LM (2006) O conceito de bioma. Acta Bot Brasilica 20:13–23. https://doi.org/10.1590/s0102-33062006000100002
Da-Gloria P, Hubbe M, Neves WA (2018) Lagoa Santa's contribution to the origins and life of early Americans. Evol Anthropol 27:121–133. https://doi.org/10.1002/evan.21587g
Daibes LFF, Zupo T, Silveira FAOFAO, Fidelis A (2017) A field perspective on effects of fire and temperature fluctuation on Cerrado legume seeds. Seed Sci Res 27:1–10. https://doi.org/10.1017/S096025851700006X
Daibes LF, Pausas JG, Bonani N, Nunes J, Silveira FAO, Fidelis A (2019) Fire and legume germination in a tropical savanna: ecological and historical factors. Ann Bot 123:1219–1229. https://doi.org/10.1093/aob/mcz028
Dairel M, Fidelis A (2020) How does fire affect germination of grasses in the Cerrado? Seed Sci Res 30:275–283. https://doi.org/10.1017/S0960258520000094

Damasceno G, Fidelis A (2020) Abundance of invasive grasses is dependent on fire regime and climatic conditions in tropical savannas. J Environ Manag 271:111016. https://doi.org/10.1016/j.jenvman.2020.111016
De Souza O, Albuquerque LB, Tonello VM, Pinto LP, Reis-Jr R (2003) Effects of fire on termite generic richness in a savanna-like ecosystem ('Cerrado') of Central Brazil. Sociobiology 42:1–11
De Souza GF, Ferreira MC, Munhoz CBR (2022) Decrease in species richness and diversity, and shrub encroachment in Cerrado grasslands: a 20 years study. Appl Veg Sci 25:e12668. https://doi.org/10.1111/avsc.12668
Delitti WBC, Pausas JG, Burger DM (2001) Belowground biomass seasonal variation in two Neotropical savannahs (Brazilian Cerrados) with different fire histories. Ann For Sci 7:713–722. https://doi.org/10.1051/Forest:2001158
Desjardins T, Carneiro Filho A, Mariotti A, Girardin C, Chauvel A (1996) Changes of the forest-savanna boundary in Brazilian Amazonia during the Holocene revealed by stable isotope ratios of soil organic carbon. Oecologia 108:749–756. https://doi.org/10.1007/Bf00329051
Díaz-Ricaurte JC, Serrano FC, Guevara-Molina EC, Araujo C, Martins M (2020) Does behavioral thermal tolerance predict distribution pattern and habitat use in two sympatric Neotropical frogs? PLoS One 15:e0239485. https://doi.org/10.1371/Journal.Pone.0239485
Díaz-Ricaurte JC, Serrano FC, Martins M (2022) VTMaxHerp: a data set of Voluntary Thermal Maximum temperatures of amphibians and reptiles from two Brazilian hotspots. Ecology:–e3602. https://doi.org/10.1002/Ecy.3602
Driscoll DA, Lindenmayer DB, Bennett AF, Bode M, Bradstock RA et al (2010) Fire management for biodiversity conservation: key research questions and our capacity to answer them. Biol Conserv 143:1928–1939. https://doi.org/10.1016/j.biocon.2010.05.026
Duleba S, Ferreira V (2014) Herpetofauna associated with termite mounds in a pasture, Mato Grosso do Sul State, Brazil. Herpetol Bull 127:10–16
Durigan G, Ratter JA (2016) The need for a consistent fire policy for Cerrado conservation. J Appl Ecol 53:11–15. https://doi.org/10.1111/1365-2664.12559
Durigan G, Pilon NAL, Abreu RCR, Hoffmann WA, Martins M et al (2020) No net loss of species diversity after prescribed fires in the Brazilian savanna. Front For Glob Change 3:13. https://doi.org/10.3389/Ffgc.2020.00013
Durigan G, Pilon NAL, Souza FM, Melo ACG, Ré DC, Souza SCPM (2022) Low intensity cattle grazing is better than cattle exclusion to drive secondary savannas toward the features of native Cerrado vegetation. Biotropica. https://doi.org/10.1111/Btp.13105
Edwards EJ, Osborne CP, Stromberg CAE, Smith SA (2010) The origins of C-4 grasslands: integrating evolutionary and ecosystem science. Science 328:587e591. https://doi.org/10.2307/40655827
Eiten G (1978) Delimitation of the Cerrado concept. Vegetatio 36:169–178. https://doi.org/10.1007/Bf02342599
Eloy L, Schmidt IB, Borges SL, Ferreira MC, dos Santos TA (2019) Seasonal fire management by traditional cattle ranchers prevents the spread of wildfire in the Brazilian Cerrado. Ambio 48:890–899. https://doi.org/10.1007/s13280-018-1118-8
Engstrom RT (2010) First-order fire effects on animals: review and recommendations. Fire Ecol 6:115–130. https://doi.org/10.4996/Fireecology.0601115
Falleiro RM, Steil L, de Oliveira MS, Lando I, Machado LOR et al (2021) Histórico, avaliação, oportunidades e desafios do manejo integrado do fogo nas terras indígenas brasileiras. Biodivers Bras 11:75–98. https://doi.org/10.37002/biodiversidadebrasileira.v11i2.1742
FAN WCS (2021) Incendios forestales en Bolivia-Análisis de impactos de los incendios forestales sobre los valores de conservación en Bolivia, 2020. Fundación Amigos de la Naturaleza, Santa Cruz de la Sierra
Faria AS, Lima AP, Magnusson WE (2004) The effects of fire on behaviour and relative abundance of three lizard species in an Amazonian savanna. J Trop Ecol 20:591–594. https://doi.org/10.1017/S0266467404001798

Fichino BS, Dombroski JRG, Pivello VR, Fidelis A (2016) Does fire trigger seed germination in the Neotropical savannas? Experimental tests with six Cerrado species. Biotropica 48:181–187. https://doi.org/10.1111/btp.12276

Fidelis A, Blanco C (2014) Does fire induce flowering in Brazilian subtropical grasslands? Appl Veg Sci:17:690–17:699. https://doi.org/10.1111/avsc.12098

Fidelis A, Zirondi Hl (2021) And after fire, the Cerrado flowers: a review of post-fire flowering in a tropical savanna. Flora 280:151849. https://doi.org/10.1016/j.Flora.2021.151849

Fidelis A, Lyra MFS, Pivello VR (2013) Above- and below-ground biomass and carbon dynamics in Brazilian Cerrado wet grasslands. J Veg Sci 24:356–364. https://doi.org/10.1111/j.1654-1103.2012.01465.x

Fidelis A, St A, Acs B, Pivello VR (2018) The year 2017: megafires and management in the Cerrado. Fire 1:49. https://doi.org/10.3390/fire1030049

Fidelis A, Rosalem P, Zanzarini V, Camargos LS, Martins AR (2019) From ashes to flowers: a savanna sedge initiates flowers 24 h after fire. Ecology 100:e02648. https://doi.org/10.1002/ecy.2648

Fidelis A, Zirondi HL, Rossatto DR, Zanzarini V (2022) Fire stimulates grass flowering in the Cerrado independent of season. J Veg Sci 33:e13125. http://hdl.handle.net/11449/240886

Fiorillo BF, Tozetti AM, Martins M (2020) Habitat use by five species of sympatric pitvipers (*Bothrops*, *Crotalus*) in a Brazilian savannah. Herpetol Notes 13:951–960

Fiorillo BF, Maciel JH, Martins M (2021) Composition and natural history of a snake community from the southern Cerrado, southeastern Brazil. Zookeys 1056:95. https://doi.org/10.3897/Zookeys.1056.63733

Flores BM, Dechoum MS, Schmidt IB, Hirota M, Abrahão A et al (2020) Tropical riparian forests in danger from large savanna wildfires. J Appl Ecol 58:419–430. https://doi.org/10.1111/1365-2664.13794

Follador GLPP, Cassini RF, Varajão AFDC, Bittencourt JS (2023) Paleofires and vegetation in a Late Pleistocene paleolake (>43 ka bp) of the savannas of Central Brazil. J Quat Sci 38:1305–1312. https://doi.org/10.1002/jqs.3551

Fonseca SN, Moura CC, Farnesi MM, Pivello VR, Machado ELM (2022) Caracterização ambiental de campos úmidos submetidos a queima prescrita no Parque Nacional das Sempre-Vivas (PNSV), MG. Rev Bras Geogr Fis 15:3229–3243. https://doi.org/10.26848/Rbgf.v15.6.p3229-3243

Forzza RC, Baumgratz JFA, Bicudo CEM, Canhos DA, Carvalho-Jr AA et al (2012) New Brazilian floristic list highlights conservation challenges. Bioscience 62:39–45. https://doi.org/10.1525/bio.2012.62.1.8

França H, Ramos Neto MB, Setzer A (2007) O Fogo no Parque Nacional das Emas. MMA, Biodiversidade, p 27

Friend GR (1993) Impact of fire on small vertebrates in mallee woodlands and heathlands of temperate Australia: a review. Biol Conserv 65:99–114. https://doi.org/10.1016/0006-3207(93)90439-8

Frizzo TLM, Bonizário C, Borges MP, Vasconcelos HL (2011) Revisão dos efeitos do fogo sobre a fauna de formações savânicas do Brasil. Oecol Australis 15:365–379. https://doi.org/10.4257/Oeco.2011.1502.13

Frizzo TLM, Campos RI, Vasconcelos HL (2012) Contrasting effects of fire on arboreal and ground—dwelling ant communities of a neotropical savanna. Biotropica 44:254–261. https://doi.org/10.1111/j.1744-7429.2011.00797.x

Furley PA, Ratter JA (1988) Soil resources and plant communities of the Central Brazilian Cerrado and their development. J Biogeogr 15:97–108. https://doi.org/10.2307/2845050

Furtado LO, Felicio GR, Lemos PR, Christianini AV (2021) Winners and losers: how woody encroachment is changing the small mammal community structure in a Neotropical savanna. Front Ecol Evol 9:774744. https://doi.org/10.3389/Fevo.2021.774744

Fuzaro L, Xavier NL, Carvalho FJ, Silva FAN, Carvalho SM, Andaló V (2018) Influence of pollination on canola seed production in the Cerrado of Uberlândia, Minas Gerais State, Brazil. Acta Sci Agron 40:e39315. https://doi.org/10.4025/Actasciagron.v40i1.39315

Geiger EL, Gotsch SG, Damasco G, Haridasan M, Franco AC, Hoffmann WA (2011) Distinct roles of savanna and forest tree species in regeneration under fire suppression in a Brazilian savanna. J Veg Sci 22:312–321. https://doi.org/10.1111/j.1654-1103.2011.01252.x

Ghizoni IR, Layme VMG, Lima AP, Magnusson WE (2005) Spatially explicit population dynamics in a declining population of the tropical rodent Bolomys lasiurus. J Mammal 86:677–682. https://doi.org/10.1644/1545-1542(2005)086[0677:Sepdia]2.0.Co;2

Giacomo AG, Giacomo AJ, Reboreda JC (2011) Effects of grassland burning on reproductive success of globally threatened Strange-tailed Tyrants Alectrurus risora. Bird Cons Int 21:411–422. https://doi.org/10.1017/S0959270910000584

Gibb H, Parr CL (2010) How does habitat complexity affect ant foraging success? A test using functional measures on three continents. Oecologia 164:1061–1073. https://doi.org/10.1007/s00442-010-1703-4

Giorgis MA, Zeballos SR, Carbone L, Zimmermann H (2021) A review of fire effects across South American ecosystems: the role of climate and time since fire. Fire Ecol 17:11. https://doi.org/10.1186/s42408-021-00100-9

Goedert WJ (1987) Solos dos Cerrados. EMBRAPA/CPAC, Planaltina

Gomes L, Miranda HS, Bustamante MMC (2018) How can we advance the knowledge on the behavior and effects of fire in the Cerrado biome? For Ecol Manag 417:281–290. https://doi.org/10.1016/j.foreco.2018.02.032

Gomes L, Miranda HS, Soares-Filho B, Rodrigues L, Oliveira U, Bustamante MMC (2020) Responses of plant biomass in the Brazilian savanna to frequent fires. Front For Glob Chang 3:116. https://doi.org/10.3389/ffgc.2020.507710

Gomez Isaza DF, Cramp RL, Franklin CE (2022) Fire and rain: a systematic review of the impacts of wildfire and associated runoff on aquatic fauna. Glob Chang Biol. https://doi.org/10.1111/gcb.16088

González TM, González-Trujillo JD, Muñoz A, Armenteras D (2021) Differential effects of fire on the occupancy of small mammals in neotropical savanna-gallery forests. Perspect Ecol Conserv 19:179–188. https://doi.org/10.1016/j.Pecon.2021.03.005

González TM, González-Trujillo JD, Muñoz A, Armenteras D (2022) Effects of fire history on animal communities: a systematic review. Ecol Process 11:11. https://doi.org/10.1186/s13717-021-00357-7

Goodland R, Pollard R (1973) The Brazilian cerrado vegetation: a fertility gradient. J Ecol 61:219–224. https://doi.org/10.2307/2258929

Gorgone-Barbosa E, Pivello VR, Bautista S, Zupo T, Rissi MN, Fidelis A (2015) How can an invasive grass affect fire behavior in a tropical savanna? A community and individual plant level approach. Biol Invasions 17:423–431. https://doi.org/10.1007/s10530-014-0740-z

Gottsberger G, Silberbauer-Gottsberger I (2006) Life in the Cerrado, a South American tropical seasonal ecosystem—origin, structure, dynamics and plant use, vol 1. Reta Verlag, Ulm

Gross DR, Eiten G, Flowers NM, Leoi FM (1979) Ecology and acculturation among native peoples of Central Brazil. Science 206:1043–1050. https://doi.org/10.1126/Science.206.4422.1043

Guidon N, Delibrias G (1986) Carbon-14 dates point to man in the Americas 32,000 years ago. Nature 321:769e771. https://doi.org/10.1038/321769a0

Haase R (1992) Physical and chemical properties of savanna soils in Northern Bolivia. Catena 19:119–134. https://doi.org/10.1016/0341-8162(92)90020-C

Haridasan M (2008) Nutritional adaptations of native plants of the Cerrado biome in acid soils. Braz J Plant Physiol 20:183–195. https://doi.org/10.1590/S1677-04202008000300003

Henriques RPB, Briani DC, Palma ART, Vieira EM (2006) A simple graphical model of small mammal succession after fire in the Brazilian Cerrado. Mammalia 70:226–230. https://doi.org/10.1515/Mamm.2006.044

Hoffmann WA (1996) The effects of fire and cover on seedling establishment in a neotropical savanna. J Ecol 84:383–393. https://doi.org/10.2307/2261200

Hoffmann WA (1998) Post-burn reproduction of woody plants in a Neotropical savanna: the relative importance of sexual and vegetative reproduction. J Appl Ecol 35:422–433. https://doi.org/10.1046/j.1365-2664.1998.00321.x

Hoffmann WA (1999) Fire and population dynamics of woody plants in a Neotropical savanna: matrix model projections. Ecology 80:1354–1369. https://doi.org/10.2307/177080
Hoffmann WA, Geiger EL, Gotsch SG, Rossatto DR (2012) Ecological thresholds at the savanna-forest boundary: how plant traits, resources and fire govern the distribution of tropical biomes. Ecol Lett 15:759–768. https://doi.org/10.1111/j.1461-0248.2012.01789.x
Hofmann GS, Cardoso MF, Alves RJV et al (2021) The Brazilian Cerrado is becoming hotter and drier. Glob Chang Biol 27:4060–4073. https://doi.org/10.1111/Gcb.15712
IBGE- Instituto Brasileiro de Geografia e Estatística (2019) Biomas e Sistema Costeiro-Marinho do Brasil 1:250.000. IBGE/Coordenação de Recursos Naturais e Estudos Ambientais. https://geoftp.ibge.gov.br/informacoes_ambientais/estudos_ambientais/biomas/mapas/biomas_e_sistema_costeiro_marinho_250mil.pdf. Accessed 16 Apr 2023
Infante J, Novoa FJ, Ibarra JT, Melnick DJ, Griffin KL, Bonacic C (2021) Altered fire regimes modify lizard communities in globally endangered Araucaria forests of the southern Andes. Sci Rep 11:22709. https://doi.org/10.1038/s41598-021-02169-3
Jaramillo C (2023) The evolution of extant South American tropical biomes. New Phytol 239:477–493. https://doi.org/10.1111/nph.18931
Jolly CJ, Dickman CR, Doherty TS, van Eeden LM, Geary WL et al (2022) Animal mortality during fire. Glob Chang Biol 28:2053–2065. https://doi.org/10.1111/Gcb.16044
Kanegae MF, Levy G, Freitas SR (2012) Habitat use by sharp-tailed tyrant (*Culicivora caudacuta*), and cock-tailed tyrant (*Alectrurus tricolor*) in the Cerrado of southeastern Brazil. Rev Bras Ornitol 20:52–58
Kauffman JB, Cummings DL, Ward DE (1994) Relationships of fire, biomass and nutrient dynamics along a vegetation gradient in the Brazilian Cerrado. J Ecol 82:519–531. https://doi.org/10.2307/2261261
Keeley JE, Pausas JG, Rundel PW, Bond WJ, Bradstock RA (2011) Fire as an evolutionary pressure shaping plant traits. Trends Plant Sci 16:406–411. https://doi.org/10.1016/j.tplants.2011.04.002
Kirkland GL, Snoddy HW, Amsler TL (1996) Impact of fire on small mammals and amphibians in a central Appalachian deciduous forest. Am Midl Nat 135:253–260. https://doi.org/10.2307/2426707
Klink CA, Machado R (2005) A conservação do Cerrado brasileiro. Megadiversidade 1:147–155
Klink CA, Moreira A (2002) Past and current human occupation and land use. In: Oliveira PS, Marquis RJ (eds) The Cerrados of Brazil—ecology and natural history of a Neotropical savanna. Columbia Univ. Press, New York, pp 69–88
Klink CA, Sato MN, Cordeiro GG, Ramos MIM (2020) The role of vegetation on the dynamics of water and fire in the Cerrado ecosystems: implications for management and conservation. Plan Theory 9:1803. https://doi.org/10.3390/plants9121803
Lahaye C, Hernandez M, Boëda E, Felice GD, Guidon N et al (2013) Human occupation in South America by 20,000 BC: the Toca da Tira Peia site, Piauí, Brazil. J Archaeol Sci 40:2840–2847. https://doi.org/10.1016/j.Jas.2013.02.019
Laris P (2002) Burning the seasonal mosaic: preventative burning strategies in the wooded savanna of Southern Mali. Hum Ecol 30:155–186. https://doi.org/10.1023/A:1015685529180
Lazarina M, Sgardelis SP, Tscheulin T, Kallimanis AS, Devalez J, Petanidou T (2016) Bee response to fire regimes in Mediterranean pine forests: the role of nesting preference, trophic specialization, and body size. Basic Appl Ecol 17:308–320. https://doi.org/10.1016/j.Baae.2016.02.001
Le Stradic S, Roumet C, Durigan G, Cancian L, Fidelis A (2021) Variation in biomass allocation and root functional parameters in response to fire history in Brazilian savannas. J Ecol 109:4143–4157. https://doi.org/10.1111/1365-2745.13786
Ledru MP (2002) Late Quaternary history and evolution of the Cerrados as revealed by palynological records. In: The Cerrados of Brazil—ecology and natural history of a *Neotropical savanna*. Columbia Univ. Press, New York, pp 33–50
Lehmann CER, Anderson TM, Sankaran M, Higgins SI, Archibald S et al (2014) Savanna vegetation-fire-climate relationships differ among continents. Science 343:548–552. https://doi.org/10.1126/science.1247355

Lilienfein J, Wilcke W, Zimmermann R, Gerstberger P, Araújo G, Zech W (2001) Nutrient storage in soil and biomass of native Brazilian Cerrado. J Plant Nutr Soil Sci 164:487–495. https://doi.org/10.1002/1522-2624(200110)164:5<487::Aid-Jpln487>3.0.Co;2-I
Lima JEFW, Silva EM (2008) Recursos hídricos do bioma Cerrado- importância e situação. In: Sano SM, Almeida SP, Ribeiro JF (eds) Cerrado: ecologia e flora. Embrapa, Brasília, pp 89–106
Lima JM, Castro AB, Lima AP, Magnusson WE, Landeiro VL, Fadini RF (2020) Influence of the fire regime on the species richness and floristic composition of an isolated Amazonian savanna - PELD oeste do Pará. Oecologia Aust 24:301–316. https://doi.org/10.4257/oeco.2020.2402.06
Lira-Martins D, Nascimento DL, Abrahão A, Costa PB, D'Angioli AM et al (2022) Soil properties and geomorphic processes influence vegetation composition, structure, and function in the Cerrado Domain. Plant Soil 476:549–588. https://doi.org/10.1007/s11104-022-05517-y
Loeb SC, Blakey RV (2021) Bats and fire: a global review. Fire Ecol. https://doi.org/10.1186/s42408-021-00109-0
Lopes LE, Malacco GB, Alteff EF, Vasconcelos MF, Hoffmann D, Silveeira LF (2010) Range extensions and conservation of some threatened or little known Brazilian grassland birds. Bird Cons Int 20:84–94. https://doi.org/10.1017/S0959270909990190
Lopes CT, Vasconcelos HL (2011) Fire increases insect herbivory in a Neotropical savanna. Biotropica 43:612–618. https://doi.org/10.1111/j.1744-7429.2011.00757.x
Lopes LE, Meireles RC, Peixoto HJC, Teixeira JPG, Macgado TLSS, Lombardi VT (2023) Movement ecology of the threatened Campo Miner Geositta poeciloptera and its implications for the conservation of tropical open grassland birds. Bird Cons Int 33 e38:1-11. https://doi.org/10.1017/S0959270922000417
Lucindo AS, Antunes AZ, Kanashiro MM, Dias MM (2015) Birds at Santa Bárbara Ecological Station, one of the last Cerrado remnants in the state of São Paulo, Brazil. Biota Neotrop 15:e0155. https://doi.org/10.1590/1676-06032015015514
Machida WS, Gomes L, Moser P, Castro IP (2021) Long term post-fire recovery of woody plants in savannas of Central Brazil. For Ecol Manag 493:119255. https://doi.org/10.1016/j.foreco.2021.119255
Maillard O, Anívarro R, Vides-Almonacid R, Torres W (2018) Estado de conservación de los ecosistemas de las serranías chiquitanas: un caso de estudio de la Lista Roja de Ecosistemas de la UICN en Bolivia. Ecol Boliv 53:128–149
Maillard O, Herzog SK, Soria-Auza RW, Vides-Almonacid R (2022) Impact of fires on key biodiversity areas (Kbas) and priority bird species for conservation in Bolivia. Fire 5:4. https://doi.org/10.3390/Fire5010004
Maravalhas J, Vasconcelos HL (2014) Revisiting the pyrodiversity–biodiversity hypothesis: long-term fire regimes and the structure of ant communities in a Neotropical savanna hotspot. J Appl Ecol 51:1661–1668. https://doi.org/10.1111/1365-2664.12338
Marini-Filho OJ (2000) Distance-limited recolonization of burned Cerrado by leaf-miners and gallers in Central Brazil. Environ Entomol 29:901–906. https://doi.org/10.1603/0046-225X-29.5.901
Marques NCS, Machado RB, Aguiar LMS, Mendonça-Galvão L, Tidon R et al (2022) Drivers of change in tropical protected areas: long-term monitoring of a Brazilian biodiversity hotspot. Perspect Ecol Conserv 20:69–78. https://doi.org/10.1016/j.pecon.2022.02.001
Martins LC, Paiva EAS (2016) Flammable resin in *Vellozia variabilis* (Velloziaceae): gland structure and chemical composition. Flora 219:94–100. https://doi.org/10.1016/j.flora.2016.01.004
Medeiros, MB (2002) Efeitos do fogo nos padrões de rebrotamento em plantas lenhosas, em campo sujo. PhD Thesis, Universidade de Brasília. 120 p
Medeiros MB, Miranda HS (2005) Mortalidade pós-fogo em espécies lenhosas de campo sujo submetido a três queimadas prescritas anuais. Acta Bot Brasilica 19:493–500. https://doi.org/10.1590/S0102-33062005000300009
Mendonça AF, Armond T, Camargo ACL, Camargo NF, Ribeiro JF et al (2015) Effects of an extensive fire on arboreal small mammal populations in a Neotropical savanna woodland. J Mammal 96:368–379. https://doi.org/10.1093/Jmammal/Gyv038

Mereles F (2007) La diversidad vegetal en el Paraguay. In: Salas-Dueñas DA, Facetti JF (eds) Biodiversidad del Paraguay: una aproximación a sus realidades. 1° Ed. Fundación Moisés Bertoni. USAID,GEF/BM. 89–100

Miranda AC, Miranda HS, Dias IDFO, Dias BFS (1993) Soil and air temperatures during prescribed cerated fires in Central Brazil. J Trop Ecol 9:313–320. https://doi.org/10.1017/S0266467400007367

Miranda H, Rocha e Silva E, Miranda A (1996) Comportamento do fogo em queimadas de campo sujo. In Impactos de Queimadas Em Áreas de Cerrado e Restinga, eds. H Miranda, C Saito, B Dias, pp. 1–10. Brasília: Universidade de Brasília

Miranda HS, Sato MN, Nascimento-Neto W, Aires FS (2009) Fires in the Cerrado, the Brazilian savanna. In: Cochrane MA (ed) Tropical fire ecology—climate change, land use, and ecosystem dynamics. Springer, Chichester, pp 427–450

Mistry J, Berardi A, Andrade V, Krahô T, Krahô P, Leonardos O (2005) Indigenous fire management in the Cerrado of Brazil: the case of the Krahô of Tocantins. Hum Ecol 33:4141–4148. https://doi.org/10.1007/s10745-005-4143-8

Mokany K, Raison RJ, Prokushkin AS (2006) Critical analysis of root:shoot ratios in terrestrial biomes. Glob Change Biol 12:84–96. https://doi.org/10.1111/j.1365-2486.2005.001043.x

Molinas-González CR, Florentín P (2021) Incendios forestales en Paraguay: implicaciones para su prevención y manejo. Rev Investy Est(UNA) 12:47–57. https://doi.org/10.47133/IEUNA2115

Montgomery RF, Askew GP (1983) Soils of tropical savannas. In: Bourlière F (ed) Tropical savannas. Elsevier, Amsterdam, pp 63–78

Morais HC, Benson WW (1988) Recolonização de vegetação de Cerrado após queimada, por formigas arborícolas. Rev Brasil Biol 48:459–466

Morais AR, Côrtes LG, Bastos RP (2011) Queimadas podem alterar as assembléias de anuros? O caso das veredas na Estação Ecológica Serra Geral do Tocantins. Rev Biol Neotrop 8:32–39. https://doi.org/10.5216/rbn.v8i2.16856

Moreira LA, Fenolio DB, Silva HLR, da Silva NJ (2009) A preliminary list of the Herpetofauna from termite mounds of the cerrado in the Upper Tocantins river valley. Pap Avulsos Zool 49:183–189. https://doi.org/10.1590/s0031-10492009001500001

Moreira B, Tormo J, Estrelles E, Pausas JG (2010) Disentangling the role of heat and smoke as germination cues in Mediterranean Basin flora. Ann Bot 105:627–635. https://doi.org/10.1093/aob/mcq017

Moretti M, Obrist MK, Duelli P (2004) Arthropod biodiversity after forest fires: winners and losers in the winter fire regime of the southern Alps. Ecography 27:173–186. https://doi.org/10.1111/j.0906-7590.2004.03660.x

Motta GS, Pilon N, Fidelis A, Kolb RM (2024) Smoke effects on the germination of Cerrado species. Plant Ecol 225:1–9. https://link.springer.com/article/10.1007/s11258-024-01427-4

Motta-Junior JC, Granzinolli MAM, Develey PF (2008) Aves da Estação Ecológica de Itirapina, estado de São Paulo, Brasi. Biota Neotrop 8(3):207–227. https://doi.org/10.1590/S1676-06032008000300019

Nascente JC, Ferreira ME, Nunes GM (2022) Integrated fire management as a renewing agent of native vegetation and inhibitor of invasive plants in Vereda habitats: diagnosis by remotely piloted aircraft systems. Remote Sens 14:1040. https://doi.org/10.3390/Rs14041040

Nascimento DTF, Novais GT (2020) Clima do Cerrado: dinâmica atmosférica e características, variabilidades e tipologias climáticas. Élisée Rev Geo 9:e922021

Nascimento AR, Malinov IK, Freire G, Freitas AVL, Diniz IR (2020) The temporal dynamics of two *Morpho fabricius*, 1807 species (Lepidoptera: Nymphalidae) are affected differently by fire in the Brazilian savanna. Environ Entomol 49:1449–1454. https://doi.org/10.1093/Ee/Nvaa128

Newberry BM, Power CR, Abreu RCR, Durigan G, Rossatto DR, Hoffmann WA (2020) Flammability thresholds or flammability gradients? Determinants of fire across savanna–forest transitions. New Phytol 228:910–921. https://doi.org/10.1111/Nph.16742

Nunes BA, Cunha A, Vieira EM (2023) Effects of a large wildfire on the community composition of medium and large mammals in a Neotropical savannah. Int J Wildland Fire. https://doi.org/10.1071/Wf22023

Oliveira HFM, Aguiar LMS (2015) The response of bats (Mammalia: Chiroptera) to an incidental fire on a gallery forest at a Neotropical savanna. Biota Neotrop 15:e–0091. https://doi.org/10.1590/1676-0611-Bn-2015-0091

Oliveira MNS, Dias BAS, Andrade GC, Tanaka MK (2015) Harvest times of *Comanthera elegans*, a worldwide traded Brazilian species of everlasting flower: implications on seed production, germination, and on species management. Braz J Bot 38:795–808. https://doi.org/10.1007/s40415-015-0179-1

Oliveira U, Soares-Filho B, Rodrigues H, Figueira D, Gomes L et al (2023) A near real-time web-system for predicting fire spread across the Cerrado biome. Sci Rep 13:4829. https://doi.org/10.1038/s41598-023-30560-9

Oliveira-Filho AT (1992a) Floodplain "murundus" of Central Brazil: evidence for the termite-origin hypothesis. J Trop Ecol 8:1–19. https://doi.org/10.1017/S0266467400006027

Oliveira-Filho AT (1992b) The vegetation of Brazilian "murundus": the Island-effect on the plant community. J Trop Ecol 8:465–486. https://doi.org/10.1017/S0266467400006817

Oliveira-Filho AT, Ratter JA (2002) Vegetation physiognomies and woody flora of the Cerrado biome. In: Oliveira PS, Marquis RJ (eds) The Cerrados of Brazil: ecology and natural history of a Neotropical savanna. Columbia University Press, New York, pp 91–120

Oliveira-Filho EC, Brito DQ, Dias ZMB, Guarieiro M, Carvalho EL et al (2018) Effects of ashes from a Brazilian savanna wildfire on water, soil and biota: an ecotoxicological approach. Sci Total Environ 618:101–111. https://doi.org/10.1016/j.Scitotenv.2017.11.051

Olson DM, Dinerstein E, Wikramanayake ED, Burgess ND, Powell GVN et al (2001) Terrestrial ecoregions of the world: a new map of life on Earth. Bioscience 51:933–938. https://doi.org/10.1641/0006-3568(2001)051[0933:TEOTWA]2.0.CO;2

Ørsted M, Jørgensen LB, Overgaard J (2022) Finding the right thermal limit: a framework to reconcile ecological, physiological and methodological aspects of CTmax in ectotherms. J Exp Biol 225: jeb244514. https://doi.org/10.1242/jeb.244514

Overbeck GE, Vélez-Martin E, Menezes LS, Anand M (2022) Placing Brazil's grasslands and savannas on the map of science and conservation. Perspect Plant Ecol Evol Syst 56:125687. https://doi.org/10.1016/j.ppees.2022.125687

Owen RD (2013) Ecology of small terrestrial mammals in an isolated Cerrado patch, eastern Paraguay: communities, species, and effects of ENSO, precipitation, and fire. Mastozool Neotrop 20:97–112

Pacheco S, Ribon R, Silva NF, Simon JE, Pinheiro RT (1994) Efeito do manejo do cerrado sobre as populações de alguns Tinamidae em Três Marias, estado de Minas Gerais. Rev Bras Biol 54:435–441

Paraguay (2010) Plan Nacional de Manejo Integrado del Fuego: 2012–2016. USAID /OFDA /LAC

Parker TA, Willis EO (1997) Notes on three tiny grassland flycatchers, with comments on the disappearance of South American fire-diversified savannas. Ornithol Monogr. https://doi.org/10.2307/40157552

Pausas JG (2019) Generalized fire response strategies in plants and animals. Oikos 128:147–153. https://doi.org/10.1111/Oik.05907

Pausas JG, Bond WJ (2020) Alternative biome states in terrestrial ecosystems. Trends Plant Sci 25:250–263. https://doi.org/10.1016/j.tplants.2019.11.003

Pausas JG, Parr CL (2018) Towards an understanding of the evolutionary role of fire in animals. Evol Ecol 32:113–125. https://doi.org/10.1007/s10682-018-9927-6

Pausas JG, Lamont BB, Paula S, Appezzato-da-Glória B, Fidelis A (2018) Unearthing belowground bud banks in fire-prone ecosystems. New Phytol 217:1435–1448. https://doi.org/10.1111/nph.14982

Pellegrini AFA, Ahlström A, Hobbie SE, Reich PB (2018) Fire frequency drives decadal changes in soil carbon and nitrogen and ecosystem productivity. Nature 553:194–198. https://doi.org/10.1038/Nature24668

Peralta G, Stevani EL, Chacoff NP, Dorado J, Vázquez DP (2017) Fire influences the structure of plant–bee networks. J Anim Ecol 82:1372–1379. https://doi.org/10.1111/1365-2656.12731

Pessenda LCR, Gomes BM, Aravena R, Ribeiro AS, Boulet R, Gouveia SEM (1998) The carbon isotope record in soils along a forest-cerrado ecosystem transect: implications for vegetation changes in the Rondonia state, southwestern Brazilian Amazon region. The Holocene 8:599–603. https://doi.org/10.1191/095968398673187182
Pilon NAL, Hoffmann WA, Abreu RCR, Durigan G (2018) Quantifying the short-term flowering after fire in some plant communities of a cerrado grassland. Plant Ecol Divers 11:259–266. https://doi.org/10.1080/17550874.2018.1517396
Pilon NAL, Cava MGB, Hoffmann WA, Abreu RCR (2021) The diversity of post-fire regeneration strategies in the cerrado ground layer. J Ecol 109:154–166. https://doi.org/10.1111/1365-2745.13456
Pilon NAL, Freire CTR, Oliveira-Alves MJ, Oliveira RS (2023) Speedy blooming in Cerrado after fire is not uncommon: new records of Cyperaceae species flowering 24 h after burning. Austral Ecol 48(5). https://doi.org/10.1111/aec.13326
Pinheiro E, Durigan G (2009) Dinâmica espaço-temporal (1962-2006) das fitofisionomias em unidade de conservação do Cerrado no sudeste do Brasil. Rev Bras Botânica 3:441–454
Pinho FF, Ferreira GB, Paglia AP (2017) Influence of vegetation physiognomy, elevation and fire frequency on medium and large mammals in two protected areas of the Espinhaço Range. Fortschr Zool 34:1–11. https://doi.org/10.3897/zoologia.34.e11921
Pinter N, Fiedel S, Keeley JE (2011) Fire and vegetation shifts in the Americas at the vanguard of Paleoindian migration. Quat Sci Rev 30:269–272. https://doi.org/10.1016/j.Quascirev.2010.12.010
Pivello VR (2011) The use of fire in the Cerrado and Amazonian rainforests of Brazil: past and present. Fire Ecol 7:24–39. https://doi.org/10.4996/Fireecology.0701024
Pivello VR, Coutinho LM (1992) Transfer of macro-nutrients to the atmosphere during experimental burnings in an open Cerrado (Brazilian savanna). J Trop Ecol 8:487–497. https://doi.org/10.1017/S0266467400006829
Pivello VR, Norton GR (1996) Firetool: an expert system for the use of prescribed fires in Brazilian savannas. J Appl Ecol 33:348–356. https://doi.org/10.2307/2404756
Pivello VR, Shida CN, Meirelles ST (1999) Alien grasses in Brazilian savannas: a threat to the biodiversity. Biodivers Conserv 8:1281–1294. https://doi.org/10.1023/A:1008933305857
Pivello VR, Oliveras I, Miranda HS, Mundayatan H, Sato MN, Meirelles ST (2010) Effect of fires on soil nutrient availability in an open savanna in Central Brazil. Plant and Soil 337:111-123. https://doi.org/10.1007/s11104-010-0508-x
Pivello VR, Vieira I, Christianini AV, Ribeiro DB, Menezes LS et al (2021) Understanding Brazil's catastrophic fires: causes, consequences and policy needed to prevent future tragedies. Perspect Ecol Conserv 19:233–255. https://doi.org/10.1016/j.pecon.2021.06.005
Prada M (2001) Effects of fire on the abundance of large mammalian herbivores in Mato Grosso, Brazil. Mammalia 65:55–62. https://doi.org/10.1515/Mamm.2001.65.1.55
Prada M, Marinho-Filho J (2004) Effects of fire on the abundance of Xenarthrans in Mato Grosso, Brazil. Austral Ecol 29:568–573. https://doi.org/10.1111/j.1442-9993.2004.01391.x
Prada M, Marini-Filho OJ, Price PW (1995) Insects in flower heads of *Aspilia foliacea* (Asteraceae) after a fire in a Central Brazilian savanna: evidence for the plant vigor hypothesis. Biotropica 27:513–518. https://doi.org/10.2307/2388965
Prance GT (1996) Islands in Amazonia. Phil Trans R Soc Lond B 351:823–833. https://doi.org/10.1098/rstb.1996.0077
Prous A (1991) Arqueologia brasileira. Brasília, Ed. Univ. de Brasília, Brasília, Brazil
Queiroz-Neto P (1982) Solos da região dos cerrados e suas interpretações. Rev Bras Cienc Solo 6:1–12
Rachid-Edwards M (1956) Alguns dispositivos para proteção de plantas contra a seca eo fogo. Bol Fac Filos Ciênc Let Univ São Paulo, Botânica: 35–68
Ramos DM, Liaffa ABS, Diniz P, Munhoz CBR (2016) Seed tolerance to heating is better predicted by seed dormancy than by habitat type in Neotropical savanna grasses. Int J Wildl Fire 25:1273–1280. https://doi.org/10.1071/WF16085

Ramos DL, Bustamante MMC, Silva e Silva FD, Carvalheiro LG (2018) Crop fertilization affects pollination service provision—common bean as a case study. PLoS One. https://doi.org/10.1371/Journal.Pone.0204460

Ramos-Neto MB, Pivello VR (2000) Lightning fires in a Brazilian savanna national park: rethinking management strategies. Environ Manag 26:675–684. https://doi.org/10.1007/s002670010124

Rawitscher F (1948) The water economy of the vegetation of the 'Campos cerrados' in southern Brazil. J Ecol 36:237–268. https://doi.org/10.2307/2256669

Reatto A, Correia JR, Spera ST (2008) Solos do bioma Cerrado: aspectos pedológicos. In: Sano SM, Almeida SP, Ribeiro JF (orgs.) Cerrado: ecologia e flora. Embrapa, Brasília, pp 107–149

Recalde FC, Breviglieri CP, Kersch-Becker MF, Romero GQ (2021) Contribution of emergent aquatic insects to the trophic variation of tropical birds and bats. Food Webs 29: e00209. https://doi.org/10.1016/j.fooweb.2021.e00209

Rede Cerrado (2024) Pelos povos e comunidades tradicionais. https://redecerrado.org.br/nossa-atuacao/defesa-de-povos-e-comunidades-tradicionais/#:~:text=S%C3%A3o%20mais%20de%2080%20etnias,%2C%20ciganos%2C%20entre%20tantos%20outros. Accessed 2 Apr 2024

Rede de Sementes do Cerrado (2002) Mercado de sementes e restauração: Provendo serviços ambientais e biodiversidade.do Cerrado. Relatório 2018–2021. https://www.rsc.org.br/files/projeto-mercado-de-sementes-e-restauracao-2018-2021.pdf. Accessed 2 Apr 2024

Reis AF, Schmiele M (2019) Characteristics and potentialities of Savanna fruits in the food industry. Braz J Food Technol. https://doi.org/10.1590/1981-6723.15017

Reis MG, Fieker CZ, Dias MM (2016) The influence of fire on the assemblage structure of foraging birds in grasslands of the Serra da Canastra National Park, Brazil. Ann Acad Bras Cienc 88:891–901. https://doi.org/10.1590/0001-3765201620150177

Ribeiro JF, Walter BMT (2008) As principais fitofisionomias do bioma Cerrado. In: Sano SM, Almeida SP de, Ribeiro JF (orgs.). Cerrado: ecologia e flora. Embrapa, Brasília, pp 151–212

Ribeiro J, Colli GR, Soares AMVM (2020) The anurofauna of a vanishing savanna: the case of the Brazilian Cerrado. Biodivers Conserv 29:1993–2015. https://doi.org/10.1007/s10531-017-1468-8

Rissi MN, Baeza MJ, Gorgone-Barbosa E, Zupo T, Fidelis A (2017) Does season affect fire behaviour in the Cerrado? Int J Wildland Fire 26:427–433. https://doi.org/10.1071/WF14210

Rizzini CT (1997) Tratado de fitogeografia do Brasil: aspectos ecológicos, sociológicos e florísticos. Âmbito Cultural Edições, Rio de Janeiro

Rocha ABA, Barbosa VRF, Faria KMS, Martins ES, Soares Neto GB (2022) Geomorphologic map of the Brazilian Cerrado by geomorphometric archetypes. Rev Bras Geomorfol 23:1674–1684. https://doi.org/10.20502/rbg.v23i3.2132

Rodrigues CA, Fidelis A (2022) Should we burn the Cerrado? Effects of fire frequency on open savanna plant communities. J Veg Sci 33:e13159. https://doi.org/10.1111/jvs.13159

Rodrigues CA, Zirondi HL, Fidelis A (2021) Fire frequency affects fire behavior in open savannas of the Cerrado. For Ecol Manag 482:118850. https://doi.org/10.1016/j.Foreco.2020.118850

Rodriguez-Montellano AM (2014) Incendios y quemas en Bolivia, análisis histórico desde 2000 a 2013. FAN-Fundación Amigos de la Naturaleza, Santa Cruz de la Sierra

Roitman I, Felfili JM, Rezende AV (2008) Tree dynamics of a fire-protected cerrado sensu stricto surrounded by forest plantations, over a 13-year period (1991–2004) in Bahia, Brazil. Plant Ecol 197:255–267. https://doi.org/10.1007/s11258-007-9375-9

Rosa TF, Camarota F, Zuanon LA, Tito R, Maravalhas JB et al (2021) The effects of high-severity fires on the arboreal ant community of a Neotropical savanna. Oecologia 196:951–961. https://doi.org/10.1007/s00442-021-04922-x

Rossi RD, Martins CR, Viana PL, Rodrigues EL, Figueira JEC (2014) Impact of invasion by molasses grass (Melinis minutiflora P. Beauv.) on native species and on fires in areas of campo-cerrado in Brazil. Acta Bot Brasilica 28:631–637. https://doi.org/10.1590/0102-33062014abb3390

Ruggiero PGC, Pivello VR, Sparovek G, Teramoto E, Pires-Neto AG (2006) Relação entre solo, vegetação e topografia em área de cerrado (Parque Estadual de Vassununga, SP): como se expressa em mapeamentos? Acta Bot Brasilica 20:383–394. https://doi.org/10.1590/S0102-33062006000200013

Saint-Hilaire AD (1937) Viagem Às Nascentes Do Rio São Francisco e Pela Província de Goyaz. Companhia Editora Nacional, São Paulo

Salgado AAR, Bueno GT, Diniz AD, Marent B (2015) Long-term geomorphological evolution of the Brazilian territory. In: Vieira BC, Salgado AAR, Santos LJC (eds) Landscapes and landforms of Brazil. Springer, Dordrecht, pp 19–31

Salgado-Labouriau ML (2005) Alguns aspectos sobre a paleoecologia dos Cerrados. In: Scariot A, Sousa-Silva JC, Felfili JM (orgs.) Cerrado: ecologia, biodiversidade e conservação. Ministério do Meio Ambiente, Brasília, pp 107–118

Sanaiotti TM, Magnusson WE (1995) Effects of annual fires on the production of fleshy fruits eaten by birds in a Brazilian Amazonian savanna. J Trop Ecol 11:53–65. https://doi.org/10.1017/S0266467400008397

Sanaiotti TM, Martinelli LA, Victoria RL, Trumbore SE, Camargo PB (2002) Past vegetation changes in Amazon savannas determined using carbon isotopes of soil organic matter. Biotropica 34:2–16. https://doi.org/10.1111/j.1744-7429.2002.Tb00237.x

Sano EE, Rosa R, Brito JLS, Ferreira LG (2010) Land cover mapping of the tropical savanna region in Brazil. Environ Monit Assess 166:113–124. https://doi.org/10.1007/s10661-009-0988-4

Santos AJB, Silva GTDA, Miranda HS, Miranda AC, Lloyd J (2003) Effects of fire on surface carbon, energy and water vapour fluxes over campo sujo savanna in central Brazil. Funct. Ecol. 17:711–19. https://doi.org/10.1111/j.1365-2435.2003.00790.x

Santos RC (2013) Efeitos do fogo sobre a estrutura de comunidades de abelhas e sobre redes de interações abelhas-planta em áreas de canga no Parque Estadual do Rola Moça. Universidade Federal de Minas Gerais, Brazil

Santos FLM, Nogueira J, Souza RAF, Falleiro RM, Schmidt IB, Libonati R (2021) Prescribed burning reduces large, high-intensity wildfires and emissions in the Brazilian savanna. Fire 4:56. https://doi.org/10.3390/Fire4030056

Santos JL, Sitters H, Keith DA, Geary WL, Tingley R, Kelly LT (2022) A demographic framework for understanding fire-driven reptile declines in the 'land of the lizards'. Glob Ecol Biogeogr 31:2105–2119. https://doi.org/10.1111/geb.13520

Sato MN, Miranda HS, Maia JMF (2010) O fogo e o estrato arbóreo do Cerrado: efeitos imediatos e longo prazo. In: Miranda SM (org.). Efeitos do regime do fogo sobre a estrutura de comunidades de cerrado: Resultados do Projeto Fogo. IBAMA/MMA, Brasília, Brazil, pp: 77–91

Schmidt IB, Fidelis A, Miranda HS, Ticktin T (2017) How do the wets burn? Fire behavior and intensity in wet grasslands in the Brazilian savanna. Brazilian J. Bot. 40:167–75. https://doi.org/10.1007/s40415-016-0330-7

Schmidt IB, Eloy L (2020) Fire regime in the Brazilian savanna: recent changes, policy and management. Flora 268:151613. https://doi.org/10.1016/j.Flora.2020.151613

Schmidt IB, Sampaio MB, Figueiredo IB, Ticktin T (2011) Fogo e artesanato de capim-dourado no Jalapão—usos tradicionais e consequências ecológicas. Biodivers Bras 1:67–85. https://doi.org/10.37002/biodiversidadebrasileira.v1i2.116

Schmidt IB, Moura LC, Ferreira MC, Eloy L Sampaio AB et al (2018) Fire management in the Brazilian savanna: first steps and the way forward. J Appl Ecol 2094–2101. doi:https://doi.org/10.1111/1365-2664.13118

Silva FAM, Assad ED, Evangelista BA (2008) Caracterização climática do bioma cerrado. In: Sano SM, Almeida SP de, Ribeiro JF (orgs.). Cerrado: ecologia e flora. Embrapa, Brasília, pp 69–88

Silva PS, Bastos A, Libonati R, Rodrigues JA, DaCamara CC (2019) Impacts of the 1.5°C global warming target on future burned area in the Brazilian Cerrado. For Ecol Manag 446:193–203. https://doi.org/10.1016/j.Foreco.2019.05.047

Silva PS, Nogueira J, Rodrigues JA, Santos FLM, Pereira JMC et al (2021) Putting fire on the map of Brazilian savanna ecoregions. J Environ Manag 296:113098. https://doi.org/10.1016/j.jenvman.2021.113098

Silveira L, Rodrigues FHG, Jácomo ATA, Diniz Filho JA (1999) Impact of wildfires on the megafauna of Emas National Park, Central Brazil. Oryx 33:108–114. https://doi.org/10.1046/j.1365-3008.1999.00039.x

Simon M, Pennington T (2012) Evidence for adaptation to fire regimes in the tropical savannas of the Brazilian Cerrado. Int J Plant Sci 173:711–723. https://doi.org/10.1086/665973

Simon MF, Grether R, Queiroz LP, Skema C, Pennington RT, Hughes CE (2009) Recent assembly of the Cerrado, a neotropical plant diversity hotspot, by *in situ* evolution of adaptations to fire. PNAS 106:20359–20364. https://doi.org/10.1073/Pnas.0903410106

Simpson KJ, Ripley BS, Christin PA, Belcher CM (2016) Determinants of flammability in savanna grass species. J Ecol 104:138–148. https://doi.org/10.1111/1365-2745.12503

Sousa HC, Soares AHS, Costa BM, Pantoja DL, Caetano GH, de Queiroz TA, Colli GR (2015) Fire regimes and the demography of the lizard *Micrablepharus atticolus* (Squamata, Gymnophthalmidae) in a biodiversity hotspot. S Am J Herpetol 10:143–156. https://doi.org/10.2994/SAJH-D-15-00011.1

Souza E, Lima AP, Magnusson WE, Kawashita-Ribeiro RA, Fadini RF et al (2021) Short- and long-term effects of fire and vegetation cover on four lizard species in Amazonian savannas. Can J Zool 99:173–182. https://doi.org/10.1139/Cjz-2020-0224

Souza CV, Lourenço A, Vieira EM (2023a) Species-specific responses of medium and large mammals to fire regime attributes in a fire-prone Neotropical savanna. Fire 6:110. https://doi.org/10.3390/Fire6030110

Souza PCG, Mota SLL, Corrêa RS, Pinto JRR (2023b) How substrate type and elevation drive woody communities and influence species ecological strategies in the Brazilian savanna at local scale. Ecol Res 38:815–827. https://doi.org/10.1111/1440-1703.12413

Spriggs EL, Christin PA, Edwards EJ (2014) C4 photosynthesis promoted species diversification during the Miocene grassland expansion. PLoS One 9:e105923. https://doi.org/10.1371/journal.pone.0105923

Stevens N, Lehmann CER, Murphy BP, Durigan G (2017) Savanna woody encroachment is widespread across three continents. Glob Chang Biol 23:235–244. https://doi.org/10.1111/gcb.13409

Thomé MTC, Martins M, Brasileiro CA (2021) Higher diversity in open Cerrado supports the role of regional processes in shaping an anuran assemblage in southeastern Brazil. South Am J Herpetol 19:1–7. https://doi.org/10.2994/Sajh-D-18-00039.1

Tomas WM, Berlinck CN, Chiaravalloti RM, Paganini G, Faggioni CS et al (2021) Distance sampling surveys reveal 17 million vertebrates directly killed by the 2020's wildfires in the Pantanal, Brazil. Sci Rep 11:23547. https://doi.org/10.1038/s41598-021-02844-5

Tomas WM, Baggio R, Berlinck CN, Camilo AR, Cunha CN et al (2024) Área de Reserva Legal em ecossistemas campestres e savânicos: desafios de manejo e conservação frente às mudanças climáticas globais. Pesq Agropecu Bras 59:e03491. https://doi.org/10.1590/S1678-3921.pab2024.v59.03491p

Tubelis DP (2019) Fire management and aspects of the nesting biology of the red-legged Seriema (*Cariama cristata*) in woodlands at Parque Nacional das Emas, central Cerrado. Rev Bras Ornitol 27:230–237. https://doi.org/10.1007/Bf03546067

Tubelis DP, Delitti WBC (2010) Fire management and the nesting of *Athene cunicularia* (Aves, Strigidae) in grasslands in central Cerrado, Brazil. Biota Neotrop. https://doi.org/10.1590/s1676-06032010000200012

Uehara-Prado M, Bello AM, Fernandes JO, Santos AJ, Silva IA, Ciancaruso MV (2010) Abundance of epigaeic arthropods in a Brazilian savanna under different fire frequencies. Fortschr Zool 27:718–724. https://doi.org/10.1590/S1984-46702010000500008

Valdujo PH, Nogueira CC, Martins M (2002) Ecology of Bothrops neuwiedi pauloensis (Serpentes: Viperidae: Crotalinae) in the Brazilian Cerrado. J Herpetol 36:169. https://doi.org/10.1670/0022-1511(2002)036[0169:EOBNPS]2.0.CO;2

Van Der Hammen T (1991) Palaeoecology of the Neotropics: an overview of the state of affairs. Bol IG–USP 8:35–55. https://doi.org/10.11606/Issn.2317-8078.v0i8p35-55

Van Der Hammen T, Hooghiemstra H (2000) Neogene and Quaternary history of vegetation, climate, and plant diversity in Amazonia. Quat Sci Rev 19:725–742. https://doi.org/10.1016/S0277-3791(99)00024-4
Vasconcelos HL, Pacheco R, Silva RC, Vasconcelos PB, Lopes CT et al (2009) Dynamics of the leaf-litter arthropod fauna following fire in a neotropical woodland savanna. PLoS One 4:e7762. https://doi.org/10.1371/Journal.Pone.0007762
Vasconcelos H, Maravalhas JB, Cornelissen T (2017) Effects of fire disturbance on ant abundance and diversity: a global meta-analysis. Biodivers Conserv 26:177–188. https://doi.org/10.1007/s10531-016-1234-3
Veloso HP (1966) Atlas florestal do Brasil. Ministério da Agricultura, Rio de Janeiro
Vieira EM (1999) Small mammal communities and fire in the Brazilian Cerrado. J Zool 249:75–81. https://doi.org/10.1017/S0952836999009073
Vieira EM, Briani DC (2013) Short-term effects of fire on small rodents in the Brazilian Cerrado and their relation with feeding habits. Int J Wildland Fire 22:1063–1071. https://doi.org/10.1071/Wf12153
Vieira EM, Marinho-Filho J (1998) Pre- and post-fire habitat utilization by rodents of Cerrado from Central Brazil. Biotropica 30:491–496. https://doi.org/10.1111/j.1744-7429.1998.Tb00086.x
Vieira EM, Andrade I, Price PW (1996) Fire effects on a *Palicourea rigida* (Rubiaceae) gall midge: a test of the plant vigor hypothesis. Biotropica 28:210–217. https://doi.org/10.2307/2389075
Vieira EM, Iob G, Briani DC, Palma ART (2005) Microhabitat selection and daily movements of two rodents (*Necromys lasiurus* and *Oryzomys scotti*) in Brazilian Cerrado, as revealed by a spool-and-line device. Mamm Biol 70:359–365. https://doi.org/10.1016/j.Mambio.2005.08.002
Villarroel D, Munhoz CBR, Proença CEB (2016) Campos y sabanas del Cerrado en Bolivia: delimitación, síntesis terminológica y sus características fisionómicas. Kempffiana 12:47–80
Vitt LJ, Caldwell JP (1993) Ecological observations on Cerrado lizards in Rondônia. Brazil J Herpetol 27:46. https://doi.org/10.2307/1564904
Ward DE, Susott RA, Kauffman JB, Babbitt RE, Cummings DL, Dias B, Holben, DE, Kaufman YJ, Rasmussen, RA, Setzer, AW (1992) Smoke and fire characteristics for Cerrado and deforestation burns in Brazil: BASE-B experiment. J. Geophys. Res. 97:14601–19. https://doi.org/10.1029/92jd01218
Warming E (1892) Lagoa Santa. EDUSP, São Paulo
Welch JR (2014) Xavante ritual hunting: anthropogenic fire, reciprocity, and collective landscape management in the Brazilian Cerrado. Hum Ecol 42:47–59. https://doi.org/10.1007/s10745-013-9637-1
Welch JR, Brondízio ES, Hetrick SS, Coimbra-Jr CEA (2013) Indigenous burning as conservation practice: Neotropical savanna recovery amid agribusiness deforestation in Central Brazil. PLoS One 8:e81226. https://doi.org/10.1371/Journal.Pone.0081226
Werneck FP (2011) The diversification of eastern South American open vegetation biomes: historical biogeography and perspectives. Quat Sci Rev 30:1630–1648. https://doi.org/10.1016/j.Quascirev.2011.03.009
Whelan RJ (1995) The ecology of fire. Cambridge University Press, Cambridge
Wieczorkowski JD, Lehmann CER (2022) Encroachment diminishes herbaceous plant diversity in grassy ecosystems worldwide. Glob Chang Biol 00:1–15. https://doi.org/10.1111/gcb.16300
Woinarski JCZ, Legge S (2013) The impacts of fire on birds in Australia's tropical savannas. Emu 113(4):319–352. https://doi.org/10.1071/MU12109
Wragg PD, Mielke T, Tilman D (2018) Forbs, grasses, and grassland fire behavior. J Ecol 106:1983–2001. https://doi.org/10.1111/1365-2745.12980
Zanzarini V, Andersen AN, Fidelis A (2022) Flammability in tropical savannas: variation among growth forms and seasons in Cerrado. Bioptropica 54:979–987. https://doi.org/10.1111/btp.13121
Zirondi HL, de Pinho JH, Daibes LF, Fidelis A (2019) Heat and smoke affect the germination of flammable resprouters: Vellozia species in the Cerrado. Folia Geobot 54:65–72

Zirondi HL, Ooi MKJ, Fidelis A (2021) Fire-triggered flowering is the dominant post-fire strategy in a tropical savanna. J Veg Sci 32:e12995. https://doi.org/10.1111/jvs.12995
Zupo T, Daibes LF, Pausas JG, Fidelis A (2021) Post-fire regeneration strategies in a frequently burned Cerrado community. J Veg Sci 32:e12968. https://doi.org/10.1111/jvs.12968
Zupo T, Gorgone-Barbosa E, Rissi MN, Daibes LF (2022) Experimental burns in an open savanna: greater fuel loads result in hotter fires. Austral Ecol 47:1101–1112. https://doi.org/10.1111/aec.13202

Fire in the Llanos of the Orinoco River

6

Juan F. Silva and Bibiana A. Bilbao

Abstract

The Llanos of the Orinoco River is a vast and unique biodiverse area in northern South America, covered by mosaics of various ecosystems, such as savanna, wetlands, and different forest types, on infertile soils. Here, we highlight fire's critical and complex role in shaping the Llanos ecosystems, underscoring how fire regimes impact landscape dynamics, biodiversity, and carbon storage. Fire promotes the survival of fire-adapted C4 grasses with regrowth capacity and pyrophytic trees with fire-resistant bark and enlarged belowground structures. Periodic fires help preserve the savanna's grass-dominated landscapes but pose challenges to forests, which are crucial for carbon sequestration. However, some fire regimes may enhance forest cover in savannas, pointing to a complex interaction of factors beyond fire. Climate change may alter carbon assimilation and vegetation recovery. On the other hand, land-use changes, including agricultural expansion and intensified livestock farming, could diminish fire occurrence, threatening biodiversity. This management dilemma requires balancing fire's ecological advantages in savannas with its forest risks, aiming for a strategy that integrates biodiversity preservation with carbon sequestration efforts and a long history of traditional fire management from pre-Columbian times. Despite the

J. F. Silva (✉)
Instituto de Ciencias Ambientales y Ecológicas ICAE, Universidad de Los Andes, Mérida, Venezuela

Fondo Verde Institute, Huancayo, Peru
e-mail: jsilva@ula.ve

B. A. Bilbao
Departamento de Estudios Ambientales, Universidad Simón Bolívar, Caracas, Venezuela

Université Paul Valéry-Montpellier III, Montpellier, France
e-mail: bbilbao@usb.ve

A. Fidelis, V. R. Pivello (eds.), *Fire in the South American Ecosystems*, Ecological Studies 250, https://doi.org/10.1007/978-3-031-89372-8_6

Llanos' economic marginality, deforestation, ecosystem fragmentation, and inadequate fire management remain critical environmental threats.

Resumen

Los Llanos del Orinoco comprenden una vasta área biodiversa cubierta por mosaicos de varios ecosistemas, como sabanas, humedales y diferentes tipos de bosques, en suelos infértiles. El fuego promueve la supervivencia de gramíneas C4 con alta capacidad de rebrote y de árboles pirófitos con corteza resistente al fuego y estructuras subterráneas agrandadas, jugando un papel crítico en la configuración, la dinámica del paisaje, la biodiversidad y el almacenamiento de carbono. Las quemas periódicas ayudan a preservar los paisajes de sabana pero plantean desafíos a los bosques, aunque algunos regímenes de quema pueden mejorar la cubierta forestal de las sabanas revelando una compleja interacción de factores. Mientras el cambio climático puede favorecer la ocurrencia de fuego, los cambios en el uso del suelo pueden restringirla. Este dilema requiere una estrategia que integre la preservación de la biodiversidad con los esfuerzos de secuestro de carbono y una larga historia de gestión tradicional del fuego. A pesar de la marginalidad económica de los Llanos, la deforestación, la fragmentación de los ecosistemas y la gestión inadecuada de los incendios siguen siendo amenazas medioambientales críticas.

Resumo

As planícies do Orinoco são uma área vasta e biodiversa coberta por mosaicos de vários ecossistemas, como savanas, zonas húmidas e florestas de diferentes tipos, sobre solos inférteis. O fogo promove a sobrevivência de gramíneas C4, com elevada capacidade de rebrota, e árvores pirofíticas, com casca resistente ao fogo e estruturas subterrâneas alargadas, desempenhando um papel crítico na configuração e dinâmica da paisagem, na biodiversidade e no armazenamento de carbono. As queimadas periódicas ajudam a preservar as paisagens de savanas, mas colocam desafios às florestas, embora alguns regimes de queima possam melhorar a cobertura florestal das savanas, revelando uma complexa interação de fatores. Enquanto as mudanças climáticas podem favorecer a ocorrência de incêndios, as alterações no uso das terras podem restringi-los. Este dilema exige uma estratégia que integre a preservação da biodiversidade com esforços de sequestro de carbono e uma longa história de manejo tradicional do fogo. Apesar da marginalidade económica dos Llanos, o desmatamento, a fragmentação dos ecossistemas e o manejo inadequado dos incêndios continuam a ser ameaças ambientais críticas.

6.1 The Orinoco Llanos

The Orinoco Llanos is an enormous plain of about 450,000 km^2 that extends between Colombia and Venezuela (Etter et al. 2010). It has a significant diversity of landscapes and ecosystems, from grasslands to humid forests, mostly known for their extensive savanna ecosystems. This region is notorious for the annual occurrence of vegetation-sweeping fires that play a role in the ecology of ecosystems and populations, the balance of carbon, and climate change.

The Orinoco Plains extend from the Guaviare River in southeastern Colombia to the mouths of the Orinoco River in the Atlantic Ocean. The Andes and Caribbean mountains range border to the west and north, and the Orinoco River borders to the south (Fig. 6.1). The materials on which the Llanos lie are sediments deposited during the Upper Tertiary and Quaternary periods. The relief is flat, but there are notable differences in elevation between older and higher plains, with coarser materials ("mesas") and lower, more recent areas, with finer materials. Geomorphology and pedogenesis have produced diverse habitats that manifest themselves at different scales, from landforms to landscapes. These can be large, relatively homogeneous units (Sarmiento and Monasterio 1971; Sarmiento 1984) or mosaic of smaller units characterized by the dominant soil, vegetation and flora (Silva et al. 1971; Silva and Sarmiento 1976).

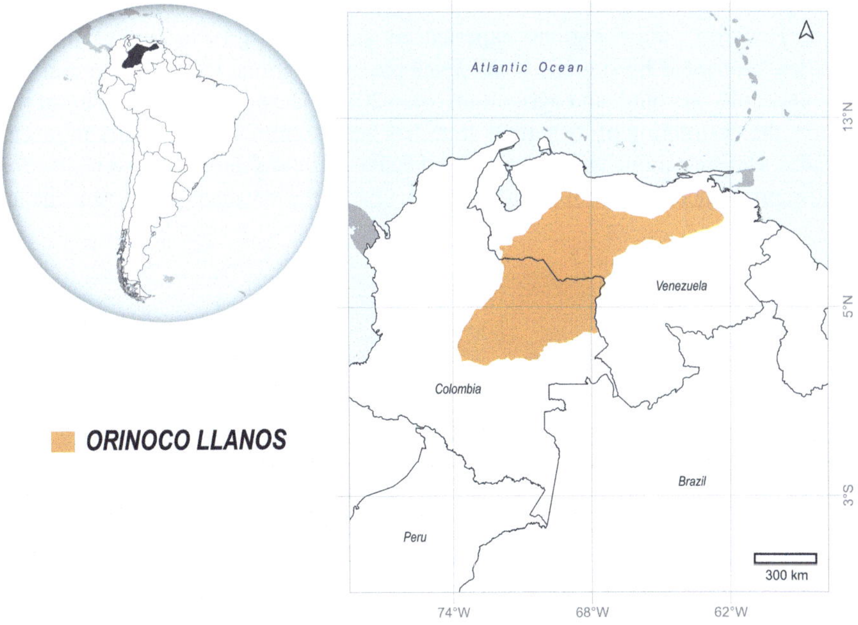

Fig. 6.1 Distribution of the Orinoco Llanos. Source: Olson et al. (2001), modified according to Sarmiento (1983) and Schargel (2007). Produced by Carlos Pérez, Bibiana Bilbao, Juan Silva and Julia Abrantes Rodrigues

The Llanos region has a macrothermal, isothermal climate, with annual rainfall between 700 and 2500 mm, in a gradient that increases from the east to the west in Venezuela and southward in Colombia. Rainfall is seasonal, generated by fluctuations in the intertropical convergence zone (ITCZ). It rains between May and November, and the dry season between December and April is dominated by trade winds blowing from the northeast.

Geomorphology and rainfall determine a wide variety of soils ranging from Ferralsols (Oxisols) to Fluvisols and Gleysols, generally of low fertility. The water regime of these soils can vary from seasonal (one season with good moisture supply and good drainage and another with water deficit) to hyper-seasonal (changing from dry to flooded in an annual cycle); the semi-seasonal savannas do not experience a dry season (Sarmiento 1996). Geomorphological and edaphic diversity is directly related to the variety of vegetation cover.

The vegetation presents a broad spectrum of variation, strongly altered by the development of industrial agriculture and other human activities. There are remnants of various types of forests, from evergreen, seasonally flooded mosaics of semi-evergreen and semi-deciduous forests to dry forests and gallery forests. Savanna cover varies from well-drained, seasonal savannas to flooded savannas dominated by grasslands; in some swampy groves, palm communities grow (Fig. 6.2). Although extensive savanna areas remain, ranching pastures and plantations increasingly replace the natural communities.

With less than 10% of the two countries' population and a very low population density, the extensive plains have been marginal to social and economic development. However, some regions experienced significant growth during the last decades, motivated by specific local development programs in oil and natural gas mining, cattle ranching, and agriculture (Siso 2012). Despite the marginality of the region, the destruction of its natural areas has been significant, especially in recent decades. Deforestation, agriculture and extensive cattle ranching are responsible for the disappearance of 40% of the forests and 32% of the savannas in the Venezuelan

Fig. 6.2 Ecosystems of the Orinoco Llanos. (**a**) Capybaras (*Hydrochoerus hydrochaeris)* grazing in the "esteros" (flooded plains) of Apure state in the Western Llanos, Venezuela (Photo: Río Verde: @rioverdeoficial); (**b**) Well-drained seasonal savanna on a dissected landscape at the edge of Mesa Becerra highland; below is the Orituco River valley with savannas and gallery forests. Parque Nacional Aguaro-Guariquito. (Photo: Juan F. Silva)

Llanos between 1988 and 2010 (Morales et al. 2021). In Colombia, clearance for agriculture (rice), plantations, and pastures has increased exponentially since 1970 and is progressing at an annual rate of more than 100,000 ha (Etter et al. 2010).

In this chapter, we will describe how fire has been an active participant in the evolution of the Llanos landscape and what we know of how it affects the different ecosystems and their biodiversity. The role of fire is a subject of discussion among academics, ecologists, and conservationists, and some consider fire to be a driver of biodiversity (He et al. 2019).

6.2 Fire and the Major Ecosystems in the Llanos

There is consensus that humans, either directly or indirectly, cause most burns in the Llanos. The occurrence, frequency and effects of fire depend to a large extent on the type of vegetation cover. A distinction can be made between the fire in savannas and the fire in forests. In forests, fire commonly comes together with deforestation and changes in land use. In the savanna, on the other hand, fire is mainly used to promote the growth of the herbaceous layer and thus generate food for livestock; in the Llanos, it is also used for hunting. Although fire induces changes in the savanna ecosystems, they soon return to a condition close to the previous one.

Vegetation response (resistance and recovery) to fire depends on several factors. Among them, the time of the year in which the burning occurs, the composition of the plant community (resistant, resilient, sensitive species), the conditions of the vegetation (moisture content, growth rates), the severity of the burn (intensity, temperatures, speed of displacement), which in turn is related to the amount of accumulated combustible material, air humidity and wind speed. The severity also has to do with the frequency of burning (Medina and Silva 1990; Silva 1996a).

6.2.1 Fire and the Savannas

Savannas occupy most of the Orinoco Llanos, sometimes with patches of groves, sometimes interrupted by the course of rivers with narrow strips of gallery forests and, less frequently, giving way to more significant extensions of forests or forming savanna-forest mosaics. The savanna ecosystem is characterized by a low, open herbaceous layer, dominated by grasses and a variably discontinuous tree layer, dominated by stunted, fire-resistant species. It has been described by numerous authors, with emphasis on different aspects such as its ecological dynamics (Solbrig et al. 1992), floristic diversity, and structural diversity (Sarmiento and Monasterio 1975; Sarmiento 1983, 1984, 1996; Ponce et al. 1994; Solbrig et al. 1996; Montes et al. 2013). Four main functional savanna types exist in the Orinoco Llanos: semi-seasonal, seasonal, hyper-seasonal, and flooded savanna (Sarmiento 1983, 1984). Similarly, up to five physiognomic types can be distinguished: savanna-grassland, savanna-open, savanna-closed, savanna-forested and savanna-parkland, which are distinguished based on the degree of cover by woody elements (Sarmiento 1983).

Fig. 6.3 Burning Llano savanna at Estación Biológica de Los Llanos (EBLL) in Calabozo, Guárico State, Western Llanos, Venezuela (Photo: Juan F. Silva)

Fire is an essential and determinant factor in the dynamics of these savannas (Medina and Silva 1990; Armenteras et al. 2005) (Fig. 6.3). It occurs more frequently towards the end of the dry season and the transition to the rainy season. There are only two studies on the physical characteristics of fire in the Orinoco Llanos, by Vareschi (1962) and Bilbao and Medina (1996). Vareschi (1962) reported temperatures between 300 and 400 °C in the grassy layer during a burn, while the soil surface barely reached 90 °C. Bilbao and Medina (1996) obtained similar temperature profiles in *Trachypogon* savannas but recorded higher temperatures up to 500 °C in savannas invaded by *Hyparrhenia rufa,* a taller African grass in the Llanos of Calabozo. The maximum temperatures in *Hyparrhenia rufa* were recorded above 1 m, compared to 0.3 m in the smaller grassy layer of *Trachypogon* savannas, indicating the interaction of fuel distribution and fire behavior. Vareschi (1962) also points out that these burns rarely result in the death of trees of typical savanna species (*Curatella americana, Byrsonima crassifolia, Bowdichia virgilioides*), either because they are protected by thick bark or because their functional state prevents the effect of fire (Vareschi 1962). However, Bilbao (1995) indicates that the high flame lengths and temperatures produced by *Hyparrhenia rufa* fires create a ladder effect, reaching and damaging tree foliage (see also Sect. 6.5).

Annual fires occupy large areas in the Llanos, although estimates differ significantly; this is partly due to differences in methodology, scale and years of study. Armenteras et al. (2005) reported 48.8 km^2 of savannas burned in 2000; Romero-Ruiz et al. (2010) estimated an average of 275 km^2 burned each year between 2000 and 2008, with a sizable interannual variation. An important detail of this study is that 195 km^2 burned more than four times, while 370 km^2 never burned in the period. Llano et al. (2018) detected 445 km^2 burned between 2000 and 2014. They identified over 19 thousand fire footprints, most pixel-sized (0.237 km^2). The largest footprint identified has an area of 880.5 km^2. These authors provide information on the duration of the burns, which is related to their extent.

Fire intensity is determined, among other factors, by the amount of accumulated dry biomass, which in turn depends on biomass production during the growing season (Silva 1996a). Biomass production relies on the length of the period with available water, which can vary depending on the edaphic characteristics and the amount of precipitation during a given year (Medina and Silva 1990). Aboveground biomass production in the herbaceous layer should also be related to its floristic composition. The herbaceous communities in these savannas vary in composition depending on the edaphic characteristics, which differ even within the same large relief unit (Silva and Sarmiento 1976). In disturbed savanna communities, invasive species can become established and change the structure of these communities and the fire's characteristics, as discussed later.

No systematic studies document changes in the savanna subjected to different fire regimes in the Orinoco Llanos. As a counterpart to the lack of fire experiments, monitoring in an area excluded from fire has been used. The hypothesis behind this approach is that since fire is the cause of forest destruction in the area, its exclusion would be sufficient to restore forest cover. In the case of the Llanos, the only fire exclusion trial available is the "Estación Biológica de Los Llanos" (EBLL), initiated by Blydenstein in 1962 on a three-hectare plot. This trial yields considerable information on species and community response to fire exclusion (Blydenstein 1963). The "experiment" excluded fire and cattle grazing and ended in 1991 with a large fire that swept the entire EBLL. San José and Fariñas documented changes in the composition of the excluded area several times between 1969 and 1986 (San José and Fariñas 1971, 1983, 1991; Fariñas and San José 1985, 1987; Fariñas 2013). These authors documented changes in the composition and structure of the savanna during the exclusion period. Woody species increased from four to 24 in 21 years without fire, and the number of stems increased from 281 to 9273; however, after almost three decades without fire, the EBLL's savanna never reached the forest structure (Fariñas 2013). Changes in woody savanna density also occurred in areas adjacent to the EBLL, managed with frequent grazing and burning (Silva et al. 2001), suggesting that factors other than fire were at work. The increase in CO_2 concentration in the atmosphere is likely a driver of these changes (Bond and Midgley 2000; Higgins and Scheiter 2012).

Despite their apparent homogeneity, grasses (as well as trees) differ in many aspects of their biology, and their responses to fire may be complex (Medina 1996; Silva 1996b). Fire represents an ecological driver (in the short term) and a selective force (in the long term) acting on these species. Fire is a cause of direct mortality and affects plant growth, but on the other hand, it removes the dry standing biomass, facilitating the access of light to the soil and foliage and mineralizes the nutrient content of the dry foliage, producing a sudden influx of nutrients to the soil (Sarmiento 1992; Silva 1996a). Some species seem to have evolved depending on fire for flowering (e.g. *Sporobolus cubensis* and *Elyonurus adustus*); however, seedling mortality by fire is high, and adult mortality is low (Canales and Silva 1987).

Sporobolus cubensis is not affected by fire exclusion (Canales and Silva 1987; Silva et al. 1990). These conflicting effects show that fire has different effects on different stages and transitions of the life cycle of the plant. *Andropogon semiberbis* experiences lower mortality with burning than under fire exclusion (Silva and Castro 1989). *Trachypogon* spp. also seems to depend on fire occurrence, whereas fire is detrimental to *Axonopus canescens*, as shown by the fire exclusion project in the EBLL (San José and Fariñas 1983). Other factors such as interspecific competition (Raventós and Silva 1995), soil moisture and changes in nutrient availability may operate (Medina and Silva 1990; Fariñas 2013).

Andropogon brevifolius, an annual grass that grows primarily in disturbed savannas, experiences high seedling mortality by fire; however, the population requires one out of three years of burning to persist (Canales et al. 1994). A study on the introduced legume, *Stylosanthes capitata*, showed that even though the fire destroys its aerial biomass, this species resists the passage of fire. It sprouts from the crown and replenishes its aerial biomass; likewise, the soil seed bank showed higher germination in the burned plots than in the control plot (Alejo et al. 1987).

The characteristic woody elements of the Llanos savannas are a few species of evergreen trees of low height and twisted trunks, namely *Curatella americana* (Dilleniaceae), *Byrsonima crassifolia, Byrsonima coccolobifolia* (Malpighiaceae) and *Bowdichia virgilioides* (Fabaceae). Other species may accompany them in some savannas, such as *Casearia sylvetris* (Salicaceae) and *Palicourea rigida* (Rubiaceae) in the western savannas and *Roupala complicata* (Proteaceae) in the eastern savannas of Venezuela. Fire is an agent with opposing effects on the growth and survival of woody savanna seedlings. It removes foliage and leaf litter from the herbaceous layer, allowing sufficient light for the rapid growth of seedlings; but, on the other hand, it represents a mortality risk. The few existing studies reveal that seed production varies widely among species (García-Núñez and Azócar 2004). Highly fecund species, such as *Curatella americana*, produce more than 170,000 seeds per individual, with a high viable fraction; another species, *Byrsonima crassifolia*, can produce an average of more than 80,000 seeds, half of which are viable; only 1% of its annual seed harvest is found in the soil seed bank (García-Núñez et al. (2001). Pérez and Santiago (2001), in their study of the seed bank in a seasonal savanna of the Eastern Llanos of Venezuela, reported an average of 1702 seeds m^{-2}, belonging to 124 species of 20 plant families. The only savanna tree detected was *Curatella americana*, with a density of 9 seeds m^{-2}.

Initial seedling growth rates under the herbaceous canopy are low (García-Núñez and Azócar 2004). Most growth is in the roots and the underground stem (xylopodium). The increased underground growth facilitates regrowth after dry-season fire destroys the standing biomass and is, therefore, essential for the individual's survival. Thus, for example, most legume species in savannas are perennial and store significant biomass, energy, and nutrients in underground xylopodia organs (Medina and Bilbao 1991). These reserves enable regrowth after fire, herbivory, or drought. However, their sparsely branched root systems make soil occupation less effective than the fibrous, highly branched roots of grasses. This suggests grasses have a higher capacity to extract water and nutrients, giving them a competitive advantage over legumes, especially in post-fire regrowth.

Seedlings that reach their first dry season face two crucial sources of stress: lack of water and fire; mortality in *Palicourea rigida* seedlings during their first dry season was similar (40%) with an early burn or without burning and decreased significantly under both conditions for 3 years; similar results but higher survival showed *Casearia sylvestris* seedlings (García-Núñez 2003).

Savanna trees survive in savannas with frequent burning, aided by a sufficiently thick bark, sprout ability, and significant understory biomass (Silva 2013). The development of the protective bark is a function of age (and size), so individuals that do not reach a certain height suffer fire damage that can destroy their aerial biomass and even kill the plant (topkill). Trees that experience this top kill effect will regrow from sprouts during the wet season. This is the so-called "fire trap" (Sarmiento et al. 1985), from which they emerge only when they exceed the height necessary to escape the top kill effect (Bond 2008). Several authors have investigated the relationship between bark thickness and the effect of fire in savannas in other regions (Hoffmann and Solbrig 2003; Hoffmann et al. 2003; Lawes et al. 2011). However, such research has not been conducted in the Orinoco savannas. Another little-studied characteristic of woody species in these savannas is the development of below-ground stems capable of resprouting, the nature of which is still poorly known.

Since savanna burns are recurrent, the density of savanna woody trees depends on the balance between their annual growth rate and the rate of return of fire. The Orinoco Llanos burning is almost annual; therefore, the savannas tend to be very open, and small trees dominate the woody populations (bank of saplings). Of course, annual growth rates of woody trees depend on various factors such as competition, water and nutrient availability (Bond 2008; Silva 2013).

6.2.2 Fire and the Deciduous Forests

Deciduous forests are characterized by the predominance of deciduous trees, a pronounced dry season, rainfall between 700–1200 mm and fertile soils (Gonzalez 2013). In the Llanos, deciduous forests are on the northern fringe, close to the Cordillera de la Costa, and in the central and eastern Llanos. These forests have one or two strata of trees 10–20 m high, dominated by some legume species such as *Lonchocarpus* spp., *Acacia* spp., *Caesalpinia mollis*, and *Cassia moschata* (Aymard 2015; Gonzalez 2013).

What remains of these forests is highly fragmented and under constant pressure from fire and deforestation. Different estimates of their current distribution seem close to 40,000 km^2 and about 5% is found in protected areas (Fajardo et al. 2005).

In some areas of the Llanos, deciduous forests occur as scattered islands in a savanna matrix. These formations have been designated "savanna park" (Sarmiento and Monasterio 1971). The islands of deciduous forest (locally called "matas") are generally small and, in their contact zone with the savanna, present an ecotone with numerous trees of savanna species (*Byrsonima crassifolia, Curatella americana*) which protects from savanna fires. The larger islands have a tall evergreen tree of *Copaifera oficinalis* (Fabaceae) that dominates the canopy of the "mata" in height. These islands of deciduous forest are interpreted as relics of the forest that was supposed to cover the whole area (Vareschi 1960) or as advances of the deciduous forest over the savanna (Sarmiento and Monasterio 1975). The savanna-forest system is a dynamic system of coexistence of these two physiognomies in a balance dependent on several factors, one of which is fire. Evidence supports this hypothesis, showing synergistic interactions involving woodland trees, leaf-cutting ants (*Atta laegivata*), and deciduous forest trees, which modify local conditions and promote the growth of forest islands. Other factors (fire and trampling) act to reverse direction (Farji-Brenner and Silva 1995).

These studies show that the savanna-forest dilemma is not attributable to fire as a single factor, but rather it is a dynamic and complex equilibrium where several factors act; savannas and forests are alternative stable states (Beckett and Bond 2019). Some evidence shows that islands of deciduous forests persist (and even increase in number and cover) in savannas under frequent fires (Thielen 2003).

6.2.3 Fire and the Gallery Forests

Numerous tributary rivers cross the Orinoco Llanos, accompanied by narrow strips called gallery or riparian forests (Naiman et al. 1998). Various riparian vegetation types, distinguished by dominant tree species, result from the physiographic modelling and the climate (Cabrera-Amaya and Rivera-Díaz 2016). The gallery forests are located on recent alluvial deposits on banks and terraces exposed to frequent flooding during the rainy season. The rest of the year, they have a relatively high-water table that allows water supply to dominant species of evergreen trees.

The galleries can be found in a forest matrix, surrounded by the forests that grow in the adjacent terraces and cuvettes, or they may be in a savanna matrix; in this case, the abrupt limit between the forest cover and the savanna vegetation is notorious.

Sandy textures dominate the soil in the banks, facilitating drainage. There is a gradient towards finer textures from the top down, with sandy loam, loam, silt loam, and even clayey soils (Berroterán 1988). These banks exhibit a gradient of environmental conditions ranging from well-drained conditions to periods of flooding reaching a meter or more in height. The gradient depends on the bank's relative position, the fluvial current's magnitude, and its seasonal runoff (Veneklaas et al. 2005).

These gallery forests represent a vital vegetation cover that serves as a habitat and ecological corridor for the movement of fauna (Correa-Gómez and Stevenson 2010). They are mixed forests, with trees of medium height, evergreen, deciduous and semi-deciduous, and several species of palms (Aymard 2015).

Gallery forests in a savanna matrix are under pressure from the frequent burning of the savanna, but unaided fire does not enter the gallery protected by fire-resistant species at its edge (Mceowen 2020; Armenteras et al. 2021). Kellman and Meave

(1997) claim these ecotonal areas may eventually serve as refugia for some forest species. Other factors may contribute to preventing savanna fire from penetrating the gallery forest, such as less fuel or better moisture conditions and the different quality of fuel materials within the forest (Biddulph and Kellman 1998). However, recurrent fires may allow savanna grasses to enter the forest and progressively fragment the canopy, killing trees and saplings. Also, high-intensity fires may break the defenses of the gallery and penetrate within the forest. Fire pressure may intensify as climate change influences savanna burning, making it more extensive, frequent and intense (Armenteras et al. 2021; González et al. 2021).

Although the gallery forests have been significantly destroyed and fragmented during the twentieth century and so far in the twenty-first century, their cover has also increased. Thielen (2003), in the National Park Aguaro-Guariquito, recorded significant increases in cover over 59 years (1938–1997) from 151.5 ha to 526.9 ha; longitudinally, the gallery increased from 9.6 km in 1938 to 27.9 km in 1997, representing an average increase of 326 m per year.

In summary, there is little information on the effect of burning and degradation of gallery forests from the Llanos and the functioning of these ecosystems. These aspects, as well as those of forest restoration, are still pending on the Llanos research agenda.

6.2.4 Fire and the Semideciduous Forests of the Western Llanos

The high, semideciduous forests are restricted to the western Llanos in a wide strip attached to the foothills from the Portuguesa State to the Apure State in Venezuela. In this vast region, the relative position of the Quaternary sediments attached to the Andean Piedmont and its subsequent geomorphological dynamics make an adequate supply of humidity possible despite the seasonality of precipitation. They represent forest mosaics that cover from higher positions (banks) with better drainage to lower positions ("bajios") that experience seasonal flooding (Monasterio et al. 1971). They are tall to intermediate forests, relatively dense, with sufficient moisture to make unassisted fire entry difficult.

Between 1950 and 1960, four large "forest reserves" were created: Turén, Ticoporo, Caparo and San Camilo. Together they occupied about 10,000 km^2. Since their creation, timber extraction began, followed by encroachment and deforestation. Today, the destruction of these forests and their replacement by agricultural and grazing systems is almost complete. According to Pacheco et al. (2011), only 16% of all forests in the Western Llanos remain.

These forests vary in composition depending on local characteristics (Huber et al. 2006; Aymard 2015). Before being deforested, it contained fine woods, with species such as *Swietenia macrophylla, Cedrela odorata and Pachira quinata.*

Slash and burn are used to replace these forests with pasture fields for cattle ranching. The process of deforestation of these forests began with the invasion by poor peasants from the Andes (of Venezuela and Colombia). This process intensified after 1984, and by the end of the twentieth century and the beginning of the twenty-first century, the forests had mainly been replaced by grasslands (Chuvieco et al. 2002; Pozzobon and Osorio 2002). Some studies point to the eradication of malaria, the construction of the Barinas-San Cristobal highway, and the Agrarian Reform of 1960, among other factors contributing to the population growth and waves of invasions into the forest reserves (Kammesheidt 1999; Rojas-López 2007). In recent decades, populist policies favored smallholder farmers' occupation of the last remnants of rainforests (Lozada 2007). This dynamic leads to small farmers selling their land after a few years and the development of large estates dedicated to intensive cattle raising (Rojas-López 1993).

The replacement of these extensive forests by pastures and agriculture undoubtedly implies an enormous loss of biodiversity. In addition to the local extinction of numerous tree species, the loss of habitats affects the survival of numerous animal species inhabiting these forests. Land use change also affects soils with progressive impoverishment of their nutrient content, compaction and erosion.

6.2.5 Fire and Other Forest Ecosystems

The "morichales" (or "cananguchales") are humid ecosystems linked to particular geomorphological, edaphic and hydric conditions, with the emblematic palm species *Mauritia flexuosa* (moriche). Morichales are located in tableland landscapes at the headwaters of streams that spring from the water table, with a persistent flow that keeps the substrate humid. The soils are predominantly loam, oligotrophic, acidic, and grey, with a generally deep organic horizon (González 1987). *Trachypogon* savannas with scattered dwarf trees of woody species surround the morichales (Fernández 2007).

In Venezuela, the morichales are found in a strip located south of the Llanos, near the Orinoco River, in the States of Monagas, Anzoátegui, Guárico and Apure (Machado-Allison et al. 2014). In Colombia, they are found mainly in the Meta highlands but are also in the Departments of Arauca, Casanare, Meta, and Vichada (Fernández 2007).

The morichales are variable communities in terms of floristic composition, dominance and structure, which, according to González (1987, 2016), represent successional stages leading to evergreen swamp forests with reduced presence of the palm (González 1987; Fernández 2007).

Annual surface fires have little consequences on morichales because of their little effect on the dominant palms; however, the crown fires, which start as surface fires and occur on the slopes of nearby gullies stimulated by the wind, can reach temperatures above 400 °C and are lethal to the vegetation (González and Rial 2011). Fire can also be especially damaging when combined with oil spills, causing high mortality rates (Bevilacqua and González 1994). This combination is the method used by oil companies to try to eliminate oil spill contamination in the eastern plains of Venezuela. Since accidents are frequent, they represent a real threat to the morichales (González and Rial 2011). Some projects work on restoring morichales through passive strategies with promising results (Narváez-Ortiz et al. 2021).

Another woody ecosystem dominated by palms is the *Copernicia tectorum* swamp, locally called "palmar llanero". These communities are quite different, ecologically and floristically, from the morichales explained above. They are located in wide fluvial terraces with hydromorphic soils, subject to seasonal flooding. The conditions are extreme since the soils dry out entirely in the dry season, while 4–6 weeks after the onset of the rainy season, they flood for the next 4–6 months. *Copernicia tectorum* is the dominant and almost unique woody element, while the herbaceous layer is dominated by grasses of the genera *Andropogon* and *Mesosetum* (Sarmiento and Monasterio 1971; Sarmiento 1983). These communities burn during the dry season, as do the surrounding savannas. *Copernicia tectorum* is said to be resistant to fire, although no studies prove it.

Other flood evergreen forest formations are also found in the Llanos. They are mainly in seasonal lagoons in the confluence of tributaries to the major rivers (such as the Orinoco). When they are of white water, as commonly occurs north of the Orinoco, they are called "rebalses". Some studies have documented these forests' floristic composition and structure (Colonnello et al. 1986; Colonnello 1990a, 1990b; Díaz and Rosales 2008), but information on their ecological dynamics is scarce. Studies on the effects of dry season burning on these ecosystems are also unavailable.

6.3 Fire and the Fauna

The different ecosystems of the Llanos are inhabited by diverse fauna, including invertebrates, amphibians, reptiles, mammals and birds (Pérez and Ojasti 1996; Vilella and Baldassarre 2010). Studies on fire-fauna interaction in these ecosystems are scarce and only on a few taxa (Giorgis et al. 2021). Despite this lack, the different faunal inventories point to fire as one of the main threats to faunal groups because burning alters their habitats, destroys food sources or drastically disrupts food chains. Habitat loss and forest fragmentation are the most critical threats to the fauna of the Llanos (Rodríguez and Rojas-Suárez 2010). The interactions between fire, fragmentation and biota are complex and related to the species, habitat, ecosystem and scale (Driscoll et al. 2021).

Savanna fires are lethal for ants and termites whose nests are found near the border of gallery forests. These and other fire-sensitive insects are food for armadillos and bats; thus, burning the savanna would harm populations of these predators (Aya-Cuero et al. 2019; Azofeifa et al. 2019). However, it is necessary to distinguish between high-intensity fires, which can occur after prolonged exclusion, and surface fires, or controlled burns in agricultural fields, which are part of ecosystem management and do not imply significant harmful effects on fauna. Furthermore, fire exclusion in savannas harms fauna more than burning under controlled conditions (Ojasti 2000).

Animal mortality, medium-term responses of fauna, and recolonization processes of burned areas in the Llanos are scarcely known (González et al. 2021). We know that the structure and composition of plant communities are fundamental for the existence of a particular fauna in the ecosystem. This close relationship is altered once fire changes the structure and composition of the vegetation. In the case of gallery forests, fire can open up the canopy, creating more open habitats. This opening favors species with a more omnivorous diet and a more flexible life cycle, while it harms specialist species that see their food and habitat supply reduced. On the other hand, some species of small mammals seem indifferent to fire. By disrupting the functioning of gallery forests, fire represents a threat to the survival of small mammal populations (González et al. 2021).

The scarce information on the effects of fire on the fauna of the Llanos is related to the low level of sampling and studies of the fauna in this region (Trujillo-Gonzalez et al. 2010; Mora-Fernandez et al. 2015). Although some authors call attention to the adverse effects of anthropogenic activities and the increasing land use change upon animal populations in the Llanos (Castillo-Figueroa et al. 2021) do not pay special attention to the recurrent burning in the savannas. Several investigations on the effect of fire on Llanos fauna are in progress in the Colombian Llanos, and it is expected that these studies will generate new publications on such a strategic topic (Armenteras et al. 2019).

6.4 Fire, Savanna Soils and Nutrient Balance

Soils in the Llanos are the product of a long geological history (dominated by sedimentary deposits from the upper Tertiary and Quaternary) under the effects of Quaternary climatic changes. They have low fertility with low to deficient organic matter and nitrogen, low pH levels (<5), low exchangeable bases, high P-fixation capacity and consequently low P availability, high concentration of Fe oxide and exchangeable Al levels reaching high saturation and toxicity (Berroterán 1988; Cochrane 1990; Rippstein et al. 2001).

In poorly drained savannas of the more recent alluvial plains, soils are Inceptisols, Entisols, Alfisols and, in some cases, Mollisols, which present clay sub-horizons that slow water movement down in their profiles, becoming waterlogged with the onset of the rainy season (May–June to November–December), so most fires occur in the dry season (Armenteras et al. 2005; Rondón et al. 2006). In well-drained savannas are Oxisols and Ultisols, beside Entisols and spots of Vertisols, Inceptisols and Alfisols (Berroterán 1988; Cochrane 1990; Rippstein et al. 2001).

Some studies in savannas of the Llanos show that savanna burning affects the soil's nutrient balance and chemical characteristics (pH, nutrient and organic matter content) (Medina and Silva 1990; Bilbao 1995; López-Hernández 2013). Most soil physical changes occur above 600 °C, temperatures very unlikely to reach from shallow fires in savannas (Frost and Robertson 1987; Bilbao and Medina 1996). The long-term effects of fire on soils physical properties may come from vegetation changes or exposure to the erosive effects of wind and water.

Fire plays a crucial role in nutrient cycling and contributes to the rapid mineralization and loss of some nutrients in savanna ecosystems. In the case of gaseous elements, like N and S, losses occur by volatilization and the spread of incompletely burned ashes, whereas in the case of sedimentary elements, like P and bases, losses by fire are associated with the spread of ashes (Medina 1982a). The direct effects of fire on soil generally occur at temperatures above 690–700 °C, which destroys the entire surface litter layer and 99% of soil organic carbon (Frost and Robertson 1987). However, as explained above, savannas, especially with low tree density, experience relatively low soil surface temperatures during burning (generally not exceeding 200 °C), which only affect functional groups of O_2-containing humic polymers particularly labile to heat treatment (Almendros et al. 2000). Thus, in general, the results of studies on the long-term effects of fire in the in tropical savannas are pretty inconsistent in terms of organic matter content in the soil (Bilbao 1995; Rondón et al. 2006). Those results are associated with the low surface temperatures of combustion, low concentration ranges and the outstanding natural heterogeneity of organic matter in savanna soils, together with the different types of burning histories, fire behavior and vegetation type. However, experimental burns of greater intensity in savannas of *Trachypogon* sp. and *Hyparrhenia rufa* in the Llanos of Venezuela increased organic matter content in the first 5 cm of the soil (Bilbao 1995). According to this author, these differences could be associated with the necromass incorporation or organic carbon residues in the ashes. Also, Bilbao registered an increase in pH (from one-tenth to 1 unit), associated with the increase in the availability of exchangeable cations in the soil and the contribution of K, Ca and Mg in the ashes (Bilbao 1995). All these changes disappeared after 90 days in the rainy season. Another study showed an impoverishment of soil organic matter, P and K, with burning, while no significant differences in total N, Ca and Mg content were found between protected savanna and burned savanna soils of Venezuelan Llanos (Hernández-Valencia and López-Hernández 2002).

According to Sanhueza and Crutzen (1998), the herbaceous layer loses 90–95% of nitrogen (as NO and NO_2) with the passage of fire. In the *Trachypogon* savannas of the Llanos, some studies have reported losses of 8.2 kg N ha^{-1} $yr.^{-1}$ (López-Hernández 2013), while Bilbao and Medina (1996) distinguish between the effect of fire on savannas invaded by *Hyparrhenia rufa* and native *Trachypogon* savannas. In the first case, volatilization losses reach 10 g m^{-2}, while in the native savanna, there are no significant differences between the burned and unburned areas. These differences probably result from more intense burning in the savanna invaded by *Hyparrhenia rufa* due to the quality of the fuel material and the homogeneity of plant cover (Bilbao and Medina 1996). Several authors suggest that losses from burning can be recovered by wet deposition, with 2.6–5 kg N ha^{-1} y^{-1} (Medina 1982b), 2.2–6 kg N ha^{-1} y^{-1} (López-Hernández 2013) and 10.5 kg N ha^{-1} y^{-1} by dry deposition (Sanhueza and Crutzen 1998). Symbiotic nitrogen fixation in the soil-plant system (13.7 and 7.8 kg N ha^{-1} y^{-1}) and microbial crust (2.4 and 4 kg N ha^{-1} y^{-1}) is another essential N input to the *Trachypogon* savannas in the Llanos (Bilbao 1995). Considering the additional losses of N ~7.5 kg N ha^{-1} y^{-1} as soil emissions of NO (N_2O, N_2 and NO), the budget is even positive since losses seem to be balanced by rainfall and biological fixation inputs (Medina 1982b; Medina and Izaguirre 2004; López-Hernández 2013). Also, these estimated amounts seem to be enough to maintain plant production.

Concerning phosphorous, Hernández-Valencia and López-Hernández (1999) determined that a burn produces phosphorus losses of 1.1 kg P ha^{-1} y^{-1} in a *Trachypogon* savanna; only half of these losses are recovered in rainfall. Savanna soil protected from fire and grazing for 32 years showed a higher phosphorus content than burned savanna. Although losses from burning represent the highest phosphorus output from the system, they are too low to alter the phosphorus cycle in this ecosystem (Hernández-Valencia and López-Hernández 1999).

Traditional savanna management practices do not affect the biomass of soil macrofauna but slightly decrease its taxonomic richness (Decaëns et al. 2001). The presence of cattle and the practice of burning favor earthworms but are detrimental to termites. In savannas of the Eastern Plains of Colombia, burning in the rainy season causes an immediate and short-lived negative effect on the soil macrofauna, although the communities recover in <6 months. During the dry season, macroinvertebrates are mainly found in the deep horizons of the soil, far from the reach of temperature and humidity changes induced by burning or by the reduction of vegetation cover (Decaëns et al. 2001).

In conclusion, the data about the effects of fire on the nutrient balance of Llanos savannas support the hypothesis that the vegetation in the natural savanna has evolved under fire constraints. However, uncertainties in flux measures indicate that more work on those aspects is required (Medina 1982b; López-Hernández 2013).

6.5 Fire and Invasive Species

Until approximately the mid-1950s, the Orinoco Llanos of Venezuela and Colombia were exclusively used for low-intensity cattle ranching, with cattle feeding on native low-forage quality grasses with low nutritive and palatability value and productive capacity, with stocking rates of <0.5 heads per hectare. In the Llanos of Venezuela in the 1960s, a new agropastoral livestock system was intensified and implemented by introducing more palatable African grass species. Similarly, in Colombia in the 1970s, research led by CIAT and ICA encouraged the intensification of livestock activities by introducing improved grasses from Africa (Table 6.1). The bases for selection of these newly introduced species are, among others, higher growth capacity and productivity, higher nutritional and palatability value, better nutrient use efficiency and resource allocation, higher reproduction rates, ability to establish through seeds with higher germination potential, great capacity for regrowth after grazing, and tolerance to defoliation (Bilbao and Medina 1990; Bilbao 1995; Baruch 1996; Baruch and Bilbao 1999). Furthermore, exotic tree species were introduced for plantations, either timber-producing, such as the Caribbean pine, or oil-producing, such as the African oil palm (Table 6.1). These exotic species are considered threats to the biodiversity of savanna communities, and their effects on fire regimes are variable (Cárdenas-López et al. 2010; Salgado-Negret et al. 2022).

Table 6.1 Some of the exotic plant species introduced in the Orinoco Llanos

Species	Family
Andropogon bicornis L	Poaceae
Cynodon dactylon L.,	Poaceae
Cenchrus purpureus (Schumach.) Morrone	Poaceae
Cynodon dactylon (L.) Pers.	Poaceae
Digitaria decumbens stent	Poaceae
Hyparrhenia rufa (Nees) Stapf,	Poaceae
Megathyrsus maximus (Jacq.) B.K. Simon & S.W.L. Jacobs	Poaceae
Melinis minutiflora P. Beauv.	Poaceae
Urochloa brizantha (Hochst. Ex A.Rich.) R.D. Webster	Poaceae
Urochloa decumbens (Stapf) R.D. Webster	Poaceae
Rottboellia cochinchinensis (Lour.) Clayton	Poaceae
Elaeis guineensis Jacq.	Arecaceae
Leucaena leucocephala (lam.) deWit.	Fabaceae
Pinus caribaea Morelet	Pinaceae

The introduction of African grasses in the Llanos and their subsequent spread into natural areas markedly decreased local vegetation diversity and profoundly affected ecosystem function and stability, particularly impacting fire dynamics in savanna ecosystems (Bilbao 1995; Baruch 1996; Bilbao and Medina 1996; Medina and Bilbao 1996; Baruch and Bilbao 1999). Several studies revealed critical morphological and physiological characteristics of *Hyparrhenia rufa* that contributed to its colonization's success, its impact on the fire behaviour, the post-fire response, and the savanna function and dynamics (Bilbao 1995; Baruch and Bilbao 1999). The opening of the herbaceous layer by savanna fires allows *Hyparrhenia rufa* colonization, promoted by its higher germination potential, fast seedling growth, and vegetative growing capacity. In one growth season, *Hyparrhenia rufa* produced bigger and taller individuals (2.5 m) than the taller native grasses (~2 m in the flowering period), which markedly increased the savanna production and carbon fluxes. *Hyparrhenia rufa* plots produced 100% more biomass than the native *Trachypogon* savanna plots; moreover, native species maintain a fraction of standing green biomass during the dry season. Thus, the supply of combustible material of *Hyparrhenia rufa* was higher than those of the native grassland, both in height and biomass, allowing more intense dry-season fires and opening more spaces in the community (Baruch and Bilbao 1999; Bilbao 1995). Although fires cause some damage to *Hyparrhenia rufa*, they allow further colonization, constituting a re/invasion cycle (D'Antonio and Vitousek 1992; Bilbao 1995). The plant architecture defines the vertical distribution of fuel: taller grasses produce higher flames and greater damage to the tree canopy than lower grasses (Frost and Robertson 1987). Consequently, in areas dominated by *Hyparrhenia rufa*, the ladder effect of fire affects the canopy of savanna native dominant trees characterized by short stature (Bilbao 1995; Bilbao and Medina 1996).

The exclusion of fire in areas dominated by *Hyparrhenia rufa* produced the death of this species, leaving a cover of dead plant mass. In contrast, in burnt savannas, *Hyparrhenia rufa* becomes dominant (100% of absolute frequency), whether by regrowth or recruitment by seed germination (Medina and Bilbao 1996). Burning in areas of *Trachypogon* savannas previously subjected to fire exclusion increased the absolute frequency of *Hyparrhenia rufa* by 36%. After a second and a third consecutive year of burning, the increase was 50% and 90%, respectively (Bilbao 1995). These results show the high dependence of *Hyparrhenia rufa* on fire for its persistence and invasion in the savannas of the Llanos.

Plots dominated by *Hyparrhenia rufa* showed a significant reduction in species diversity and composition (15 species and 0.72–0.86 diversity indexes) compared to native savanna plots (34–41 species and 1.35–1.44 diversity index) (Baruch and Bilbao 1999). Current tendencies in the Llanos are to replace native savannas with introduced pastures to improve cattle production. This practice compromises the permanence and function of native savannas. The challenge remains to determine the fate of the low diversity substituted savannas and its effect on ecosystem function and stability (Baruch and Bilbao 1999).

6.6 Fire, Climate Change and Land Use Change

Climate change predictions suggest that fire in the Llanos will increase in frequency, intensity and extent. These increases will have a more significant effect on the ecosystems of the Llanos, especially their biodiversity. On the other hand, the increase in land use change, with more significant extensions dedicated to agriculture and plantations and the intensification of livestock farming with introduced pastures, may reduce the extent of the area burned in the region (Rondón et al. 2006). Savanna burning emits a wide range of volatile and particulate materials, with a magnitude comparable to global industrial emissions (Grace et al. 2006). Forty-one percent of methane emissions and 35% of nitrogen oxide emissions in the savannas of the Colombian highlands come from burning.

The appraisal of carbon emissions to the atmosphere in the Llanos is not straightforward because of the variety of values reported and the types of assumptions made. San José et al. (1998) estimated emissions from burning native vegetation at 20 Tg C yr^{-1}, whereas San José and Montes (2001) estimated 4.85 Tg C yr^{-1}. These estimates represent 0.7% and 3.7% of annual emissions from burning the world's savannas.

Assessments of carbon stocks in the different ecosystems of the Llanos are essential information for the design of management strategies related to fire and climate change mitigation. However, the available studies present varied estimates (Table 6.2), making it difficult to conclude. These variations are primarily due to the remarkable spatial and temporal heterogeneity of the vegetation cover in the Llanos. In addition, no direct methods are available to measure carbon stocks in vegetation

Table 6.2 Total carbon content estimated for different vegetation types in the Orinoco Llanos

Author	Vegetation type	Mg C ha^{-1}
San José et al. 1998	Burned savanna	70.8
	Protected savanna (25 yr)	106.7
	Semideciduous forest	214.1
San José et al. 2008	Native savanna	17.02
	Dense savanna (28% tree cover)	25.35
	Woodland (78% tree cover)	48.66
	Savanna with *Andropogon gayanus*	38.86
Calderón-Balcázar et al. (2023)	Gallery forest	73.36
	Morichal	100.31
	Oil palm plantation	11.8–24.61

Table 6.3 Carbon fluxes estimated for different vegetation types in the Orinoco Llanos

Author	Vegetation type	MgC ha^{-1} yr.$^{-1}$
Morales-Rincón et al. (2021)	Savannas in Colombia	0.35–2.73
San José et al. (2008)	Open savanna	0.06
	Dense savanna (28% tree cover)	1.16–1.39
	Woodland (78% tree cover)	3.36
	Savanna with *Andropogon gayanus*	
San José (1995) in Grace et al. (2006)	Burned savanna	3.4
San José (1995) in Grace et al. (2006)	Fire-excluded savanna	8.1

and soil; indirect methods tend to be unrepresentative of the areas studied, making emissions estimates difficult (Gibbs et al. 2007). Other sources of variation are the different indirect estimation methods used.

San José et al. (1998) reported estimates two to ten times higher than the values reported by San José et al. (2008). The latter values range from 17.02 mg C Ha^{-1} in a native savanna to 48.66 mg C ha^{-1} in a woodland savanna (78% tree cover). Calderón-Balcázar et al. (2023) estimated values of 73.36 mg C ha^{-1} for gallery forest and 100.3 mg C ha^{-1} for morichal; these authors also estimated the carbon stock of an oil palm plantation for three consecutive years, reporting values in a range of 11.8–24.6 mg C ha^{-1}. Rainford et al. (2021), using a model fed with data from CIAT (International Center for Tropical Agriculture), predict that the Eastern Plains of Colombia contain an organic carbon stock of 3.7 Tg C ha^{-1}.

Carbon fluxes in savannas seem to depend more on the extent of drought than on the magnitude of precipitation, determining whether the ecosystem functions as a carbon source or Sink (Morales et al. 2021). Reported values for carbon fluxes vary widely from 0.06 mg C ha^{-1} yr.$^{-1}$ (San José et al. 2008) in an open savanna to 8.1 mg C ha^{-1} yr^{-1} (Grace et al. 2006) in a fire-excluded savanna (Table 6.3). According to these authors, the savannas of the Orinoco Llanos act as carbon sinks. In their model of carbon stocks and fluxes, San José and Montes (2001) concluded that the Orinoco Llanos are a carbon sink of approximately 17.5 Tg C yr.$^{-1}$. Their reported carbon sequestration values in savannas are within the range of estimates for other savanna ecosystems worldwide (Grace et al. 2006).

Fire suppression in savannas effectively contributes to climate change mitigation due to increased woody biomass (San José et al. 2008). However, it is necessary to consider other effects of exclusion, specifically the decrease in biodiversity and local extinction of savanna specialist species (Abreu et al. 2017).

About 30% of the area of the Llanos in Venezuela is under intensive agricultural and livestock use, a total of approximately nine million hectares (Hernández-Hernández et al. 2017). In the Eastern Plains of Colombia, although the replacement of native savannas by crops and plantations has advanced notably in this century, there is a remarkable dynamic of changes between different types of use (Vargas-Serna 2022). According to Romero-Ruiz et al. (2011), by 2007, 55.5% of the Eastern Plains remained as native savannas and 18.2% as native forests, mainly gallery forests; 20.6% were crops and pastures, while only 0.97% consisted of oil palm Plantations. This last use has advanced notably; by 2022, oil palm plantations reached an extension of 450,000 Ha (Vargas-Serna 2022), representing an Increase of 55% per year.

Oil palm promoters argue that the impact of these plantations on biodiversity is minimal since the planted areas are savannas that have already been transformed into introduced pastures. Furthermore, they add that plantations replace unproductive savannas with no biodiversity value (Diaz Moreno 2016). This argument ignores the natural richness of the savannas, the existence of specialized species, the endemic species and the ecosystem services. Both plantations and other agricultural land uses, such as intensive agriculture of corn, rice and other crops, not only imply the destruction of habitats and, therefore, an impact on plant and animal biodiversity but also an exclusion of fire, which is essential in the maintenance of the savanna ecosystem (Abreu et al. 2017).

6.7 Fire Management

In the Llanos savannas of Venezuela and Colombia, fire is a crucial tool for the "Llaneros" (people of mixed European and native descent), used to regrow grasses at the onset of the dry season, extending the cattle foraging period. This tradition dates back to the seventeenth century when Andalusian Spaniards introduced cattle and has evolved with the integration of improved cattle breeds and productive African forage species (Medina and Bilbao 1996; Hétier and López 2003; Duno et al. 2007; Sibaja-Madera 2023). However, changes in land ownership, invasive species, and cultural loss have confined fire use to uncultivated areas, leading to significant, unmanaged fires exacerbated by climate change with its longer droughts and increased temperatures. Additionally, efforts to expand agricultural and oil frontiers over the last century, coupled with socio-political disruptions from armed conflict and drug trafficking, have transformed land use, resulting in both organized and chaotic deforestation using uncontrollable fires for savanna expansion (Bilbao 1995; Hétier and López 2003; Romero-Ruiz et al. 2010; Barreto and Armenteras 2020; Meza 2023; Sibaja-Madera 2023).

Paleoecological studies indicate that early human settlements managed savanna-forest mosaics with fire as early as 9200 cal BP, showcasing a deep historical knowledge of fire management to balance agriculture with natural ecosystems during pre-Columbian Holocene times (Berrio et al. 2002; Leal and Bilbao 2011; Hooghiemstra and Flantua 2019; Leal et al. 2019; Piraquive-Bermúdez and Behling 2022). Today, Indigenous Peoples, despite enduring historical devastations, manage a small portion of land within national parks or marginalized territories. Admirably, groups like the Pumé in Venezuela and the Saliva in Colombia continue traditional fire practices in savanna and forest settings, employing fire in hunting, gathering, and swidden agriculture, which improves the fertility of tropical soils. These practices seem to maintain ecological balance and enhance biodiversity by creating varied environmental niches (Saturno and Zent 2016; Huertas et al. 2019; Sibaja-Madera 2023).

The need for revised fire management policies is critical, focusing on promoting prevention practices based on fire management over suppression. While Colombia and Venezuela have established fire management laws, enforcement is weak and often contradicts Indigenous traditions, underscoring the need for fire management strategies that align with traditional knowledge and ecological needs (Bilbao et al. 2022).

6.8 Conclusions

There is consensus among savanna scientists that fire is a fundamental factor in the functioning of savanna ecosystems in the Orinoco Llanos.

Concerning extension, most burns do not reach 1 km^2, while a few can reach up to 10 km^2 and, exceptionally, a few hundred km^2. We have little information on the characteristics of the fires in terms of frequency, extent and duration.

At the level of species and plant populations of the Llanos savannas, the responses to fire are as varied as the flora is diverse. Some species seem to depend on fire for functioning and do not survive in the prolonged absence of fire.

Fire damages forest wildlife by destroying habitats, but little information exists on the relationship between animal species and forest fires. The relationship between savanna animal populations and recurrent fires is even less clear, to which plant biodiversity seems to be well adapted.

Land use change towards intensive agriculture and farming and the establishment of extensive forest plantations exclude fire from large savanna areas. This management is detrimental to the conservation of savanna biodiversity despite being promoted by some conservationists.

Climate change promotes environmental conditions favorable to the occurrence of burning, and some predict that it will increase the extension of savannas to the detriment of forests (Salazar et al. 2007). By eliminating forest cover, fire favors emissions and reduces carbon sequestration capacity. On the other hand, savanna burning does not seem important in this context since emissions from burning could be offset by CO_2 sequestration during the growing season.

The lack of knowledge about fire as an ecological factor in the Orinoco Llanos is remarkable. This deficiency is even more severe if we consider this region's tremendous environmental and biotic heterogeneity, and its current dynamics oriented towards the disappearance of native ecosystems. A few ecosystems and some species, such as the morichales and native grasses, have received more attention. However, there is still a significant gap in information on a large part of the interactions between fire and the animal and plant biodiversity of the Llanos.

References

Abreu RCR, Hoffmann WA, Vasconcelos HL, Pilon NA, Rossatto DR, Durigan G (2017) The biodiversity cost of carbon sequestration in tropical savanna. Sci Adv 3:e1701284. https://doi.org/10.1126/sciadv.1701284

Alejo AP, Rodríguez AG, Fisher MJ (1987) Tolerancia de Stylosanthes capitata a la quema en los Llanos Orientales de Colombia. Pasturas Trop 10:2–7. https://www.tropicalgrasslands.info/public/journals/4/Elements/DOCUMENTS/1988-vol10-rev1-2-3/Vol10_rev1_88_art2.pdf

Almendros G, Giampaolo S, Pard MT (2000) Laboratory appraisal of carbon sequestration and nutrient availability after different organic matter inputs in virgin and cultivated Zimbabwean soils. Commun Soil Sci Plant Anal 32:877–894

Armenteras D, Romero M, Galindo G (2005) Vegetation fire in the savannas of the llanos Orientales of Colombia. World Resource Rev 17:531–543. https://citeseerx.ist.psu.edu/document?repid=rep1&type=pdf&doi=5c3b0f6866ea040eb6fba1811b5d85fca2c516a2

Armenteras D, Meza MC, González TM, Salazar N, Barreto S et al (2019) Estado del conocimiento de la ecología del fuego en Colombia: síntesis de hallazgos y aplicaciones. Ecolmod Universidad Nacional de Colombia. https://www.researchgate.net/publication/337404240_Estado_del_conocimiento_de_la_ecologia_del_fuego_en_Colombia_sintesis_de_hallazgos_y_aplicaciones

Armenteras D, Meza MC, González TM, Oliveras I, Balch JK, Retana J (2021) Fire threatens the diversity and structure of tropical gallery forests. Ecosphere 12. https://doi.org/10.1002/ecs2.3347

Aya-Cuero C, Trujillo F, Mosquera-Guerra F, Chacón-Pacheco J, Caicedo D (2019) Distribution of armadillos in Colombia, with an analysis of ecoregions and protected areas inhabited. Mammal Res 64:569–580. https://doi.org/10.1007/s13364-019-00435-4

Aymard G (2015) Bosques de los Llanos de Venezuela: estructura, composición florística, diversidad y estado actual de conservación. In: López-Falcón R, Hétier JM, López HD, Schargel R, Zinck A (eds) Tierras llaneras de Venezuela. IRD-CIDIAT, Mérida, Venezuela, pp 241–268

Azofeifa Y, Estrada-Villegas S, Mavárez NJM (2019) Activity of aerial insectivorous bats in two rice fields in the northwestern llanos of Venezuela. Acta Chiropterol 21:149–163. https://hal.science/hal-02464094

Barreto JS, Armenteras D (2020) Open data and machine learning to model the occurrence of fire in the ecoregion of "llanos Colombo–Venezolanos". Remote Sens 12:3921. https://doi.org/10.3390/rs12233921

Baruch Z (1996) Ecophysiological aspects of the invasion by African grasses and their impact on biodiversity and function of Neotropical savannas. In: Solbrig OT, Medina E, Silva J (eds) Biodiversity and savanna ecosystem processes: a global perspective ecological studies, vol 121. Springer, Berlin, pp 79–93. https://doi.org/10.1007/978-3-642-78969-4

Baruch Z, Bilbao B (1999) Efects of fire and defoliation on the life history of native and invader C4 grasses in Neotropical savannas. Oecologia 119:510–520. https://doi.org/10.1007/s004420050814
Beckett H, Bond WJ (2019) Fire refugia facilitate forest and savanna co-existence as alternative stable states. J Biogeogr 46:2800–2810. https://doi.org/10.1111/jbi.13707
Berrio JC, Hooghiemstra H, Behling H, Botero P, Van der Borg K (2002) Late-quaternary savanna history of the Colombian llanos orientales from lagunas Chenevo and Mozambique: a transect synthesis. The Holocene 12:35–48. https://doi.org/10.1191/0959683602h1518rp
Berroterán JL (1988) Paisajes ecológicos de sabanas en Llanos Altos Centrales de Venezuela. Ecotropicos 1:92–108. http://www.saber.ula.ve/handle/123456789/25709
Bevilacqua M, González V (1994) Consecuencias de derrames de petróleo y acción del fuego sobre la fisionomía y composición florística de una comunidad de morichal. Ecotropicos 7:23–34. http://www.saber.ula.ve/handle/123456789/25687
Biddulph J, Kellman M (1998) Fuels and fire at savanna-gallery forest boundaries in southeastern Venezuela. J Trop Ecol 14:445–461. https://doi.org/10.1017/S0266467498000339
Bilbao B (1995) Impacto del régimen de quemas en las características edáficas, producción de materia orgánica y biodiversidad de sabanas tropicales en Calabozo, Venezuela. Dissertation, IVIC, Caracas
Bilbao B, Medina E (1990) Nitrogen-use efficiency for growth in a cultivated African grass and a native south American pasture grass. J Biogeogr 17:421–425. https://doi.org/10.2307/2845372
Bilbao B, Medina E (1996) Types of grassland fires and nitrogen volatilization in tropical savannas, Calabozo, Venezuela. In: Levine J (ed) Biomass burning and global change. MIT, Cambridge, MA, pp 569–574. https://www.researchgate.net/publication/301677302_Types_of_Grassland_Fires_and_Nitrogen_Volatilization_in_Tropical_Savannas_of_Calabozo_Venezuela
Bilbao BA, Millán A, Matany Luque M, Mistry J, Gómez-Martínez R et al (2022) An intercultural vision for integrated fire management in Venezuela. Trop Forest Issues 61:39–46. https://doi.org/10.55515/CNUU7417
Blydenstein J (1963) Cambios de la vegetación después de protección contra el fuego. II. Análisis de una parcela de la Estación Biológica después de un año sin quema. Bol Soc Ven Cienc Nat 103:239–244
Bond WJ (2008) What limits trees in C4 grasslands and savannas? Annu Rev Ecol Evol Syst 39:641–659. https://doi.org/10.1146/annurev.ecolsys.39.110707.173411
Bond WJ, Midgley JJ (2000) A proposed CO2-controlled mechanism of woody plant invasion in grasslands and savannas. Glob Change Biol 6:865–869. https://doi.org/10.1046/j.1365-2486.2000.00365.x
Cabrera-Amaya D, Rivera-Díaz O (2016) Composición florística y estructura de los bosques ribereños de la cuenca baja del río Pauto, Casanare, Colombia. Caldasia 38:53–85. https://doi.org/10.15446/caldasia.v38n1.57829
Calderón-Balcázar A, Cárdenas CD, Díaz-Vasco O, Fandiño E, Márquez T, Pizano C (2023) Biomass and carbon stocks of four vegetation types in the llanos Orientales of Colombia (Mapiripán, meta). Trees For People 12:100380. https://doi.org/10.1016/j.tfp.2023.100380
Canales MJ, Silva J (1987) Efecto de una quema sobre el crecimiento y demografía de vástagos en Sporobolus cubensis. Acta Oecol-Oec Gen 8:391–401. https://www.ciencias.ula.ve/icae/publicaciones/sabanas/pdf/canales1987.pdf
Canales MJ, Trevisan MC, Silva JF, Caswell H (1994) A demographic study of an annual grass (Andropogon brevifolius Schwarz) in burnt and unburnt savanna. Acta Oecol 15:261–273. http://www.ciens.ula.ve/icae/publicaciones/sabanas/pdf/canales1994.pdf
Cárdenas-López D, Castaño N, Cárdenas-Toro J (2010) Análisis de riesgo de especies de plantas introducidas para Colombia. In: Baptiste MP, Castaño N, Cárdenas D et al (eds) Análisis de riesgo y propuesta de categorización de especies introducidas para Colombia. Instituto de Investigación de Recursos Biológicos Alexander von Humboldt, Bogotá, Colombia, pp 51–68. http://hdl.handle.net/20.500.11761/31384

Castillo-Figueroa D, Martínez-Medina D, Rodríguez-Posada ME (2021) Activity patterns of medium and large mammals in two savanna ecosystems in the Colombian llanos. Revista Acad Colomb Ci Exact 45:1071–1083. https://doi.org/10.18257/raccefyn.1461

Chuvieco E, Salas J, Meza E, Vargas F (2002) Empleo de la teledetección en análisis de la deforestación forestal tropical: el caso de la reserva forestal de Ticoporo (Venezuela). Serie Geogr 10:55–76. http://hdl.handle.net/10017/1160

Cochrane TT (1990) Regional soil differentiation in Neotropical savannas. In: Sarmiento G (ed) Las Sabanas Americanas Aspectos de su biogeografía, ecología y utilización. Fondo Editorial Acta Cient Venez, Caracas, pp 99–124

Colonnello G (1990a) A Venezuelan flood plain study on the Orinoco River. Forest Ecol Manag 33:103–124. https://doi.org/10.1016/0378-1127(90)90187-G

Colonnello G (1990b) Elementos fisiográficos y ecológicos de la Cuenca del Río Orinoco y sus rebalses. Interciencia 15:476–485

Colonnello G, Castroviejo S, López G (1986) Comunidades vegetales asociadas al Río Orinoco en el sur desde Monagas y Anzoategui (Venezuela). Mem Soc Cienc Nat 46:127–165

Correa-Gómez D, Stevenson P (2010) Estructura y diversidad de bosques de galería en una sabana estacional de los llanos orientales colombianos (Reserva Tomo Grande, Vichada). Orinoquía 14:31–48. http://www.scielo.org.co/pdf/rori/v14s1/v14s1a04.pdf

D'Antonio CM, Vitousek PM (1992) Biological invasions by exotic grasses, the grass/fire cycle and global change. Annu Rev Ecol Syst 23:63-87

Decaëns T, Jiménez JJ, Rangel AF, Cepeda A, Moreno AG, Lavelle P (2001) La Macrofauna del Suelo en la Sabana Bien Drenada de los Llanos Orientales. In: Rippstein G, Escobar G, Motta F (eds) Agroecología y biodiversidad de las sabanas en los Llanos Orientales de Colombia. CIAT, pp 111–137. https://cgspace.cgiar.org/bitstream/handle/10568/55159/agroecologia_y_biodiversidad.pdf

Díaz W, Rosales J (2008) Análisis fitosociológico y estructural del bosque inundable de varzea. Kuawäi 1:13–39

Diaz Moreno I (2016) Palma, Estado y Región en los Llanos colombianos (1960–2015). In: Ulloa A, Coronado S (eds) Extractivismos y posconflicto en Colombia: retos para la paz territorial. Universidad Nacional de Colombia, pp 167–200. https://www.academia.edu/35175985/2016_Ulloa_A_Coronado_S_Eds_Extractivismos_posconflicto_pdf

Driscoll DA, Armenteras D, Bennett AF, Brotons L, Clarke MF (2021) How fire interacts with habitat loss and fragmentation. Biol Rev Camb Philos Soc 96:976–998. https://doi.org/10.1111/brv.12687

Duno DR, Aymard G, Huber O (eds) (2007) Catálogo anotado e ilustrado de la flora vascular de los llanos de Venezuela, Caracas, Fudena.

Etter A, Sarmiento A, Romero M (2010) Land use changes (1970–2020) and the carbon emissions in the Colombian llanos. In: Hill M, Hannan NP (eds) Ecosystem function in savannas: measurement and modeling at landscape to global scales. CRC Press, pp 383–402. https://doi.org/10.1201/b10275-26

Fajardo L, Gonzalez V, Nassar J, Lacabana M, Portillo PQ, Carrasquel CAF, Rodríguez JP (2005) Tropical dry forests of Venezuela: characterization and current conservation status. Biotropica 37:531–546. https://doi.org/10.1111/j.1744-7429.2005.00071.x

Fariñas M (2013) Cambios en la sabana de la Estación Biológica de los Llanos «Francisco Tamayo», Sociedad Venezolana de Ciencias Naturales, Calabozo, Venezuela. In: Medina E, Huber O, Nassar JM, Navarro P (eds) Recorriendo el paisaje vegetal de Venezuela. IVIC, Caracas, pp 231–248. http://www.saber.ula.ve/bitstream/handle/123456789/39306/Farinas2013.pdf?sequence=1&isAllowed=y

Fariñas MR, San José JJ (1985) Cambios en el estrato herbáceo de una parcela de sabana protegida del fuego y el pastoreo durante 23 años. Calabozo, Venezuela Acta Cient Venez 36:199–200

Fariñas MR, San José JJ (1987) Efectos de la supresión del fuego y el pastoreo sobre la composición de una sabana de Trachypogon en los Llanos del Orinoco. In: San José JJ, Montes R (eds) La Capacidad Bioproductiva de las Sabanas. CIET-Unesco-MAB-IVIC-USB-IUPC, Caracas, pp 513–545. http://www.saber.ula.ve/handle/123456789/37653

Farji-Brenner A, Silva JF (1995) Leaf-cutting ants and forest groves in tropical savanna: facilitated succession? J Trop Ecol 11:651–669. https://doi.org/10.1017/S0266467400009202
Fernández A (2007) Los morichales de los Llanos de Venezuela. In: Duno de Stefano R, Aymard G, Huber H (eds) Catálogo anotado e ilustrado de la flora vascular de los llanos de Venezuela. FUDENA-Fund Polar-FIBV, Caracas, pp 91–98. https://www.academia.edu/21395929/Flora_and_Vegetation_of_the_Venezuelan_Llanos
Frost PGH, Robertson F (1987) The ecological effects of fire in savannas. In: Walker BH (ed) Determinants of tropical savannas. IRL Press, Oxford, UK, pp 93–140
García-Núñez C (2003) Aspectos ecofisiológicos del establecimiento de dos especies de árboles siempreverdes de las sabanas estacionales. Universidad de Los Andes, Mérida, Venezuela. https://saber.ula.ve/bitstream/handle/123456789/35511/17.GarciaNuñez%2cCarlos.2003.pdf?sequence=1&isAllowed=y
García-Núñez C, Azócar A (2004) Ecología de la regeneración de árboles de la sabana. Ecotropicos 17:1–24. http://ecotropicos.saber.ula.ve
García-Núñez C, Azócar A, Silva JF (2001) Seed production and soil seed bank in three evergreen woody species from a neotropical savanna. J Trop Ecol 17:563–576. https://doi.org/10.1017/S0266467401001420
Gibbs HK, Brown S, Niles JO, Foley JA (2007) Monitoring and estimating tropical forest carbon stocks: making REDD a reality. Environ Res Lett 2:045023. https://doi.org/10.1088/1748-9326/2/4/045023
Giorgis MA, Zeballos SR, Carbone L, Zimmermann H, von Wehrden H et al (2021) A review of fire effects across south American ecosystems: the role of climate and time since fire. Fire Ecol 17:11. https://doi.org/10.1186/s42408-021-00100-9
González V (1987) Los morichales de los llanos orientales. Un enfoque ecológico. Ediciones CORPOVEN, Caracas
González V (2016) Los palmares de pantano de Mauritia flexuosa en suramerica: una revisión. In: Lasso CA, Colonnello G, Moraes M (eds) Morichales y canaguales de la Orinoquia y Amazonia: Colombia-Venezuela: Parte 2. Instituto de Investigación de Recursos Biológicos Alexander von Humboldt, Bogotá, Colombia, pp 45–83. http://hdl.handle.net/20.500.11761/31423
González V, Rial A (2011) Las comunidades morichal en los llanos orientales de Venezuela, Colombia y el delta del Orinoco: impactos de la actividad humana sobre su integridad y funcionamiento, vol 17. CO-BAC, Bogotá. https://doi.org/10.1186/s42408-021-00100-9
González TM, González-Trujillo JD, Muñoz A, Armenteras D (2021) Differential effects of fire on the occupancy of small mammals in neotropical savanna-gallery forests. Perspect Ecol Conserv 19:179–188. https://doi.org/10.1016/j.pecon.2021.03.005
González V (2013) La vegetación de Venezuela al norte del Río Orinoco. Fundación Instituto Botánico de Venezuela Dr. Tobías Lasser (FIBV), Caracas, Venezuela
Grace J, San Jose J, Meir P, Miranda HS, Montes RA (2006) Productivity and carbon fluxes of tropical savannas. J Biogeogr 33:387–400. https://doi.org/10.1111/j.1365-2699.2005.01448.x
He T, Lamont BB, Pausas JG (2019) Fire as a key driver of Earth's biodiversity. Biol Rev 94:1983–2010. https://doi.org/10.1111/brv.12544
Hernández-Hernández RM, Acevedo D, Sarmiento L (2017) Carbon sequestration potential in the savannas ecosystems of Venezuelan flatlands. In: Proceedings of the global symposium on soil organic carbon. Food and Agriculture Organization of the United Nations, Rome, pp 444–448. https://www.fao.org/3/i7565en/i7565en.pdf
Hernández-Valencia I, López-Hernández D (1999) Efectos de la quema sobre el ciclo del fósforo en una sabana de Trachypogon. Ecotropicos 12:3–7. http://www.saber.ula.ve/handle/123456789/25487?show=full
Hernández-Valencia I, López-Hernández D (2002) Pérdida de nutrimentos por la quema de vegetación en una sabana de Trachypogon. Rev Biol Trop 50:1013–1019. https://www.scielo.sa.cr/scielo.php?script=sci_arttext&pid=S0034-77442002000300019
Hétier JM, López FR (2003) Tierras Llaneras de Venezuela. CIDIAT, Mérida, Venezuela
Higgins SI, Scheiter S (2012) Atmospheric CO2 forces abrupt vegetation shifts locally, but not globally. Nature 488:209–212. https://doi.org/10.1038/nature11238

Hoffmann WA, Solbrig OT (2003) The role of topkill in the differential response of savanna woody species to fire. Forest Ecol Manag 180:273–286. https://doi.org/10.1016/S0378-1127(02)00566-2

Hoffmann WA, Orthen B, Do Nascimento PKV (2003) Comparative fire ecology of tropical savanna and forest trees. Funct Ecol 17:720–726. https://doi.org/10.1111/j.1365-2435.2003.00796.x

Hooghiemstra H, Flantua SGA (2019) Colombia in the Quaternary: an overview of environmental and climatic change. In: Gómez J, Pinilla-Pachon AO (eds) The geology of Colombia. Publicaciones Geológicas Especiales, Servicio Geológico Colombiano, pp 43–95. https://doi.org/10.32685/pub.esp.38.2019.02

Huber O, de Stefano RD, Aymard G, Riina R (2006) Flora and vegetation of the Venezuelan llanos: a review. In: Pennington RT, Ratter JA (eds) Neotropical savannas and dry forests: diversity, biogeography, and conservation, The systematics association special, vol 69, pp 95–120. https://doi.org/10.1201/9781420004496

Huertas A, Baptiste B, Toro M, Huertas H (2019) Manejo de la quema de pastizales de sabana inundable: una mirada de pueblo originario Saliva en Colombia. Chungara Rev Antropo Chilena 51:167–176

Kammesheidt L (1999) The role of population growth and land-use policy in deforestation: a case study in the western Venezuelan plains. In: Deutscher Tropentag. http://ftp2.de.freebsd.org/pub/tropentag/proceedings/1999/referate/RME4.pdf. Accessed May 13, 2023

Kellman M, Meave J (1997) Fire in the tropical gallery forests of Belize. J Biogeogr 24:23–34. https://doi.org/10.1111/j.1365-2699.1997.tb00047.x

Lawes MJ, Richards A, Dathe J, Midgley JJ (2011) Bark thickness determines fire resistance of selected tree species from fire-prone tropical savanna in North Australia. Plant Ecol 212:2057–2069. https://doi.org/10.1007/s11258-011-9954-7

Leal A, Bilbao B (2011) Cambios de vegetación durante el holoceno tardío en un morichal de los Llanos del Orinoco Venezuela. Acta Bot Venez 34:237–256

Leal A, Gassón R, Behling H, Sánchez F (2019) Human-made fires and forest clearance as evidence for late Holocene landscape domestication in the Orinoco llanos (Venezuela). Veget Hist Archaeobot 28:545–557. https://doi.org/10.1007/s00334-019-00713-w

Llano XC, Pascual DA, Niño LF (2018) ≌Cuán grande es un incendio en las sabanas del norte de Suramérica? Perspect Rurales 16:25–40. https://doi.org/10.15359/prne.16-31.2

López-Hernández D (2013) N biogeochemistry and cycling in two well-drained savannas: a comparison between the Orinoco Basin (llanos, Venezuela) and Ivory Coast (western Africa). Chem Ecol 29:280–295. https://doi.org/10.1080/02757540.2012.744830

Lozada JR (2007) Situación actual y perspectivas del manejo de recursos forestales en Venezuela. Rev Fores Venez 51:195–218. http://bdigital.ula.ve/storage/pdf/fores/v51n2/articulo7.pdf

Machado-Allison A, Rial A, Lasso CA (2014) Morichales. In: Lasso CA, Rial A, Collonnello G et al (eds) XI. Humedales de la Orinoquia (Colombia-Venezuela), pp 207–209. Serie Recursos hidrobiológicos y pesqueros continentales de Colombia. Instituto de Investigación de Recursos Biológicos, Alexander von Humboldt, Bogotá. http://hdl.handle.net/20.500.11761/31361

Mceowen K (2020) Managing the gallery forests of the Colombian Llanos for resistance and resilience. https://doi.org/10.13140/RG.2.2.33015.55202

Medina E (1982a) Nitrogen balance in the Trachypogon grasslands of Central Venezuela. Plant Soil 67:305–314. https://doi.org/10.1007/BF02182777

Medina E (1982b) Physiological ecology of neotropical savanna plants. In: Huntley BJ, Walker BH (eds) Ecology of tropical savannas. Springer, pp 308–335. https://doi.org/10.1007/978-3-642-68786-0_15

Medina E (1996) Diversidad morfológica y funcional del grupo de productores primarios en sabanas. Interciencia 21:193–202. https://www.researchgate.net/publication/286328552_Diversidad_morfologica_y_funcional_del_grupo_de_productores_primarios_en_sabanas

Medina E, Bilbao BA (1991) Significance of nutrient relations and symbiosis for the competitive interaction between grasses and legumes in tropical savanna. In: Esser G, Overdieck D (eds) Modern ecology: basic and applied aspects. Elsevier Science, Amsterdam, The Netherlands, pp 24–36. https://doi.org/10.1016/B978-0-444-89183-9.50019-8

Medina E, Bilbao BA (1996) Relaciones nutricionales e impacto de incendios en la produccion de materia organica de pastizales naturales y cultivados. Capítulo 13. In: Sarmiento G, Cabido M (eds) Biodiversidad y Funcionamiento de Pastizales y Sabanas en America Latina, CYTED-CIELAT, pp 245–260

Medina E, Izaguirre ML (2004) N2-fixation in tropical American savannas evaluated by the natural abundance of 15N in plant tissues and soil organic matter. Trop Ecol 45:87–96. https://www.researchgate.net/publication/238107885_N2-fixation_in_tropical_American_savannas_evaluated_by_the_natural_abundance_of_15N_in_plant_tissues_and_soil_organic_matter

Medina E, Silva JF (1990) The savannas of northern South America: a steady state regulated by water-fire interactions on a background of low nutrient availability. J Biogeogr 17:403–413. https://doi.org/10.2307/2845370

Meza MC (2023) Efectos de los Incendios forestales sobre la resiliencia de bosques tropicales de tierras bajas. Universidad Nacional de Colombia, Bogotá, Colombia

Monasterio M, Sarmiento G, Silva JF (1971) Reconocimiento ecológico de los Llanos Occidentales. III. El sur del Estado Barinas. Acta Cient Venez 22:153–169. http://www.saber.ula.ve/bitstream/handle/123456789/37669/monasterio1971.pdf?sequence=1&isAllowed=y

Montes R, San José JJ, Aymard GA (2013) Flora y Vegetación características de la altiplanicie de Mesa y planicie eólica del Parque Nacional Aguaro-Guariquito. Caldasia 35:219–240. http://www.scielo.org.co/pdf/cal/v35n2/v35n2a1.pdf

Mora-Fernandez C, Peñuela-Recio L, Castro-Lima F (2015) El medio natural: los llanos de Casanare y área de estudio. Orinoquia 19:253–271. http://www.scielo.org.co/pdf/rori/v19n2/v19n2a12.pdf

Morales OFM, Albarrán AJT, Gutierrez JJC (2021) Estudio espacio temporal de la cobertura de la tierra: Norte del río Orinoco-Venezuela, periodo 1988-2010. GeoFocus 28:95–119. https://doi.org/10.21138/GF.686

Morales-Rincon LA, Hernandez AJ, Rodriguez-Hernandez NS, Jimenez R (2021) Carbon exchange and accumulation in an Orinoco high plains native savanna ecosystem as measured by eddy covariance. Front Environ Sci 9:673932. https://doi.org/10.3389/fenvs.2021.673932

Naiman RJ, Fetherston K, McKay SJ, Chen J (1998) Riparian forests. In: Naiman RJ, Bilby RE (eds) River ecology and management: lessons from the Pacific coastal ecoregion, pp 289–323. https://doi.org/10.2307/177387

Narváez-Ortiz I, Ortíz-Paz RA, López Patarroyo E (2021) Regeneración pasiva de morichales (Mauritia flexuosa L.f.) en los llanos orientales de Colombia. Ecosistemas 30:2230. https://doi.org/10.7818/ECOS.2230

Ojasti J (2000) Manejo de Fauna Silvestre Neotropical. In: SI/MAB Series # 5. Smithsonian Institution/MAB Biodiversity Program, Washington DC. https://bibliotecavirtualaserena.files.wordpress.com/2017/11/libro-de-manejo-de-fauna-de-ojasti.pdf

Olson DM, Dinerstein E, Wikramanayake ED, Burgess ND, Powell GVN et al (2001) Terrestrial ecoregions of the world: a new map of life on earth. Bioscience 51:933–938. https://doi.org/10.1641/00063568(2001)051[0933:TEOTWA]2.0.CO;2

Pacheco C, Aguado I, Mollicone D (2011) Dinámica de la deforestación en Venezuela: Análisis de los cambios a partir de mapas históricos. Interciencia 36:566–586

Pérez EM, Ojasti J (1996) La utilización de la fauna silvestre en la América tropical y recomendaciones para su manejo sustentable en las sabanas. Ecotropicos 9:71–82. http://www.saber.ula.ve/bitstream/handle/123456789/25731/articulo3.pdf?sequence=2

Pérez EH, Santiago ET (2001) Dinámica estacional del banco de semillas en una sabana en los Llanos Centro-Orientales de Venezuela. Biotropica 33:435–446. https://www.jstor.org/stable/3593188

Piraquive-Bermúdez D, Behling H (2022) Holocene paleoecology in the Neotropical savannas of northern South America (llanos of the Orinoquia ecoregion, Colombia and Venezuela): what do we know and on what should we focus in the future? Front Ecol Evol 10:824873. https://doi.org/10.3389/fevo.2022.824873

Ponce ME, González V, Brandin J, Ponce M (1994) Vegetation analysis associated to a soil toposequence in the central eastern llanos of Venezuela. Ecotropicos 7:11–22. http://erevistas.saber.ula.ve/index.php/ecotropicos/article/view/10293/10233

Pozzobon E, Osorio R (2002) Evaluación de las deforestaciones en la reserva forestal de Ticoporo, estado Barinas-Venezuela, en base al análisis multitemporal de imágenes de percepción remota. Rev Geogr Venez 43:215–235. http://www.saber.ula.ve/bitstream/handle/123456789/24581/articulo43-2-3.pdf?sequence=2andisAllowed=y

Rainford S, Martín-López JM, Da Silva MM (2021) Approximating soil organic carbon stock in the Eastern Plains of Colombia. Front Environ Sci 9:685819. https://doi.org/10.3389/fenvs.2021.685819

Raventós J, Silva JF (1995) Competition effects and responses to variable number of neighbors in two tropical savannas grasses. J Trop Ecol 11:39–52. https://doi.org/10.1017/S0266467400008385

Rippstein G, Amézquita E, Escobar G, Grollier C (2001) Condiciones Naturales de la Sabana. In: Rippstein G, Escobar G, Motta F (eds) Agroecología y biodiversidad de las sabanas en los Llanos Orientales de Colombia. CIAT, Cali, Colombia, pp 1–21. https://cgspace.cgiar.org/bitstream/handle/10568/55159/agroecologia_y_biodiversidad.pdf

Rodríguez JP, Rojas-Suárez F (2010) Libro Rojo de la Fauna Venezolana: Actualización Periódica de la Situación de las Especies Amenazadas Del País. In: Machado-Allison, A. et al. (eds), Investigación y Manejo de Fauna Silvestre en Venezuela en Homenaje al " Dr. Juhani Ojasti ". ISBN: 978-980-6195-14-1. pp. 121-132.

Rojas-López J (1993) La colonización agraria de las reservas forestales: ≃un proceso sin solución? Cuadernos Geográficos 10. ULA Mérida-Venezuela. https://bdigital.uncu.edu.ar/objetos_digitales/9541/1.pdf

Rojas-López J (2007) Regulación ambiental y colonización agraria en reservas de bosque. El drama de Ticoporo, estado Barinas-Venezuela. Rev Geogr Venez 48:129–141. https://www.redalyc.org/articulo.oa?id=347730365008

Romero-Ruiz M, Etter A, Sarmiento A, Tansey K (2010) Spatial and temporal variability of fires in relation to ecosystems, land tenure and rainfall in savannas of northern South America. Glob Change Biol 16:2013–2023. https://doi.org/10.1111/j.1365-2486.2009.02081.x

Romero-Ruiz MH, Flantua SGA, Tansey K, Berrio JC (2011) Landscape transformations in savannas of northern South America: land use/ cover changes since 1987 in the llanos orientales of Colombia. Appl Geogr 32:766–776. https://doi.org/10.1016/j.apgeog.2011.08.010

Rondón MA, Acevedo D, Hernandez RM, Rubiano Y, Rivera M (2006) Carbon sequestration potential of neotropical savannas of Colombia and Venezuela. In: Lal R, Cerri C, Bernoux M et al (eds) . Haworth Press, Carbon sequestration in soils of Latin America, pp 213–243. http://www.saber.ula.ve/bitstream/handle/123456789/39315/rondon2006.pdf?sequence=1&isAllowed=y

Salazar LF, Nobre CA, Oyama MD (2007) Climate change consequences on the biome distribution in tropical South America. Geophys Res Lett 34:L09708. https://doi.org/10.1029/2007GL029695

Salgado-Negret B, Urbina-Cardona N, Noguera-Urbano EA, Cruz-Rodríguez CA, Baptiste MP (2022) Predicting the risk of exotic plant invasions in the Orinoco region: importance of distribution models, climatic niche and functional richness. Front Environ Sci 10:687723. https://doi.org/10.3389/fenvs.2022.687723

San José JJ, Fariñas MR (1971) Estudios sobre los cambios de la vegetación protegida de la quema y el pastoreo en la Estación Biológica de los Llanos. Bol Soc Ven Cs Nat 119-120:136–147

San José JJ, Fariñas MR (1983) Changes in tree density and species composition in a protected Trachypogon savanna, Venezuela. Ecology 64:447–453. https://doi.org/10.2307/1939963

San José JJ, Fariñas MR (1991) Temporal changes in the structure of a Trachypogon savanna protected for 25 years. Acta Oecol 12:237–247. http://www.ciencias.ula.ve/icae/publicaciones/sabanas/pdf/san_jose1991a.pdf

San José JJ, Montes RA (2001) Management effects on carbon stocks and fluxes across the Orinoco llanos. Forest Ecol Manag 150:293–311. https://doi.org/10.1016/S0378-1127(00)00588-0

San José JJ, Montes RA, Fariñas MR (1998) Carbon stocks and fluxes in a temporal scaling from a savanna to a semi-deciduous forest. Forest Ecol Manag 105:251–262. https://doi.org/10.1016/S0378-1127(97)00288-0

San José JJ, Montes R, Grace J, Nikonova N (2008) Land-use changes Alter CO2 flux patterns of a tall-grass Andropogon field and a savanna-woodland continuum in the Orinoco lowlands. Tree Physiol 28:437–450. https://doi.org/10.1093/treephys/28.3.437

Sanhueza E, Crutzen PJ (1998) Budgets of fixed nitrogen in the Orinoco savannah region. Glob Biogeochem Cycles 12:653–666. https://doi.org/10.1029/98GB02314

Sarmiento G (1983) Patterns of specific and phenological diversity in the grass community of the Venezuelan tropical savannas. J Biogeogr 10:373–391. https://doi.org/10.2307/2844747

Sarmiento G (1984) The neotropical savannas. Harvard University Press, Cambridge, Massachussetts. https://www.hup.harvard.edu/catalog.php?isbn=9780674418554

Sarmiento G (1992) A conceptual model relating environmental factors and vegetation formations in the lowlands of tropical South America. In: Furley PA, Proctor J, Ratter JA (eds) Nature and dynamics of forest-savanna boundaries. Chapman and Hall, London, pp 583–601. http://www.ciencias.ula.ve/icae/publicaciones/sabanas/pdf/sarmiento_g1992b.pdf

Sarmiento G (1996) Ecología de pastizales y sabanas en América Latina. In: Sarmiento G, Cabido M (eds) Biodiversidad y funcionamiento de pastizales y sabanas en América Latina. CYTED-CIELAT, Mérida, pp 15–24. http://www.ciencias.ula.ve/icae/publicaciones/sabanas/pdf/sarmiento_g1996b.pdf

Sarmiento G, Monasterio M (1971) Ecología de las sabanas de América tropical. I. Análisis macroecológico de los Llanos de Calabozo. Venezuela Cuadernos Geogr 4:1–126. http://www.ciencias.ula.ve/icae/publicaciones/sabanas/pdf/sarmiento_g1971b.pdf

Sarmiento G, Monasterio M (1975) A critical consideration of the environmental conditions associated with the occurrence of savanna ecosystems in tropical America. In: Golley F, Medina E (eds) . Springer, Tropical ecological systems, pp 223–250. http://www.ciencias.ula.ve/icae/publicaciones/sabanas/pdf/sarmiento_g1975b.pdf

Sarmiento G, Goldstein G, Meinzer F (1985) Adaptive strategies of woody species in neotropical savannas. Biol Rev 60:315–355. http://www.ciencias.ula.ve/icae/publicaciones/sabanas/pdf/sarmiento_g1985.pdf

Saturno S, Zent S (2016) Aspectos etnoecológicos de la agricultura entre los Pumé. Bol Mus Para Emílio Goeld 11:653–676. https://doi.org/10.1590/1981.81222016000300007

Schargel R (2007) Geomorfología y suelos. In: Duno de Stefano R, Aymard G, Huber O (eds) Catálogo anotado e ilustrado de la flora vascular de los Llanos de Venezuela. FUDENA-Fundación Polar-FIBV, Caracas, pp 15–31. https://www.academia.edu/21395929/Flora_and_Vegetation_of_the_Venezuelan_Llanos

Sibaja-Madera FJ (2023) Fuego y ganadería en el valle de los ríos Sinú y San Jorge (Caribe colombiano), en el siglo XX. HAAL 4:90–109. https://doi.org/10.53077/haal.v4i02.176

Silva JF (1996a) Biodiversity and stability in tropical savannas. In: Solbrig OT, Medina E, Silva JF (eds) Biodiversity and savanna ecosystem processes: a global perspective. Springer, pp 161-171. doi: https://doi.org/10.1007/978-3-642-78969-4_9

Silva JF (1996b) Fire as a factor influencing the diversity of life forms in neotropical savannas. In: Pereira RC, Nasser L (eds) Proceedings of the 1st international symposium of tropical savannas and VIII Simposium on Cerrado. EMBRAPA-CPAC, Brasília, pp 6–9. http://www.saber.ula.ve/handle/123456789/16885

Silva JF (2013) El fuego y los árboles de las sabanas. In: Medina E, Huber O, Nassar JM, Navarro M (eds) Recorriendo el paisaje vegetal de Venezuela. IVIC, Caracas, pp 157–117. http://www.saber.ula.ve/bitstream/handle/123456789/39232/Silva2013.pdf?sequence=2&isAllowed=y

Silva JF, Castro F (1989) Fire, growth and survivorship in a Neotropical savanna grass (Andropogon semiberbis) in Venezuela. J Trop Ecol 5:387–400. https://doi.org/10.1017/S0266467400003849

Silva JF, Sarmiento G (1976) Influencia de factores edáficos en la diferenciación de las sabanas. Análisis de componentes principales y su interpretación ecológica. Acta Cient Venez 27:68–78. http://www.ciencias.ula.ve/icae/publicaciones/sabanas/pdf/silva1971.pdf

Silva JF, Monasterio M, Sarmiento G (1971) Reconocimiento ecológico de los Llanos Occidentales. II. El Norte del Estado Barinas. Acta Cient Venez 22:61–72. http://www.ciencias.ula.ve/icae/publicaciones/sabanas/pdf/silva1971.pdf

Silva JF, Raventós J, Caswell H (1990) Fire and fire exclusion effects on the growth and survival of two savanna grasses. Acta Oecol 11:783–800. http://www.ciencias.ula.ve/icae/publicaciones/sabanas/pdf/silva1990.pdf

Silva JF, Zambrano A, Fariñas MR (2001) Increase in the woody component of seasonal savannas under different fire regimes in Calabozo, Venezuela. J Biogeogr 28:977–983. http://www.ciencias.ula.ve/icae/publicaciones/sabanas/pdf/silva2001.pdf

Siso GJ (2012) La población de Venezuela: evolución, crecimiento y distribución geográfica. Terrain 28:109–140. https://ve.scielo.org/scielo.php?script=sci_arttext&pid=S1012-70892012000100006&lng=es&nrm=iso&tlng=es

Solbrig OT, Goldstein G, Medina E, Sarmiento G, Silva JF (1992) Responses of tropical savannas to stress and disturbance: a research approach. In: Mohan KW (ed) Ecosystem rehabilitation, Ecosystem analysis and synthesis, vol 2. SPB Academic Publishing, The Hague, Netherlands, pp 63–73. http://www.saber.ula.ve/handle/123456789/37661

Solbrig OT, Medina E, Silva JF (1996) Biodiversity and tropical savanna properties: a global view. In: Mooney HA, Cushman JH, Medina E et al (eds) Biodiversity and global change, vol 55. SCOPE, John Wiley, pp 185–205. http://www.ciencias.ula.ve/icae/publicaciones/sabanas/pdf/solbrig1996a.pdf

Thielen DR (2003) Tendencias de la precipitación y la dinámica espacial y temporal de las leñosas en una sabana Neotropical. Universidad de los Andes, Mérida. http://www.ciencias.ula.ve/icae/postgrado/doctorado_tesis.php

Trujillo-Gonzalez F, Beltrán Gutiérrez M, Diaz-Pulido FPA, Payan Garrido E (2010) Mamiferos. In: Lasso CA, Usma JS, Trulillo F et al (eds) Biodiversidad de la cuenca del Orinoco: bases científicas para la identificación de áreas prioritarias para la conservación y uso sostenible de la biodiversidad. Instituto de Investigación de Recursos Biológicos Alexander von Humboldt, WWF Colombia, Fundación Omacha, Fundación La Salle e Instituto de Estudios de la Orinoquia (Universidad Nacional de Colombia), Bogotá, D. C., Colombia, pp 147–172

Vareschi V (1960) La Estación Biológica de los Llanos de la Sociedad Venezolana de Ciencias Naturales y su tarea. Publicación 1,. Estación Biológica de Los Llanos, Venezuela

Vareschi V (1962) La quema como factor ecológico en los Llanos. Bol Soc Ven Cs Nat 23:9–31

Vargas-Serna J (2022) Análisis del uso de la tierra en las sabanas de los Llanos Orientales de Colombia, la región de la Orinoquía, desde 1999 hasta 2022. Universidad de Antioquia, Colombia. https://bibliotecadigital.udea.edu.co/bitstream/10495/28560/2/VargasJeronimo_2022_clasificacionUsosSuelo.pdf

Veneklaas EJ, Fajardo A, Obregon S, Lozano J (2005) Gallery forest types and their environmental correlates in a Colombian savanna landscape. Ecography 28:236–252. https://doi.org/10.1111/j.0906-7590.2005.03934.x

Vilella FJ, Baldassarre GA (2010) Abundance and distribution of water birds in the llanos of Venezuela. Wilson J Ornithol 122:102–115. https://doi.org/10.1676/09-070.1

Fire Ecology on Megadiverse Montane Grasslands

7

Soizig le Stradic, Swanni T. Alvarado, Orlando Vargas Ríos, and Fernando A. O. Silveira

Abstract

South American montane grasslands are hotspots of biodiversity, characterized by unusually high levels of endemism. They are the birthplace of the massive rivers and watersheds, shaping the geography and biota of the continent and providing water to millions of people. Here, we provide an overview of fire ecology in two megadiverse South American montane grasslands, the *Páramo* in the Andes and the C*ampo Rupestre* in eastern Brazil. For the *Páramo*, we discuss ecological and evolutionary aspects of fire ecology with a focus on vegetation, given the apparent lack of knowledge for both animal and microbial communities. For the C*ampo Rupestre*, in addition to a brief discussion on fire effects on vegetation, we also attempt to synthesize fire effects on seed germination, invertebrates, vertebrates and microbial communities. After laying out the state-of-art on fire ecology on montane grasslands, we identify knowledge gaps and suggest research priorities that can enhance our capacity to manage fire sustainably.

S. le Stradic
Biodiversité, Génes & Communautés-BIOGECO, UMR INRAE/Université de Bordeaux, Pessac Cedex, France
e-mail: soizig.le-stradic@inrae.fr

S. T. Alvarado · O. V. Ríos
Departamento de Biología, Facultad de Ciencias, Universidad Nacional de Colombia, Bogotá, Colombia
e-mail: stalvarador@unal.edu.co; jovargasr@unal.edu.co

F. A. O. Silveira (✉)
Departamento de Genética, Ecologia e Evolução, Universidade Federal de Minas Gerais, Belo Horizonte, Brazil
e-mail: faosilveira@icb.ufmg.br

A. Fidelis, V. R. Pivello (eds.), *Fire in the South American Ecosystems*, Ecological Studies 250, https://doi.org/10.1007/978-3-031-89372-8_7

Resumen

Los pastizales de montaña de América del Sur son centros críticos de biodiversidad y presentan niveles inusualmente altos de endemismo. Son el lugar del nacimiento de enormes ríos y cuencas hidrográficas, que dan forma a la geografía y la biota del continente y proporcionan agua a millones de personas. Aquí ofrecemos una visión general de la ecología del fuego en dos pastizales montanos megadiversos de América del Sur, el *Páramo* en los Andes y el Campo Rupestre en el este de Brasil. Para el *Páramo*, discutimos aspectos ecológicos y evolutivos de la ecología del fuego con un enfoque en la vegetación, dada la aparente falta de conocimiento e información tanto de las comunidades animales como microbianas. Para el Campo Rupestre, además de una breve discusión sobre los efectos del fuego en la vegetación, también intentamos sintetizar los efectos del fuego en la germinación de semillas, los invertebrados, los vertebrados y las comunidades microbianas. Después de exponer los últimos avances en ecología del fuego en pastizales montanos, identificamos lagunas de conocimiento y sugerimos prioridades de investigación que pueden mejorar nuestra capacidad para gestionar el fuego de manera sostenible.

Resumo

Os campos montanhosos da América do Sul são centros de biodiversidade e apresentam níveis elevados de endemismo. Estes campos são as nascentes de enormes rios e bacias hidrográficas, moldando a geografia e a biota do continente e fornecendo recursos hídricos a milhões de pessoas. Aqui, fornecemos uma visão geral da ecologia do fogo em dois campos montanhosos megadiversos da América do Sul, o *Páramo,* nos Andes, e o Campo Rupestre, no leste do Brasil. Para o *Páramo*, discutimos aspectos ecológicos e evolutivos da ecologia do fogo com foco na vegetação, dada a aparente falta de conhecimento das comunidades animais e microbianas. Para o Campo Rupestre, além de uma breve discussão sobre os efeitos do fogo na vegetação, também tentamos sintetizar os efeitos do fogo na germinação de sementes, invertebrados, vertebrados e comunidades microbianas. Após apresentar o estado da arte sobre a ecologia do fogo nas vegetações campestres montanhosas, identificamos lacunas de conhecimento e sugerimos prioridades de pesquisa que podem melhorar a nossa capacidade de gerir o fogo de forma sustentável.

7.1 Origin, Ecology and Biogeography of Montane Grasslands

Grasslands are ancient ecosystems that cover around 40% of the terrestrial land and occur on all continents except Antarctica (Gibson 2009; Veldman et al. 2015; Dinerstein et al. 2017). Characterized by a dominance of a continuous layer of grasses and herbs, grasslands are often present where climatic and edaphic conditions and disturbance regimes constraint trees establishment (Bond 2019). Grasslands are long recognized by their roles in shaping rangelands and agricultural production systems (Gibson 2009). As such, ancient tropical grasslands have long been misunderstood as by-products of forest degradation, and consequently, undervalued and underprotected by ecologists and conservation biologists (Parr et al. 2014; Silveira et al. 2022). However, multiple lines of evidence suggest the tropical grasslands arose in the Miocene, millions of years before humans have walked the Earth (Edwards et al. 2010), so that their ecological importance can now be fully appreciated (Strömberg and Staver 2022). Grasslands not only provide key ecosystem services, particularly water resources, forage, food and belowground carbon storage (Bengtsson et al. 2019; Zhao et al. 2020), but also, contrary to previously thought, are home to biodiversity comparable to that of tropical rainforests in many taxa (Murphy et al. 2016).

South American grasslands occur across tropical and temperate regions. These grasslands can be divided into tropical/subtropical grasslands and savannas, temperate grasslands and montane grasslands (Olson et al. 2001). South American grasslands are also home to extinct and extant large mammal herbivores which have left significant imprints on plant traits and functioning of current ecosystems (Strömberg et al. 2013; Dantas and Pausas 2022). In seasonal montane grasslands and savannas, the biomass of the herbaceous layer built-up during the rainy season dries out during the dry season providing abundant fuel for natural fires ignited by lightning at the onset of the rainy season (Bond 2019). However, fire activity on tropical montane grasslands can be fuel-limited or moisture-limited, depending on soil conditions and the amount and distribution of rainfall (Alvarado et al. 2020). As a consequence, most tropical grasslands have evolved under the influence of fire and grazing 10-8 million years ago (Buisson et al. 2019).

Grass dominated systems are naturally flammable (Zanzarini et al. 2022), and fire has been playing a major role shaping their biodiversity, vegetation structure, and functional trait composition (Maurin et al. 2014; Simon et al. 2009; Pilon et al. 2021). Here, we provide an updated account on fire effects on the *Páramo* and *Campo Rupestre*, two montane grasslands in South America. Our goals are to explore the state-of-art on fire ecology on these montane grasslands and to identify knowledge gaps and research priorities that can enhance our capacity to manage fire sustainably. First, we briefly describe the montane grasslands that we address and provide a literature overview of the fire-related topics studied in these grasslands. Then, we discuss ecological and evolutionary aspects of fire ecology with a focus on plants and vegetation, given the apparent lack of knowledge for both animal and microbial communities. We close the chapter by suggesting avenues for future research.

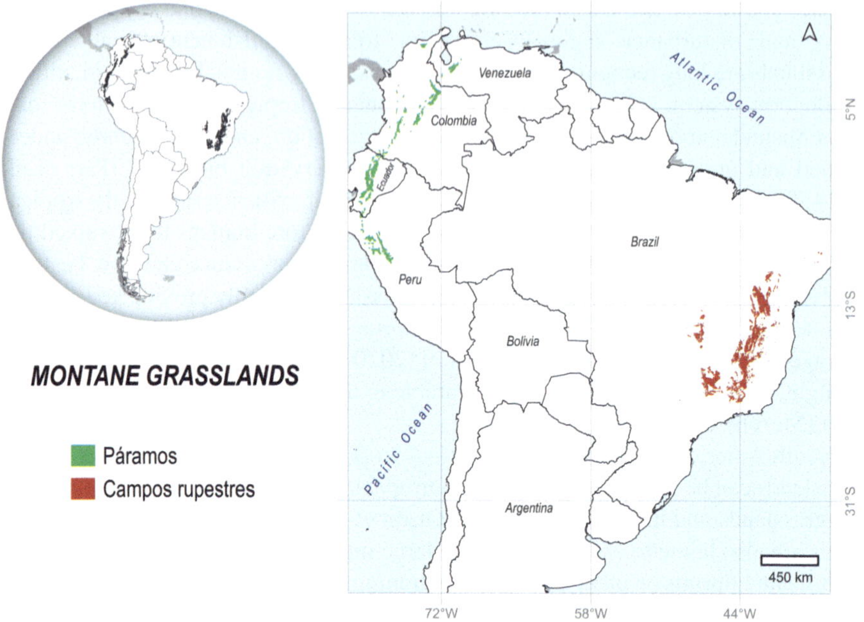

Fig. 7.1 Distribution of the major montane grasslands ecosystems in South America following Olson et al. (2001) and Silveira et al. (2016). Detailed maps for highland grasslands are not available but they concentrate as sky islands along the Brazilian Atlantic Forest. Produced by S.T. Alvarado and Júlia Abrantes Rodrigues

7.2 South American Montane Grasslands

South America is home to a wide diversity of montane grasslands. Montane grasslands are mainly found on the Andes, but also in southern Brazil, northern Argentina, Uruguay as well as in eastern Brazil along the Espinhaço Range, the Mantiqueira Range and Serras Gerais. Olson et al. (2001) described ecoregions dominated by montane grasslands (Fig. 7.1, Table 7.1), but the relationships between fire and vegetation have been very poorly studied in most of them. These highland vegetation types occur across gradients of elevation and precipitation, under different climatic regimes and are common throughout the continent. Here, we focus on the *Páramo* and *Campo Rupestre*, two montane grasslands where knowledge on fire ecology is more advanced, despite persistent knowledge gaps shared by both.

7.2.1 *Páramo*

The *Páramo* is a Neotropical, high-elevation grassland considered an alpine ecosystem. The *Páramo* is mostly located in the northern region of the geologically recent Andes mountain range, but it also stretches to small portions of Central America.

Table 7.1 Diversity of climate and vegetation in major montane grasslands in South America

Ecosystem	Elevation (m)	Climate	Rainfall (mm)	Vegetation	Distribution
Campo rupestre	900–2000	Dry winters and wet summers	800–1800	Tropical grasslands	Brazil, Bolivia, Guyana, Venezuela
Espinal	1625	Warm temperate, fully humid, hot summer	~800	Temperate grasslands and savannas	Argentina
High Monte	1500–3000	Temperate and arid to semi-arid	100–450	Montane grassland-shrublands	Argentina
Humid pampas	1116	Humid, temperate with dry winters and wet summers	~900	Temperate grasslands and savannas	Argentina
Páramo	2900–4800	Cold and humid with daily fluctuations in temperature and humidity	600–3000	Alpine grasslands	Venezuela, Colombia, Ecuador, Peru
Puna grasslands	3200–5000	Variable, depending on elevation, but generally cold with precipitation	400–2000	Montane grassland-scrublands	Peru, Bolivia, Chile, Argentina
Southern Andean steppe	2000–5000	Precipitation falls mainly in winter. Dry, with snowstorms and hailstorms common and frequent frosts year-round	200–600	Montane grassland-scrublands	Chile, Argentina

Modified from Olson et al. (2001)

This ecosystem extends across Peru (4200 km^2), Ecuador (12,600 km^2), Colombia (14,430 km^2) and Venezuela (3990 km^2) and patches in Panama and Costa Rica (Kappelle and Horn 2016; Podwojewski and Poulenard 2004). It usually covers Andean mountaintops ranging from 2900 to 4800 m above sea level, between the tree line of Andean montane forests and the permanent snow line (Madriñán et al. 2013). The climate of the *Páramo* is windy, cold and wet, with mean annual temperatures of 9 °C, with decreases of 0.6 °C for each 100 m upward change in elevation, and a day-night amplitude over 15 °C where temperatures often drop below 0 °C with increasing elevation (Podwojewski and Poulenard 2004).

The vegetation is dominated by grasslands, locally known as *pajonales*, and a mosaic of wetlands and peat bogs. The *Páramo* is known to be a refuge of highly endemic and diverse plant communities which are typically adapted to high daily temperature variation, fogs, low atmospheric pressure and high solar irradiation (Hofstede and Llambí 2019; Madriñán et al. 2013; Peyre et al. 2019; Cruz and Lasso 2021). The vegetation of the *Páramos* can be divided according to the elevational range: *superpáramo*, *páramo sensu stricto* (*páramo s.s.*) and *subpáramo*

Fig. 7.2 (**a, b**) Typical landscape of the *Páramo* under mist. (**c, d**) Abundant water resources and the iconic *Espeletia* (Asteraceae). (**e, f**) *Páramo* on fire and recently burned, showing habitat heterogeneity that affects fire spread. (Photos: **a**, **b**, **c**, **d**, **f** S.T. Alvarado; **e** O. Vargas Rios)

(Fig. 7.2). The *superpáramo* (4200–4800 m above sea level) is a periglacial environment characterized by a sparse vegetation cover and species adapted to nocturnal freezing and high solar radiation by leaf xeromorphic adaptation (reduced surfaces, folded, densely hairy, whitish leaves). The *Páramo s.s.* (3500–4200 m) is an open-canopy vegetation characterized by the presence of herbaceous vegetation dominated by bunchgrasses (*Calamagrostis* and *Festuca*) and other C3 grasses, dwarf bamboos (*Chusquea*), mosses, rosettes and small shrubs the Asteraceae, Ericaceae, Melastomataceae, and Hypericaceae that collectively cover 70% of the soil surface (Cleef 1981; Rangel 2000). The giant rosette-plants *Espeletia* (*frailejones*; Asteraceae) and *Puya* (Bromeliaceae), acaulescent rosettes, cushion-like vegetation (especially in swampy areas) are emblematic examples of morphological and physiological adaptations to extreme environmental conditions (Madriñán et al. 2013; Cruz and Lasso 2021; Fig. 7.2). Patches of *Polylepis* forest grow up to 4200 m. Finally, the *subpáramo* (2900–3500 m) represents a transitional zone between the *Páramo* and the Andean forest, which is characterized by a cover of small trees and scrubs taller than in the *Páramo*.

The *Páramo* is estimated to host between 3000 and 4000 species of vascular plants (Luteyn 1999; Podwojewski and Poulenard 2004; Hofstede and Llambí 2019), with endemism levels reaching 60% (Mena et al. 2001; León-Yánez 2011), confirming its importance as global biodiversity hotspot (Myers et al. 2000). Paleoecological evidence dating to 5 Ma has demonstrated that *Páramo* high diversity has a recent origin and is the result of a fast diversification (Hooghiemstra et al. 2006; Torres et al. 2013; Perez-Escobar et al. 2022). This high diversity level has been shaped by two main drivers: (1) the strong daily weather fluctuations which cause stressful conditions to plants (daily temperature variation, fogs, low atmospheric pressure and high solar irradiation), shaping morphological and physiological adaptations (Madriñán et al. 2013; Hofstede and Llambí 2019; Peyre et al. 2019; Cruz and Lasso 2021); and (2) the Pleistocene glacial and interglacial fluctuations that resulted in a dynamic environment shifting its distribution across a large elevational range (from 1500 to 4900 m), and forming cycles of ecosystem expansion and retraction among mountaintop *Páramo* patches, facilitating allopatric speciation and secondary contact (sympatric speciation) (Hooghiemstra and van der Hammen 2004; Nevado et al. 2019; Flantua and Hooghiemstra 2018; Flantua et al. 2019).

The *Páramo* corresponds to wetter areas in the Northern Andes in comparison to the drier areas in southern parts of the Andean range (Román-Cuesta et al. 2014). As a result, it is vital for water resources management providing water supply to millions of people, as well as other collection, regulation and maintenance ecosystem services (Chuncho and Chuncho 2019). Nevertheless, the *Páramo* is currently threatened by climate change with consequences for the hydrological cycles and carbon storage capacity (Buytaer et al. 2011; Yuste et al. 2017), but also by changes in land use related to agricultural expansion (Buytaert et al. 2005), and other contemporary issues such as the introduction of exotic plants, increasing fire frequency and mining activities (Chuncho and Chuncho 2019).

7.2.2 *Campo Rupestre*

Campo Rupestre can be defined as a montane, grassy–shrubby, fire-prone vegetation mosaic with nutrient-impoverished quartzite, sandstone or ironstone rocky outcrops along with white sands, stony soils and sometimes subjected to seasonal waterlogging (Miola et al. 2021) (Fig. 7.3). Small patches of *Campo Rupestre* occur as sky islands across many biomes in South America, but they are especially dominant on mountaintops above 900 m a.s.l. at the ancient and nutrient-poor soils along the Espinhaço Range, the largest, uninterrupted Precambrian orogenic belt in eastern South America, dating back 2.5–1.8 Gya (Silveira et al. 2016).

Fig. 7.3 (**a**) Typical heterogeneous landscape of the *Campo Rupestre*. (**b**) Rocky outcrops often support a woody flora which is typically isolated from fires. (**c, d**) *Campo rupestre* is a major provider of water resources and home to endemic biodiversity. (**e**) Fire-induced flowering in *Bulbostylis paradoxa* (Cyperaceae). (**f, g**) *Vellozia* spp. also has flower-stimulated flowering in many cases but fire is not essential for flowering in many other. (Photos: F.A.O. Silveira)

The unique location of the Espinhaço Range places it at the ecotone of three major biogeographic domains, the Atlantic Forest (tropical rainforest) to the east, the Cerrado (tropical savanna) to the west, and the Caatinga (tropical dry forest) to the north. Consequently, the Espinhaço Range is a unique center of biodiversity, endemism and plant diversification (Colli-Silva et al. 2019; Silveira et al. 2020; Vasconcelos et al. 2020; Barros-Souza and Borges 2023; Dantas-Queiroz et al.

2023). The area covered by *Campo Rupestre* is less than 0.8% of Brazil's land surface, yet it is home to more than 5300 species, which corresponds to >15% of the country's plant diversity. More than 40% of those species are endemic to the *Campo Rupestre* (Miola et al. 2021).

The climate is usually wet subtropical with two well-defined seasons: a rainy season from October to April and a dry season from May to September. Annual precipitation is ~1400 mm, and the mean temperature is 21 °C. Man-set fires are recurrent during the dry season, so that natural, lightning-ignited fires at the onset of the rainy season have become rare (Fernandes et al. 2020).

The plant communities established on different types of geological substrate, including quartzitic, ferruginous (*canga*) and sedimentary rocks, that are all shallow and extremely nutrient-poor soils with successive periods of water logging during the rainy season and water deficit during the dry season. The *Campo Rupestre* vegetation is characterized by an open grassy–shrubby mosaic with patches of rocky outcrop vegetation sustaining woody vegetation (Loiola et al. 2023). The flora is dominated by Poaceae, Cyperaceae, Asteraceae, Eriocaulaceae, Xyridaceae and Velloziaceae (Silveira et al. 2016).

Ecosystem services provided by the *Campo Rupestre* extends far beyond its small area, reaching some of the most populated areas of Brazil. Tens of millions of people benefit from the water supply, hydroelectric energy generation, ecotourism, medicinal plants, native fodder production, and belowground carbon storage. Unfortunately, ill-planned road construction and urban growth, uncontrolled mining, artisanal mining and ecotourism, afforestation, cattle overgrazing, overharvesting of ornamental species, and uncontrolled fires poses significant threats to biodiversity and ecosystem services provided by *Campo Rupestre* (Fernandes et al. 2020).

7.3 Fire in Montane Grasslands

Our aim here is to provide a brief account on the state of the art of fire ecology and evolution in the *Páramo* and in *Campo Rupestre*, rather than providing a systematic literature survey. In August 2023, we have searched the Web of Science using the terms fire (all fields) AND "campo rupestre" OR "campos rupestres" OR "rupestrian field" OR "rupestrian grassland" OR "canga" OR "rocky grassland" OR paramo OR paramos (all fields). We screened all 162 papers and retained fire-related empirical studies, thus excluding review papers. We broadly classified papers into vegetation-level (those looking at plant communities' composition and diversity, restoration, palaeoecology, and ecosystem functioning), regeneration (population-level studies examining fire effects on demography, flowering, resprouting, germination, seedling ecology and physiology), species interactions, animal communities, and microbial communities. After excluding non-related papers, we found only 32 papers for the *Páramo* and 19 for the *Campo Rupestre*. The results summarized in Fig. 7.4 clearly show that fire-related studies are rare in both ecosystems, but a closer look reveals some differences between them.

The effect of fire on plant communities of montane grasslands in South America is a topic still poorly addressed. In the last decades, the importance of these ecosystems, and the services they offered have been highlighted (Silveira et al. 2016; Overbeck et al. 2022), but very few scientific papers dealt with the effect of fire on plant regeneration strategies, effect of different fire regimes on community structure, the evolution of fire-related plant traits, fire-triggered phenology, effects of fire on carbon emission and carbon storage or the impact of fire on hydrological cycle. This may be surprising as these ecosystems are often clearly qualified as fire-prone environments. In a recent review about the history of research in *Campo Rupestre*, fire is rarely mentioned within the bibliographic survey (Miola et al. 2021), and for the *Páramo* only a few manipulative studies using experimental or prescribed fire in the *Páramo* have been carried out (Keating 1998; Ramsay and Oxley 1996), illustrating this gap of knowledge in montane grassland ecology in South America.

Fire-related studies in the *Páramo* have started in 1990, and there was no apparent trend in the literature (Fig. 7.4). Most studies for the *Páramo* were at the vegetation-level, notably palaeoecological studies looking at fire history in the Holocene (e.g. Frederick et al. 2018). There were few studies examining fire effects on population level in the *Páramo*, only one study linking fire and animal ecology (Raczka

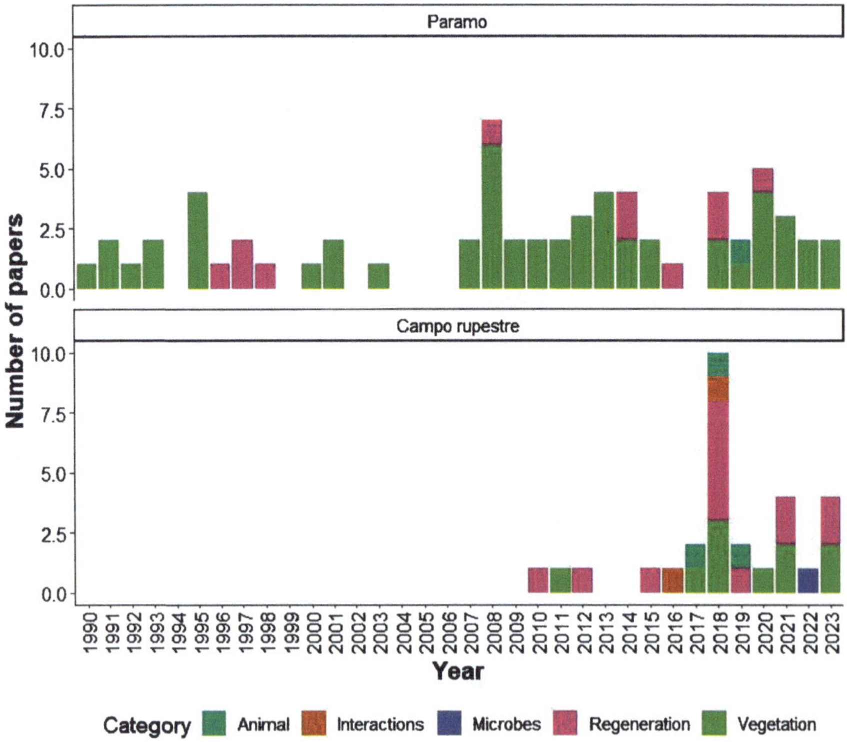

Fig. 7.4 State-of-the-art of fire ecology in the *Páramo* (**a**) and *Campo Rupestre* (**b**) according to the literature survey (details in the text)

et al. 2019), and none looking at ecological interactions or microbial communities (Fig. 7.4). In contrast, for the *Campo Rupestre*, the publication history is much more recent and scarcer, but more diverse. Apparently, there is a trend for an increasing number of publications more recently (Fig. 7.4). More than half of studies have looked at regeneration, and there were also studies examining fire effects on invertebrates (e.g. Nunes et al. 2019), plant-animal interactions (Costa et al. 2018) and microbial communities (Stürmer et al. 2022).

Whereas the number of studies related to fire especially in savannas increased during the last decade in South America, the effect of fire on montane grasslands still did not attracted much attention, despite the fact that several clues point out to the fact that fire is an integral part of these ecosystems (Espinoza et al. 2022), that fire frequency increased with human activities (Ledru et al. 2022) and also the increase of studies on the Integrated Fire Management as a conservation policy for fire-prone vegetation (Neger and Rosa-Paz 2022). Overall, the role of fire in the evolution, development and maintenance of these montane grasslands is still unclear.

7.4 Fire in the *Páramo*

Paleoenvironmental reconstructions of the Holocene through pollen and carbon records in northern South America have shown changes in climate and fire history. Jantz and Behling (2012) recorded hot and dry conditions in the early to mid-Holocene (~8000 to 3900 BP) in a *Páramo* in Ecuador, with the frequent presence of coal particles from early to mid-Holocene (~7600 B.P.). There is parallel evidence of fires throughout the Holocene in the Eastern Mountain range of Colombia, with a slight increase during mid-Holocene driven mainly by human activities (Espinoza et al. 2022). While the palaeoecological records suggest that human populations have been influencing the Ecuadorian *Páramo* since ~8000 B.P., Jantz and Behling (2012) observed that changes in fire activity since 4100 B.P. were likely related to more severe ENSO events, which were subsequently intensified by human activities after 3800 B.P., thus promoting the reorganization of plant communities since then (Espinoza et al. 2022).

7.4.1 Fire Regimes

Understanding the role of fire in the *Páramo* requires knowledge on fire regime. The potential fire frequency is determined by the amount and distribution of precipitation, coupled with fuel availability (mainly standing necromass) which is equal to the amount of biomass remaining after consumption by animals. *Páramo* fires generally occur during the dry season, between December and January, and can extend until February or March. Although 2-day long drought spells may be enough to dry out the vegetation in an atmospherically dry *Páramo*, in an atmospherically humid *Páramo*, drying periods are usually longer, especially for bamboo-type vegetation

(*Chusquea tessellata,* Poaceae) on poorly drained soils (Vargas-Ríos 2013). Likewise, the heterogeneity of the landscape is the main factor shaping the spread of *Páramo* fires, including the presence of natural (e.g. watercourses) or artificial barriers (such as roads). Indeed, the heterogeneity of vegetation associated with soils or topography is relevant to explain fire behavior and distribution because specific vegetation types can act as barriers, for example, peat bogs. Since vegetation distribution is highly dependent on the disturbance regime of the area, the extent and patchy patterns of fire events are the consequence of spatial patterns of past fires. A site that has been recently burned is unlikely to accumulate an adequate amount of fuel to support intense fires.

Very dissimilar intervals of natural fire frequency in equatorial tropical high mountain ecosystems have been reported, making it difficult to establish general patterns. Fire return intervals are estimated to be 6–30 years for the *Páramo* of Costa Rica (Horn 1989), 1–1.6 for the *Páramo* at in Los Nevados National Natural Park (Central Mountain range of Colombia; Verweij 1995) and 7–11 years for grassy *Páramos* (*pajonal*) of the Eastern Mountain range in Colombia (Vargas-Ríos 2000). Historical reconstructions based on paleoecological records, or the reconstruction of the most recent fire regime based on remote sensing data are still missing to better explain fire patterns in these ecosystems.

Fire history of a site can have a marked effect on fire intensity, via fuel availability. There is a close relationship between fire intensity and frequency, with fire temperature depending on biomass accumulation and its vertical distribution. For example, in experimental clump burnings, differences in fire intensity were found depending on the area of the plant involved (Ramsay and Oxley 1996). In the upper part of the leaves, at heights above 1 m, as well as in the middle levels of the clumps (35–65 cm), temperatures ranged between 400 and 500 °C, while at the base of the tufts and 2 cm belowground, the temperature was close to 65 °C. While in fire experiments in Chingaza National Natural Park in Colombia, in a plant community of *Espeletia grandiflora—Calamagrostis efusa* on well-drained soil, the increase of soil temperature after fire reaches no more than 8 °C (Cardenas 2013). However, more studies are needed to study the increase of temperatures during fires and its intensities in a gradient of different plant communities and landscape heterogeneity.

7.4.2 Fire Effects on Vegetation

Páramo fires generally can completely destroy the aboveground biomass (biomass + necromass), and only small ground-level plants can avoid being burned depending on fire severity and intensity (Vargas-Ríos 2000, 2013). The susceptibility of the *Páramo* vegetation to fires is related to the following aspects:

1. Amount of accumulated fuel: the *Páramo* comprises ecosystems where standing necromass accumulates mainly in the clumps of grasses (e.g. *Calamagrostis* spp.) and dead rosette leaves. In unburned *Páramo*, a continuous matrix of nec-

romass is formed in *Espeletia*-dominated vegetation. The main source of fuel is dead leaves, mainly from *Calamagrostis effusa*. About 80% of the epigeal phytomass of the clumps may consist of dead material (Cardoso and Schnetter 1976; Hofstede and Witte 1993; Hofstede et al. 1995; Vargas-Ríos 2013).
2. Low moisture content: *Páramo* fires occur in the dry season when the moisture content in the vegetation and soil are both low.
3. Low decomposition rates: the low decomposition rates of organic matter caused by low soils and environment temperatures origin aboveground necromass to accumulate which is the main source of fuel for fire.
4. Vegetation spatial distribution: the development of a continuous matrix of clumps is the most important factor for the spread of fires in the *Páramo*; additionally, high densities of *frailejones* (*Espeletia* spp.; Asteraceae) with their standing necromass in areas with low fire activity contribute significantly to fire spread.
5. Low levels of herbivory: Low herbivory in the *Páramo* contributes to the accumulation of organic matter. Unlike the *puna*, in the central Andes, large mammals have not evolved in the *Páramo*, insects are the main herbivores.

Climatic constraints of this ecosystem and the consequently low growth rate of the *Páramo* species are expected to be the main drivers of vegetation dynamics. However, data on fire frequency and return interval provides insights into vegetation resilience to fire and suggest that fire has significant effects on vegetation structure and functioning even at low frequency. The low growth rate of the *Páramo* species and the climatic constraints of this ecosystem also reduce the recovery rate after a fire. For example, the recovery period after a fire has been estimated to be 3–3.5 years at the community level, but the complete regeneration of the initial *Espeletia*-dominated communities took place only after 10 years in Los Nevados National Natural Park in Colombia (Verweij 1995). Total vegetation recovery has been estimated to range between 3 and 10 years elsewhere (Horn 1989; Verweij and Budde 1992; Ramsay and Oxley 1996). Recovery after fires takes 8–15 years in terms of vegetation structure (height, biomass and necromass) (Ramsay 1992; Vargas-Ríos 2002). However, the time needed to recover species composition (richness, life-forms and diversity of habitats for the fauna) is more difficult to determine. Estimates indicate that even after 20 years without disturbance, the vegetation is still not similar to pre-disturbance levels (Keating 1998; Verweij 1995). The species composition will depend on fire frequency and cattle grazing intensity, with woody species tending to be more negatively affected by fire than the herbaceous ones (Hofstede 1995; Vargas-Ríos 2002; Verweij 1995). Thus, vegetation response depends on the capacity for regeneration and reproduction, with resprouters regenerating immediately from the base when fires do not overcome the first few centimeters of the ground, and seeders establishing through sexual reproduction (Hofstede et al. 2014). In the *Páramo*, the major regeneration strategies can be summarized as (1) expression of a bud bank, mainly of clumps-type grasses; (2) shrub resprouting; (3) seed dispersal; and (4) expression of seed banks (Vargas-Ríos 1997, 2000).

Contrary to other ecosystems, ashes from fires do not have a fertilizing effect in the *Páramo*. Consequently, the vegetation does not grow faster after a fire, and productivity tends to be reduced because vegetative buds are exposed to the extreme climatic conditions characteristic of high elevation (Hofstede 1995; Vargas-Ríos 2002). Fire takes away vegetation insulating layer, exposing it to even further temperature and humidity fluctuations (Hofstede et al. 2014). High air temperatures also cause an increase in evapotranspiration and the decomposition of organic matter, increasing soil drought (Hofstede 1995; Ramsay and Oxley 1996).

Considering the fire effects on soils, fire drastically reduced the saturation of bases in inceptisol soils (organic) which means that these soils often lost bases (Vargas-Ríos 2002). The total P content both in inceptisoles and histosols increases significantly in the first month after fire compared to unburned areas, fluctuating during the recovery. pH varies only in the inceptisoles and decreases during the first month after fire but increases from the third month onwards. The levels of Ca and Mg increase whereas the level of K decreases in the first month after fire reaching the nutrient levels of unburned vegetation only 1 year after fire. The C/N rate in the vegetation drops drastically after a fire and even after a year of recovery it does not reach pre-disturbance levels.

7.4.3 Fire-Related Plant Traits

Páramo species are adapted to low temperatures, water and nutrient stress, rather than possessing fire-related traits (Monasterio and Sarmiento 1991). The *Páramo* can be too cold, wet or dry to burn (Hardesty et al. 2005), in some cases they lack a sufficient source of fuel or ignition to consider fire as an evolutionary force. Therefore, the *superpáramo* located at elevations ranging between 4200 and 4800 m can be classified as fire-independent since they correspond to ecosystems that have low biomass production as a result of stress caused by low temperatures and/or hydric stress. However, the *subpáramo* and *páramo sensu stricto* can be classified as fire-sensitive ecosystems given their apparent lack of fire-related adaptations coupled with high individual mortality even under low intensity fires (Vargas-Ríos et al. 2021). Only few shrub species present belowground organs such as tubers or xylopodia. The *Páramo* has evolved as biogeographic sky islands that, contrary to many open ecosystems (e.g. savannas and tropical grasslands), it was not subjected to a continuous natural disturbance regime. Consequently, *Páramo* species have not developed particular disturbance-related adaptations (such as fire and grazing/browsing by large herbivores). Their plants are not adapted to resist recurring fires, just as their soils and vegetation are not adapted to resist trampling or permanent grazing by animals. According to Young et al. (2011), the *Páramo* is sensitive to disturbances for several reasons: (1) their distribution in relatively small patches (islands) makes them especially vulnerable to edge effects; (2) due to high endemism levels, extinction risk increases with increasing *Páramo* degradation and the likelihood of among-patch migration decreases; and (3) they are very sensitive to increasingly frequent fires.

7.4.4 Fire Interactions with Environmental Drivers

Livestock practices are closely linked with the use of fire, which is why many ecosystems globally present interactions between both disturbances (Vargas-Ríos 2002). The use of fire as an agricultural tool to create more productive systems is a recent practice in the *Páramo*, where people set fires to renew pastures for cattle and prepare the land for crops such as green peas, potato and broad bean. The effects of these combined disturbances are evident at different scales, from population- to the landscape-level (Guhl 1968; Laegaard 1992; Hofstede and Rossenaar 1995). Fire has a strong impact at the population-level mainly on the mortality of small individuals (Verweij and Kok 1992; Premauer 1999). Cattle trampling also affects the recruitment and establishment of species by affecting seedlings and saplings. The combination of intensive grazing and fire produces open vegetation with a lot of exposed soil, low organic matter content, and low water retention capacity, making it more sensitive to trampling and not suitable for recolonization by natural vegetation adapted to germinate in the presence of taller vegetation (Hofstede 1995; Hofstede et al. 1995). This can halt or arrest natural succession and prevent regeneration under the combined effects of both disturbances (Vargas-Ríos 2000).

The indiscriminate use of fire results in vegetation patches with different degrees of alteration, changes the population structure of plants and animals, and changes the composition of plant communities (Premauer 1999). In a relatively small area, it is possible to find a mosaic of sites that have not been burned for several years, sites with moderate and high fire frequency, sites recently burned, and sites with a strong impact of burning and grazing, showing a pattern of patches with different disturbance intensities (Verweij 1995).

In *Páramos* highly disturbed by recurrent fires and intensive grazing, the shrubby and herbaceous layers are reduced with direct consequences on the characteristic species (e.g. *Chusquea tessellata*). A decrease in vegetation cover of common life-forms such as grass clumps can disappear completely when fragmented by trampling, leaving the meristems exposed to intense UV-B light and below 0 °C temperatures (Posada and Cárdenas 1999). Likewise, a decline in perennial species rather than annual ones (Hofstede 1995; Verweij 1995) reduces vegetation to a carpet of short grasses and herbs (especially eudicots) capable of tolerating continuous trampling and browsing, with high total cover and high proportion of live biomass (foliage especially) (Hofstede et al. 1995; Hofstede and Rossenaar 1995).

Under moderate frequency of both fire and cattle grazing, the primary productivity can be higher than in undisturbed vegetation (Hofstede et al. 1995; Hofstede and Rossenaar 1995). Premauer (1999) found that with increasing disturbance, the total amount of phytomass decreased. There was also a decrease in the height, richness and evenness of the vertical structure, and an increase in vegetation clumpness, possibly related to species reproducing vegetatively. Finally, fire disturbances also promote rapid colonization of invasive species such as *Ulex europeus* (Vargas-Ríos 2013). Unfortunately, data on fire effects on animals and microbes are largely missing.

7.5 Fire in *Campo Rupestre*

The *Campo Rupestre* is a fire-prone ecosystem, and studies on community structure, regeneration, restoration, and conservation should address natural and anthropogenic fires. Natural fires are ignited by lighting usually at the dry-to-rainy season transition. Currently, the highest number of natural fires occurs in August, but fires in September are those covering larger areas (Figueira et al. 2016). Natural fires are becoming increasingly rare because anthropogenic fires are becoming so widespread and frequent that they extend to far-reaching areas and consume most of the dry biomass before the arrival of the first rains and lightning that would otherwise trigger natural fires. Fire frequencies range from twice a year to a peak frequency of 7 years depending on the site. Large-scale fires occur every 6–8 years, in sites accumulating fine fuel biomass above 9 ton/ha (Figueira et al. 2016). Unfortunately, long-term fire regimes in *Campo Rupestre* are largely unknown, but the use of fire to support livestock pastures management and arson has dominated the fire dynamics of these ecosystems since post-European colonization times (Figueira et al. 2016).

7.5.1 Fire Effects on Vegetation and Plant Populations

Present evidence suggests that *Campo Rupestre* communities are fire-resilient. Biomass recovery quickly after wildfires (Alberton et al. 2023), with no significant or even higher species richness in recently burnt patches (Le Stradic et al. 2018a). Burned grasslands tended to return to a pre-fire state in terms of richness, plant cover, and life-form diversity, but each community presented different trajectories in floristic transitions and different degrees of species similarity in relation to adjacent unburned areas. Despite these floristic differences, plant cover is similar to unburned areas (Araujo and Conceição 2021). Changes seem to be faster in herbaceous communities and slower on rocky outcrops (Neves and Conceição 2010; Alberton et al. 2023). As discussed above, the *Campo Rupestre* is a highly heterogeneous ecosystem with patches of mountaintop forest islands immersed in a matrix of grasslands. In contrast to the fire-resilient grasslands, these forest islands are fire-sensitive, thus creating a dilemma for fire management policies (Coelho et al. 2018).

Several traits of *Campo Rupestre* species allow them to survive natural fire regimes and/or take advantage of the post-fire environment, suggesting an ancient role of fire in their evolution (Figueira et al. 2016). The main dominant species resprout from underground storage organs (USOs) and from aerial organs (Neves and Conceição 2010) indicating that resprouting is the most common post-fire regeneration strategy. The mechanisms shaping regeneration are related to the evolution of USOs, rather than colonization from seeds (see section below). USOs are common in the *Campo Rupestre* and allow for quick resprouting after disturbances (Simon et al. 2009; Brum et al. 2017). A diversity of carbohydrate-based structures in USOs provides tolerance to environmental disturbances such as fire and support the mobilization of resources to aboveground parts (Joaquim et al. 2018).

Mass flowering after fire is observed in several *Campo Rupestre* species (Conceição and Orr 2012; Conceição 2018; Le Stradic et al. 2018b; Souza et al. 2018), that usually show high seed germinability (Le Stradic et al. 2015), indicating a fine-tuned relationship between fire and plant species reproduction. Most *Campo Rupestre* species are classified as fire-stimulated (Fidelis and Zirondi 2021), and few species have been shown to have fire-dependent flowering such as the sedge *Bulbostylis paradoxa* (Fidelis et al. 2019).

Uncontrolled fires and species overharvesting pose significant threats to endemic *Campo Rupestre* species, thus a major challenge is to reconcile knowledge of fire ecology with the sustainable harvest of species with economic value that support Local Communities. Current extractive practices interfere with the recruitment dynamics in *Comanthera elegantula*, an everlasting species (Bedê et al. 2018). An intermittent fire regime appears to favor seedling establishment and growth, while periodic fire episodes negatively impact the progression of seedling cohorts and the regrowth of the herbaceous cover. Souza et al. (2018) also show that synchronized flowering of *Vellozia sincorana* is triggered by fire; no flowers were recorded in the absence of recent fires. The long-term effects of two common harvesting techniques and different fire intervals indicate that adult survival is the most important life-history variable for population persistence, and recent fires are primarily responsible for increases in reproductive rates. Harvesting may become sustainable, as long as a balance between harvesting and fire use is achieved.

7.5.2 Fire Effects on Regeneration from Seed

Fire plays multiple roles in seed germination including stimulating seed production, alleviating dormancy and promoting germination. Fire-related germination has been especially studied in fire-prone ecosystems (Pausas and Lamont 2022; Rosbakh et al. 2023). Several papers have been published on fire-related germination in *Campo Rupestre*, but these papers have focused on a handful species, mostly on a few selected dominant families.

Recently, Ordóñez-Parra et al. (2025) have used meta-analyses techniques to synthesize the effects of fire (heat and smoke) on germination percentage and germination time for all available species. When all data are pooled together, fire-related cues (heat shock and smoke) reduce germination percentage and time. The effect of fire-related cues, however, varies according to the treatment applied and seed traits. For example, when decomposing analyses by each heat shock treatments, they found a negative effect on germination percentage in seeds exposed by 1-min heat shocks at 200 °C. When this treatment is excluded, heat shocks did not affect germination, regardless of species growth-form, distribution or microhabitat. Seed size was a moderate predictor of fire-effects on seed germination. Fire-related cues increased germination percentage in relatively heavier seeds, and smaller seeds remain more sensitive to heat.

There is no evidence that smoke affects germination percentage, but overall, there is moderate evidence that it reduces germination time in herbaceous and endemic species. Faster germination after fires suggests that species optimize establishment in post-fire environments when competition with neighbors is relaxed and resources are more readily available. This agrees with a previous study showing faster germination in post-fire fruiting herbs compared to fire-independent fruiting ones. Overall, the current literature indicates a relatively small effect of fire-related cues in *campo rupestre*, as shown more broadly for the lowland cerrado (Daibes et al. 2022).

7.5.3 Fire Effects Beyond Plants

As discussed above, research in *Campo Rupestre* historically focused on fire effect on vegetation, plant populations and seed germination (Fig. 7.4). Consequently, knowledge on fire effects on organisms beyond plants is scarce. Here, we attempt to summarize the few studies on fire effects on invertebrates, vertebrates and microorganisms. Three independent studies on invertebrate communities have been conducted in a *campo rupestre* at Serra do Cipó, southeastern Brazil. Ant community structure in response to fire were examined in two elevations (800 m above sea level and 1300 m a.s.l.) 1, 4, and 10 months after a fire event (Neves et al. 2016). In both elevations, species richness did not differ between burned and unburned plots, but species composition was different at the highest elevation. Although data suggest that ant communities are resilient to fires, fires represent a disturbance shaping the structure of ant communities. A second study examined the effects of fire on dung beetle richness, abundance, mean community biomass and composition (Nunes et al. 2019). Consistently, with the results from ants, no effect of fire occurrence, time since last fire and fire frequency on any community variable was found. These results suggest that dung beetle communities are resilient to fire, but contrary to ants, fire seem to have no effect on the community structure of dung beetles. Finally, Fernandes et al. (2023) evaluated the short-term effects of fire on termites across an elevational gradient. Consistently, with the previous studies, no effect of fire on average species richness and bait colonization was found for any elevation belt.

For vertebrates, it appears that both amphibian and mammal communities are also fire-resilient. Drummond et al. (2018) monitored anurans up to 7 years after fire events. An increase in anuran species richness was found immediately after the fire event, with five species recorded at low densities exclusively in the post-fire period. Aligned with this finding, fire passage had no effect on anuran phylogenetic diversity (Anjos et al. 2024). Using camera traps to monitor medium and large mammal occurrence and activity in two *Campo Rupestre* sites, Pinho et al. (2017) found that fire frequency did not affect species richness or use intensity.

Stürmer et al. (2022) compared arbuscular mycorrhizal fungi (AMF) species composition, spore numbers, species richness, and mycorrhizal inoculum potential in burned and unburned *Campo Rupestre* in the Amazon. Again, no consistent difference was found in any parameter between burned and unburned plots, suggesting that AMF spore communities are not affected by burning.

Finally, using network metrics, Costa et al. (2018) examined to what extent fires disrupt ant-plant interactions at Serra do Cipó, southeastern Brazil. Fire has led to short-term negative effects on the diversity and frequency of interactions, but network metrics in burned plots returned to pre-disturbance levels in 6 months, highlighting the resilience of ant-plant interactions after fire disturbances. Altogether, current evidence suggests that communities of terrestrial invertebrates, vertebrates and plant-animal interactions in *Campo Rupestre* are resilient to fires, but no data are available on the effects of changing fire regimes on terrestrial biota.

7.6 Knowledge Gaps and Future Research

To improve conservation outcomes, fire management must simultaneously address habitat heterogeneity and socioeconomic complexities of megadiverse montane grasslands landscapes and be supported by sound scientific evidence. Here, we found that research in fire evolutionary ecology is still emerging and remains insufficient to provide a solid basis for fire management in both *Páramo* and *Campo Rupestre*. Despite the recent flourishing of savanna fire ecology, similar advances have not yet climbed up the mountains. We identified several knowledge gaps that persist for decades and preclude the development of evidence-based fire management programs.

Such knowledge gaps include (1) evolutionary perspectives on fire-related traits, (2) animal and microbial fire ecology, (3) alternative stable states mediated by fire and climate change; (4) palaeoecological studies to reconstruct fire history; (5) effective ways to reconcile people's needs and fire management, (6) drawing knowledge on use of fire by Indigenous Peoples, (7) implications for integrative fire management.

In fire-prone montane grasslands, recent changes in fire regime pose a major threat to endangered species (Monteiro et al. 2018). Thus, understanding the effects of fire on vegetation requires experimental burning studies similarly as those developed for the Cerrado (Fidelis 2020). This is particularly true in a context of climate change. We also need to understand how habitat heterogeneity modulates fire effects on vegetation. For example, comparing fire effects on dry and humid *páramos* (*páramos* de pajonal y *páramos* de chuscal) and fire effects on forest islands vs grasslands in *Campo Rupestre* will shed some light on why some of these ecosystems seem to be fire-sensitive despite fire being part of these ecosystems. Finally, the role of fire in the evolution, maintenance and restoration of these montane grasslands remains to be investigated. Despite the lack of long-term research and specific legislation and frameworks for the *Campo Rupestre*, adapting the integrative fire management plants for the Cerrado seems like a significant step forward. We hope the overview provided here fuels novel research on fire ecology that will help protecting megadiverse montane grasslands through specific fire management policies that can be applied in the Andes, the Espinhaço Range and in other mountains where fire is becoming increasingly important from the conservation perspective. Unfortunately, slow progress in legislation has prevented the development of specific fire management use and policy in montane grasslands. Enhancing our understanding of fire effects on montane grassland biota and functioning is expected to support IFM as developed for lowland fire-prone savannas.

References

Alberton A, Alvarado ST, Torres RS, Fernandes GW, Morellato LPC (2023) Monitoring immediate post-fire vegetation dynamics of tropical mountain grasslands using phenocameras. Ecol Inform 78:102341. https://doi.org/10.1016/j.ecoinf.2023.102341

Alvarado ST, Andela N, Silva TSF, Archibald S (2020) Global thresholds of transition between moisture-limited and fuel-limited fire regimes in tropical savannas and grasslands. Glob Ecol Biogeor 29:331–344. https://doi.org/10.1111/geb.13034

Anjos AG, Alvarado ST, Solé M, Benchimol M (2024) Influence of fire regime on the taxonomic and phylogenetic diversity of frog communities in a Brazilian flammable ecosystem. For Ecol Manag 551:121556. https://doi.org/10.1016/j.foreco.2023.121556

Araujo T, Conceição AA (2021) High functional redundancy drives vegetation recovery in Campo rupestre affected by wildfires. Flora 281:151866. https://doi.org/10.1016/j.flora.2021.151866

Barros-Souza Y, Borges LM (2023) Spatial-and lineage-dependent processes underpin floristic assembly in the megadiverse Eastern South American mountains. J Biogeogr 50:302–315. https://doi.org/10.1111/jbi.14527

Bedê LC, Paglia AP, Neves ACO, Martins RP (2018) Effects of traditional extractive management on the seedling recruitment dynamics of *Comanthera elegantula* (Eriocaulaceae) in Espinhaço mountain range, SE Brazil. Flora 238:216–224. https://doi.org/10.1016/j.flora.2017.06.011

Bengtsson J, Bullock JM, Egoh B, Everson C, Everson T, O'Connor T et al (2019) Grasslands—more important for ecosystem services than you might think. Ecosphere 10:e02582. https://doi.org/10.1002/ecs2.2582

Bond WJ (2019) Open ecosystems: ecology and evolution beyond the forest edge. Oxford University Press, Oxford

Brum M, Teodoro GS, Abrahão A, Oliveira RS (2017) Coordination of rooting depth and leaf hydraulic traits defines drought-related strategies in the Campos rupestres, a tropical montane biodiversity hotspot. Plant Soil 420:467–480. https://doi.org/10.1007/s11104-017-3330-x

Buisson E, Le Stradic S, Silveira FAO, Durigan G, Overbeck GE et al (2019) Resilience and restoration of tropical and subtropical grasslands, savannas, and grassy woodlands. Biol Rev 94:590–609. https://doi.org/10.1111/brv.12470

Buytaer W, Cuesta F, Tobón F (2011) Potential impacts of climate change on the environmental services of humid tropical alpine. Glob Ecol Biogeor 20:19–33. https://doi.org/10.1111/j.1466-8238.2010.00585.x

Buytaert W, Wyseure G, de Bievre B, Deckers J (2005) The effect of land-use changes on the hydrological behaviour of histic andosols in South Ecuador. Hydrol Proc 19:3985–3997. https://doi.org/10.1002/hyp.5867

Cardoso GH, Schnetter ML (1976) Estudios ecológicos en el Páramo de Cruz Verde, Colombia III. La biomasa de tres asociaciones vegetales y la productividad de *Calamagrostis effusa* (HBK) Steud y *Paepalanthus columbiensis* Ruhl. en comparación con la concentración de clorofila. Caldasia 54:85–91

Cárdenas CD, LÁ (2013) El fuego y el pastoreo en el páramo húmedo de Chingaza (Colombia): efectos de la perturbación y respuestas de la vegetación. Universitat Autònoma de Barcelona

Chuncho C, Chuncho G (2019) Páramos del Ecuador, importancia y afectaciones: una revisión. Bosques Latitud Cero 9:71–83

Cleef AM (1981) The vegetation of the paramos of the Colombian Cordillera Oriental. Mededelingen van het Botanisch Museum en Herbarium van de Rijksuniversiteit te Utrecht 481:1–321

Coelho MS, Neves FS, Perillo LN, Morellato LPC, Fernandes GW (2018) Forest archipelagos: a natural model of metacommunity under the threat of fire. Flora 238:244–249. https://doi.org/10.1016/j.flora.2017.03.013

Colli-Silva M, Vasconcelos TN, Pirani JR (2019) Outstanding plant endemism levels strongly support the recognition of campo rupestre provinces in mountaintops of eastern South America. J Biogeogr 46:1723–1733. https://doi.org/10.1111/jbi.13585

Conceição AA (2018) A hot case for conservation: Candombá (*Vellozia pyrantha*), a flammable plant endemic to a national park is used to make a fire and threatened by fire suppression policy. J Nat Conserv 45:118–121. https://doi.org/10.1016/j.jnc.2018.08.011

Conceição AA, Orr BJ (2012) Post-fire flowering and fruiting in *Vellozia sincorana*, a caulescent rosette plant endemic to Northeast Brazil. Acta Bot Bras 26:94–100. https://doi.org/10.1590/S0102-33062012000100011

Costa FV, Blüthgen N, Viana-Junior AB, Guerra TJ, Di Spirito L, Neves FS (2018) Resilience to fire and climate seasonality drive the temporal dynamics of ant-plant interactions in a fire-prone ecosystem. Ecol Indic 93:247–255. https://doi.org/10.1016/j.ecolind.2018.05.001

Cruz M, Lasso E (2021) Insights into the functional ecology of páramo plants in Colombia. Biotropica 53:1415–1431. https://doi.org/10.1111/btp.12992

Daibes LF, Ordóñez-Parra CA, Dayrell RL, Silveira FAO (2022) Regeneration from seeds in South American savannas, in particular the Brazilian Cerrado. In: Baskin CC, Baskin JM (eds) Plant regeneration from seeds. Academic, pp 183–197. https://doi.org/10.1016/B978-0-12-823731-1.00002-0

Dantas VL, Pausas JG (2022) The legacy of the extinct Neotropical megafauna on plants and biomes. Nat Commun 13:129. https://doi.org/10.1038/s41467-021-27749-9

Dantas-Queiroz MV, Hurbath F, Godoy FMR, Lanna FM, Versieux LM, Palma-Silva C (2023) Comparative phylogeography reveals the demographic patterns of neotropical ancient mountain species. Mol Ecol 32:3165–3181. https://doi.org/10.1111/mec.16929

Dinerstein E, Olson D, Joshi A, Vynne C, Burgess ND, Wikramanayake E, Hahn N et al (2017) An ecoregion-based approach to protecting half the terrestrial realm. Bioscience 67:534–545. https://doi.org/10.1093/biosci/bix014

Drummond LO, Moura FR, Silverio Pires MR (2018) Impact of fire on anurans of rupestrian grasslands (Campos rupestres): a case study in the Serra do Espinhaço, Brazil. Salamandra 54:1–10

Edwards EJ, Osborne CP, CAE S, Smith SA, Consortium CG, Bond WJ et al (2010) The origins of C4 grasslands: integrating evolutionary and ecosystem science. Science 328:587–591. https://doi.org/10.1126/science.1177216

Espinoza IG, Franco-Gaviria F, Castañeda I, Robinson C, Room A et al (2022) Holocene fires and ecological novelty in the high Colombian Cordillera oriental. Front Ecol Evol 10:895152. https://doi.org/10.3389/fevo.2022.895152

Fernandes GW, Arantes-Garcia L, Barbosa M, Barbosa NP, Batista EK, Beiroz W et al (2020) Biodiversity and ecosystem services in the Campo Rupestre: a road map for the sustainability of the hottest Brazilian biodiversity hotspot. Perspect Ecol Conserv 18:213–222. https://doi.org/10.1016/j.pecon.2020.10.004

Fernandes GW, Oki Y, Negreiros D, Constantino R, Novais S (2023) No short-term effects of fire on termite diversity in a tropical mountain. Insect Soc 70:203–211. https://doi.org/10.1007/s00040-023-00909-4

Fidelis A (2020) Is fire always the "bad guy"? Flora 268:151611. https://doi.org/10.1016/j.flora.2020.151611

Fidelis A, Zirondi HL (2021) And after fire, the Cerrado flowers: a review of post-fire flowering in a tropical savanna. Flora 280:151849. https://doi.org/10.1016/j.flora.2021.151849

Fidelis A, Rosalem P, Zanzarini V, Camargos LS, Martins AR (2019) From ashes to flowers: a savanna sedge initiates flowers 24 h after fire. Ecology 100:1–4. https://doi.org/10.1002/ecy.2648

Figueira JEC, Ribeiro KT, Ribeiro MC, Jacobi CM, França H, Neves ACO et al (2016) Fire in rupestrian grasslands: plant response and management. In: Fernandes GW (ed) Ecology and conservation of mountaintop grasslands in Brazil. Springer, Cham, pp 415–448. https://doi.org/10.1007/978-3-319-29808-5_18

Flantua SG, Hooghiemstra H (2018) Historical connectivity and mountain biodiversity. In: Hoorn C, Perrigo A, Antonelli A (eds) Mountains, climate and biodiversity. Wiley–Blackwell, pp 171–185

Flantua SG, O'Dea A, Onstein RE, Giraldo C, Hooghiemstra H (2019) The flickering connectivity system of the north Andean Páramos. J Biogeogr 46:1808–1825. https://doi.org/10.1111/jbi.13607

Frederick L, Brunelle A, Morrison M, Crespo P, Johnson W (2018) Reconstruction of the mid-Holocene paleoclimate of the Ecuadorian Andean páramo at Tres Lagunas, Ecuador. The Holocene 28:1131–1140. https://doi.org/10.1177/0959683618761547

Gibson DJ (2009) Grasses and grassland ecology. Oxford University Press, Oxford

Guhl E (1968) Los páramos circundantes de la Sabana de Bogotá. Su ecología y su importancia para el régimen hidrológico de la misma. Colombia Geográfica 9:195–212

Hardesty J, Myers RL, Fulks W (2005) Fire, ecosystems, and people: a preliminary assessment of fire as a global conservation issue. The GeorgeWright Forum 22:78–87

Hofstede RGM (1995) Effects or burning and grazing on a Colombian páramo ecosystem. PhD dissertation. University of Amsterdam, Amsterdam

Hofstede R, Llambí L (2019) Plant diversity in Páramo—neotropical high mountain humid grasslands. Encyclopedia of the World's Biomes:362–372. https://doi.org/10.1016/B978-0-12-409548-9.11858-5

Hofstede RG, Rossenaar J (1995) Biomass of grazed, burned and undisturbed páramo grasslands, Colombia. II. Root mass and aboveground:belowground ratio. Artic Alpine Res 27:13–18

Hofstede RGM, Witte HJL (1993) An evaluation of the use of the dry-weight-rank and the comparative yield biomass estimation methods in páramo ecosystem research. Caldasia 17:205–210

Hofstede RGM, Mondragón X, Rocha CM (1995) Biomass of grazed, burned and undisturbed páramo grasslands, Colombia. I. Aboveground vegetation. Artic Alpine Res 27:1–12

Hofstede RGM, Calles J, López V, Polanco R, Torres F, Ulloa J, Vásquez A, Cerra M (2014) Los Páramos Andinos ≚Qué sabemos? Estado de conocimiento sobre el impacto del cambio climático en el ecosistema páramo. IUCN, Quito

Hooghiemstra H, van der Hammen T (2004) Quaternary Ice-Age dynamics in the Colombian Andes: developing an understanding of our legacy. Phil Trans R Soc Lond B 359:173–181. https://doi.org/10.1098/rstb.2003.1420

Hooghiemstra H, Wijninga V, Cleef A (2006) The palaeobotanical record of Colombia: implications for biogeography and biodiversity. Ann Missouri Bot Gard 93:297–324. https://doi.org/10.3417/0026-6493(2006)93[297:TPROCI]2.0.CO;2

Horn SP (1989) Prehistoric fires in the Chirripó highlands of Costa Rica: sedimentary charcoal evidence. Rev Biol Trop 37:139–148

Jantz N, Behling H (2012) A Holocene environmental record reflecting vegetation, climate, and fire variability at the Páramo of Quimsacocha, southwestern Ecuadorian Andes. Veg Hist Archaeobotany 21:169–185. https://doi.org/10.1007/s00334-011-0327-x

Joaquim EO, Silva TM, Figueiredo-Ribeiro RDCL, Moraes MG, Carvalho MAM (2018) Diversity of reserve carbohydrates in herbaceous species from Brazilian campo rupestre reveals similar functional traits to endure environmental stresses. Flora 238:201–209. https://doi.org/10.1016/j.flora.2017.01.001

Kappelle M, Horn SP (2016) The Paramo ecosystem of Costa Rica's highlands. In: Kappelle M (ed) Costa Rican ecosystems. The University of Chicago Press, Chicago, pp 492–523

Keating PL (1998) Effects of anthropogenic disturbance on paramo vegetation in Podocarpus National Park, Ecuador. Phys Geogr 19:221–238. https://doi.org/10.1080/02723646.1998.10642648

Laegaard S (1992) Influence of fire in the grass páramo vegetation of Ecuador. In: Balslev H, Luteyn J (eds) Páramo an Andean ecosystem under human influence. Academic, London

Le Stradic S, Silveira FAO, Buisson E, Cazelles K, Carvalho V, Fernandes GW (2015) Diversity of germination strategies and seed dormancy in herbaceous species of campo rupestre grasslands. Austral Ecol 40:537–546. https://doi.org/10.1111/aec.12221

Le Stradic S, Hernandez P, Fernandes GW, Buisson E (2018a) Regeneration after fire in campo rupestre: short-and long-term vegetation dynamics. Flora 238:191–200. https://doi.org/10.1016/j.flora.2016.12.001

Le Stradic S, Buisson E, Fernandes GW, Morellato LPC (2018b) Reproductive phenology of two co-occurring Neotropical mountain grasslands. J Veg Sci 29:15–24. https://doi.org/10.1111/jvs.12596

Ledru MP, Aquino-Alfonso O, Finsinger W, Samaniego P, Hidalgo S (2022) Changes in the vegetation and water cycle of the Ecuadorian páramo during the last 5000 years. The Holocene 32:950–963. https://doi.org/10.1177/09596836221101251

León-Yánez S (2011) La flora de los páramos ecuatorianos. In: Mena-Vásconez P et al (eds) Páramo. Paisaje estudiado, habitado, manejado e institucionalizado. EcoCiencia/Abya Yala/ECOBONA, Quito, pp 25–40

Loiola PP, Morellato LPC, Camargo MGG, Kamimura VA, Mattos JS et al (2023) Shared-role of vegetation types, elevation and soil affecting plant diversity in an old-tropical mountain hotspot. J Mount Sci 20:1842–1853. https://doi.org/10.1007/s11629-022-7838-

Luteyn JL (1999) Paramos: a checklist of plant diversity, geographic distribution and botanical literature. Mem New York Bot Garden 84:1–278

Madriñán S, Cortés AJ, Richardson JE (2013) Páramo is the world's fastest evolving and coolest biodiversity hotspot. Front Genet 4:56606. https://doi.org/10.3389/fgene.2013.00192

Maurin T, Davies TJ, Burrows JE, Daru BH, Yessoufou Muasya AM, van der Bank M, Bond WJ (2014) Savanna fire and the origins of the 'underground forests' of Africa. New Phytologist 204, 201–214

Mena P, Medina G, Hofstede RGM (2001) Los Páramos del Ecuador: particularidades, problemas y perspectivas. Abya Yala, Quito

Miola DT, Ramos VD, Silveira FAO (2021) A brief history of research in campo rupestre: identifying research priorities and revisiting the geographical distribution of an ancient, widespread Neotropical biome. Biol J Linn Soc 133:464–480. https://doi.org/10.1093/biolinnean/blaa175

Monasterio M, Sarmiento L (1991) Adaptive radiation of *Espeletia* in the cold Andean tropics. Trends Ecol Evol 6:387–391

Monteiro L, Machado N, Martins E, Pougy N, Verdi M, Martinelli G, Loyola R (2018) Conservation priorities for the threatened flora of mountaintop grasslands in Brazil. Flora 238:234–243. https://doi.org/10.1016/j.flora.2017.03.007

Murphy BP, Andersen AN, Parr CL (2016) The underestimated biodiversity of tropical grassy biomes. Philos Trans R Soc 371:20150319. https://doi.org/10.1098/rstb.2015.0319

Myers N, Mittermeier RA, Mittermeier CG, Fonseca GAB, Kent J (2000) Biodiversity hotspots for conservation priorities. Nature 403:853–858. https://doi.org/10.1038/35002501

Neger C, Rosas-Paz LD (2022) A characterization of fire-management research: A bibliometric review of global networks and themes. Fire, 5(4), 89

Nevado B, Contreras-Ortiz N, Hughes C, Filatov DA (2019) Pleistocene glacial cycles drive isolation, gene flow and speciation in the high-elevation Andes. New Phytol 219:779–793. https://doi.org/10.1111/nph.15243

Neves SPS, Conceição AA (2010) Campo rupestre recém-queimado na Chapada Diamantina, Bahia, Brasil: plantas de rebrota e sementes, com espécies endêmicas na rocha. Acta Bot Bras 24:697–707. https://doi.org/10.1590/S0102-33062010000300013

Neves FS, Lana TC, Anjos MC, Reis AC, Fernandes GW (2016) Ant community in burned and unburned sites in Campos rupestres ecosystem. Sociobiology 63:628–636. https://doi.org/10.13102/sociobiology.v63i1.779

Nunes CA, Beiroz W, Silva PG, Braga RF, Fernandes GW, Neves FS (2019) Fire? They don't give a dung! The resilience of dung beetles to fire in a tropical savanna. Ecol Entomol 44:315–323. https://resjournals.onlinelibrary.wiley.com/doi/abs/10.1111/een.12705

Olson DM, Dinerstein E, Wikramanayake ED, Burgess ND, Powell GV et al (2001) Terrestrial ecoregions of the world: a new map of life on Earth. Bioscience 51:933–938. https://doi.org/10.1641/0006-3568(2001)051[0933:TEOTWA]2.0.CO;2

Ordóñez-Parra CA, Medeiros NF, Dayrell RLC, Le Stradic SL, Negreiros D, Cornelissen T, Silveira FAO (2025) Seed functional ecology in Brazilian rock outcrop vegetation: an integrative synthesis. Annals of Botany 135, 371–385

Overbeck GE, Vélez-Martin E, Menezes LS, Anand M, Baeza S et al (2022) Placing Brazil's grasslands and savannas on the map of science and conservation. Perspect Plant Ecol Evol Syst 56:125687. https://doi.org/10.1016/j.ppees.2022.125687

Parr CL, Lehmann CER, Bond WJ, Hoffmann WA, Andersen AN (2014) Tropical grassy biomes: misunderstood, neglected, and under threat. Trends Ecol Evol 29:205–213. https://doi.org/10.1016/j.tree.2014.02.004

Pausas JG, Lamont BB (2022) Fire-released seed dormancy-a global synthesis. Biol Rev 97:1612–1639. https://doi.org/10.1111/brv.12855

Perez-Escobar OA, Zizka A, Bermúdez MA, Meseguer AS, Condamine FL et al (2022) The Andes through time: evolution and distribution of Andean floras. Trends Plant Sci 27:364–378. https://doi.org/10.1016/j.tplants.2021.09.010

Peyre G, Balslev H, Font X, Tello JS (2019) Fine-scale plant richness mapping of the Andean Páramo according to macroclimate. Front Ecol Evol 7:377. https://doi.org/10.3389/fevo.2019.00377

Pilon NA, Cava MG, Hoffmann WA et al (2021) The diversity of post-fire regeneration strategies in the cerrado ground layer. J Ecol 109:154–166. https://doi.org/10.1111/1365-2745.13456

Pinho FFD, Ferreira GB, Paglia AP (2017) Influence of vegetation physiognomy, elevation and fire frequency on medium and large mammals in two protected areas of the Espinhaço Range. Fortschr Zool 34:e11921. https://doi.org/10.3897/zoologia.34.e11921

Podwojewski P, Poulenard J (2004) Páramos soils. Encyclopedia of Soil Science:1–4

Posada C, Cárdenas C (1999) Banco de semillas germinable de una comunidad vegetal de páramo sometida a quema y pastoreo (Parque Nacional Natural Chingaza). Tesis, Departamento de Biología, Universidad Nacional

Premauer JM (1999) Efecto de diferentes regímenes de disturbio por quema y pastoreo sobre la estructura horizontal y vertical de la vegetación del páramo (Parque Nacional Natural Chingaza). Universidad Nacional, Departamento de Biología

Raczka MF, Mosblech NA, Giosan L, Valencia BG, Folcik AM et al (2019) A human role in Andean megafaunal extinction? Quat Sci Rev 205:154–165. https://doi.org/10.1016/j.quascirev.2018.12.005

Ramsay P (1992) The paramo vegetation of Ecuador: the community ecology, dynamics and productivity of tropical grasslands in Andes. PhD thesis, Bangor

Ramsay PM, Oxley ERB (1996) Fire temperatures and post fire plant community dynamics in Ecuatorian grass paramo. Vegetatio 124:129–144. https://doi.org/10.1007/BF00045489

Rangel C (2000) La región de vida paramuna. Colombia Diversidad Biótica III. Universidad Nacional de Colombia, Instituto de Ciencias Naturales, Bogotá

Román-Cuesta RM, Carmona-Moreno C, Lizcano G, New M, Silman M et al (2014) Synchronous fire activity in the tropical high Andes: an indication of regional climate forcing. Glob Chang Biol 20:1929–1942. https://doi.org/10.1111/gcb.12538

Rosbakh S, Carta A, Fernández-Pascual E, Phartyal SS, Dayrell RL et al (2023) Global seed dormancy patterns are driven by macroclimate but not fire regime. New Phytol 240:555–564. https://doi.org/10.1111/nph.19173

Silveira FAO, Negreiros D, Barbosa NP, Buisson E, Carmo FF et al (2016) Ecology and evolution of plant diversity in the endangered campo rupestre: a neglected conservation priority. Plant Soil 403:129–152. https://doi.org/10.1007/s11104-015-2637-8

Silveira FAO, Dayrell RL, Fiorini CF, Negreiros D, Borba EL (2020) Diversification in ancient and nutrient-poor neotropical ecosystems: how geological and climatic buffering shaped plant diversity in some of the world's neglected hotspots. In: Rull V, Carnaval AC (eds) Neotropical diversification: patterns and processes. Springer, pp 329–368. https://doi.org/10.1007/978-3-030-31167-4_14

Silveira FAO, Ordóñez-Parra CA, Moura LC, Schmidt IB, Andersen AN et al (2022) Biome Awareness Disparity is BAD for tropical ecosystem conservation and restoration. J Appl Ecol 59:1967–1975. https://doi.org/10.1111/1365-2664.14060

Simon MF, Grether R, de Queiroz LP, Skema C, Pennington RT et al (2009) Recent assembly of the Cerrado, a Neotropical plant diversity hotspot, by in situ evolution of adaptations to fire. Proc Natl Acad Sci 106:20359–20364. https://doi.org/10.1073/pnas.0903410106

Souza JM, Schmidt IB, Conceicao AA (2018) How do fire and harvesting affect the population dynamics of a dominant endemic Velloziaceae species in campo rupestre? Flora 238:225–233. https://doi.org/10.1016/j.flora.2017.02.007

Strömberg CA, Staver AC (2022) The history and challenge of grassy biomes. Science 377:592–593. https://doi.org/10.1126/science.add1347

Strömberg CAE, Dunn RE, Madden RH, Kohn MJ, Carlini AA (2013) Decoupling the spread of grasslands from the evolution of grazer-type herbivores in South America. Nat Commun 4:1478. https://doi.org/10.1038/ncomms2508

Stürmer SL, Heinz KGH, Marascalchi MN, Giongo A, Siqueira JO (2022) Wildfire does not affect spore abundance, species richness, and inoculum potential of arbuscular mycorrhizal fungi (Glomeromycota) in ferruginous Canga ecosystems. Acta Bot Bras 36:e2021abb0218. https://doi.org/10.1590/0102-33062021abb0218

Torres V, Hooghiemstra H, Lourens L, Tzedakis PC (2013) Astronomical tuning of long pollen records reveals the dynamic history of montane biomes and lake levels in the tropical high Andes during the Quaternary. Quat Sci Rev 63:59–72. https://doi.org/10.1016/j.quascirev.2012.11.004

Vargas-Ríos O (1997) Un modelo de sucesión regeneración de los páramos después de quemas. Caldasia 19:331–345

Vargas-Ríos O (2000) Sucesión—Regeneración del páramo después de quemas. Universidad Nacional de Colombia

Vargas-Ríos O (2002) Disturbios, patrones sucesionales y grupos funcionales de especies en la interpretación de matrices de paisaje en los páramos. Perez-Arbelaezia 13:73–89

Vargas-Ríos O (2013) Distúrbios en los páramos andinos. In: Cortés J, Sarmiento C (eds) Visión socioecosistémica de los páramos y la alta montaña colombiana. Instituto Humboldt

Vargas-Ríos O, Melgarejo LM, Rojas-Zamora Ó, Ávila-Rodríguez LA, Pérez-Martínez LV, Insuasty-Torres J., ... & Armero-Estrada M (2021) Bases ecológicas y sociales para la restauración de los páramos

Vasconcelos TN, Alcantara S, Andrino CO, Forest F, Reginato M, Simon MF, Pirani JR (2020) Fast diversification through a mosaic of evolutionary histories characterizes the endemic flora of ancient Neotropical mountains. Proc Roy Soc B 287:20192933. https://doi.org/10.1098/rspb.2019.2933

Veldman JW, Buisson E, Durigan G, Fernandes GW, Le Stradic S et al (2015) Toward an old-growth concept for grasslands, savannas, and woodlands. Front Ecol Environ 13:154–162. https://doi.org/10.1890/140270

Verweij PA (1995) Spatial and temporal modelling of vegetation patterns: burning and grazing in the paramo of Los Nevados National Park. University of Amsterdam

Verweij PA, Budde PE (1992) Burning and grazing gradient in páramo vegetation. In: Balslev H, Luteyn JL (eds) Paramo: an andean ecosystem under human influence. Academic, pp 177–195

Verweij PA, Kok K (1992) Effects of fire and grazing of *Espeletia hartweiana* populations. In: Balslev H, Luteyn JL (eds) Paramo: an andean ecosystem under human influence. Academic, pp 215–229

Young B, Young, KR, Josse C (2011) Vulnerability of tropical Andean ecosystems to climate change. In: Climate change and biodiversity in the tropical Andes. SCOPE IAI, pp 170–181

Yuste JC, Hereş AM, Ojeda G, Paz A, Pizano C, García-Angulo D, Lasso E (2017) Soil heterotrophic CO2 emissions from tropical high-elevation ecosystems (Páramos) and their sensitivity to temperature and moisture fluctuations. Soil Biol Biochem 110:8–11. https://doi.org/10.1016/j.soilbio.2017.02.016

Zanzarini V, Andersen AN, Fidelis A (2022) Flammability in tropical savannas: variation among growth forms and seasons in Cerrado. Biotropica 54:979–987. https://doi.org/10.1111/btp.13121

Zhao Y, Liu Z, Wu J (2020) Grassland ecosystem services: a systematic review of research advances and future directions. Landsc Ecol 35:793–814. https://doi.org/10.1007/s10980-020-00980-3

Fire in the Chaco Region: Ecological Aspects and Land Management

8

Sandra Bravo, Roxana Ledesma, Darío Coria, and Dante Loto

Abstract

The Chaco region is one of the most important and diverse forestry ecoregions in South America, where fire represents an ecological event recurrent, indispensable for the natural dynamic of the open areas such as grasslands and savannas. Native species of forests and savanna trees have different strategies to survive fire by vegetative regeneration. In this chapter, we discuss antecedents related to the different topics of the fire ecology of the Chaco region: climatic and land use factors influencing the fire frequency to regional scale, plant flammability and functional traits related to fire resistance and tolerance of woody plants, and the effects of fire in soil seed banks. Due to the fire being used in this region as a vegetation management tool, some aspects related to prescribed fire plans for grasslands and savanna restoration/rehabilitation are also discussed in this chapter.

S. Bravo (✉)
Complejo Edilicio Jardín Botánico, Facultad de Ciencias Forestales, Universidad Nacional del Estero, Santiago del Estero, Argentina
e-mail: ssjbravo@unse.edu.ar

R. Ledesma · D. Coria
Estación Experimental Agropecuaria Santiago del Estero, Instituto Nacional de Tecnología Agropecuaria, Santiago del Estero, Argentina
e-mail: ledesma.roxana@inta.gob.ar; coria.ruben@inta.gob.ar

D. Loto
Complejo Edilicio Jardín Botánico, Facultad de Ciencias Forestales, Universidad Nacional del Estero, Santiago del Estero, Argentina

Consejo Nacional de Investigaciones Científicas y Técnicas, El Zanjón, Argentina
e-mail: danteloto@unse.edu.ar

A. Fidelis, V. R. Pivello (eds.), *Fire in the South American Ecosystems*, Ecological Studies 250, https://doi.org/10.1007/978-3-031-89372-8_8

Resumen

La Región Chaqueña es una de las ecorregiones forestales más importantes y diversas de Sudamérica, donde el fuego representa un evento ecológico recurrente e indispensable para la dinámica natural de áreas abiertas como pastizales y sabanas. Las especies leñosas de bosques nativos y los árboles de sabanas tienen diferentes estrategias para sobrevivir al fuego, principalmente por regeneración vegetativa. En este capítulo se discuten antecedentes relacionados a diferentes aspectos de la ecología de fuego de la Región Chaqueña: los factores climáticos y de uso de la tierra que influencian la frecuencia del fuego a escala regional, la inflamabilidad de la vegetación y los rasgos funcionales relacionados a la resistencia y tolerancia al fuego, y los efectos del fuego sobre los bancos de semillas. Debido a que el fuego se usa en la Región Chaqueña como una herramienta de manejo de la vegetación en ambientes productivos, algunos aspectos relacionados a los planes de quemas prescriptas para restauración/rehabilitación de pastizales y sabanas, se discuten también en este capítulo.

Resumo

A região do Chaco é uma das ecorregiões florestais mais importantes e diversificadas da América do Sul, onde o fogo representa um evento ecológico recorrente, indispensável para a dinâmica natural das áreas abertas, como campos e savanas. As espécies lenhosas nativas de florestas e savana têm estratégias diferentes para sobreviver ao fogo, principalmente por meio da regeneração vegetativa. Neste capítulo, discutimos os antecedentes relacionados aos diferentes tópicos da ecologia do fogo para a região do Chaco: os fatores climáticos e de uso da terra que influenciam a frequência do fogo em escala regional, a inflamabilidade da vegetação e os efeitos do fogo nos bancos de sementes do solo. Devido ao fato do fogo ser usado na região do Chaco como uma ferramenta de manejo da vegetação, alguns aspectos relacionados ao planejamento de queimadas prescritas para pastagens e restauração/reabilitação de savanas também são discutidos neste capítulo.

8.1 Introduction

The Chaco region is an extensive sedimentary plain integrating the arid and semiarid diagonal extending with direction northeast to Southwestern in South America together to the Caatinga and the Cerrado biomes, from Brazil (Fig. 8.1). It spreads over 1,000,000 Km^2 among Bolivia, Paraguay, Argentina, and a small sector of Brazil. The name "Chaco" in the native language Quechua means "Hunting Land". This region has a particular biogeography identity in South America, with a tropical semiarid to tropical wet climate (Iriondo 1993). The mean annual rainfall oscillates from 400 mm in the west to 1200 mm in the east. The mean annual temperature is 20 °C, and the mean of the warmest (January) and coldest (July) months are 26 °C

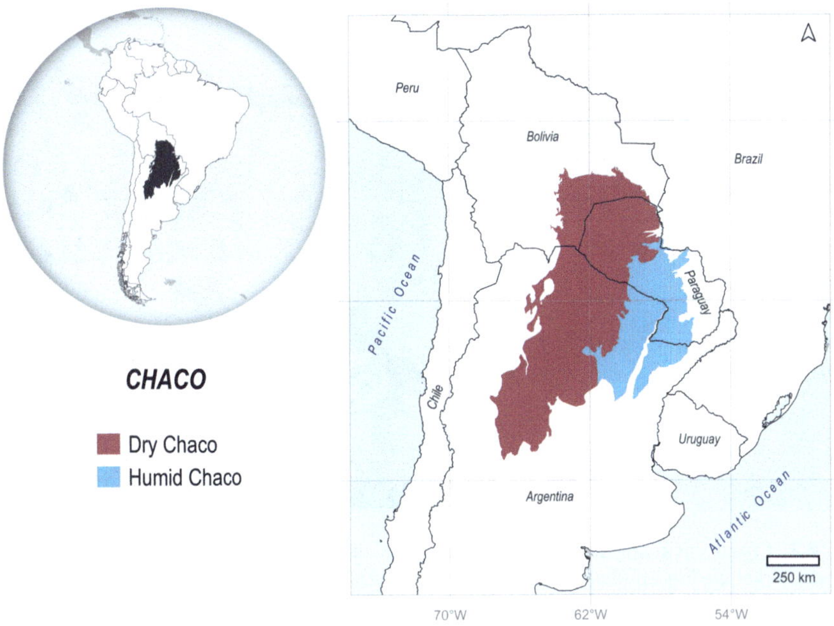

Fig. 8.1 The Gran Chaco region in South America with an identification of Dry and Humid ecoregions, following Olson et al. (2001). (Produced by Júlia Abrantes Rodrigues)

and 11 °C, respectively. The Gran Chaco limits on the west by the Andes mountains ranges and on the east by the Paraguay and Paraná rivers. Due to the accentuated climatic gradient, the vegetation units vary broadly depending on soils, altitude, water availability, and disturbance history. In the driest areas (west), predominate open sclerophyllous forests, edaphic and pyrogenic grasslands, and savannas, whereas in the humid sectors (east), highly diverse forests, tallgrass lands, swamps with palms are present. Iriondo (1993) postulated that the Chaco region's biogeographic identity has been established in the Miocene.

In the Dry Chaco of Argentina, the large rivers crossing the region (Pilcomayo, Bermejo, and Salado) delimitate alluvial terraces between river channels and other geomorphological agents. This ecoregion experimented on two dry climate periods in the late Quaternary that influenced the formation of loess soils and dunes. In the east Chaco, rivers formed mainly alluvial fans and swamps because the silty clay soils have difficult infiltration, and the low or nil slope avoids the runoff of the water excess. Temporary marshes are named "bañados" (which are 30–40 cm deep), and permanent marshes (less than 1 m deep) "esteros". In both, fires are common disturbances promoted by high fuel loads. Old trees remain on the edges of "esteros" due to the anaerobiosis conditions created by flooding (Fig. 8.2).

Chaco region forests play a significant role at a global scale since they are the second most forested area in South America. As well, grasslands and savannas are other valuable areas for conservation, and in the Dry Chaco they have shown

Fig. 8.2 Chaco environments. (**a**) Semiarid forests of "dos quebrachos" with a canopy of *Aspidosperma quebracho-blanco* and *Schinopsis lorentzii* (Photo: S. Bravo) (**b**) Native semiarid savanna in Winter (Photo: S. Bravo). (**c**) Semiarid forest with under silvopastoral management. (Photo: S. Bravo) (**d**) Savanna of *Copernicia alba* in Parque Nacional Río Pilcomayo, Formosa, Argentina (Photo: C. Spagarino)

retraction of their natural areas due to overgrazing, excessive use of fires, agricultural frontier advance, and climatic factors that altered the natural fire regime (Fernández et al. 2019; Coria et al. 2021). In the countries where the region is spread, the different land cover types (forests, grasslands/savannas, shrublands, and wetlands) have particularities in species composition and conservation states even in the same land cover type therefore, references to particular sites are needed to comprise the discussed information. For example, the tall grasslands of *Spartina* sp. alternate with forests and shrublands in the more humid sectors of the Argentine Chaco region, whereas the *Elionorus muticus* savannas are representative in the drier. On the other hand, the forests from the Central Chaco region of Argentina show different structural attributes mainly related to the density of trees of dominant species. However, the species composition does not vary significantly. In these forests, the shrubby and intermediate layer are dense and highly diverse, and changes in species composition in response to disturbance regimes are mainly evident in them (Loto and Bravo 2020). All these ecological aspects influence directly or indirectly the response to fires of native vegetation and their effects on natural dynamics.

The over-exploitation, deforestation, and livestock activities altered forest structure and species composition in vast natural areas of the Chaco region contributing to the fragmentation of native forests, contraction of open areas, and losses of different ecosystem services (Britos and Barchuk 2008; Grau et al. 2015; Coria et al. 2021). The Chaco Region is one of the most fire-affected areas of Argentina and contains valuable areas with high conservation interest because of the endemic species presence of the genus *Schinopsis* and *Gonopterodendron* (Torrella et al. 2015; Anuario de Estadística Forestal 2022). The highest percentage of burnt areas corresponds to grasslands and shrublands where fine fuel availability and desiccation rates promote the ignition and spread of fires (Fisher et al. 2007; Landi et al. 2017). In Paraguay, the most surface from burnt areas corresponds to the Wet Chaco (59%) located in the Eastern sector (Coronel et al. 2021). In the Bolivian Chaco region, most affected areas correspond to grasslands and shrublands, mainly related to agricultural use of fires (Rodríguez-Montellano and Camargo-Arce 2021). The objective of this chapter is to gather information about fire ecology and fire regimes in the Grand South American Chaco to contribute to sustainable management of the natural resources in the region, as well as with restoration and mitigation activities in the current scenario of climate change where increasing fire frequency and severity are expected.

8.2 Fire History and Climatic Changes in the Chaco Region: Paleoecological Records and Recent History

Fire regime studies are essential to comprise the vegetation dynamics in areas with seasonal climates where fine fuel accumulation and desiccation pulses promote the occurrence and spread of fires (Allen 2008). The parameters of fire regimes are frequency, seasonality, the extension of burnt areas, and severity (Keeley 2009). In the Chaco region, antecedents about paleoclimatic changes and their effects on native vegetation are still scarce, as well as studies on fire regimes. The more ancient data from tales of travelers and the first ecologist's reports referred to the use of fire by Indigenous Peoples and rural Local Communities for clearing lands for agriculture, for war, and for hunting wildlife (Bucher 1982). The fire season in the Chaco region coincides with the dryness period starting in autumn and ending in late spring when aborigines and local farmers use fire to promote grass regrowth and improve the forage quality (Kunst and Bravo 2003; Kunst et al. 2016; Coronel et al. 2021). Fires typically occur from April to October and along this prolonged dry and cool period, native vegetation experiments several phenological changes and variations in the moisture content of plants and soils, which originate different fire behavior along the same fire season (Bravo et al. 2014). However, there is even a substantial information gap about the relationship between climate changes and vegetation dynamics over the past centuries, and the potential role of disturbances such as wildfires, volcanic activities, and flooding, among others, as drivers of environmental changes. The geomorphological features of Chaco plains allowed sediment studies from lakes and rivers, and micro, meso, and macro-charcoals in soils and sediments of rivers that constituted an alternative source of information about past climatic conditions (Lindskoug 2016; Speranza 2020). The pollen analysis in soil and identification of variations in phytolith density indicated changing climatic conditions over the last centuries that could have particular effects on fire regimes (Lindskoug 2016). The microcharcoals were frequent in the soil profiles from the Mountain Chaco region of Argentina, with different ratios of presence/absence suggesting the fire regime variability over the last millennium. The radiocarbon dating allowed analyzing the frequency, extent, and severity of past fires, concluding that fires were recurrent events in Ambato Valley from 4500 years B.P, even before the arrival of humans to this area (Lindskoug 2016). According to Lindskoug (2016), there have been periods with higher severity fires than others, identified by peaks of charcoal in the soils of all sampling sites.

On the other hand, the dendrochronological studies in native species allow us to obtain data on fire recurrence with an annual resolution, enhancing the perception of climatic and environmental changes related to fire regimes over the last centuries (Speer 2010). The pollen and dendrochronological studies indicated fire frequency oscillations over dry and wet periods from 1765 to 1945 in the Argentine Chaco region, with greater water availability from the ending period (Speranza 2020). These climatic oscillations could influence the frequency, extension, and severity of the past fires coinciding with those observed in broader temporal scales by Lindskoug (2016). Periods with greater water availability probably promoted higher fine fuel fields than lower water availability periods, increasing fire recurrence. This fire regime variability was also detected through more recent fire chronologies. Bravo et al. (2001) built the first fire chronology (1925–2000) in ecotones among savannas and native forests in the Argentine semiarid Chaco region (600 mm of annual mean precipitation), assessing a temporal change in fire frequency and a relationship between rainfall and some fire regime parameters (Bravo et al. 2010). In this study, a fire frequency (FF) of 0.18 fire. year.$^{-1}$ and Mean Fire Interval (MFI) of 3.3 years were determined for the entire chronology. The period post-1971 showed a fire frequency more than two times higher than the period 1925–1970 ($p < 0.05$) accompanied by 22% more rainfall than the first period. A Superposed Epoch Analysis allowed establishing that the fire events occurred in years with rainfall above the mean annual and the more extensive fires (affecting more than 50% of sampling sites), occurred 2 years after a year with rainfall above the annual mean followed by a dry year. These results highlighted the effect of water availability on the fine fuel field and the role of fine fuel accumulation and desiccation in promoting fires in these environments coinciding with those observed in other semiarid and arid regions of the world (Allen 2008).

However, this first fire chronology from semiarid Chaco savannas allowed estimating the severity of past fires using dendroecological indicators such as the height of fire wounds and the number of trees dating each fire event (Bravo 2006; Bravo et al. 2001, 2008; Fig. 8.3a–c). Fire wound height is an estimator positively related to the flame length, and this last parameter of fire behavior reflects accurately the fire intensity (Cochrane 2009). Both indicators showed that fires in semiarid Chaco savannas increased in severity after 1970, and most of them were identified in higher classes of fire wound height (≥1.30 m). Results indicated a fire regime of a high frequency of fires with medium fire severity, with an increase in frequency, extension, and severity from 1970 (Bravo et al. 2001). More recently, a new fire chronology from 1943 to 2007 of the arid Chaco, Los Llanos, La Rioja, Argentina (400 mm annual precipitation) was obtained using dendrochronology techniques (Bravo et al. 2021, Fig. 8.3d). The FF determined in the dry open forest was 0.40 fires. year $^{-1}$,

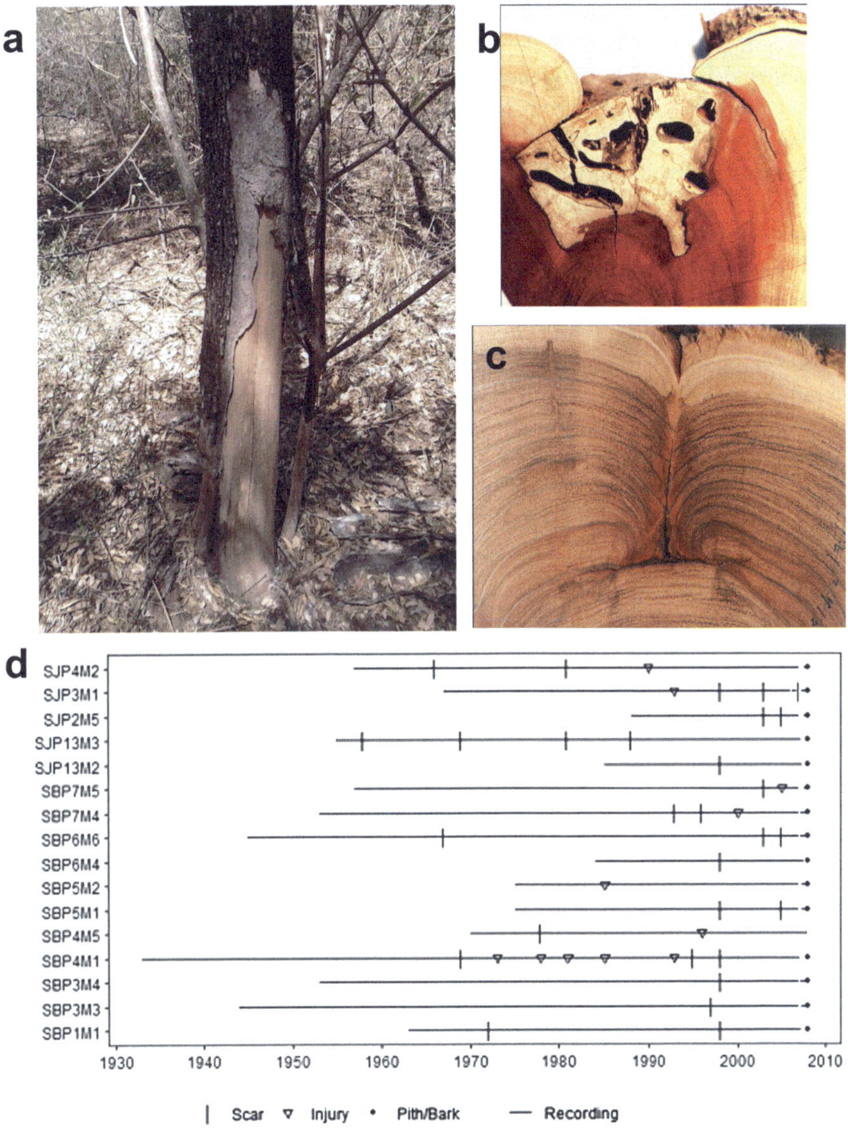

Fig. 8.3 Fire scars and dendrochronological studies about parameters of fire regimes in native woody species from Chaco. (**a**) Fire scar in *Schinopsis lorentzii*. (**b**, **c**) Macroscopy of wood samples with fire scars of respectively *Schinopsis lorentzii* and *Neltuma nigra*. (**d**) Fire chronology in *Neltuma pugionata* in a Dry Chaco forest (Bravo et al. 2021). (Photos: S. Bravo)

and an MFI of 2.5 years, with an increased FF from 1980. Despite the high FF, the height of fire wounds in the arid Chaco suggested low-severity fires ($\leq$1.30 m). These results allowed assessing a fire regime of high frequency but low severity, which could be attributed to the minor fine fuel load due to lower water availability than in the semiarid Chaco region.

The use of satellite images has increased the comprehension of temporal changes in fire regimes in the Chaco region at the landscape scale. These studies allowed us to establish different fire frequencies in different land cover types related to different fuel models and the flammability of species composing plant communities. Landi et al. (2017) identified greater fire frequency in grasslands and savannas than in forests attributed to higher fine fuel availability and lower desiccation thresholds in open areas. Also, these authors assessed higher fire frequency and burnt areas in semiarid land covers than arid Chaco region attributing these results to higher fine fuel availability in the former. The time required to build a more or less continuous canopy has been considered a threshold of forest resistance to fires in other semiarid regions of the world due to the effect of plant cover in fuel desiccation.

Climate change and land use changes in the Gran Chaco region may introduce alterations in the natural fire regimes mediated by different fuel fields, accumulation rates, and desiccation patterns (Kunst et al. 2014; Argañaraz et al. 2016) as in other multiple regions of South America and the world. The increase in livestock activities and silvopastoral systems in the Gran Chaco region added other ignition sources to others of climate nature. Under the current climate change scenario, expect accentuated La Niña years for the next two decades allowing supposing a standing increase in the risk of wildland fires at the regional level.

Besides the productive areas, several tourist regions and urbanizations represent hot pots of fires due to the human activity role as ignition sources. Naval Fernández et al. (2023) mentioned a mean fire recurrence of 3–9 years in the Sierras Chicas, Mountain Chaco region, Argentina, considering the period from 2020 to today. These authors remarked that fire seasons had the second largest burnt area in 34 years and the occurrence of 5 mega fires (≥10,000 ha) in this period, affecting public land, protected areas, and valuable rural areas with protected forests by federal laws. Most countries composing the Grand South American Chaco have laws regulating the use of fires for livestock and other activities. However, the scarcity of human and financial resources limits the success of its application.

8.3 Role of Fire in Chaco Forests and Savannas: Adaptive Traits of Native Species

The role of fire as a factor in modeling landscapes is closely related to the fire history of the geographic area, the species composition of vegetation, and plant traits determining the resistance of different species to physical damage during fires and the ability to recover and recruit after fires. In fire-prone ecosystems plant species are flammable, have traits conferring fire resistance, fire tolerance (resprouting ability) to respond after fires recovering lost or damaged structures, and reproductive syndromes such as fire-stimulating flowering and heat shock tolerant seeds and/or germination stimulated by fire, smoke or charred wood. Acquisition of these traits requires evolution under environmental pressures generated by fires acting within particular parameters (fire regimes). For that, data about historic fire regimes are essential to assess the role of fire in ecosystems.

8.3.1 Plant Flammability in Chaco Environments

The Chaco region vegetation shows different plant flammability in forests, grassland/savannas, and shrublands attributed to the different types and loads of fuels and species composition (Landi et al. 2017). The different proneness to the fire of different land cover is mainly due to the coexistence of fuel models (grasslands, savannas, shrubland, and forests) which experiment progressive desiccation over the fire season. The growth habit (grass, shrub, and tree), plant architecture (ramification degree), leaf and twig dry and wet matter contents, and foliar persistence are functional traits related to the flammability degree of species (Santacruz-García et al. 2019). Savannas and grasslands have a continuous grass layer with scattered shrubs and trees that promote fast desiccation due to exposition to full radiation and winds. Rainfall exerts a remarkable climatic control on fine fuel production and desiccation levels, and the ending of fire season determines the time with the highest plant flammability (Bravo et al. 2010, 2014). Both natural (lighting) and anthropic factors are associated with ignitions influencing the fire frequency (Bertone et al. 2022). In the Gran South American Chaco region, the period without rainfall is up to 6 months (from autumn to the middle spring), representing the period with more susceptibility to fire occurrence in grasslands and savannas. The temporal window for fire event occurrence is shorter in most humid areas than in those drier areas coinciding with the period when vegetation reaches the desiccation threshold to spread fires.

Santacruz-García et al. (2019) studied the plant flammability of 11 woody species of native forests Argentine Chaco region, identifying that 63% have elevated to a very elevated flammability degree. Shrubby species have a higher flammability degree than tree species, promoted by plant architecture, ramification degree, and perennial foliage that contribute to the sustainability of combustion. In addition, Santacruz-García et al. (2021) mentioned a better biochemical response to experimental burns in shrubby than tree species from Argentine Chaco which gives them higher survival and potential growth after fires building a plant architecture that promotes fires. In the Chaco forests, species of shrubby layer represent a natural upstairs of fuel to the canopy when wildfires are incoming to forests under extreme environmental conditions such as droughts, desiccant, and steady winds or heat waves. The high flammability of shrubs seems to explain the high intensity of fires in open areas heavily encroached where fires consume fine, medium, and coarse fuels, burning vast areas for several days, and sometimes with extreme fire behavior. Argañaraz et al. (2016) mentioned the relationship between live fuel moisture content and the Fire Risk Index in all plant cover types in the Mountain Chaco Region, remarking on the effect of water availability besides fuel loads in these environments.

8.3.2 Fire Resistance: The Role of Barks and Growth Habit

Fire resistance in plants mainly depends on traits related to the protection of meristematic tissues (buds, cambium) by bark, bracts, and hairs, among other structures

(Bravo et al. 2008; Pausas and Keeley 2014). Both the growth rates and the growth habit of species influence the plant age at which they reach fire resistance thresholds (plant height, diameter, bark thickness) (Hoffmann et al. 2012). The resprouting ability is a functional trait widely distributed among plant clades representing an adaption to herbivory, wildfires, and insect attacks, and it is considered a *fire tolerance* trait that allows to plants recover from damaged organs and/or to form adventitious buds to resprout after fires. The fire resistance of cambial tissue and aerial buds is a determining factor in the maintenance of live plant structures indispensable to epicormic resprouting after fires (Bravo 2010; Pausas and Keeley 2014). When fire resistance of species is overcome by fire intensity, plants lose all aerial structures and resprout from basal (plant collar) and undergrowth organs (roots, lignotubers, xylopods) representing reservoirs of buds.

The resprouting strategy seems to be the most important regeneration mechanism of woody species from Gran Chaco (Gurvich et al. 2005; Ledesma et al. 2018). The aerial bud banks differ among woody species, and they are related to changes in architecture and plant size (Bravo et al. 2019). The size of the aerial bud bank could be estimated by growing modules expressed as branches, spines and thorns on the main bole. In juvenile individuals of four native woody species from the Argentine Chaco aerial bud bank decreased with the increase in plant height and diameter, suggesting the partition of carbohydrates between resprout production and plant growth. That strategy was especially remarkable in juvenile individuals of tree species, which prioritize height growth to expense resprouting production. Conversely, the shrubby species promote basal resprouting to expense its height growth even in low to medium-intensity fires (Bravo et al. 2014, 2019; Ledesma et al. 2018). High-intensity fires provoke loss of aerial structure in shrubs and juvenile individuals of tree species resprouting from adventitious buds o buds located in underground organs, even in the seedling stage (Barchuck et al. 2006; Ledesma et al. 2018). According to that, *Schinopsis lorentzii* and *Sarcomphalus mistol* produced a higher number of epicormic resprouts after the low-intensity early experimental burns, than high-intensity late burns, which eliminate most shoots and aerial buds in juvenile individuals (Bravo et al. 2014, 2018). The studies about the vegetative regeneration strategy in native woody species from Mountain Chaco region forests mentioned basal resprout as the most common type of resprouting, suggesting medium to high-intensity fires occur in this region (Herrero et al. 2016). The resprouting from aerial organs represents a fast alternative for recovering biomass and nutrient recycling in burnt areas but basal resprouting could give higher opportunities for new shoots survival because of better environmental conditions increasing with the time from fire.

The vegetation responses after fires include competition among species to capture resources released by disturbance, different asexual and sexual regeneration strategies, and adaptability to changing environmental conditions such as occur in burnt areas (Ocampo-Zuleta and Bravo 2019). Species from the Chaco region represent a flora remarkably adapted to adverse environmental conditions imposed by prolonged dry seasons, freezing, hot summers, and disturbance regimes characterizing the semiarid climate (Tálamo and Caziani 2003; Tálamo et al. 2012). Wildfires, flooding, and herbivory are considered the most common disturbances in this region (Bucher 1982). The plant adaptations to this environmental context seem to cooperate with the resilience of native flora to fires (Santacruz-García et al. 2021). The discussions concerning if the Chaco region is a fire-prone ecosystem are yet open, and more studies are desirable to assess the tolerance/dependence of native vegetation to fire. The relationship between reproductive processes and fire occurrence is fundamental for assigning the category of "fire-prone ecosystems" since these ecosystems require the evolution of plant functional traits flora under a determined fire regime (Pausas and Keeley 2014).

8.3.3 Resistance to Fire in Native Woody Species

In woody species, bark thickness (inner, outermost, and total) is considered the main plant trait related to fire resistance since it constitutes the main isolating structure of the cambial tissue, phellogen, and phloem during fires (Resco de Dios 2020). The rate of bark thickness increase with age varies substantially among woody species, and the acquisition of the minimum bark thickness allowing survival to main meristematic tissue represents a resistance threshold to survive recurrent fires. Most native woody species from Chaco Region survive by resprouting after fires (Bravo et al. 2008, 2018; Loto and Bravo 2020; Santacruz-García et al. 2021), they have a wide gradient of bark thickness and bark densities which seem to influence fire resistance even in juvenile individuals (Bravo et al. 2014). The inner bark proportion and bark density have been mentioned as other determining features of fire resistance because both influence the time to ignition and heat transference to growth tissues (Bravo et al. 2014, 2021). The inner bark contains higher moisture content than the outermost bark, which diminishes the heat shock on cambial cells due to the heat liberation by vaporization. On the other hand, high bark density increases the time to ignition since the heat transference is slower than in dead tissues such as cork (Pinard and Huffman 1997; Dickinson and Johnson 2001). The dominant tree species of the primary and secondary Argentine Chaco region forests, *Aspidosperma quebracho-blanco*, *Schinopsis lorentzii*, *Neltuma alba*, and *Neltuma nigra*, in a mature state, have bark thicknesses that overcome or approximate the minimum 1.5 cm threshold considered necessary to resist low to medium fire intensity, whereas shrubby species usually are characterized by thin barks (Bravo et al. 2008). However, juvenile individuals of tree species have inner bark proportions above or near to 50% of total bark thickness, which allows them to maintain alive the shoots after low to medium-intensity fires (Bravo et al. 2014, 2021).

Similarly, the variability in bark thickness and density of woody species has been observed in the Bolivian Chaco forests. Pinard and Huffman (1997) stated that bark thickness has a stronger effect on cambial temperatures than bark density and moisture content in Bolivian Chaco species. Mostacedo et al. (1999) referred to a higher affectation by fire among sub-humid forest species than species from Bolivian dry forest, in most diameter classes suggesting a great adaptation to fire. Some palm species such as *Attalea speciose*, and valuable tree species such as *Aspidosperma cylindrocarpon*, and *Casearia gossypiosperma* have greater basal areas in burnt areas, suggesting a high aptitude to survive fires. Several species of dry Bolivian Chaco are congener with Argentine Chaco region species such as *Aspidosperma rigidum*, *Caesalpinia floribunda*, *Acacia polyphylla*, *Ceiba samauma*, or shared species as *Astronium urundeva*, *Phyllostylon rhamnoides*, and *Tabebuia impetiginosa* in transitional areas with Cerrado from Brazil. These results seem to indicate that tree species from Gran Chaco have bark thicknesses and density that allow them to resist fire damage in medium to low-intensity fires, which is important to maintain tree growth habits and resprout from aerial organs. After more severe fires that eliminate aerial structures, trees resprout from underground organs, acquiring the shrubby growth habit that characterizes environments with high-frequency fires.

8.3.4 The Role of Seed Banks in Post-Fire Recruitment

The aerial and soil seed banks are the source of viable sexual propagules indispensable for the recruitment of new individuals and the maintenance of the natural dynamic of plant communities. It represents a source of genetic diversity that determine the potential of adaptability to changing environments. The seed bank's persistence is a key issue in the analysis of the role of fire in the post-disturbance dynamic of vegetation. The recruitment depends on the seed viability of species composing plant communities, their heat shock tolerances, and the edaphic, biological, and climatic conditions influencing burnt areas (Ocampo-Zuleta and Bravo 2019; Bravo et al. 2022). Besides, the success of post-fire recruitment varies among different land covers and the potential of species to form persistent soil seed banks (SSB).

Fire is an essential component in the natural dynamic of grasslands and savannas affecting the biomass field, the dead biomass elimination, and the litter amount (Carilla et al. 2011). Alterations in natural fire regimes could affect SSBs of grasslands through changes in the species composition and dominance, in its resprouting capacity, and in soil properties. Funes et al. (2003) informed a low similarity of standing vegetation and species composition of SSB in native grasslands from Central Argentina exposed to high fire frequency, which indicates that a low number of species could form a reservoir of viable seeds in the soil under this fire regimes, although most of them form persistent SSBs. On the other hand, Coria et al. (2021) determined a significant and positive association between fire frequency and palatable species composition in Chaco region savannas, highlighting the role of natural fire regimes in the maintenance of diversity and forage aptitude of these vegetation units. Nevertheless, a higher fire frequency than that characterizes the natural fire regimes, diminishing the vigor of plants to resprout, reach the reproductive phase and re-load the SSBs (Funes et al. 2003). Added to that, overgrazing and drought periods could affect synergistically the natural dynamic of SSBs in grasslands and savannas. Arcamone and Jaureguiberry (2018) determined in a set of 10 broad-leaved annual and perennial herbaceous species, a positive response to fire which give them the advantage to colonize burnt areas and become dominant in scenarios with frequent and low to medium fire severity fires. On the other hand, a particular species group as Cacti, considered typical in arid and semiarid fire-prone environments, exhibited a high diversity of responses to heat shock and several species have the ability to colonize Chaco Serrano environments where wildfires are frequent (Roca et al. 2021). Both broad-leaved herbaceous species and Cacti species diminish the forage value of grasslands and savannas by competition with grass seedlings by space, nutrients, and water.

Lipoma et al. (2021) considered that the natural regeneration of Chaco region forests is mainly dependent on the resprouting strategy and, the recruitment of woody species from SSBs does not seem the central resilience mechanism. There is not even evidence of a positive linkage between reproductive traits and stimulation by fires, smoke, and charred wood among native species. Jaureguiberry and Díaz (2015) and Ibañez Moro et al. (2021) identified stimulated, tolerant, and sensible species of native vegetation from the Chaco region when tested for heat shock tolerance in the laboratory. The first study informed that 65% of studied species (n = 26) showed heat-tolerant germination to temperatures between 70 and 100 °C (including only graminoid species); 27% were heat-stimulated germination, and 8% had heat-sensible germination. However, these thermal ranges are overcome at leaf litter level during fires (Kunst and Bravo 2003; Ibañez Moro 2022), suggesting that heat tolerance and stimulation only occur if seeds are buried into soils. Ibañez Moro et al. (2021) stated that representative native tree species from the semiarid Chaco region (n = 6) showed heat shock sensible seeds (n = 4), a low number of species have tolerant (n = 1) or stimulated (n = 1) seeds by heat shock temperatures from 70 to 110 °C. The heat shock did not break seed dormancy; reaffirming that incorporation into the soil is essential to avoid fire damage and recruit after fires (Ibañez Moro 2022). However, *Vachellia aroma* (stimulated) and *Neltuma nigra* (tolerant) can colonize savanna and grasslands, perhaps favored by the heat shock tolerance of seeds and the fruit wall protection, and aptitude to incorporate into the soil (Ibañez Moro et al. 2021, 2024).

Besides wildfires incoming to forests from savannas and grasslands, the native Chaco forests experiment anthropic disturbances related to productive activities such as logging, roller chopping, prescribed fires, and the tropical grass sown to increase the forage for livestock in silvopastoral systems (Kunst et al. 2016). The combination of disturbances that remove native vegetation as roller chopping and prescribed fires reduce the seed density in SSBs and the tree species representativeness, which could cause difficulty in recruitment after these management practices (Bravo et al. 2023). The regulation of intensity and frequency of use of these management practices are relevant to maintaining the sustainability of natural resources in the Gran Chaco since it may alter the recruitment rates. The synergic biomass removal by fire and grazing could affect the reproductive phases and hamper or limit the load of SSB. Data about resprouting potential, seedlings growth, and time required to reload SSBs of the native tree species are indispensable data to achieve sustainable forest management in silvopastoral systems and genetic diversity conservation of native forests.

8.3.5 Vegetation Changes Induced by Fire and Synergism Among Disturbances

Primary productivity of vegetation is one of the most used environmental metrics to characterize site quality and the effects of disturbances on the functionality of ecosystems. The disturbances such as fires, forest logging, and grazing have effects on forest structures, biomass storage, regeneration patterns, and the return period of the pre-disturbance condition (Erb et al. 2018; Lasslop et al. 2020). The intensity of these disturbances depends on the human pressure on natural ecosystems and land-use history. However, the increasing demand for knowledge about disturbance effects in plant communities, mainly in forest persistence is related to the novel intensity and scaling of forest deforestation and degradation in many ecosystems of the world (Hansen et al. 2013). The global challenge is finding a framework where disturbance intensity allows the production without loss of forest functioning.

Disturbances such as forest logging, livestock, and fires have a long history in the Dry Chaco region as land management practices (Bucher and Huszar 1999; Bravo et al. 2001; Grau et al. 2015). Indigenous Peoples managed the fire especially after colonization and the introduction of domestic cattle (Bucher and Huszar 1999; Kunst et al. 2014). At present, Dry Chaco forests have diversified land uses from traditional charcoal and firewood production and extensive livestock to production under silvopastoral systems, agroecological practices for the subsistence of rural inhabits, and hardwood logging from native forests. Since 2000, the Dry Chaco region become a global hotspot of deforestation, and forest degradation where extensive forest areas have been transformed especially for agriculture (Hansen et al. 2013; Gasparri et al. 2008), increasing the carbon emissions to the atmosphere. Biomass production and carbon storage of the remaining native forests of the Dry Chaco region have been scarcely studied regardless of the different intensities of fires, livestock, and forest logging. Previous studies comparing carbon stored in aboveground biomass in Dry Chaco forests found decreased carbon storage due to lower tree density and tree cover in secondary than primary forests (Bonino 2006; Conti et al. 2014). The rate of recovery of forests can vary after different disturbance combinations and among natural and anthropogenic disturbances (De Marzo et al. 2021). The previous forest state (vigor, health) and structure could also affect the potential of forest and grasslands/savanna structures recovery (Landi et al. 2020; Coria et al. 2021).

Loto (2021) evaluated the effects of increasing fire frequency, the intensity of livestock, and forest logging on the carbon storage in native forests from Dry Chaco using standard methods for forest inventory, allometric equations to estimate carbon storage, and MODIS remote sensing products for thermal anomalies for estimations of frequency and extension of burnt areas. The intensity of livestock was estimated through bovine vaccination records of SENASA (Registro Nacional de Sanidad de Productores Pecuarios), and geographic layers of climate, land cover, soils, and economic variables for the entire region of the Argentine Dry Chaco (Fernández et al. 2020). Results allowed determining an average of 46.7 ± 19.0 Mg ha^{-1} of total carbon in forest plots, with a minimum of 4.3 and a maximum of 128.4 Mg ha^{-1}.

Forty-five trees native species were identified in all studied plots where *A. quebracho-blanco*, *Schinopsis lorentzii*, *Sarcomphalus mistol*, *Gonopterodendron sarmientoi,* and *Neltuma nigra* were dominant species with the most accumulated carbon in their biomasses. General linear models showed a decreasing relationship between total carbon and the year of fires and livestock intensity in the studied forest plots (Fig. 8.4a, b). Total carbon increased with longer distances from routes and railways (Fig. 8.4c, d).

Results obtained until today indicated that fire as a disturbance can alter vegetation community structure and species composition through the differential fire tolerance and fire frequency. As well, the combined effects of other disturbances act synergically generating changes in the valuable ecosystem services that can be attributed to differences in the attributes of functional traits related to the main processes (Bravo et al. 2018; Loto and Bravo 2020).

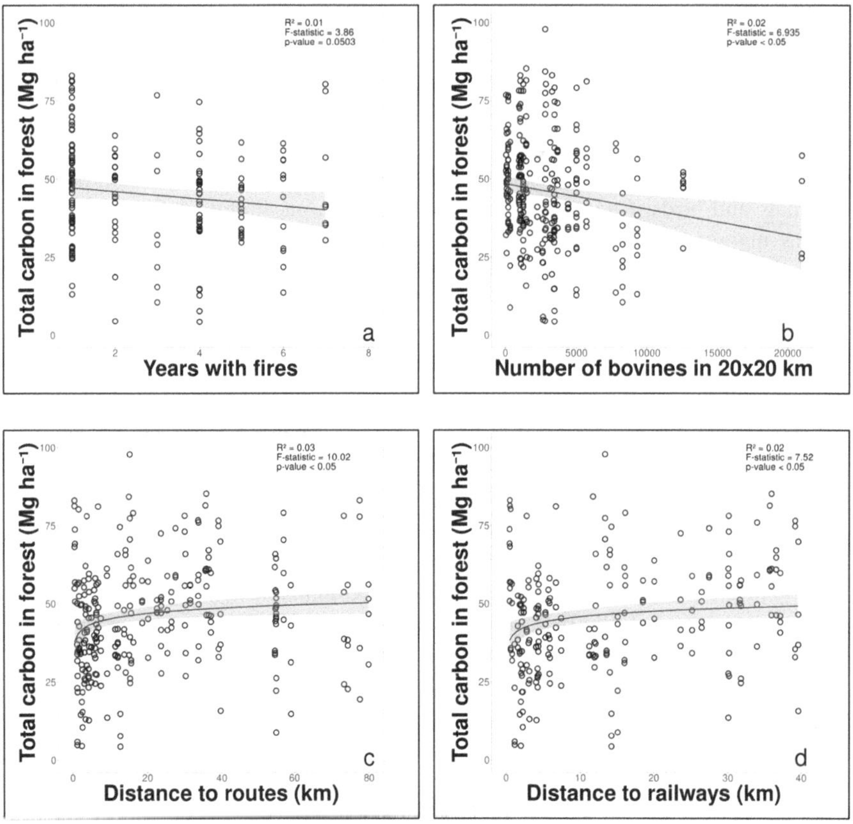

Fig. 8.4 Synergisms among disturbances and effects on total carbon in Dry Chaco Region of Argentina. (**a**) Number of years with fires. (**b**) Livestock in the study area represented by bovines in 400 km^2. (**c**, **d**) Access to native forests is estimated as distance to routes and railways. (Source: Loto (2021))

8.3.6 Fire Effects on Animals

Disturbances affect wildlife through changes in the refuge and food availability, nesting areas, alteration in the physical environment, and species interactions. The Gran Chaco region represents the distribution area of significant carnivores such as the *Panthera onca*, family (Felidae), also called the American tiger, "yaguareté", and *Puma concolor* (puma) which need an extended area of well-conserved forests for hunting and breeding (Quiroga et al. 2013; Di Blanco et al. 2020). Other native mammals such as *Lycalopex gymnocercus (pampean fox)*, *Tapirus terrestris* (tapir), *Priodontes maximus* (giant armadillo), and *Myrmecophaga trydactila* (spiny anteater), are included in the Red List of Endangered Species of IUCN (Di Blanco et al. 2020). The most threats to biodiversity at the regional scale are deforestation, extreme wildfires, and the lack of wise restoration activities in perturbed or modified natural areas. Wildfires contribute to forest fragmentation and the progressive diminishing of connectivity of forest patches whereby the restoration of natural fire regimes is a priority action for biodiversity conservation mainly of the big mammals. The abundance and diversity of small mammals in the Chaco environments are mainly dependent on grassland and savanna cover (Marás et al. 2022), which suggests that prescribed fires could represent a habitat restoration tool in encroached open areas. Some protected areas from the Chaco region are developing fire regime restoration programs of natural fire regime to recover habitat, refuge, and forage for mammals and birds. *Rhea americana* (ñandú), a large, flightless South American bird, as many others, depends of these challenges since the open areas are their major refuges and nesting spaces.

8.3.7 Fire Effects on Soils

Soil heating during vegetation fires could produce changes in physical, chemical, and biological soil properties some of which are poorly understood (Coronel et al. 2021). As well, soil temperatures can vary broadly during fires so avoid generalizations of how soil properties can be altered (Brady et al. 2022). In the Chaco region, there is scarce information about how fires transform the biological, chemical, and physical properties of soils. Due to the Chaco region fire season coinciding with the winter dryness period, the first and more evident effect of fires is the accentuated erosion risk favored by soil desiccation. The fire intensity and the residence time of fires are the main factors to predict its severity on the soil, as the greater fire residence time in the area, as more severe the effect on the vegetation and soil (Wright and Bailey 1982; Stoof et al. 2013). In the Dry Chaco forests, fires do not affect significantly the carbon balance of soils in areas do not expose to grazing which suggests the resilience of native vegetation when fire intensity is not elevated (Abril et al. 2005). The high nitrate availability after fires promotes plant growth, principally grasses and herbs which initiates the recovery of nitrogen mineralization processes. However, the recovery of soil nutrient levels after fires could slow in arid and semiarid regions where the decomposition rates are also slow. The magnitude of physical, chemical, and biological changes in soils after fires depends on temperature ranks reached during fire and exposition times (Neary 2004).

To develop burning prescriptions to control shrubby species in the Chaco, experimental burns were carried out on different dates with different fuel loads to evaluate the fire intensity and the air and soil temperatures (Ledesma et al. 2018). Thermocouples placed at leaf litter level recorded an average maximum soil temperature of 310 °C in experimental plots (SD = 178 °C; maximum = 504 °C; minimum = 79 °C). Experimental burns in the middle fire season (July) showed lower fire intensities (2407 K Wm^{-1}) than in October probably by lower fuel desiccation and air temperatures than late fire season. On the other hand, in the late fire season often occurred high-intensity fires (4320 $KWm^{-1}s$) due to the higher air temperatures and fuel desiccation. The fuel loads had a greater effect on fire intensity than the burning season; and higher fine fuel loads generated the major fire intensities than lower fuel loads (4343 KWm^{-1}; 2383 KWm^{-1}, respectively). These data allow us to advise some aspects of fire behavior but information about heat transference patterns in the soil is not available for the Chaco region, at least to our knowledge. Antecedents about the heat shock effect on soils indicated that temperatures about 300 °C characterize low-intensity fires, however, at this threshold occur the formation of hydrophobic organic compounds in superficial horizons that favors surface runoff (Neary 2004). González et al. (2001) determined in Chaco region savannas that total soil organic carbon, total soil nitrogen content, soil respiration, and soil microbial biomass nitrogen reduced under annual fires. This fire frequency reduced the soil biological activity which may increasing soil susceptibility to erosion processes. However, the fire frequency of one fire every 3–4 years, and one fire every 10 years do not generate significant changes indicating a relative soil stability to this fire recurrence.

8.4 Proposed Protocols to Use Fire as a Management Tool

The comprehension of fire behavior in different vegetation units is a central issue when environmental managers promote the use of fire as a tool for vegetation management. The parameters defining the fire behavior include the time to ignition, the rate of spread, the characteristics of the front fire, and its intensity. The time to ignition mainly depends on the moisture content of the grasses (Cheney and Sullivan 2008; Bianchi et al. 2014), and the meteorological conditions during fires. According to Kunst et al. (2015), the seasonal variation of the moisture content of grass fuels reflects the annual rainfall distribution in the Chaco region. The period from autumn to middle spring without rainfall (up to 6 months) represents the time-lapse with more susceptibility to fire occurrence since the live grass moisture content is 5–10%. Indeed, environmental managers using prescribed fires consider air humidity as the most potent indicator of the moisture content of grass fuels.

The first step to delineate fire management protocols is to identify critical areas where fires start and areas with great fuel accumulation. Most fire events in the Chaco region start in grasslands and savannas under livestock activities, where farmers use fire to promote the regrowth of pastures and to control shrub encroachment (Bravo et al. 2014; Ledesma et al. 2018). The fire application with management goal requires well-established objectives, data about fire regimes in areas to manage, and recent history of fire occurrence at regional and local levels (satellite images). This knowledge could contribute to establishing more conservative and efficient fire management plans that simulate the natural fire regime and allow the conservation and provision of multiple ecosystem services (Coria et al. 2021). This analysis must consider changes expected under current climate change such as an increasing incidence of thunderstorms as ignition sources and heat waves (Bertone et al. 2022).

The horizontal and vertical continuity of fuels influencing fire spread in the Chaco region environments, and the scarcity of fine fuels under tree plant cover represents a natural limit to fire spread in forests. Therefore, the assessment of fuel models, the design of firewalls, and the improvement of burning plans (date, ignition techniques, and personnel, among others) represent a second step for delineating protocols for fire management. The seasonal monitoring of fuel loads and plant flammability in different fuel models could allow for improving the use of resources during prevention and extinction activities (Santacruz-García et al. 2019; Bravo et al. 2022). The reduction of fuel loads in critical areas is the main objective of preventive forestry management since it influences flame height, flame depth, fire intensity, and the difficulty of suppression.

8.4.1 Fire in Grasslands and Savannas

In native grasslands from the Dry Chaco, fine fuel accumulation varies between 4000 and 5000 kg DM.year^{-1}.ha^{-1} (Kunst et al. 2015; Ledesma et al. 2018), and fires have medium to high intensity (with flame length of 3–4 m). Pastures of exotic tropical grasses such as *Megathyrsus maximus* (Gatton panic), *Pennisetum ciliare* (buffel grass), and *Chloris gayana* (grama rhodes) are commonly implanted in the Chaco region to improve the forage availability in open areas and in silvopastoral systems, increasing the fine fuel biomass to 8000–10,000 kg DM.year^{-1}.ha^{-1} (Kunst et al. 2016). These fine fuel loads can generate fire behavior modifications, increasing its intensity and diminishing the possibilities of extinction. Firewalls are the main strategies to avoid out-control fire events, and the width of these free-vegetation areas can vary regarding the ignition techniques (Fig. 8.5a; Kunst and Rodriguez 2003). The use of fire in these environmental contexts requires considerable expertise and knowledge about fire behavior to avoid accidents in operators and give more security during fire control activities when there are fire escapes or during firewall building. The firewalls are mandatory for ecotones between grassland/savannas and forests and between silvopastoral system plots and native forests.

Fig. 8.5 Fuel treatments: (**a**) Emplacement of the firewall with mechanical treatment and prescribed fire in grasslands from the Chaco region (Photo: R. Ledesma); (**b**) Measurement of types and loads of fuels (Photo: S. Bravo)

8.4.2 Fire in Forests

The fire behavior in Chaco forests has been scarcely studied yet because experimental burns are difficult to carry out due to the dense shrubby layer and the high flammability degree (Santacruz-García et al. 2019). Wildland fires are opportunities to evaluate fire behavior and fire effects but provide partial data about the fuels and vegetation responses since the meteorological conditions at the fire moment; and other related to fire behavior are unknown. The use of prescribed fires in forests should consider fuel conditions (load, moisture, horizontal and vertical continuity), wind speed, and slope (Rothermel 1983).

The fuel model in the Chaco region forests varies according to land use history and forest structure. In scarcely disturbed native forests, the amount of live grasses could varies from 0.1 to 500 kg DM.year^{-1} (Kunst et al. 2015), due to the dense shrubby layer. Prescribed fires are applied to reduce volume of shrubby species, and to facilitate the accessibility to the animals and personal in the paddock. In this type of fuel model with absence of fine fuels, slow-burning ground fires with low flame heights are the rule. However, fire can begin at the ecotones and rapidly move forward the forest generating high severity crown fires under severe weather conditions (high temperatures, low relative humidity, and high winds) (Rothermel 1983).

In order to improve the forage productivity for the cattle in silvopastoral systems farmers apply the roller chopping as a mechanical disturbance to create a park physiognomy, which is more appropriate for livestock operations. Forage yields and stocking rates usually increase sowing exotic grass species simultaneously with the roller chopping. *Megathyrsus maximus* yield under tree cover is estimated 5000–8000 kg DM.ha^{-1}, a magnitude higher than the standing crop of native pastures, increasing significantly stocking rates (Kunst et al. 2015). This exotic grass modifies the fire behavior due to the high amount of biomass (fuel) and the plant arquitecture. Table 8.1 shows some characteristics of fuels in forests and grasslands sites from Chaco region, and different meteorological and fire behavior parameters measured in experimental fires. This information belongs to a specific study carried out in two experimental sites (grasslands and forests) and it provides data for designing vegetation management practices and prevention (firewalls). These data highlight the different fuels and fire behavior in silvopastoral systems and grasslands/savannas. Woody debris burning should be carefully developed because fire could reach high intensity degrading soil and eliminating standing trees. The evaluation of the size of the fuel particles allows determining loads and therefore making predictions about the fire intensity and severity (Fig. 8.5b). In grasslands and open areas in the Chaco region, the fire front moves quickly, with a mean flame length of 3–4 m, indicating high fire intensity, the need for firefighter assistance, and to apply prescriptions to avoid fire escapes such as the emplacement of firebreaks with a width more than 30 m (Wright and Bailey 1982). Fire spread is associated with the floristic composition of the vegetal community, this attribute influences mainly on the fuel load and its distribution in the space. The amount of released heat is related to the presence of chemical compounds in the plants, dry matter content, and their inflammability (Bravo et al. 2022). The design of protocols for fire application under different fuel models could contribute to generating plans that minimize the environmental damage and maintain ecosystem services of forests, grasslands, and other vegetation units in the Chaco region.

Table 8.1 Characteristics of fuels in forests and encroached grasslands from Dry Chaco, Argentina, shrub and grass loads (kgDM* ha^{-1}), and meteorological and fire behavior parameters during experimental fires

Site	Forest	Grassland
Grass fine fuel (kg ha^{-1})	2805	6635
Tree density (individuals. m^{-2})	0.91	0.96
Shrub richness	10	10
Woody residues (kgDM ha^{-1})		
Fine (0–0.5 cm)	891	347
Medium (0.5–2.5 cm)	8972	1772
Large (2.5–7.5 cm)	24,858	2644
Shrub individual mean volume (m^3)	0.69	1.33
Shrub individual height (m)	1.37	1.62
Shrub sprouts per plant (number)	3	4.89
Meteorological conditions during an experimental fire		
Temperature (°C)	32–36.2	19.3–24.8
Mean wind speed (km.h^{-1})	2.5	1.1
Air humidity (%)	45–32	45–34
Fire behavior parameters		
Flame length (m)	1.8	5
Fire intensity (kW m^{-1})	932.5	8595.0

8.5 Opportunities to Use Prescribed Fires for Management and Restoration of Grasslands and Savannas

In the last 150 years, many arid and semiarid grasslands/savannahs have suffered a process of woody encroachment because of the spread of woody plants' establishment in grass-dominated areas (D'Odorico et al. 2012). Shrub cover usually increases substantially while tree cover shows variable responses (Eldridge et al. 2011). This situation alters biodiversity, hydrological cycles, and the carbon balance of ecosystems and hurts the forage value of grasslands and savannas (Archer et al. 2017). At broad temporal scales, the balance between woody and herbaceous cover is affected by global warming and the increase of CO_2 concentrations in the atmosphere; and at a spatial scale, this balance varies according to geomorphology, topography, soil types, and natural and anthropic disturbances (Archer et al. 2017). Grazing interacts with fire frequency by regulating the production of herbaceous biomass through grass consumption by animals, and fire-promoted grass fields (Allen 2008; Scott et al. 2013). The "overgrazing hypothesis" considers that chronic overgrazing drastically decreases herbaceous biomass production that, at the same time, reduces fire events and promotes woody plant spreading and soil erosion (Morello and Adamoli 1974; Van Auken 2009; D'Odorico et al. 2012). This mechanism would be the main cause of woody encroachment in different arid and semi-arid regions, with negative effects on vegetation structure, biodiversity, and soil quality (Kunst et al. 2006; Van Auken 2009; Coria et al. 2023). Nevertheless, the ecological restoration activities of grasslands/savannahs are almost non-existent in the Gran Chaco region. Instead, the main management paradigm focuses on solving the pastoral problems by reducing or eliminating woody cover using mechanical disturbances (rolling chopping) and the introduction of tropical forage grasses.

Coria et al. (2021) evaluated the association between historic grazing intensity and fire frequency in native savannas from the Chaco region with vegetation variables such as shrub and tree cover, herbaceous biomass, and abundance of grass functional types. Table 8.2 shows the significant associations found among variables (Spearman rank correlation coefficient, $P = 0.05$). Results confirmed the overgrazing hypothesis that produced the diminishing of fire frequency, the significant increase of tree and shrub cover, and the decrease of herbaceous biomass. Associated with these environmental changes it has observed a diminishing of perennial, palatable, shade-intolerant species. Therefore, this study highlights the convenience of reintroducing the normal fire regimes in grassland/savannas from the Chaco region by prescribed fires. This approach would contribute to regulating the shrub and tree cover, favor biomass productivity of palatable, fire-tolerant, and shade-intolerant grasses typical of grassland/savannas in good state and improve the forage value.

The encroached native grassland restoration could require high-intensity fires to diminish the woody plant cover. These types of fires (length flame >3 m), requires a fine fuel load greater than 3000 kg DM/ha, and severe weather conditions during burning (air temperature between 20 and 30 °C, wind speed between 12 and 25 km/h and air humidity between 25% and 35%) (Kunst et al. 2003). When this amount of fine fuel is available, rational grazing and application of prescribed fires each

Table 8.2 Correlations between structural and functional traits in grasslands/savannas from Dry Chaco, Argentina, and environmental factors

Variable (1)	Variable (2)	Spearman Coef.	P-value	Strength
Fire frequency	Historical grazing intensity	−0.71	0.000	Strong
Fire frequency	Shrub cover	−0.67	0.001	Moderate
Fire frequency	Tree cover	−0.59	0.003	Moderate
Fire frequency	Herbaceous biomass	0.63	0.001	Moderate
Fire frequency	Grasses	0.60	0.003	Moderate
Fire frequency	Perennial grasses	0.60	0.003	Moderate
Fire frequency	Palatable grasses	0.60	0.003	Moderate
Fire frequency	Shade intolerant grasses	0.51	0.012	Moderate
Fire frequency	High fire tolerant grasses	0.64	0.001	Moderate
Historical grazing intensity	Shrub cover	0.43	0.044	Moderate
Historical grazing intensity	Tree cover	0.43	0.048	Moderate
Historical grazing intensity	Herbaceous biomass	−0.71	0.000	Strong
Historical grazing intensity	Grasses	−0.71	0.000	Strong
Historical grazing intensity	Perennial grasses	−0.71	0.000	Strong
Historical grazing intensity	Palatable grasses	−0.71	0.000	Strong
Historical grazing intensity	Shade tolerant grasses	−0.72	0.000	Strong
Historical grazing intensity	Shade intolerant grasses	−0.54	0.009	Moderate
Historical grazing intensity	High fire tolerant grasses	−0.61	0.002	Moderate
Historical grazing intensity	Middle-low fire tolerant grasses	−0.50	0.018	Moderate

3–5 years should allow the progressive restitution and/or the maintenance of grasslands/savannas functionality as habitat native flora and fauna and forage source for cattle (Coria 2022). In scenarios with high woody cover (>75%) and almost null fire occurrence, the threshold of 3000 kg DM/ha to get high-intensity grassland fires could not occur. In these cases, grassland/savannah restoration could need the sequential application of mechanical plant removal methods and prescribed fires (Ledesma et al. 2018). Low-intensity rolling chopping is promoted as a mechanic removal method to decrease the shrubby cover, minimizing coarse woody debris generation (diameter >7.6 cm) (Kunst et al. 2012). Prescribed fires follow the rolling chopping application, and grazing suppression during one or two growing seasons to promote grass establishment (Ledesma et al. 2018; Coria 2022).

Grasslands/savannas respond positively with a significant herbaceous biomass accumulation (Coria et al. 2023). This increase in fine fuel biomass allows performance prescribed fires to decrease even more the shrub competence and favor again herbaceous vegetation. In grassland/savannas severely altered by woody encroachment, the superficial soil horizon could be absent and salinized, without any plant cover. In these conditions, roller chopping, fire, and grazing management are no viable alternatives for grassland/savanna restoration because to will not produce enough increases in grass biomass to conduct prescribed fires. In these cases, the restoration will need to enhance the soil quality and cultivate one or a few tolerant plant species to recover some utility value from encroached grasslands and savannas.

In the Chaco region, even though fires are commonly used for vegetation management, rarely they meet prescription criteria to maximize control possibilities and reduce negative effects on ecosystems and rural structures. However, the world uses prescribed fires as a useful and relatively cheap tool for vegetation management with the prevention goal of wildland fires and restoration of grassland/savannas, among others. The baseline for successful fire management is the formation and expertise of participants and users, and the design of specific protocols for multiple uses. The systems or programs based in rural extension, society formation and education at different levels should work together to promote the use of prescribed fires. In Argentina, the prescribed fires in ranches are permitted previous the approbation of a burn plan (National Law 26562).

8.6 Chaco Region Vegetation and Fire Resilience in a Climate Change Scenario

The native Chaco region vegetation seems to be adapted to recurrent disturbances that act synergistically, and the potential for recovery of species composition, structure, and ecosystem functioning depends on pre- and post-disturbance conditions (Landi et al. 2020; Loto and Bravo 2020; de Marzo et al. 2021). The flora adaptation to particular environmental conditions by different ecophysiological mechanisms seems to collaborate in the vegetation resilience to fire when frequency and severity do not overcome the species tolerance, and physical and biological legacies allow the natural regeneration. The resilience of native woody vegetation depends mainly on resprouting strategy, and it has been related to a positive biochemical response mediated by chlorophylls a and b, carotenes, tannins, and phenols, among other secondary metabolites that allow a fast recovery of photosynthetic apparatus and sprouts growth (Santacruz-García et al. 2021). The post-fire recruitment from SSB could have difficulties in scenarios with more intensive and extended droughts, increased average temperatures, and heat waves, such as being expected for the future due to all of them predisposing to higher severe and frequent fires. Seeds of native species showed limited heat shock tolerance (80–110 °C), and the recruitment after wildfires seem to depend on seed burial capacity and post-fire environmental conditions (Ibañez Moro et al. 2021; Bravo et al. 2023). The antecedents related to fire management and vegetation resilience from the Chaco region to fires indicate that the monitoring and treatment of fuels, and the responsible use of prescribed fires, are the best prevention actions to mitigate extreme fire events and excessive fire frequency, promoting the vegetation recovery by natural regeneration mechanisms.

References

Abril A, Barttfeld P, Bucher E (2005) The effect of fire and overgrazing disturbances on soil carbon balance in the Dry Chaco forest. For Ecol Manag 206(1–3):399–405. https://doi.org/10.1016/j.foreco.2004.11.014

Allen H (2008) Fire: plant functional types and patch mosaic burning in fire prone ecosystems. Prog Phys Geogr 32:421–437. https://doi.org/10.1177/0309133308096754

Anuario de Estadística Forestal (2022) Productos primarios, carbón y durmientes. Productos manufacturados. Incendios forestales. Comercio Exterior. Ministerio de Ambiente y desarrollo Sustentable de la Nación, Argentina

Arcamone J, Jaureguiberry P (2018) Germination response of common annual and perennial forbs to heat shock and smoke treatments in the Chaco Serrano, Central Argentina. Austral Ecol. https://doi.org/10.1111/aec.12593

Archer S, Andersen E, Predick K, Schwinning S, Steidl R, Woods S (2017) Woody plant encroachment: causes and consequences. In: Briske D (ed) Rangeland systems. Processes, management and challenges. Springer, Cham, pp 25–84

Argañaraz JP, Landi M, Bravo S, Gavier-Pizarro G, Scavuzzo C, Bellis L (2016) Estimation of live fuel moisture content from MODIS images for fire danger assessment in Southern Gran Chaco. IEEE J Sel Top Appl Earth Obs Remote Sens 9(12):5339–5349. https://doi.org/10.1109/JSTARS.2016.2575366

Bertone F, Nicora G, Vidal L (2022) Thunderstorm days over Argentina: integration between human observations of thunder and the worldwide lightning location network lightning data. J Climatol 42. https://doi.org/10.1002/joc.7800

Bianchi L, Defossé G, Dentoni M, Kunst C, Ledesma R, Bravo S (2014) Dinámica de la humedad de los combustibles y su relación con la ecología y el manejo de fuego en la región chaqueña occidental (Argentina) I: conceptos básicos. RIA 40(2):154–164. http://www.scielo.org.ar/pdf/ria/v40n2/v40n2a08.pdf

Bonino E (2006) Changes in carbon pools associated with a land-use gradient in the Dry Chaco, Argentina. For Ecol Manag 223(1–3, 183):–89. https://doi.org/10.1016/j.foreco.2005.10.069

Brady M, Dickinson M, Miesel J, Wonkka C, Kavanagh K et al (2022) Soil heating in fire (SheFire): a model and measurement method for estimating soil heating and effects during wildland fires. Ecol Appl 32(6). https://doi.org/10.1002/eap.2627

Bravo S (2006) Régimen de fuego en una sabana del Chaco Occidental y sus efectos morfológicos en especies leñosas. Facultad de Ciencias Naturales e Instituto Miguel Lillo, Universidad Nacional de Tucumán, Argentina

Bravo S (2010) Anatomical changes induced by fire-damaged cambium in two native tree species of the Chaco region, Argentina. IAWA Journal 31 (3): 283–292. https://doi.org/10.1163/22941932

Bravo S, Kunst C, Giménez A, Moglia G (2001) Fire regime of an *Elionorus muticus* Spreng. savanna, western Chaco Region, Argentina. Int J Wildland Fire 10(1). https://doi.org/10.1071/WF01014

Bravo S, Kunst C, Grau H (2008) Suitability of the native woody species of the Chaco region, Argentina, for use in dendroecological studies of fire regimes. Dendrochronologia 26:43–52. https://doi.org/10.1016/j.dendro.2007.05.004

Bravo S, Kunst C, Grau R, Aráoz E (2010) Fire–rainfall relationships in Argentine Chaco savannas. J Arid Environ 74:1319–1323. https://doi.org/10.1016/j.jaridenv.2010.04.010

Bravo S, Kunst C, Leiva M, Ledesma R (2014) Response of hardwood tree regeneration to surface fires, western Chaco region, Argentina. For Ecol Manag 326. https://doi.org/10.1016/j.foreco.2014.04.009

Bravo S, Abdala R, del Corro F, Ibañez Moro V, Santacruz García A, Loto D, Ojeda F (2018) Regeneración en especies de leñosas nativas del Chaco de Argentina y su respuesta a disturbios. In: Los bosques y el fututo Bolzón G, Giménez A (eds) Universidad Federal do Parana, Brazil/Universidad Nacional de Santiago del Estero, Argentina, pp 127–150

Bravo S, Basualdo M, Kunst C, del Corro F (2019) Aerial bud bank and structural changes of woody species from Argentine Chaco in response to disturbances. J Environ Sci Eng A 8(2). https://doi.org/10.17265/2162-5298/2019.02.002

Bravo S, Bogino S, Leiva M, Cendoya M, Kunst C, Biurrun F (2021) Wood anatomy, fire wounds and dendrochronological potential of *Prosopis pugionata* Burkart (Fabaceae) in arid Argentine Chaco. IAWA J 37(1):1–10. https://doi.org/10.1163/22941932-bja10056

Bravo S, Abdala N, Santacruz-García A, Loto D et al (2022) La vegetación chaqueña y su relación con disturbios, desde una perspectiva funcional. Respuestas de bosques chaqueños a disturbios. In: Martínez M et al (eds) Universidad Nacional de Santiago del Estero y sus aportes a la sociedad, Santiago del Estero, pp 856–880. ISBN 978-987-4456-27-4

Bravo S, Abdala N, Ibañez-Moro A (2023) Seed banks of dry tropical forests under different land management. Forests 14:3. https://doi.org/10.3390/f14010003

Britos A, Barchuk A (2008) Cambios en la cobertura y en el uso de la tierra en dos sitios del Chaco Árido del noroeste de Argentina. Agri 25:97–110. https://revistas.unc.edu.ar/index.php/agris/article/view/2746/2174

Bucher E (1982) Chaco and Caatinga, South American arid savannas, woodlands and tickets. In: Huntley B, Walker B (eds) Ecological studies 42: ecology of tropical savannas. Springer, Berlin. https://doi.org/10.1007/978-3-64268786-0_4

Bucher E, Huszar P (1999) Sustainable management of the Gran Chaco of South America: ecological promise and economic constraints. J Environ Manag 57(2):99–108. https://doi.org/10.1006/jema.1999.0290

Carilla J, Aragón R, Gurvich D (2011) Fire and grazing differentially affect aerial biomass and species composition in Andean grasslands. Acta Oecologica 37: 337–345

Cheney P, Sullivan A (2008) In: Krester A (ed) Grassfires: fuel, weather and fire behavior. CSIRO Publishing, Collingwood

Cochrane M (2009) Tropical fire ecology. Climate change, land use and ecosystem dynamics. Springer-praxis books in environmental sciences. Springer. https://books.google.com.ar/books/about/Tropical_Fire_Ecology.html?id=6J6fWSULMVEC&redir_esc=y

Conti G, Pérez-Harguindeguy N, Fabien Quètier L, Gorné et al (2014) Large changes in carbon storage under different land-use regimes in subtropical seasonally dry forests of Southern South America. Agric Ecosyst Environ 197:68–76. https://doi.org/10.1016/j.agee.2014.07.025

Coria D (2022) Restauración y manejo de pastizales/sabanas lignificados de sitios ecológicos bajos del Chaco semiárido. Facultad de Agronomía y Agroindustrias, Universidad Nacional de Santiago del Estero

Coria R, Kunst C, Bravo S (2021) A contribution to the understanding of the woody encroachment in grasslands/savannas from the South American Semiarid Chaco. Ecologia Austral 31(3):456–474. https://doi.org/10.25260/EA.21.31.3.0.1615

Coria D, Kunst C, Anriquez A, Bravo S (2023) Management of grassland/savannas encroached by woody plants in South American semiarid Chaco: ecological restoration vs. African grasses introduction. Ecologia Austral 33:489–506. https://doi.org/10.25260/EA.23.33.2.0.2156

Coronel G, Pasten M, Breuer N, Rejalaga L (2021) Wildfires in Paraguay: environmental and human impacts. In: Leal Filho W, Miranda UA, Faronni ASS (eds) Sustainability in natural resources management and land planning. Springer. https://link.springer.com/book/10.1007/978-3-030-76624-5

D'Odorico P, Okin G, Bestelmeyer B (2012) A synthetic review of feedbacks and drivers of shrub encroachment in arid grasslands. Ecohydrology 5:520–530. https://doi.org/10.1002/eco.259

de Marzo T, Pflugmacher D, Baumann M, Lambin E, Gasparri I, Kuemmerle T (2021) Characterizing forest disturbances across the Argentine Dry Chaco based on Landsat time series. Int J Appl Earth Obs 98. https://doi.org/10.1016/j.jag.2021.102310

De Marzo T, Pflugmacher D, Baumann M, Lambin E, Gasparri I, Kuemmerle (2023) Characterizing forest disturbances across the Argentine Dry Chaco based on Landsat time series. International Journal of Applied Earth Observations and Geoinformation 98. https://doi.org/10.1016/j.jag.2021.102310

Di Blanco Y, Varela D, Abba A (2020) Myrmecophaga tridactyla. In: SAyDS–SAREM (ed) Categorización 2019 de los mamíferos de Argentina según su riesgo de extinción. Lista Roja de los mamíferos de Argentina. Versión digital: http://cma.sarem.org.ar

Dickinson M, Johnson E (2001) Fire effects on trees. In: Johnson E, Mishanishi K (eds) Forest fires: behaviour and ecological effects. Academic, San Diego, pp 477–526

Eldridge D, Bowker M, Maestre F, Roger E, Reynolds J, Whitford W (2011) Impacts of shrub encroachment on ecosystem structure and functioning: towards a global synthesis. Ecol Lett 14:709–722. https://doi.org/10.1111/j.1461-0248.2011.01630

Erb K, Kastner T, Plutzar C, Bais A, Carvalhais N et al (2018) Unexpectedly large impact of forest management and grazing on global vegetation biomass. Nature 553(7686):73–76. https://doi.org/10.1038/nature25138

Fernández P, Matthias B, Baldi G, Banegas N, Bravo S et al (2019) Grasslands and open savannas of the Dry Chaco. In: Goldstein M, Della Sala D (eds) Encyclopedia of the World's biomass. Elsevier, USA. https://doi.org/10.1016/B978-0-12-409548-9.12094-9

Fernández P, De Waroux Y, Jobbágy E, Loto D, Gasparri N (2020) A hard-to-keep promise: vegetation use and aboveground carbon storage in silvopastures of the Dry Chaco. Agric Ecosyst Environ 303:107117. https://doi.org/10.1016/j.agee.2020.107117

Fisher M, di Bella C, Jobbágy E (2007) Factores que controlan la distribución espacio-temporal de los incendios en la Región Semiárida Argentina. In: Martin (ed) Teledetección—Hacia un mejor entendimiento de la dinámica global y regional. https://www.aet.org.es/congresos/xii/arg37.pdf

Funes G, Basconcelo S, Díaz S, Cabido M (2003) Seed bank dynamics in tall-tussock grasslands along an altitudinal gradient. J Veg Sci 14:253–258. http://www.jstor.org/stable/3236700

Gasparri I, Grau H, Manghi E (2008) Carbon pools and emissions from deforestation in extra-tropical forests of Northern Argentina between 1900 and 2005. Ecosystems 11:1247–1261. https://doi.org/10.1007/s10021-008-9190-8

González C, Studdert G, Kunst C, Albanesi A (2001) Comportamiento de algunas propiedades del suelo en una sabana del Chaco Semiárido Occidental bajo distintas frecuencias de fuego. Ciencia del Suelo 19:92–100

Grau HR, Torres R, Gasparri N (2015) Natural grasslands in the Chaco. A neglected ecosystem under threat by agriculture expansion and forest-oriented conservation policies. J Arid Environ 123:40–46. https://doi.org/10.1016/j.jaridenv.2014.12.006

Gurvich D, Enrico L, Cingolani A (2005) Linking plant functional traits with post-fire sprouting vigor in woody species in Central Argentina. Austral Ecol 30:789–796

Hansen M, Potapov P, Moore R, Hancher M, Turubanova A et al (2013) High-resolution global maps of 21st-century forest cover change. Science 342(6160):850–853. https://doi.org/10.1126/science.1244693

Herrero M, Torres R, Renison D (2016) Do wildfires promote woody species invasion in a fire-adapted ecosystem? Post-fire resprouting of native and non-native woody plants in Central Argentina. Environ Manag 57:308–317. https://doi.org/10.1007/s00267-015-0616-8

Hoffmann W, Geiger E, Gotsch S, Rossatto D, Silva L et al (2012) Ecological thresholds at the savanna-forest boundary: how plant traits, resources and fire govern the distribution of tropical biomes. Ecol Lett 15:759–768. https://doi.org/10.1111/j.1461-0248.2012.01789.x

Ibañez Moro A (2022) Efecto del Fuego en el Banco de Semillas de Especies Leñosas de Nativas del Chaco Occidental de Argentina. Tesis Doctoral. Facultad de Ciencias Agrarias y Forestales. Universidad Nacional de La Plata, Argentina, 166 pp. https://sedici.unlp.edu.ar/bitstream/handle/10915/148253/Documento_completo.pdf?sequence=1&isAllowed=y

Ibañez Moro A, Bravo S, Abdala N, Borghetti F, Chaib A, Galetto L (2021) Heat shock effects on germination and seed survival of five woody species from the Chaco region. Flora 275:151751. https://doi.org/10.1016/j.flora.2020.151751

Ibañez Moro V, Borghetti F, Galetto L, Cellini, J, Bravo S (2024) The influence of seed functional traits and anthropogenic disturbances on persistence and size of the soil seed bank from dry subtropical forest species

Iriondo M (1993) Geomorphology of late Quaternary of the Chaco (South America). Geomorphology 7:289–303. https://doi.org/10.1016/0169-555X(93)90059-B

Jaureguiberry P, Díaz S (2015) Post-burning regeneration of the Chaco seasonally dry forest: germination response of dominant species to experimental heat shock. Oecologia 177:689–699. https://doi.org/10.1007/s00442-014-3161-x

Keeley A (2009) Fire intensity, fire severity and burn severity: a brief review and suggested usage. Int J Wildland Fire 18:116–126. https://doi.org/10.1071/WF07049

Kunst C, Bravo S (2003) Ecología y régimen de fuego en la región Chaqueña Argentina. In: Kunst C, Bravo S, Panigatti J (eds) Fuego en ecosistemas argentinos. Ediciones INTA, Santiago del Estero, pp 109–118

Kunst C, Bravo S, Monti E, Cornacchione M Godoy J(2003) El fuego y el manejo de pasturas naturales y cultivadas en la Región Chaqueña. In: Kunst C, Bravo S, Panigatti S (eds) Fuego en los ecosistemas argentinos. Ediciones Inta, Santiago del Estero, pp: 239–248

Kunst C, Monti E, Pérez H, Godoy J (2006) Assessment of the rangelands of southwestern Santiago del Estero, Argentina, for grazing management and research. J Environ Manag 80:248–265. https://doi.org/10.1016/j.jenvman.2005.10.001

Kunst C, Ledesma R, Bravo S, Albanesi A, Anriquez A, van Meer H, Godoy J (2012) Disrupting woody steady states in the Chaco region (Argentina): responses to combined disturbance treatments. Ecol Eng 42. https://doi.org/10.1016/j.ecoleng.2012.01.025

Kunst C, Bravo S, Ledesma R, Navall M et al (2014) Ecology and management of the dry forests and savannas of the Western Chaco Region, Argentina. In: Greer E (ed) Dry forests: ecology, species diversity, and sustainable management. Nova Publishers, New York. https://www.goodreads.com/book/show/35326708, pp 133–163

Kunst C, Ledesma R, Bravo S et al (2015) Dinámica del contenido de humedad de pastos y su relación con la ecología del fuego en región chaqueña occidental (Argentina). Revista de Investigaciones Agropecuarias 41(1):83–93. http://www.scielo.org.ar/pdf/ria/v41n1/v41n1a12.pdf

Kunst C, Naval M, Ledesma R et al (2016) Silvopastoral systems in the Western Chaco region, Argentina. In: Peri P, Dube F, Varella A (eds) Silvopastoral systems in Southern South America. Advances in agroforestry11. Springer, pp 63–87. https://link.springer.com/book/10.1007/978-3-319-24109-8

Landi M, di Bella C, Argañaraz J, Bellis L (2017) Frecuencia y propagación de incendios en el paisaje del Chaco árido y semiárido, efecto de la cantidad de biomasa combustible. In: VI Jornadas y III Congreso Argentino de Ecología de Paisajes, pp 220–223. https://lpdaac.usgs.gov/

Landi M, di Bella C, Bravo S, Bellis L (2020) Structural resistance and functional resilience of the Chaco forest to wildland fires: an approach with MODIS time series. Austral Ecol. https://doi.org/10.1111/aec.12977

Lasslop G, Hantson S, Harrison S, Bachelet D, Burton CH et al (2020) Global ecosystems and fire: multi-model assessment of fire-induced tree-cover and carbon storage reduction. Glob Chang Biol 26:5027–5041. https://doi.org/10.1111/gcb.15160

Ledesma R, Kunst C, Bravo S, Leiva M, Lorea L, Godoy J, Navarrete V (2018) Developing a prescription for brush control in the Chaco region, effects of combined treatments on the canopy of three native shrub species. Arid Land Res Manag 32(3):351–366. https://doi.org/10.1080/15324982.2018.1430072

Lindskoug H (2016) Forest fires and abandonment patterns from the Aguada culture, northwest Argentina: a paleoenvironmental study based on the analysis of microcharcoals in the Ambato Valley, Catamarca Province, Argentina. British Archaeological Reports Ltd. United Kingdom. Bar International Series 2803

Lipoma M, Cabrol D, Cuchietti A, Enrico L, Gorné L, Díaz S (2021) Low resilience at the early stages of recovery of the semi-arid Chaco forest. Evidence from a field experiment. J Ecol 109:3246–3259. https://doi.org/10.1111/1365-2745.13622

Loto D (2021) Ciclo de carbono en biomasa de bosque con relación al régimen de disturbios en el Chaco seco argentino. Facultad de Ciencias Forestales, Universidad Nacional de Santiago del Estero

Loto D, Bravo S (2020) Species composition, structure, and functional traits in Argentine forests under two different disturbance histories. Ecol Indic 113. https://doi.org/10.1016/j.ecolind.2020.106232

Morello J, Adamoli J (1974) Las grandes unidades de vegetación y ambiente del Chaco argentino. Segunda parte: Vegetación y ambiente de la provincia del Chaco. Instituto Nacional de Tecnología Agropecuaria. Buenos Aires, Argentina, pp 130. https://biblio.darwin.edu.ar/meran/opac-detail.pl?id1=2895

Mostacedo B, Fredericksen T, Gould K, Toledo M (1999) Comparación de la respuesta de comunidades vegetales a los incendios forestales en los bosques tropicales secos y húmedos de Bolivia. Documenteo Técnico 83/1999, USAID: 511-0621-C-00-3027. https://pdf.usaid.gov/pdf docs/Pnacl876.pdf

Naval Fernández M, Albornoz J, Bellis L, Bald AJ, Silvetti L, Álvarez M, Argañaraz J (2023) Megaincendios 2020 en Córdoba: incidencia del fuego en áreas de valor ecológico y socioeconómico. Ecol Austral 33:136–151. https://doi.org/10.25260/EA.23.33.1.0.2120

Neary D (2004) Fire effects on soils. Rocky Mountain Research Station, Flagstaff, pp 18–19. https://doi.org/10.25260/ea.23.33.1.0.2120

Ocampo-Zuleta K, Bravo S (2019) Reclutamiento de especies leñosas en bosques tropicales expuestos a incendios: una revisión. Ecosistemas 28(1):106–117. https://doi.org/10.7818/ecos.1642

Olson DM, Dinerstein E, Wikramanayake ED, Burgess ND, Powell GV et al (2001) Terrestrial ecoregions of the world: a new map of life on Earth. Bioscience 51:933–938. https://doi.org/10.1641/0006-3568(2001)051[0933:TEOTWA]2.0.CO;2

Pausas J, Keeley J (2014) Evolutionary ecology of resprouting and seeding in fire-prone ecosystems. New Phytol 204:55–65. https://doi.org/10.1111/nph.12921

Pinard M, Huffman J (1997) Fire resistance and bark properties of trees in a seasonally dry forest in eastern Bolivia. J Trop Ecol 13:727–740. https://doi.org/10.1017/S0266467400010890

Quiroga V, Boaglio G, Noss A, Di Bitetti S (2013) Critical population status of the jaguar *Panthera onca* in the Argentine Chaco: camera-trap surveys suggest recent collapse and imminent regional extinction. Oryx. https://doi.org/10.1017/S0030605312000944

Resco de Dios V (2020) Plant-fire interactions: applying ecophysiology to wildfire management. Springer International Publishing, p 203. https://doi.org/10.1007/978-3-030-41192-3

Roca J, Jaureguiberry P, Gurvich D (2021) Are wildfires affecting seed germination in cacti? An experimental assessment. Austral Ecol 46:818–832. https://doi.org/10.1111/aec.13027

Rodríguez-Montellano A, Camargo-Arce R (2021) Reporte 2021 Bolivia. Fundación Amigos de La Naturaleza. Áreas Quemadas al Inicio de Temporada de Incendios Forestales en Bolivia 2021. https://incendios.fan-bo.org/Satrifo/2021/

Rothermel R (1983) How to predict the spread and intensity of forest and range fires. USDA Forest Service Intermountain Forest and Range Exp. Sta. GTR INT-143. https://www.fs.usda.gov/rm/pubs_int/int_gtr143.pdf

Santacruz-García A, Bravo S, del Corro F, Ojeda F (2019) A comparative assessment of plant flammability through a functional approach: the case of woody species from Argentine Chaco region. Austral Ecol 44:1416–1429. https://doi.org/10.1111/aec.12815

Santacruz-García A, Bravo S, del Corro F, García E, Molina-Terrén D, Nazareno M (2021) How do plants respond biochemically to fire? The role of photosynthetic pigments and secondary metabolites in the post-fire resprouting response. Forests 12:1–20. https://doi.org/10.3390/f12010056

Scott A, Bowman W, Bond J, Pyne S, Alexander M (2013) Fire on Earth: an introduction. Wiley, Chichester

Speer JH (2010) Fundamentals of tree-ring research. University of Arizona Press, Flagstaff, AZ.

Speranza F (2020) Variabilidad paleoclimática y cambios paleoambientales en la Región Chaqueña Semiárida , Godoy Jde Argentina. Tesis Doctoral. Universidad Autónoma de Barcelona. http://hdl.handle.net/10803/670872

Stoof C, Moore D, Fernandes P, Stoorvogel J, Fernandes R, Ferreira A, Ritsema C (2013) Hot fire, cool soil. Geophys Res Lett 40:1534–1539. https://doi.org/10.1002/grl.50299

Tálamo A, Caziani S (2003) Variation in woody vegetation among sites with different disturbance histories in the Argentine Chaco. For Ecol Manag 184:79–92. https://doi.org/10.1016/S0378-1127(03)00150-6

Tálamo A, Casenave J, Caziani S (2012) Components of woody plant diversity in semi-arid Chaco forests with heterogeneous land use and disturbance histories. J Arid Environ 85:79–85. https://doi.org/10.1111/aec.12815

Torrella S, Ginzburg R, Galetto L (2015) Forest fragmentation in the Argentine Chaco: recruitment and population patterns of dominant tree species. Plant Ecol 216:1499–1510. https://doi.org/10.1007/s11258-015-0532-2

Van Auken O (2009) Causes and consequences of woody plant encroachment into western North American grasslands. J Environ Manage 90:2931–2942. https://doi.org/10.1016/j.jenvman.2009.04.023

Wright H, Bailey A (1982) Fire ecology in the United States and Canada. Wiley, New York, p 501

Fire in Temperate Forests and Shrublands

9

Thomas Kitzberger, Mauro E. González, and Juan Paritsis

Abstract

Fire is and has been a dominant disturbance in temperate forests and shrublands of southwestern South America but current and future fire regimes pose significant threats to the long term persistence of sensitive ecosystems and to vulnerable human populations. Here we underscore the prime importance of fire weather, human ignitions and biophysical drivers that explain large-scale and long-term patterns of fire occurrence, severity and frequency. We provide a temporal line of how fire was perceived and used during different cultural/historical periods (pre-settlement, settlement and fire suppression). We review the main life history traits and functional strategies of species in relation to fire as well as the patterns of stand development following fire for the main forest types. At the landscape scale we identify several examples of fire-vegetation feedbacks and alternative stable community states modified by fire. Finally, we synthesize the current state of knowledge of how fire regimes will respond in this region to future climatic scenarios for the twenty-first century and provide some guidelines for adaptation and mitigation of fire impacts and hazards.

T. Kitzberger (✉) · J. Paritsis
INIBIOMA-Universidad Nacional del Comahue, CONICET, Bariloche, Argentina
e-mail: kitzberger@comahue-conicet.gob.ar; juanparitsis@comahue-conicet.gob.ar

M. E. González
Laboratorio de Ecología de Bosques, Instituto de Conservación, Biodiversidad y Territorio, Facultad de Ciencias Forestales y Recursos Naturales, Universidad Austral de Chile, Campus Isla Teja, Valdivia, Chile

Center for Climate and Resilience Research (CR2), Santiago, Chile
e-mail: maurogonzalez@uach.cl

A. Fidelis, V. R. Pivello (eds.), *Fire in the South American Ecosystems*, Ecological Studies 250, https://doi.org/10.1007/978-3-031-89372-8_9

Resumen

El fuego es y ha sido una perturbación dominante en los bosques y matorrales templados del suroeste de América del Sur. Sin embargo, los regímenes de incendios actuales y futuros plantean amenazas significativas a la persistencia a largo plazo de ecosistemas sensibles y a la vulnerabilidad de poblaciones humanas. Aquí resaltamos la importancia del clima, las igniciones humanas y los factores biofísicos que explican los patrones de ocurrencia, severidad y frecuencia de incendios a gran escala y a largo plazo. Proporcionamos una línea temporal de cómo se percibió y utilizó el fuego durante diferentes períodos culturales/históricos (pre-asentamiento, asentamiento y supresión). Revisamos los principales rasgos de la historia de vida y las estrategias funcionales de las especies en relación con el fuego, así como los patrones de desarrollo de rodales después del incendio para los principales tipos de bosques. A escala de paisaje, identificamos varios ejemplos de retroalimentaciones fuego-vegetación y estados comunitarios estables alternativos modificados por el fuego. Finalmente, sintetizamos el estado actual del conocimiento sobre cómo responderán los regímenes de incendios en esta región a escenarios climáticos futuros para el siglo XXI y proporcionamos algunas pautas para la adaptación y mitigación de los impactos y peligros de los incendios.

Resumo

O fogo é e tem sido um distúrbio dominante nas florestas temperadas e vegetação arbustiva do sudoeste da América do Sul, mas as tendências atuais e futuras dos regimes de incêndio representam ameaças significativas à persistência dos ecossistemas sensíveis ao fogo e às populações humanas vulneráveis, em longo prazo. Aqui, destacamos a importância primordial do clima de incêndios, das ignições humanas e dos fatores biofísicos que explicam os padrões de ocorrência, gravidade e frequência de incêndios em larga escala e em longo prazo. Apresentamos uma linha temporal de como o fogo foi percebido e usado durante diferentes períodos culturais/históricos (pré-assentamento, assentamento e supressão de incêndios). Revisamos as principais características do histórico de vida e as estratégias funcionais das espécies em relação ao fogo, bem como seus padrões de desenvolvimento após o fogo, para os principais tipos de floresta. Em escala de paisagem, identificamos vários exemplos de retroalimentação entre fogo e vegetação e estados alternativos de comunidades estáveis modificadas pelo fogo. Por fim, sintetizamos o estado atual do conhecimento de como os regimes de fogo responderão nessa região aos cenários climáticos futuros do século XXI e fornecemos algumas diretrizes para adaptação e mitigação dos impactos e riscos do fogo.

9.1 Climatic and Vegetation Setting

Influenced by the westerly circulation of moisture-laden winds from the Pacific Ocean and the N–S oriented Andean Cordillera lie the temperate forests and shrublands of South America (Fig. 9.1). Long isolated from other subtropical forests by the Arid Diagonal of South America, this continuous narrow (circa 200–300 km wide), latitudinally oriented (circa 2400 km long) forested area represents a unique biogeographical island (Armesto et al. 1996).

Along this region, rainfall seasonality decreases from north to south (Mediterranean climate in the north and year-round wet conditions in the south) due to decreasing seasonal blocking of westerly circulation caused by the South Pacific Anticyclone. Rainfall amounts decrease several-fold from west to east due to the rain shadow effect of the Andean Cordillera upon predominant Pacific circulation. Mean annual temperature decreases southwards (Armesto et al. 1996) and continentality is generally much lower than at similar latitudes of the Northern Hemisphere due to the prevailing ocean mass of the Southern Hemisphere. This maritime influence decreases across the Andes towards the Patagonian inland plains and increases southwards due to the narrowing of the South American continent. In addition, the steep Andean topography generates important altitudinal gradients resulting in

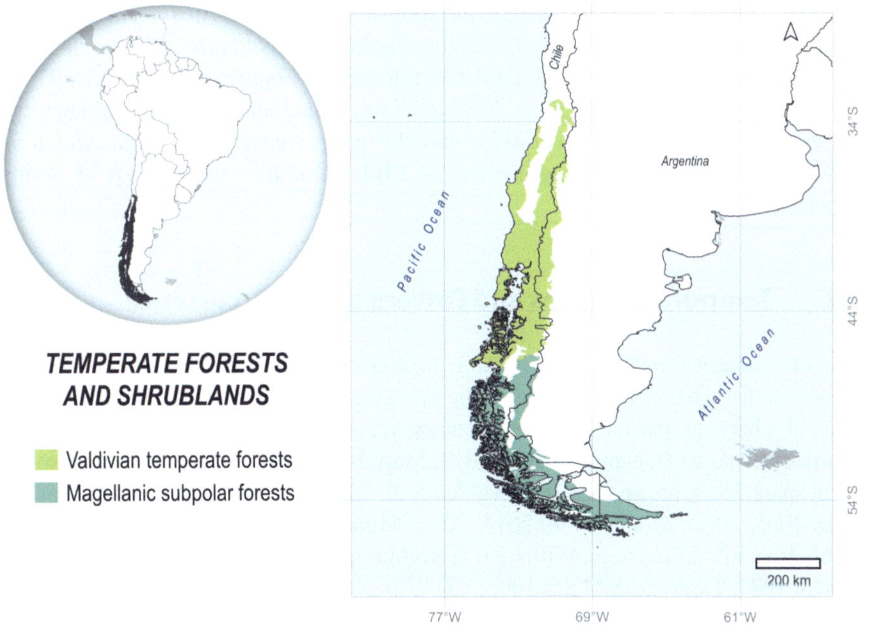

Fig. 9.1 Distribution of the temperate forests and shrublands in South America subdividing the northern Valdivian (yellow) and southern Magellanic (green) forest districts. Modified from Olson et al. (2001) and produced by Júlia Abrantes Rodrigues

sharp forest compositional changes including the disappearance of tree elements above a tree line (located at circa 2000 m a.s.l. at 35°S and 500 m a.s.l. at 55°S).

North of circa 37°S subtropical sclerophyllous forests and shrublands develop with ingressions of temperate broadleaved forests at circa 33–37°S dominated by the deciduous species *Nothofagus macrocarpa, Nothofagus obliqua, Nothofagus glauca, Nothofagus alpina,* the evergreen *Nothofagus dombeyi* and the narrow-ranged endemic *Nothofagus alessandrii*. Between circa 37–40°S, with increasing precipitation, vegetation transitions into several forest types dominated by deciduous *Nothofagus obliqua, Nothofagus alpina* and the evergreen *Nothofagus dombeyi*, and towards the Andean slopes of Chile and Argentina, into forests dominated by the conifer *Araucaria araucana*. From circa 40–47°S the lush, Valdivian rainforest develops west of the Andean divide. This species-rich forest can be composed in various combinations by *Nothofagus dombeyi,* and *Nothofagus nitida* together with several other evergreen angiosperm trees such as *Laureliopsis philippiana, Weinmania trichosperma, Aextoxicon punctatum, Eucryphia cordifolia, Drimys winteri, Gevuina avellana*, several myrtaceous trees (e.g., *Amomyrtus luma, Amomyrtus meli*), endemic conifers (e.g., *Saxegothaea conspicua, Podocarpus nubigena*) as well as the long-lived conifer *Fitzroya cupressoides*. On the drier leeward slopes of the Andes, forests become impoverished in tree elements with the evergreen *Nothofagus dombeyi* dominating the lower altitude forests and the deciduous *Nothofagus pumilio* forming monospecific subalpine forests. This latter forest type dominates forests below treeline from circa 40–55°S. Towards the dry eastern boundary with the Patagonian steppe, the xeric-tolerant conifer *Austrocedrus chilensis* and shrublands composed by the deciduous short tree *Nothofagus antarctica* and other woody elements dominate de Andean foothills. Finally, south of circa 47°S, many rainforest elements are lost, leading to the Magellanic forests dominated by the evergreen *Nothofagus betuloides* and the long-lived conifer *Pilgerodendron uviferum* that extends in coastal areas of Chile until the southern tip of South America (Donoso 1993).

9.2 Temporal Patterns and Drivers of Fire Occurrence

Fire in southern temperate forests and shrublands is and has been a variable but dominant disturbance during at least the late glacial as suggested by the pervasiveness of charcoal particles in sedimentary records throughout the region (e.g., Whitlock et al. 2007; Nanavati et al. 2019; Martel-Cea et al. 2024), charcoal in soils, burnt spars, fire scars on trees (e.g., Veblen et al. 1999; Aravena et al. 2003; González et al. 2005, 2020; Mundo et al. 2013, 2017; Muñoz et al. 2023), historically documented/recorded fires (e.g., Willis 1914; Rothkugel 1916; Elizalde 1958; Bruno and Martín 1982; Donoso and Lara 1996; CONAF 2017) and fire scars detected by satellites (Miranda et al. 2022; Kitzberger et al. 2022; Fig. 9.2).

Given the strong limitation of natural ignition sources in this region, most prehistoric and historic wildfires are assumed to have originated primarily from ignitions set by humans for different purposes since the arrival of humans in South America

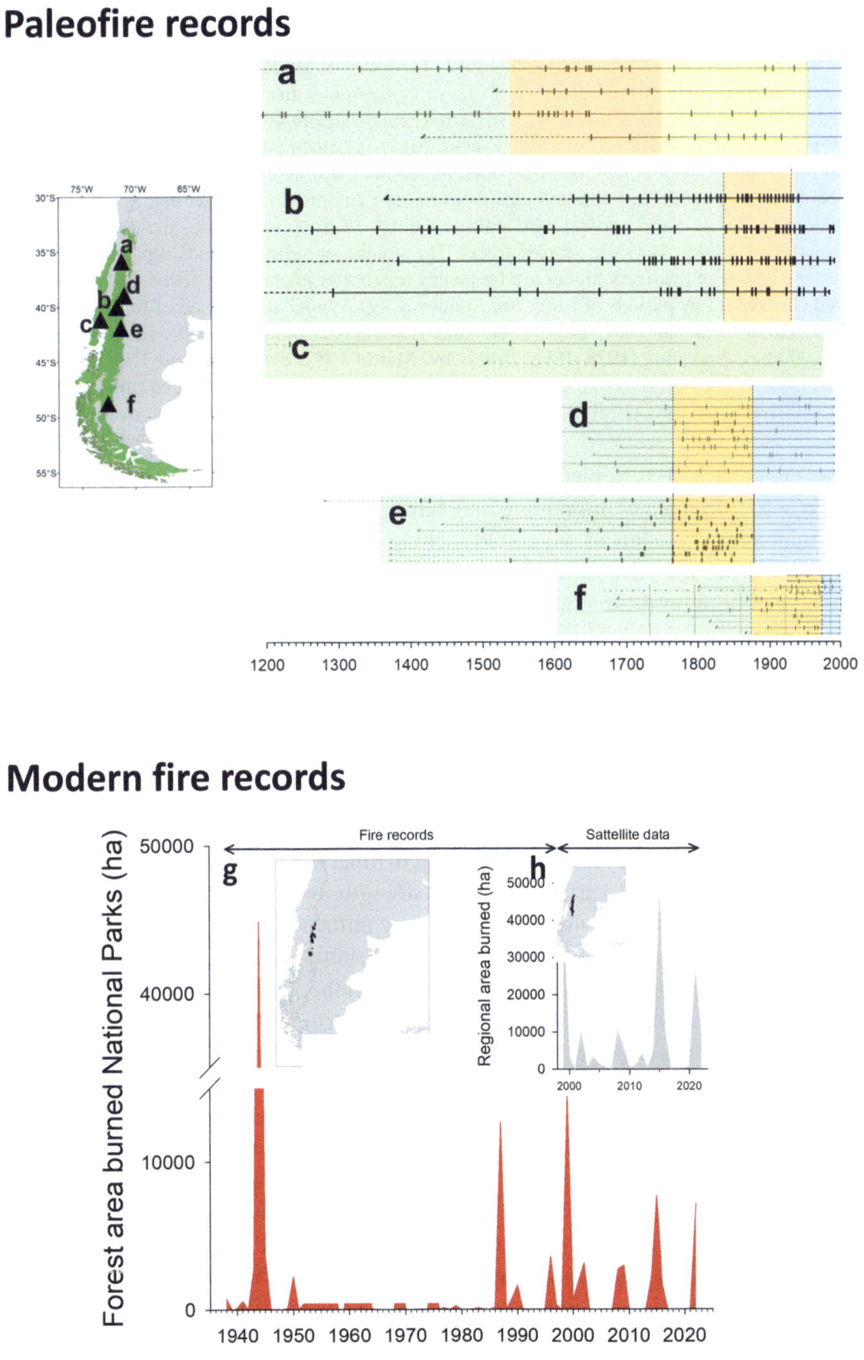

Fig. 9.2 Temporal patterns of fire activity revealed by composite (site level) dendroecological fire scar records from six studies (**a–f**) and modern fire records (**g, h**). (**a**) *Austrocedrus chilensis* fire

chronologies from the upper Cachapoal valley (modified from Rozas et al. 2018), (**b**) *Araucaria araucana* fire chronologies from Malalcahuello National Reserves and Villarrica and Tolhuaca National Parks (modified from Riquelme 2021; González et al. 2005, 2022a, b, c, for Malleco National Reserve unpublished data), (**c**) *Fitzroya cupressoides* fire chronologies from cordillera Pelada (modified from Lara et al. 1999), (**d**) *Araucaria araucana* fire chronologies from Neuquén province, Argentina (modified from Mundo et al. 2012), (**e**) *Austrocedrus chilensis* fire chronology from Lanin and Nahuel Huapi national parks, Argentina (modified from Veblen et al. 1999), and (**f**) Nothofagus pumilio fire chronologies from Santa Cruz province, Argentina (modified from Mundo et al. 2017). Short vertical bars indicate the date of fire (synchronous dating of ≥1 (a–c), and > 10% or >20% (d–f) fire-scarred trees). The orange or yellow periods separated by dates indicate a period of relatively higher fire frequency (or shorter interval), preceded (green) and followed (light blue) by periods of lower fire frequency. (**g**) Annual area burned of forest in Nahuel Huapi, Lanín National Park, Lago Puelo and Los Alerces National Parks (total area circa 1,400,000 ha), Argentina (1938–1998, Bruno and Martín 1982; data APN) and TM/MODIS fire scars (1999–2022; Barberá et al. unpubl. data), (**h**) Regional annual area burned (total area circa 2,900,000 ha) based on TM/MODIS fire scars (1999–2022; Barberá et al. unpubl. data)

(Aravena et al. 2003; Veblen et al. 2003). Prior to the arrival of humans to this region, a low-frequency fire regime may have prevailed, driven by occasional lightning and volcanic events as suggested by long sedimentary records of charcoal and pollen covering the past 40,000 years (Heusser 1994). However, the time-varying role of lightning on fire regime shifts is still not clear, particularly considering the recent warming-related increases in lightning ignited fires in northern Patagonia related to decadal changes in dominant summer circulation patterns (Veblen et al. 2008).

Despite being limited by ignitions, fire is pervasive in the region and is mostly driven by the seasonal and/or interannual drying of fuel-rich ecosystems, which enhances ignition efficiency and fire spread. Summer-dry weather and fuel conditions generated by the seasonal blocking of westerly circulation north of circa 43°S (i.e., South Pacific Anticyclone) generates a distinct fire season between November and March. Interannual variability in the strength and latitudinal position of this semi-permanent high-pressure cell has a strong influence on the annual area burned at these midlatitudes (Kitzberger et al. 1997). Rainfall and fire has been linked to tropical and extratropical climate drivers of the southern hemisphere, namely the El Niño-Southern Oscillation (ENSO) and the Southern Annular Mode (SAM). Both modern fire records (Holz et al. 2012) as well as dendroecological reconstructions spanning back several centuries (Kitzberger and Veblen 1997; Veblen et al. 1999; González and Veblen 2006; Mundo et al. 2017; Holz et al. 2017; González et al. 2005, 2020) show spatially varying and interactive effects of SAM and ENSO on fire occurrence over southwestern South America. During strong cold (La Niña) phases of ENSO, dry winters occur across southern South America, particularly over central Chile (circa 25–35°S) and dry springs occur across southern South America particularly over south-central Chile and Argentina (circa 35–42°S) while south of circa 42°S this teleconnection becomes weaker. SAM instead, has a prominently southern influence on both precipitation and temperature variability over southern South America. When SAM is in its positive mode, anomalous high pressure develops over mid-latitudes compared to high latitudes, leading to a strengthened circumpolar vortex and a poleward deflection of the westerly storm tracks (Abram et al. 2014) generating almost year-round warmer/drier conditions and increased fire activity over south of circa 38°S (Holz et al. 2012, 2017). Although these tropical and extratropical drivers of climate and fire variability are not completely independent, there is evidence from both modern fires and reconstructed fire histories that the interaction of La Niña and positive SAM produces optimal conditions for increased fire occurrence in the region of teleconnection overlap (circa 39–42°S; Veblen et al. 1999; Holz et al. 2012; Mundo et al. 2017).

While climatic variability modulating fuel desiccation seems to be the main top-down temporal driver of fire occurrence in southern temperate forests (Kitzberger et al. 2022), other (bottom-up) ecological processes may also influence fire activity. Many southern temperate forest understories are dominated by bamboos of the genus *Chusquea,* largely known to display synchronous, semelparous flowering followed by massive culm die off (González and Donoso 1999). After these large-scale reproductive events involving many hundreds square kilometers, massive amounts of fine dead fuels remain available in the forest understories until decomposition (Veblen et al. 2003). This window of increased fuel availability of circa 10 years (Holz and Palma 2012) can generate higher fire risk over entire regions and extremely large and severe individual wildfire events.

For being ignition-limited, fire regimes in southern South America have been largely modified by humans. Both dendroecological reconstructions as well as analyses of modern drivers of fire underscore the importance of human occupation patterns (both in space and time) to influence fire activity. Networks of fire chronologies derived from fire scars can distinguish regional (i.e. likely climate driven) influences from local (i.e. likely human-driven) influences on fire occurrence (Veblen et al. 1999; Mundo et al. 2013). Historic changes in land use which have varied geographically across southern South America correlate locally with sharp changes in fire regimes (Fig. 9.2a–f).

9.3 Spatial Patterns and Drivers of Fire Occurrence

Fire activity across the marked regional climatic and vegetation gradients shows a strong spatial pattern. One of the main bottom-up drivers of fire activity is ecosystem productivity as it regulates the amount and continuity of fuel that can be burned. It has been proposed that fire activity follows a humped relationship with productivity in which high and low productivity extremes have reduced fire activity due to poor fuel condition and low fuel availability, respectively, while activity is highest at intermediate portions of this gradient where enough fuels are readily burnable (Bowman et al. 2014; Krawchuk et al. 2009; Pausas and Ribeiro 2013).

Two main productivity gradients are evident in this region: a latitudinal gradient produced by the southward decreasing influence of the South Pacific Anticyclone on rainfall and a longitudinal gradient produced by the rain shadow effect of the Andean Cordillera on precipitation. Along the former gradient, fire activity follows a left-biased humped function (Holz et al. 2012) being maximal in Mediterranean shrublands that have enough fuels to carry, sustain and spread fire, but that also have abundant climatic opportunities for these fuels to become flammable (Fig. 9.3a). Fire activity drops sharply north of circa 32°S where drier semi deserts carry less, more discontinuous fuels. Southwards, fire activity decreases gradually and becomes low south of circa 44°S, a region where precipitation seasonality disappears and where fires can only burn during anomalous droughts (Fig. 9.3a). Longitudinally, the probability of fire occurrence is highest where dry forests coexist with

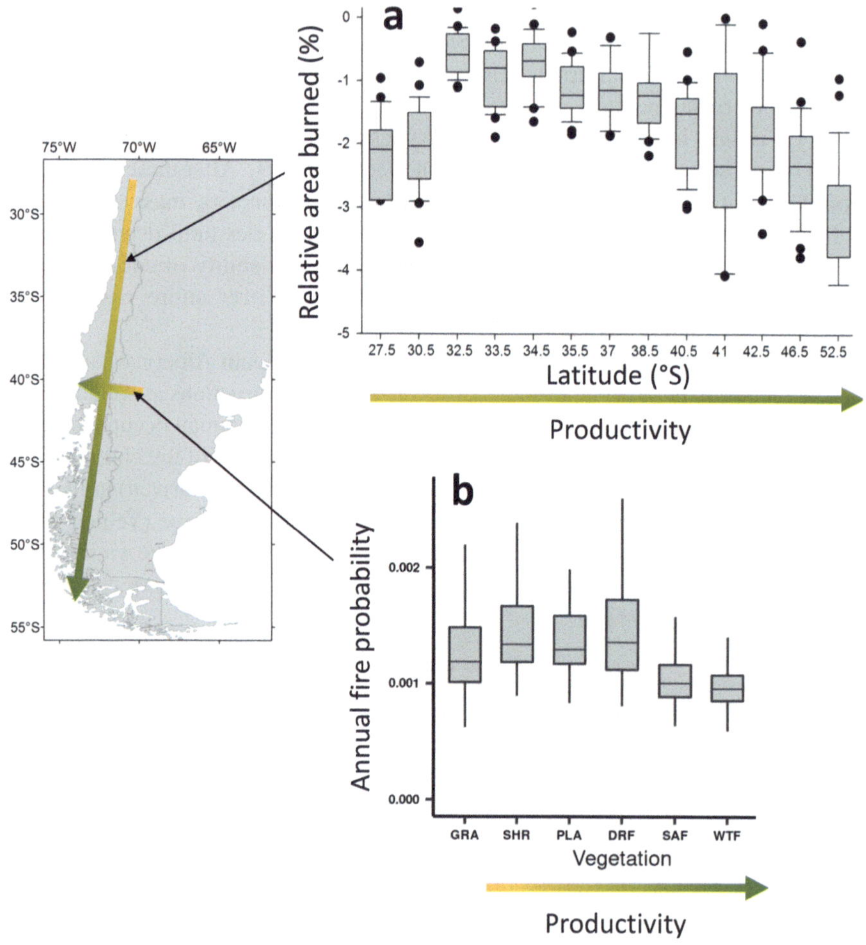

Fig. 9.3 Productivity-fire activity relations along: (**a**) a latitudinal productivity gradient (log transformed %; area burned in relation to available area per administrative region, based on CONAF 1984–2008 fire records, modified from Holz et al. (2012)), and (**b**) a trans-Andean longitudinal productivity gradient in southern South America (modeled probability from 1986–87 TM, and 1998–2018 MODIS burned area product, modified from Kitzberger et al. (2022)). *GRA* grasslands and steppes, *SHR* shrublands, *PLA* pine plantations, *DRF* dry *Austrocedrus* and *Araucaria* forests, *SAF* subalpine *Nothofagus pumilio* forests, and *WTF* wet *Nothofagus dombeyi* forests and rainforests

shrublands and exotic pine plantations (Kitzberger et al. 2022; Fig. 9.3b). Fire probabilities drop at the fuel-limited xeric Patagonian steppe and in lush less flammable lowland and coastal range rainforests and subalpine forests (Fig. 9.3b).

In line with ignition-limited fire regimes both modern fire records (Mundo et al. 2013) and satellite-derived fire scar collections (Paritsis et al. 2013; Kitzberger et al. 2022), a strong spatial association evident between fire activity and indicators of human presence (distance to roads, villages, cities; Díaz-Hormazabal and González 2016; McWethy et al. 2018; Pozo et al. 2022). The pattern of fire activity near settlements and roads is often mirrored by strong distance-dependent pattern of self-reinforced conversion from relatively non-flammable but vulnerable forest to flammable fire resilient shrublands or grasslands (Mermoz et al. 2005; Gowda et al. 2012; Paritsis et al. 2013).

9.4 Ecological Responses of Temperate Forests to Fire

9.4.1 Fire-Related Plant Traits and Individual Responses

Fire has variable effects on vegetation at the level of both the individual plants and the plant community. The effects of fire on plants may depend partly on the frequency and intensity of the fire, but also on the differential abilities or adaptations of plants to survive or recolonize (Agee 1993; Pausas et al. 2004; Pausas 2012; Keeley et al. 2012). Vegetation provides the fuel that makes fire possible, consequently fire effects on plants and their responses should be considered as the result of this interaction. Thus, to predict individual plant or vegetation responses to fire, it is necessary a comprehensive understanding of the different fire regimes that determine different potentials for recovery associated with specific plant traits (Keeley et al. 2012). This would mean that plants are not adapted to fire strictly, but to a particular fire regime (Keeley et al. 2012).

In general terms, there are five main strategies of how plant species may respond to fires (Rowe 1983; Agee 1993; Whelan 1995; Baker 2009; Pausas 2012; Keeley et al. 2012; Table 9.1). The strategies or abilities of the species are based on their life-history traits and since some of the strategies may not be exclusively related to fire, they can fit more than one category because of multiple strategies or adaptations. For example, a species can be invader (post-fire pioneer species) due to its rapid post-fire colonization, but it can also be an endurer due to its resprouting capacity (e.g., *Nothofagus obliqua, Nothofagus alpina, Nothofagus antarctica*).

Fire has long been an important natural and anthropogenic disturbance in temperate southern hemisphere forest ecosystems of Chile and Argentina. In these temperate ecosystems woody species have shown different fire-relevant functional traits to persist by either resisting fire damage, resprouting (after top kill) or through effective seed dispersal (Table 9.2; Fig. 9.4). For example, some deciduous trees of the genus *Nothofagus* commonly resprout when burned. Other tree species may be fire resisters, such as *Araucaria araucana*, showing thick bark, low-flammability foliage, low epiphyte loads, decay resistance (sealing scars avoiding diseases),

Table 9.1 Main life-history traits and functional strategies of species relative to fire responses in temperate forests and shrublands of southern South America (Agee 1993; Whelan 1995; Pausas 2012)

Plant strategy to fire	Plant response	Characteristic in relation to survival and persistence
Intolerant to fires	Avoiders or sensitive to fires	Species do not survive fires, and have essentially no abilities or adaptations to fire. These are generally shade-tolerant, late successional species that slowly reinvade burned areas, and their speed of recolonization depends on the dispersal capacity and the distance to the remnant populations. They are often slow-growing, long-lived woody plants. E.g., *Aextoxicum punctatum, Laureliopsis philippiana, Laurelia sempervirens, Podocarpus nubigena, Saxegothaea conspicua,* and *Persea lingue*
Post-fire colonizers (obligate seeders)	Invaders	Species do not resist fire nor do they generate a resistant seed bank. However, the dispersal is very efficient and they appear by germination after the fire. Seed dispersal comes from less severely burned/unburned boundaries or islands surrounded by the fire. E.g., *Nothofagus dombeyi, Nothofagus pumilio, Austrocedrus chilensis* and *Fitzroya cupressoides*
Resprouters	Endurers	Species have the ability to survive fires through the strategy of regrowth from the root crown, lateral root or aerial crown. The fire produces the reduction of the size and aerial biomass of the plants. These species are called tolerant or persistent to fire (e.g., *Nothofagus antarctica, Nothofagus alessandrii, Nothofagus obliqua, Nothofagus alpina*
Post-fire recruiters	Evaders	Species and populations can persist thanks to the presence of an abundant seed bank stored in the soil (seed banks) or canopy (serotinous structures). Fire stimulates the recruitment of these species. They are often plants with high flammability called pyrophytic species. No temperate tree seems to fit this strategy but some vines (e.g., *Muehlenbeckia hastulata*) and lower shrubs (e.g., *Fabiana imbricata*) have fire-stimulated persistent seed banks (Segura et al. 1998; Dudynski and Ghermandi 2013)
Resisters to fire	Resisters	These are thick-barked species that can survive variable intensity fires. Species of commonly less flammable plants are resisters. Two thick-bark species could be classified as resisters: *Araucaria araucana* and *N. Glauca*

self-prune lower branches, and basal and epicormic (crown) resprouting. It is important to note that these functional traits may have been selected, in an evolutionary sense not only by fire, but also by other selective forces (e.g., herbivory, drought, disease), a process commonly known as exaptation.

Despite the increasing incidence of fire, the degree of pre-adaptation, resistance and resilience vary markedly among the various temperate forest and shrubland communities that develop. Thus, highly flammable and fire-resilient communities can coexist in the landscapes in close contact with communities that are poorly adapted and sensitive to fire (Tiribelli et al. 2018). In the event of conflagrations, the former often serve as rapid and efficient propagators of fire towards the latter (Tiribelli et al. 2024) that affect desiccation and conditions of fuel ignition. Life

Table 9.2 Main traits and functional strategies of temperate tree species relative to fire resistance and recovery from fire (x = low or weak; X = high or effective)

Tree species	Thick-barked (fire resistant)	Resprouter (vegetative regrowth)	Post-fire colonizer (prolific post-fire seeding)	Observation
Araucaria araucana	**X**	**X**		Root and basal sprouts from killed seedlings and saplings; irregular recruitment from seeds
Fitzroya cupressoides	x	**X**	**X**	Large trees may resist fire; irregularly root suckers
Austrocedrus chilensis			**X**	Crown burns easily; large trees may resist fire
Nothofagus antarctica		**X**		Vigorously resprouts from lignotubers and basal buds
Nothofagus pumilio			**X**	Thin-barked easily killed by fire; large trees may resist fire
Nothofagus dombeyi			**X**	Thin-barked easily killed by fire; large trees may resist fire
Nothofagus obliqua	x	**X**	**X**	Large trees may resist fire
Nothofagus alpina	x	**X**	**X**	Large trees may resist fire
Nothofagus glauca	**X**	**X**	**X**	Crown and basal sprouts
Nothofagus alessandrii		**X**	x	Vigorous basal sprouts; low recruitment from seeds

history characteristics (longevity, reproductive age, fertility, etcirca), anatomy (bark thickness, bud protection, seed dispersal capacity) and modes of reproduction (resprouting from underground organs, resprouting from aerial organs, obligate seedlings dispersed by wind or biotic vectors) will ultimately determine the resistance of the species and resilience of the communities against fire regimes of different frequency, size or severity.

9.4.2 Stand Level Forest Recovery Following Fires

The history of volcanic eruptions and lightning as a source of forest fires over a long period have probably influenced the strategies and adaptive process of some species of the temperate regions of Chile and Argentina (Donoso 1996; Veblen et al. 1996). Although temperate southern forests of Patagonia are not dependent on fire for their survival, trees and other life forms that participate in them present different life-history traits and/or strategies to persist (i.e., resist or recover) after fire events. Here

Fig. 9.4 Traits of temperate tree species relevant to fire resistance and recovery from fire. (**a**) Postfire recruitment of Araucaria seedlings in Tolhuaca National Park (Chile); (**b**) *Araucaria araucana* bark thickness (circa 20 cm thick); (**c**) Postfire crown epicormic sprouting of *Nothofagus glauca* (Maule region, Chile); (**d** and **e**) Basal sprouts after severe fire of *Nothofagus alessandrii* and *Nothofagus glauca*, respectively; (**f**) *Nothofagus glauca* bark thickness made up of several layers showing its ability to resist intense fires. Photos: ME. González

we review the strategies, abilities, or adaptations to resist or recover from fires of the principal coniferous and broad-leaved trees of the temperate forests in the southern Andes. We focus on three conifers (*Araucaria araucana*, *Fitzroya cupressoides*, *Austrocedrus chilensis*) and on six dominant broad-leaved *Nothofagus* trees in southern Chile and Argentina.

9.4.2.1 *Araucaria-Nothofagus* Forests

Fire has been an ubiquitous ecological process influencing the *Araucaria-Nothofagus* forests. Fire history reconstructions document ca. 300- and 500-years old fire chronologies in the Andes range of Argentina and Chile, respectively (González et al. 2005, 2020; Mundo et al. 2012, 2017). Fire regimes in these forests are characterized by mixed-severity fires (low-severity surface fires to crown fires) that result in a range of fire effects and responses. *Araucaria araucana* and

Nothofagus species account for distinct strategies and adaptations to resist and/or recover from fire events (Tortorelli 1942; Montaldo 1974; Veblen 1982a, b; Burns 1991, 1993; Veblen et al. 1995, 1996; González et al. 2005, 2010a, b, 2013a, b; Assal et al. 2018; Fuentes-Ramirez et al. 2019, 2022).

Araucaria araucana is perhaps one of the few tree species in both Argentina and Chile with unequivocal fire adaptations. Some life-history traits of *Araucaria araucana* that favor its persistence capacity and adaptation to fire are its thick bark (>10 cm), natural pruning of the lower branches (which reduces the probability of fire reaching the canopy), protected terminal buds on branches, vegetative regrowth from the roots and crown, and the protection of the seeds inside large cones by layers of thick outer bracts (Tortorelli 1942; Veblen 1982a, b; Burns 1993; González et al. 2010a; Veblen et al. 1995; Table 9.2). However, heat treatments on seeds have shown negative effects on the germination of *Araucaria araucana* and *Nothofagus antarctica* (Cóbar-Carranza et al. 2015). Similarly, the *Nothofagus* species have shown certain abilities to survive or recover after fires, such as through their vegetative reproduction capacity (e.g., *Nothofagus antarctica*; Burns 1993; González et al. 2010a) or their relatively high and effective seed dispersal of the obligate seeders (e.g., *Nothofagus dombeyi, Nothofagus pumilio*; Veblen et al. 1996, 2003; González et al. 2005, 2010a; Table 9.2). Thus, in mixed *Araucaria-Nothofagus* forests affected by high-moderate severity fires, that leaves surviving trees, *Nothofagus dombeyi, Nothofagus pumilio* and *Nothofagus antarctica* can establish densely, forming even-aged forest stands (Burns 1993; González et al. 2010a). However, in northern Patagonia, failure of postfire *Nothofagus pumilio* regeneration is common on steep, north-facing slopes at high elevations (Veblen et al. 1996). Similarly, in the southern Chilean Patagonia (Aysén region), extensive burning of *Nothofagus pumilio* forests during the first half of the twenty century, resulted only in narrow bands of tree establishment under the protection of the forest margin. This regeneration failure could be explained by a combination of poor seed dispersal, herbivory, and strong drying winds (Veblen et al. 1996). In the case of *Araucaria araucana*, although the establishment of individuals from vegetative sprouts and seeds is less dense (and dependent in the latter case on the survival of female trees), it is still possible to identify pulses of postfire regeneration (González et al. 2010a).

During the last two decades, severe and extensive wildfires have affected *Araucaria-Nothofagus* forests in northern Patagonia (González and Lara 2015; Sanguinetti 2015; Fuentes-Ramírez et al. 2020; Franco et al. 2022). Recent studies of the early postfire regeneration patterns have shown the different survival and reproduction strategies of the species to cope with more frequent and high-severity fire events. After high-severity fires and/or repeated fires, early recruitment of seed-dependent species (e.g., *Nothofagus dombeyi, Nothofagus pumilio*) seems to fail or be very slow, making their persistence mostly depend on seed dispersal from unburned sites around the fire boundary (Urrutia-Estrada et al. 2018; Fuentes-Ramírez et al. 2020, 2022; González et al. 2010b, 2013b, 2020). In the case of *Araucaria*, some trees may survive with a diameter above 40–50 cm (Burns 1991; Sanguinetti 2015; González and Veblen 2007). Also, a relatively rapid sprouting capacity has been reported for burned *Araucaria* trees from both the canopy and belowground buds (González et al. 2013a; Fuentes-Ramirez et al. 2019; Sanguinetti 2015). Thus, its post-fire recruitment may establish both from seeds and vegetative sprouts from basal epicormic buds of burned young individuals (<10 cm in diameter at breast height; González et al. 2013a, b; Fuentes-Ramirez et al. 2019). Also, the rapid recruitment of *Araucaria* seedlings under mother trees implies that mother trees bearing female cones at the time of the fire, located higher in the crown and ready to disseminate their propagules, would offer protection to the seeds from the fire (González et al. 2013a, 2014).

Under a high fire severity, plant species diversity and abundance are reduced, and post-fire colonization of herbs and shrubs established mostly by vegetative sprouts (roots, bulbs and rhizomes) is relatively rapid, especially in areas burned with low and moderate severity (González and Veblen 2007; Fuentes-Ramirez et al. 2019, 2022). For example, rapid resprouting of *Chusquea culeou* results commonly in a dense cover of this bamboo species after fire (González and Veblen 2007; González et al. 2010b; Fuentes-Ramírez et al. 2022).

9.4.2.2 *Fitzroya Cupressoides* Forests

Natural or human wildfires have been an important disturbance that has shaped the structure and dynamics of *Fitzroya* forests (Veblen and Ashton 1982; Veblen et al. 1996, 1999; Lara et al. 1999; Urrutia-Estrada et al. 2018; González et al. 2022a). Fire has been the main disturbance in the central depression, the Coastal Range, and at lower elevations in the Andes. Fires in these temperate rain forests occur naturally or are set by humans during infrequent dry years (Kitzberger et al. 1997; Veblen et al. 1999; Lara et al. 1999). Fire history reconstructions document ca. 600- and 900-year-old fire chronologies in the coastal range in south-central Chile (Lara et al. 1999; Urrutia-Estrada et al. 2018) and ca. 1000-year old fire chronologies in the Argentinean Andes (Veblen et al. 1999).

Fitzroya cupressoides presents some life-history traits important to resist or recover from fires (Table 9.2). For example, the bark is typically >20-cm thick on large (>1.5 m diameter) trees, which allow *Fitzroya to survive* intense fires (Veblen et al. 2003). Also, *Fitzroya* forests affected by fires have shown the ability to recruit burned sites by seedlings and root suckers. At high altitude in the Coastal Range the abundance of seedlings and saplings in areas affected by recent, low intensity fires - where seed sources were available-, indicate adequate regeneration of *Fitzroya* (Veblen and Ashton 1982; Lara et al. 1999). However, on the most severely burned sites re-establishment of *Fitzroya* could fail or be exceedingly slow (Veblen and Ashton 1982). Furthermore, the age class structures of *Fitzroya* forest stands in the central depression and Coastal Range show single-cohort populations, suggesting establishment after stand-devastating fire disturbances (Lara et al. 1999; Silla et al. 2002). Fire promotes *Fitzroya* regeneration by seed and vegetative reproduction (root sucker sprouting and layering) being overall the latter more prevalent (Veblen and Ashton 1982; Silla et al. 2002; Lara et al. 2003).

In north Patagonia stand-replacing fires also create conditions suitable for the seedling establishment of the shade-intolerant *Fitzroya*, which would be replaced by other tree species in the absence of coarse-scale disturbance (Veblen et al. 2003).

Moreover, the resistance of large trees to burning and the establishment of new cohorts seem to ensure the persistence of *Fitzroya* in these temperate rainforests (Veblen et al. 2003).

9.4.2.3 *Austrocedrus Chilensis* Forests

Fire is the most important natural disturbance affecting the *Austrocedrus* forests of south-central Chile (1988; Le Quesne 1988; Aravena et al. 2003; Bustos-Schindler et al. 2010; Rozas et al. 2018) and those growing in northwestern Patagonia (Veblen et al. 1992a, b). In both regions, long-term history of *Austrocedrus* fire regimes have been developed documenting the fire activity over the last 450 and 800 years for northwest Patagonia and central Chile, respectively (Kitzberger et al. 1997; Kitzberger and Veblen 1997; Veblen et al. 1999; Rozas et al. 2018).

Austrocedrus is commonly killed by intense fires as a result of its relatively thin bark and because it lacks the capacity for vegetative reproduction from basal or root sprouts its post-fire regeneration depends on the dispersal of seeds to the site (Veblen et al. 1995). Infrequent stand-devastating fires are common in the dense *Nothofagus dombeyi–Austrocedrus* and pure *Austrocedrus* stands in northwest Patagonia. On the other hand, fires in open woodland habitats near the ecotone with the Patagonian steppe or in the western slopes of the Andes in central Chile are less intense as a result of the low and discontinuous fuel loads and many adult trees could survive these low-intensity fires (Veblen et al. 1995; Rozas et al. 2018). For the mesic forest zone of northwestern Patagonia, age-structure analyses indicate that, where stands of *Austrocedrus* and *Nothofagus dombeyi* are destroyed by fire, both obligate seeder species can regenerate (Veblen and Lorenz 1987). Post-fire recruitment at mesic sites is usually dominated (i.e., in density and growth rates of seedlings) during the first 20–30 years by *Nothofagus dombeyi*. However, both species develop even-aged post-fire populations (Veblen and Lorenz 1987; Veblen et al. 1992a, b).

In the Andes of central Chile, the diameter and age structure of trees in the stand clearly reflect that *Austrocedrus* regeneration is associated with frequent, low intensity fires, due to a discontinuous distribution of dead fuel. These frequent low-intensity fires originate all-aged open forests, with multiple pulses of regeneration (Barría 2007). Further south in the Andes (35–36°S) *Austrocedrus* forests presents inverted J-type diameter structure that could be interpreted as a multi-age forest, an indication that the species is self-replacing despite the human influence (i.e., high-grading; Le Quesne 1988; Donoso 1993). However, in some stands affected by logging and probably fire, the diameter structures of the trees also seem to indicate a relatively even-aged establishment of *Austrocedrus* even though its regeneration and initial growth had been limited and/or suppressed by the dense recruitment of *Nothofagus obliqua*, *Nothofagus glauca* and others common evergreen tree and shrub species (Ibarra and Mourgues 1976; Le Quesne 1988; Veblen et al. 1995). Since the age structure of these stands are not available, some doubt remains as to the true temporal establishment pattern of *Austrocedrus*, where fire and logging are major factors. Furthermore, in Alto Bío Bío (38°S at 900 m asl) the age structure suggests a regeneration pulse after a fire in 1893, with a recruitment period lasting 60 years (Aravena et al. 2003). Few individuals of *Austrocedrus* survived and showed a pronounced growth decrease in the year 1893. These surviving trees were the seed source for the new cohort established (Aravena et al. 2003). Toward southern Chile, stand-replacing fires affecting isolated Andean patches may promote regeneration pulse of *Austrocedrus*, however its establishment seems less synchronous on rocky outcrops or shallow soils (Veblen et al. 1995).

9.4.2.4 *Nothofagus* Forests (*Nothofagus Glauca, Nothofagus Alessandrii, Nothofagus Obliqua, Nothofagus Alpina*)

Fire is referred to as an important selective agent for certain reproductive strategies (i.e., sprouting capacity) and protective characteristics (i.e., thick bark) of deciduous species in these ecosystems. The history of frequent volcanic eruptions, lightning strikes and human-ignited forest fires seem to account for the role of fire in the adaptive processes and traits of some of these *Nothofagus* species (Donoso 1996; Veblen et al. 1996). For example, all these species have developed a vigorous sprouting capacity from stumps, roots, or the canopy (Donoso 1993, 1996; Veblen et al. 1996; Litton and Santelices 2002; González et al. 2015, 2022a, b; Burger 2021; Table 9.2), and some of them can develop a thick bark that provides protection from fire (Donoso 1996; Burger 2021; González et al. 2022b; Table 9.2).

In south-central Chile and Northwest Patagonia *Nothofagus* forests are dominated mostly by second-growth even-aged forests originated after fuelwood harvesting, loggins and fires during the last 150 years (Veblen 1985; Donoso 1993, 1996; Veblen et al. 1996; González et al. 2015, 2022b). Sometimes these young stands could still contain few surviving and scattered mature trees that may have served as a source of seeds.

In central Chile, second-growth stands of *Nothofagus glauca* and *Nothofagus obliqua* originated both from seeds and from stump sprouts were partly the result of a shifting cultivation practice of timber cut and burn debris to wheat cultivation. Due to this practice, high elevation Andean *Nothofagus* forests make up a mosaic of stands of different ages (Donoso 1988, 1996). Also, after clear-cuts of *Pinus radiata* plantations, second-growth stands of *Nothofagus glauca* vigorously develop from stump sprouts (Donoso 1996).

Nothofagus glauca exhibits characteristics commonly found in fire-adapted forest ecosystems. This species shows rough, coriaceous leaves and thick barks with numerous scaly layers (Donoso 1996). In the coastal range of central Chile, Litton and Santelices (2002) studied the effect of a wildfire on early secondary succession in a *Nothofagus glauca* forest, recording vigorous resprouts from the crown and the base of trees. Crown sprouts originate from dormant buds (i.e., epicormic buds) that are protected by the thick bark of the species (González et al. 2022c). Furthermore, the diameter of *Nothofagus glauca* trees determines their probability of survival, indicating that individuals with larger diameters and thicker bark have greater survival. Thus, trees with diameters >20 cm, the probability of survival is more than 75% (Burger 2021). Although it is not yet clear that these adaptations are a result of an evolutionary history to fire, this species appears well adapted to this ecological process (Litton and Santelices 2002). Due to the extensive plantation of *Pinus radiata* around these remnants' native forests, a major concern is the successful invasion of this exotic species on these postfire disturbed forests (Burger 2021; Litton and Santelices 2002; Leal-Medina et al. 2024).

In the case of *Nothofagus alessandrii*, similar human anthropogenic disturbances have prompted even-aged second growth forests (González et al. 2022b, c). Unlike *Nothofagus glauca*, this species is very susceptible to fire as it has a thin bark. Therefore, most trees survive mainly through sprouting from the epicormic buds at the bases of the trunks of the dead trees, which frequently display protuberances as evidence of continuous re-sprouting following past fire events (González et al. 2022b, c).

During the late nineteenth and early twentieth centuries, extensive areas of low and mid-elevation *Nothofagus* forests were burned in south-central Chile and north-western Patagonia. In the Valdivian rainforest (38°–42°S), large areas cut and burned promoted massive establishment of second-growth stands of *Nothofagus obliqua* and *Nothofagus alpina* (Veblen et al. 1996). In the central depression, vast areas of *Nothofagus obliqua* forests were burned and converted to farmland. After intense fires these species recruit vigorously from seeds and sprouts (Donoso 1993; Veblen et al. 1996; González et al. 2015; Table 9.2). Moreover, *Chusquea* bamboos (e.g., *Chusquea culeou, Chusquea quila*) if present in the undergrowth resprout strongly after being burned impeding the establishment of tree species (Veblen 1985; Veblen et al. 1981, 1996). Similarly, Northern Patagonia experienced extensive forest burning and clearing for livestock pastures (Veblen et al. 1996) resulting in the abundance of 70–100-year-old even-aged *Nothofagus* forests (Veblen and Lorenz 1987), where *Nothofagus alpina* and *Nothofagus dombeyi* are common dominant postfire species.

9.4.3 Landscape-Level Responses to Changing Fire Regimes

Vegetation responses to early century suppression-related drops in fire frequency were dramaticirca Settlement-related agricultural landscapes pervasive during the first half of the twentieth century in many low elevation *Nothofagus dombeyi*-dominated forests areas massively afforested into dense even-aged stands increasing overall fuel load and continuity across the landscape (Fig. 9.5a; Kitzberger and Veblen 1999). A second large-scale pattern of afforestation is the successional replacement of shrublands into *Austrocedrus chilensis* forests in areas where the existence of sufficient fire refugia in the landscape allowed for the recruitment of the relatively shade-tolerant conifer in the shrubland understory (Landesmann et al. 2015, 2016). Regional land cover change analyses spanning circa one century (Gowda et al. 2012, 2023) show net increases of forests of circa 10% (circa 0.2% $yr.^{-1}$), mostly concentrated during the 1914–1973 period, after which this trend stabilized. Concomitantly, circa 17% of the shrubland area was reduced and mostly converted into forests (Gowda et al. 2012, 2023). However, the process of afforestation was mostly restricted to lower elevation sites and/or mesic topographic microsites (SO SE-facing aspects) of higher elevations. Drier aspects particularly at higher elevations tended to remain as stable shrublands during the last century.

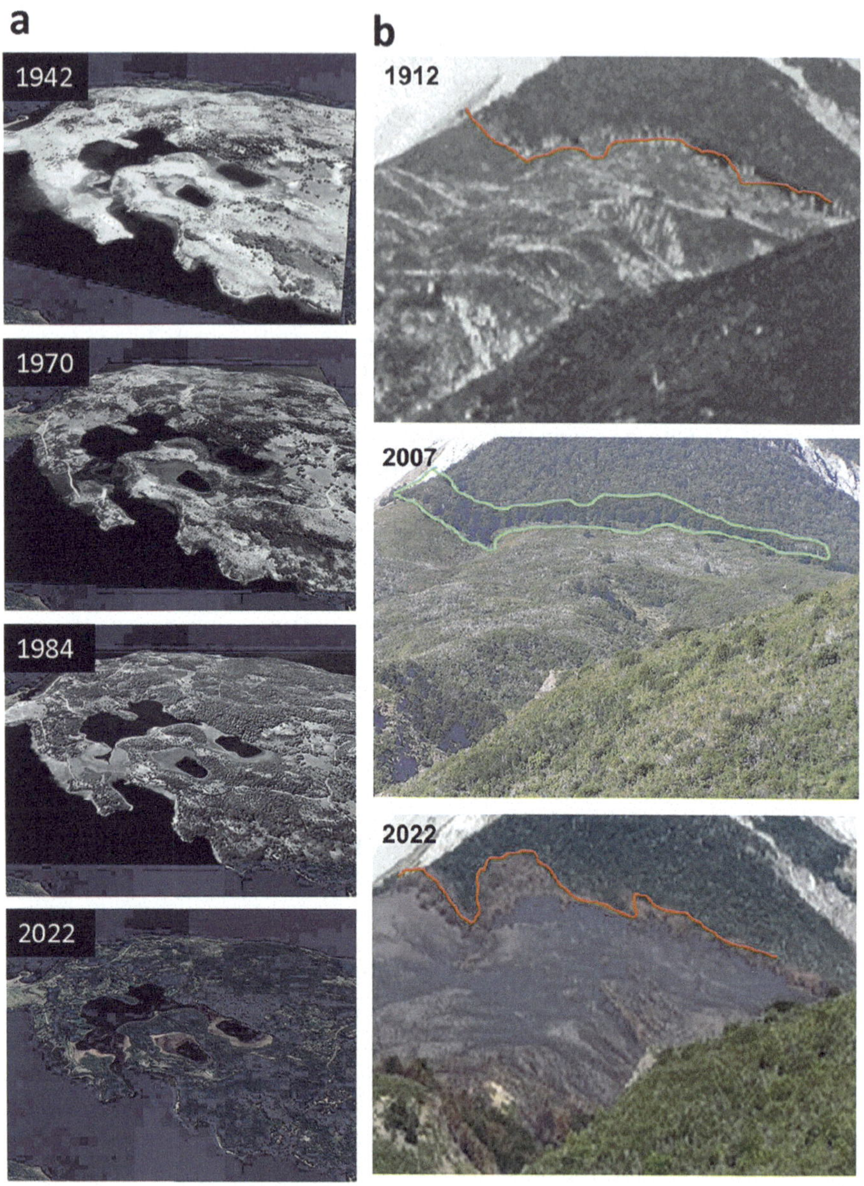

Fig. 9.5 (**a**)Tilted terrain-draped repeat air photography (1942, 1970 and 1984) and high-resolution satellite imagery of the Lago Moreno-Lago Morenito area west of the city of Bariloche, Argentina showing massive natural afforestation across the landscape as a response to declining fire frequency and rural-to-suburban/residential land use changes. (**b**) Historical photograph of Cerro Áspero, Río Negro Argentina after a severe fire occurred in 1912 (orange line) associated with an intense drought (photo: B. Willis). By 2007, a thin strip (<50 m) of *Nothofagus pumilio* young forest was evident (green polygon) suggesting dispersal-restricted post-fire regeneration (Photo: T. Veblen). In the context of a strong drought in 2022, a fire spreading circa 8500 has

(L. Martin-L. Steffen fire) reaches the slopes of Cerro Áspero and consumes the entire young forest as well as a portion of the mature (unburnt in 1912, orange line) forest, thus producing the retraction of forest (Photo: A. Premoli). The relative narrow strip of *Nothofagus pumilio* regeneration is explained by the poor dispersal capacity of *Nothofagus pumilio,* harsh microenvironmental within the shrubland and relatively frequent fires that stop at the shrubland-forest edge because mature *Nothofagus pumilio* forests tend to have high fuel moisture content. A dynamic equilibrium establishes that determines relatively stable forest-shrubland boundaries. However, under hot/dry climate conditions (e.g., fire season of 2022) severe fires can spread through the shrubland reach the forest and burn both young and mature forest due to reduced fuel moisture levels. As a result, *Nothofagus pumilio* retracts to higher elevation and/or moister microsites (S-facing slopes) at the expense of shrubland advances

Shrublands have maintained themselves stable by three main mechanisms: (1) Many permanent shrublands are located in microsites where tree species cannot develop (e.g., valley bottoms with high frost frequency), (2) Many fire-generated shrublands have been depleted from tree seed sources and thus cannot be efficiently colonized by non-resprouting obligate seeder trees, and (3) Many shrublands by being more flammable than the adjacent forests have spread more efficiently subsequent severe fires and thus maintain themselves as consequence of being "trapped" in high fire frequency/severity regime (Mermoz et al. 2005; Tiribelli et al. 2019; Landesmann et al. 2021).

9.5 Fire Effects on Animals

Knowledge on the effects of fire on animals in this region is scarce and dispersed. Changes in animal communities due to fire have been mainly studied on insects (Sackmann and Farji-Brener 2006; Sasal et al. 2010; Tello et al. 2020), birds (Grigera and Pavic 2007; Novoa et al. 2021) and lizards (Infante et al. 2021). Studies on these taxa show that fire impacts are dependent on fire severity and time since fire because these factors are strongly associated with modifications in habitat structure. Furthermore, effects of fire on these animal communities are conditional on both the taxonomic group under consideration and the burned vegetation type.

Insects show heterogeneous responses to fire in these southern temperate forests and shrublands. Ground dwelling beetle communities in *Nothofagus-Austrocedus* forests and shrublands in northwestern Patagonia in Argentina appear to decrease their abundance and change their composition after fire. Shortly after fire (i.e. 3–5 years), habitat simplification and homogenization lead to a reduced diversity in beetle communities where specialists disappear (Sasal et al. 2010). On the contrary, ant communities are much less affected than beetles by habitat changes originated by fire, probably due to their use of bare soil patches (Sackmann and Farji-Brener 2006). Similarly, within the saproxylic beetle assemblage of Chilean *Araucaria-Nothofagus* forests, high severity fire caused a decrease in mycophagous (fungi feeders) species but an increase in xylophagous (deadwood feeders) beetle species 15 years after fire (Tello et al. 2020, 2022).

Responses of vertebrates to fire in this region are also contingent on the particular guilds and vegetation types. For instance, in *Araucaria araucana* forests in Chile the lizard community varied from a dominant arboreal species in unburned and long-recovered stands (i.e. 16 years after fire), to a combination of ground-dwelling species in areas affected by two successive fires (i.e. 13 years apart, with a 3-year recovery period) (Infante et al. 2021). This response was also observed in bird communities in *Araucaria-Nothofagus* forest of southern Chile where habitat changes due to fire (1–16 years after fire) reduced the density of most avian guilds but granivores, shrub users, and migrants showed positive responses. On the other hand, understory users, foliage users, and resident species showed decreases in post-fire sites (Novoa et al. 2021). However, in northwestern Patagonia, Argentina, in a burned *Nothofagus pumilio* forest (2 and 3 years after fire) bird species diversity was higher than in the unburnt control (Grigera and Pavic 2007), which supports the idea that responses are highly heterogeneous and largely depend on the burned vegetation community. Overall, more knowledge on the impacts of fire on fauna needs to be generated as it can contribute to the understanding of mechanisms of post-fire succession (e.g., seed dispersal/predation, soil formation).

9.6 Fire-Vegetation Feedbacks

Under the current and predicted global change scenarios, temperate forests face increasing pressures of conversion to non-forest vegetation types (Enright et al. 2015; Kitzberger et al. 2022). This process takes place due to a combination of factors such as changes in the fire regime and post-fire regeneration failure. Risks of conversion can be exacerbated if positive feedbacks between fire and vegetation take place (Tepley et al. 2018). A positive feedback occurs when two contrasting communities in terms of their flammability, can develop in the same landscape and under very similar environmental conditions. There are several systems in Southern South America in which fire vegetation feedbacks occur. The most well documented feedback occurs on the eastern slopes of the Andes between deciduous *Nothofagus pumilio* forests and mixed-species shrublands (Fig. 9.5b). Two other feedbacks take place between tall evergreen humid forests and bamboo thickets in well drained areas, and between rainforests and bog vegetation in waterlogged locations. There is also an increasingly common feedback that occurs within non-native pine plantations and invasions at several locations along south-central Chile and the Patagonian Andes (Cóbar-Carranza et al. 2014; Paritsis et al. 2018; González et al. 2022b).

In the eastern portion of the Andes in northern Patagonia, there are sharp and persistent boundaries between tall *Nothofagus pumilio* forests and diverse shrublands (Fig. 9.5b). These borders do not typically correspond with the underlying abiotic conditions and therefore they are allegedly caused and maintained by fire due to differences in flammability between these communities (Paritsis et al. 2015; Kitzberger et al. 2016). Spread of fire is more likely to occur in shrublands than in tall *Nothofagus pumilio* forests. This is related to the fuel structure of shrublands compared to *Nothofagus pumilio* forests. Shrublands are composed of several species of small trees and multi-stemmed shrubs which do not form a closed canopy. Common dominant species are small trees such as *Nothofagus antarctica, Maytenus boaria, Lomatia hirsuta* and large shrubs such as *Schinus patagonicus* and *Diostea juncea* (Quinteros et al. 2010; Blackhall et al. 2015) together with a variety of climbers (e.g., *Mutisia* spp.), bamboo (*Chusquea culeou*), herbs and grasses, many of them accumulating large amounts of dead fine fuels. The relatively open upper canopy leads to a warmer and dryer microclimate compared to that of the adjacent tall forest (Paritsis et al. 2015; Barberá et al. 2023). In addition, most common species in shrublands have high foliar flammability due to their shape, chemical composition and/or moisture content (Blackhall et al. 2015; Barberá et al. 2023). Conversely, tall and closed canopies of the deciduous *Nothofagus pumilio* forest create cool and more humid microclimates which in turn result in moister fuels (Barberá et al. 2023). Furthermore, understories in these forests are species poor and individuals are typically suppressed (less than 2 m height) resulting in less overall fuels compared to shrublands. This also results in a lack of vertical fuel continuity since the lower branches of the tree canopy start at circa 7–8 m (Paritsis et al. 2015).

All these traits and conditions combined result in a strong differential flammability between these two well-represented communities which is also supported by other lines of evidence. Dendrochronological reconstructions of fire regimes at circa 40°S show higher levels of fire activity in shrublands (*Nothofagus antarctica*) compared to tall *Nothofagus pumilio* forests (Veblen et al. 1992a, b). Also, statistical models based on actual fire occurrence and spread resulted in lower probabilities of fire activity in *Nothofagus pumilio* forests than in adjacent mixed and *Nothofagus antarctica* shrublands (Mermoz et al. 2005; Paritsis et al. 2013).

The positive feedback between *Nothofagus pumilio* forests and shrublands is established by the burning of the relatively pyrophobic forest that typically occurs only during extreme drought years in which fuels lose enough moisture to ignite and spread fire (Mermoz et al. 2005; Fig. 9.5b). In most cases fires are severe and eliminate all seed sources within the burned area (Landesmann et al. 2021; Tiribelli et al. 2024). Even with the presence of seed sources, regeneration is mostly constrained to specific topographic conditions (e.g., south-facing slopes and ravines) and to a narrow belt of less than 100 m from the remnants (Landesmann et al. 2021) due to limited seed dispersal and the elevated sensitivity of *Nothofagus pumilio* seedlings to unfavorable climatic conditions (Tercero-Bucardo et al. 2007), edaphic conditions (Kitzberger et al. 2005a) and browsing and trampling by livestock (Kitzberger et al. 2005b; Raffaele et al. 2011). Areas with low or intermediate precipitation where *Nothofagus pumilio* forests have burned with mid to high severity are more susceptible to be replaced by shrublands (Landesmann et al. 2021). In these locations understory shrubs and bamboos vigorously resprout after fire or are dispersed in by birds or gravity forming dense shrublands very rapidly (Cavallero et al. 2013). Under very high livestock pressure, these environments can even maintain grassland communities (Blackhall et al. 2017). While *Nothofagus pumilio* forest may take 50 years to well over a century to reestablish, shrublands are well developed in 10–20 years (Tiribelli et al. 2018). Due to the high frequency of fires in shrublands (i.e. a few tens of years) and the resprouting ability of its dominant species, the shrubland community perpetuates itself with positive feedbacks and eliminates any potential regeneration of *Nothofagus pumilio*. While mesic and cooler conditions favor longer fire intervals and *Nothofagus pumilio* reestablishment current and projected xeric and warmer conditions promote the perpetuation or even the advance of shrubland state by favoring frequent fire (Fig. 9.5b).

Other systems where positive fire vegetation feedbacks take place are the temperate rain forests along the Andes and along the coast of southwestern South America south of circa 37°S and up to the southern end of the continent in Tierra del Fuego. These rainforests are composed of several thin-barked tree species, most of which are easily killed by fire. Furthermore, none of these species is serotinous and lack any other fire adapted traits. Species such *as Gevuina avellana* and *Weinmannia trichosperma* resprout after fire, but many others do not resprout and are obligate seeders, such as *Nothofagus nitida*, *Nothofagus dombeyi* and *Nothofagus betuloides*. Instead, these species can establish after fire, forming dense sapling stands depending on site conditions. *Chusquea* spp. bamboos a common and dominant species in the understory of these rainforests north of circa 48°S (Veblen 1982) are key species influencing fire behavior since they have abundant fine fuels with significant amounts of dry material and ignite easily (Bianchi et al. 2015). Furthermore, they flower synchronously and then die over large portions of the landscape at multidecadal intervals, which creates immense amounts of fine fuels ready to ignite (González et al. 2014). Where bamboo species grow, they are responsible for mediating one important fire feedback mechanism in these forests.

The positive fire feedback in these systems initiates with the burning of closed-canopy rainforests in areas where *Chusquea* bamboo dominates the understory (Kitzberger et al. 2016). These forests have microclimates with cooler temperatures and higher relative humidity due to less direct solar radiation which keep fuels generally moist. Also, similarly to what occurs in *Nothofagus pumilio* forests, there is a discontinuity in the vertical fine fuel profile due to the tall canopy relative to the *Chusquea*-dominated understory. In the post-fire environment, a rapid and vigorous resprouting of *Chusquea* limits growth of resprouting tree species and more so of obligate seeders. These large and dense bamboo stands (i.e. >150 Mg ha^{-1}; Veblen et al. 1980) are inherently more flammable than the closed-canopy rainforests and therefore more susceptible to burn even under less fire prone weather compared to rainforests. Thus, the combination of high loads of fine fuels with the height of erect culms forming ladder fuels makes this community very fire prone. Inhibition or delay of tree regeneration by *Chusquea* extends this stage of high flammability which may result in an alternative state dominated by bamboo. In sum, the forests are more resistant to burning (pyrophobic) and the *Chusquea*-dominated vegetation (pyrophytic) has high chances to burn again before the tall rainforests can close the canopy.

In humid forests with poor drainage and high-water tables another positive feedback may occur. After canopy tree removal by fire or other disturbances (e.g., logging) takes place, sites get waterlogged due to strong reductions in vegetation transpiration (Díaz et al. 2007). The new soil conditions impose severe restrictions to tree establishment, thus creating a mosaic of patches with species tolerant to waterlogged conditions such as *Tepualia stipularis* or *Sphagnum* spp. bogs with diverse graminoids (Holz 2009; Albornoz et al. 2013). During extreme drought conditions, fine fuels in the bogs dry more easily than the understory fuels in the rainforests facilitating fire ignition (most commonly generated by humans) and spreading in these environments. This is reflected by the tree ring record which shows high fire frequencies of low to mid severities at the edges of forest and bog environments (Holz 2009). This feedback has the potential to transform a bog, from a carbon sink to a carbon source (Nelson et al. 2021).

The last and increasingly pervasive positive feedback, which has the potential to cause massive environmental impacts, is associated with non-native pine invasions and plantations. In the ecotone between the dry Patagonian steppe and the western forests of the Andes, pine plantations (*Pinus ponderosa, Pseudotsuga menziesii, Pinus radiata,* and *Pinus contorta*) and their associated invasions generate striking changes in fuel load and structure (Taylor et al. 2017; Paritsis et al. 2018; Franzese et al. 2022). In addition, these foreign fuels are inherently more flammable than most native vegetation due to their moisture content and chemical composition (Bianchi et al. 2018; Blackhall and Raffaele 2019). Fire behavior modeling studies suggest that unmanaged pine plantations and advanced (dense) invasions have the potential to increase fire intensity nearly 30 times compared to the intensity in the native surrounding steppe (Paritsis et al. 2018). These changes in behavior, and not only the changes in pre-fire vegetation, can change post-fire trajectories likely due to soil extreme heating which can destroy seedbanks of native species but not those of serotinous pines (Taylor et al. 2017). Therefore, after burns in plantations or dense invasions, the post-fire environment is highly conducive to pine recolonization, especially if serotinous species, such as *Pinus contorta* and *Pinus radiata*, are present (Franzese and Raffaele 2017). This fire-driven mechanism favors the spread and self-perpetuation of non-native pines in these landscapes, rapidly increasing the flammability and the likelihood of severe fires across the region.

Climate change (warming and drying) is the greatest factor promoting transitions from pyrophobic forests to landscapes of more pyrophytic vegetation in southern South America (Fig. 9.5b). Other exacerbating factors, such as post-fire livestock pressure, increased human ignitions and tree invasions also favor the initiation and persistence of positive fire vegetation feedbacks in the region. Thus, as the ecological impacts of global change advance, the positive feedbacks described here will likely result in more fire activity than would be anticipated from warming and drying climate trends alone.

9.7 Historic Changes in the Perception and Use of Fire

The societal perception and the relationships between humans and fire in southern South American temperate ecosystems has changed radically throughout recent history. Before European settlement native Americans modified fire regimes by intentionally burning for hunting, signaling, and creating small clearings in the dense forest for opening traveling routes or small-scale agriculture (Veblen et al. 2003). In drier eastern forests and woodlands, early accounts by missionaries and explorers indicate that the main purpose of setting fire was to encircle and drive herds of the large camelid guanaco (*Lama guanicoe*), the lesser rhea choique (*Rhea pennata*) and, eventually, the small cervid huemul (*Hippocamelus bisulcus;* Cox 1863; Fonck 1896; Musters 1871). With the arrival of European cattle (circa seventeenth century) native inhabitants started burning open woodlands to improve pastures (Musters 1871). Fire is also mentioned as used for signaling and warfare. These ignitions may have occasionally escaped to produce large forest wildfires (Moreno 19,897). In the mesic forest, large fires ("quemazones") were observed by early explorers, commonly associated with the clearing for livestock raising and Andean travel routes used for trading (Fonck 1896) or the opening of small areas of forests for crop cultivation (Cox 1863).

By the mid-twentieth century, the white Euro-Chilean settlement culture viewed fire as a fundamental force for large-scale landscape transformation of what was conceived as an impenetrable/unusable native forest. An account of this view is offered by Vicente Perez Rosales (Pérez Rosales 1886), in charge of the colonization process since 1849 in south-central Chile. In his memories, he describes the native forests near Lago Llanquihue by 1851 as an unknown region where: "*only the Indians could provide us with useful information about this obscure region, since it is impossible to enter it in any other way than on foot, cutting a narrow path with a machete through the dense foliage…*". In order to obtain agricultural land promised to the German immigrants he continued saying: "*…as we traveled back, I offered Pichi-Juan thirty days' salary, which at that time was equivalent to 30 pesos, to burn the forests between Chanchán and the mountains to alleviate the discontent that was beginning to reign among the immigrants temporarily staying in Valdivia…*". Vast areas of forests near Osorno were burnt in a short period of time probably related to extreme drought events and/or after *Chusquea quila* flowering events. In his later revisit Perez Rosales commented: "*…when the fires started, I embarked on what turned out to be a pleasant expedition through the lands to the north of Lake Llanquihue… the average width of the burned area was five leagues by fifteen in depth*". These accounts depict the rapid transformation of the Valdivian lowland landscape. As witnesses of these times many burnt spars or scarred large trees still persist in the modern south-central Chilean agricultural landscape. Another pervasive legacy of past settlement-related burning are large treeless areas in lower western slopes of the Andes dominated by *Chusquea quila* ("quilantos"). These impenetrable masses of bamboo may have vigorously resprouted from past fires impeding further tree regeneration. Based on the same paradigm, a similar trend of extensive forest burning for cattle and agriculture occurred somewhat later (1890–1910s) on the eastern forests across the Andes. By 1916 Rothkugel (1916) mapped circa 37% of *Nothofagus*- and *Austrocedrus*-dominated forests in the provinces of Neuquén, Rio Negro and Chubut, Argentina as burned.

Most forested areas in southern South America entered, at some point of the early-mid twentieth century, into a period of declining fire activity (Fig. 9.2a–f). This was related to a cultural change in the perception of fires, from being considered genuine tools for effective landscape management and transformation to being undesirable catastrophic events that need to be suppressed. In northern Patagonia this cultural change that started during the 1920–1930s was simultaneous with the creation of large protected National Parks in Argentina and elsewhere, and with the development of state-financed fire suppression policies. Accordingly, fire occurrence started to decline during the first half of the twentieth century reaching minimum fire activity during the 1960–1970, a period coincident with relatively wet/cool conditions (Fig. 9.2a–g). Due to the sharp decline in fire frequency, many forest areas that had been cleared by large settlers set fires during the late 19th—early twentieth century started to vigorously and massively regenerate either as young even aged forests (e.g., *Nothofagus dombeyi* or *Nothofagus dombeyi-Austrocedrus chilensis* stands, Veblen and Lorenz 1987) or (in the absence of obligate seeder trees) to resprout as dense shrublands (Mermoz et al. 2005), thus increasing landscape-scale fuel loads and connectivity (Kitzberger and Veblen 1999; Gowda et al. 2012, 2023). By the 1970s-early 1980s fire was considered by many national and regional forest management authorities as a "rare phenomenon". This perception changed radically after the 1987, 1996 and 1999 very active fire seasons (Fig. 9.2g, h) that, driven by a reversal to regional warm dry conditions, triggered uncontrollable fires, the firing of a Ministry for the Environment and later creation of the current Argentinian Service for Fire Management. These fire seasons marked the beginning of a new period of high regional fire activity up until today (Fig. 9.2g, h). With more environmental education and awareness, large/severe fires affecting ecosystems and people increasingly perceive them as manifestations of climate change.

9.8 Modern Fire Trends and Anthropogenic Influences on Future Fire Regimes

Although no statistically robust trend in increasing fire activity is yet detectable (Fig. 9.2g, h), several indicators suggest a step-like qualitative change in the fire regime. Warm/dry conditions generate more severe fires particularly in wetter *Nothofagus dombeyi* forests where coarse fuels normally do not desiccate (Tiribelli et al. 2024). These conditions also promote that fires that normally stop spreading at certain elevations above certain fuel moisture thresholds, are now sweeping up until the upper treeline, frequently inducing vegetation-type conversion towards shrublands and grasslands (Mermoz et al. 2005; Kitzberger et al. 2016). Individual fire events are also becoming larger and uncontrollable (>tens of thousands of hectares) due to self-reinforced meteorological conditions. More frequent reburns (new fires on recently burned areas) are occurring due to positive fire-vegetation feedbacks that generate more flammable early postfire communities (González and Lara 2015; Tiribelli et al. 2018, 2019; Landesmann et al. 2021).

Drivers of these changes are multiple. From a fuel availability perspective, both native and exotic plant species may be responsible for the recent rise in fire activity. On one hand, the highly flammable dominant bamboo *Chusquea culeou* (not having flowered since the early 1940s) started to massively flower and die-off during the 2000s in different areas of the region, fueling very large/severe fire events. On the other hand, during the 1970s onwards the Argentinean state started to promote the plantation of exotic conifers (e.g., *Pinus radiata, Pinus contorta, Pinus ponderosa* and *Pseudotsuga menziesii*) some of them highly flammable, serotinous and invasive of the native ecosystems. Since then, the amount of land converted to plantation has steadily grown, but more importantly the surrounding areas of invasion have increased causing dramatic changes in the overall landscape flammability (Taylor et al. 2017; Paritsis et al. 2018). Moreover, many pine-invaded ecosystems that have recently burned have reburned due to a vigorous serotinous behavior of some of these species, thus self-reinforcing the invasion and the increase in flammable fuels (Raffaele et al. 2016). Also, the Chilean state subsidy for the establishment of *Pinus radiata* and eucalypt plantations has resulted in a large, planted area (>3 million ha) in south-central Chile. The homogeneity, extension and greater flammability of these plantations has favored the occurrence of large fires in recent decades and a strong invasion process on the few remnants of native forests, increasing their fuel load and risk of future fires (McWethy et al. 2018; Bowman et al. 2019; González et al. 2018; Gómez-González et al. 2019; González et al. 2022b, c; Leal-Medina et al. 2024).

From the ignition perspective, a growing trend of suburban and exurban development and increases in tourism in the region is increasing the presence of humans in areas that were formerly less used by people, thus increasing the probabilities of accidental ignitions. In addition, natural ignitions are dramatically increasing in the region due to alterations in the atmospheric circulation patterns. Since the mid-1980s a step-like circa three-fold increase in incidence of lightning ignited fires has occurred (Veblen et al. 2008) in concordance with warmer summers and the more frequent blocking high pressure features at mid latitudes and the incursion of warm northeasterly subtropical air masses that are unstable and produce dry lightning storms mostly over the eastern flanks of the Andes but eventually also crossing the Cordillera and producing lightning storms and fires over the Chilean territory.

Lightning ignited fires, although much less frequent than human-ignited fires, produce more devastating effects as the ignition and initial spread is often located in remote and inaccessible areas (e.g., steep slopes of remote mountain terrain) where detectability is lower, and the dispatch of fire suppression resources is often impossible or dangerous. In fact, the largest and ecologically most destructive fire events that occurred in northwestern Patagonia since the beginning of this century (the 2002 Malleco-Tolhuaca fire, the 2015 Cholila fire and 2022 Steffen-Martin fire) were all ignited in inaccessible territory by lightning. Current trends and General Circulation Model-predicted alterations in dominant westerly circulation patterns will likely increase lightning activity and effective lightning ignitions over the region, thus producing disproportionate effects on fire occurrence due to the difficulty in the initial suppression and combat related to these events.

From the perspective of fuel condition, warming/drying trends and mid-late-twenty-first century climatic projections and trends of main patterns of climate for the region point out for more active fire seasons with larger and more severe fires and more transitions from forest to resprouting plant communities (Holz and Veblen 2011; Kitzberger et al. 2022). SAM indices have shown since the inception of their measurements in 1948 a strong upward (positive) trend (Thompson and Solomon 2002; Marshall 2003), which correlates to reduced annual precipitation and warmer temperatures over southern South America (Garreaud et al. 2009). Furthermore, models show that this upward trend is forced by increased greenhouse gasses and reduced stratospheric ozone concentrations and suggest the positive phase is expected to continue throughout the twenty-first century (Fyfe and Saenko 2006). Although still correlative, these coincident trends suggest fire-conducive climatic conditions for the remainder of the twenty-first century (Holz and Veblen 2011). Ensembles of general circulation models (IPCC 2013) indicate that the region will experience by the late twenty-first century under the pessimistic RPC 8.5 greenhouse gas emission pathway reductions of 20% and 25% in spring and summer precipitation, respectively and increases of circa 4 °C in maximum summer temperature. These expected changes are reduced by half under the more plausible RCP 4.5 pathway. Modeling of static and dynamic drivers of fire along a large trans Andean region in northern Patagonia suggests that under the predicted RCP 8.5 climate, probabilities for active fire season will increase 2–three-fold and 7–eight-fold by the mid- and late 21st centuries, respectively, while under the RCP4.5 conditions the probabilities will double and triple for these same timeframes, respectively (Kitzberger et al. 2022). Concomitant reductions in fire intervals will likely induce important landscape-level reductions in forests and increases in shrublands and grasslands. Simulation models suggest that *Nothofagus pumilio*, the most abundant but most fire sensitive tree in Patagonia, will reduce its range extension at a rate of >1.5% century^{-1}under RCP 8.5 conditions, while the more fire resilient *Austrocedrus chilensis* and *Nothofagus dombeyi* will retract at half of that rate (Kitzberger et al. 2022).

9.9 Management Implications

Increasingly, government, academia, and the general public are viewing wildfire in this region as a complex multivariate socio-ecological phenomenon both from its causes and consequences. Better understanding of the drivers of fire is leading to more precise spatial and temporal predictions of fire occurrence. Empirical fire spread models (models trained to mimic the behavior of historic fires) such as *SimBAd* (*Simulador de Bosque AnDino-Patagónico;* Morales et al. 2015) are currently being used to develop fire hazard maps over entire regions, to identify fire-vulnerable wildland-urban interfaces and for near real-time prediction of fire spread during major fire events, a tool that has allowed planning of more efficient suppression efforts. Analyzing post-fire successional fuel dynamics of different plant communities has helped develop more ecosystem-oriented fire management plans, thus avoiding the common practice of "importing" fire paradigms of ecosystems with different fuel dynamics. Specifically, findings of positive fire-vegetation feedbacks in this region discourage applying the "pine-fire paradigm" conceived in low severity fire regimes of the SW United States coniferous forests (e.g., Covington and Moore 1994) but often misused and extrapolated to other fire regimes (Kitzberger et al. 2012). Unlike their north American counterparts, temperate forests of South America often display decreases in flammability with increasing stand age (positive fire-vegetation feedbacks) in which *fire begets fire* (Tiribelli et al. 2019; Landesmann et al. 2021) (as opposed to: *lack of fire begets fire*, also known as the Yellowstone effect). Thus, applying prescribed burning in these systems seems absolutely unadvisable as it would possibly worsen fire hazard (e.g., by promoting the resprouting of many understory species and by inducing more fuel desiccating microenvironments (Barberá et al. 2023). In fact, several recent large fire events have been found to partially overlap the perimeter of previous fires that had mostly converted forests into flammable shrublands (Mermoz et al. 2005; Tiribelli et al. 2019).

Despite advances in the knowledge of the fire ecology of southern temperate ecosystems, several facts make fire management a challenging endeavor. First, highly flammable ecosystems (shrublands, pine plantations) spatially intermingle in the landscape creating fire corridors that facilitate fire spread into fire-vulnerable forests (e.g., montane and subalpine forests). Furthermore, shrublands tend to be associated with previously disturbed areas closer to human settlements and/or roads (Mermoz et al. 2005; Paritsis et al. 2013). Therefore, there is a high likelihood the fires igniting in shrublands or plantations will effectively carry subsequent fires into the more fire-vulnerable forests. Thus, efforts should be made to reduce fuel continuity/loads in these potential hotspots of fire initiation. Proposed management actions to reduce shrubland flammability involve harvesting for firewood in these communities (Kitzberger and Grosfeld 2016). However, recent trials of this practice suggest contrasting effects as harvesting reduced fuel amount (potentially reducing fire spread) but also reduced live fuel moisture content (potentially increasing flammability) (Goldenberg et al. 2020). Clearly, promoting the transition of shrublands into forests would reduce landscape-scale flammability. Thus, efforts should be made during fire suppression to protect natural fire refugia (e.g., rocky/sandy outcrops, ravines, coasts) where obligate seeders may survive and in the long term successionally replace the shrubland (Gowda et al. 2019) and reduce its intrinsic flammability by shading of understory plant and inducing moister microclimates (Landesmann et al. 2015, 2016; Barberá et al. 2023). In addition, restoration efforts should devote efforts in planting trees or assisting the natural regeneration by eliminating barriers and threats to their growth to speeding up (freeing from competition, Letorneau et al. 2004) this successional process. Second, native ecosystems are being actively planted with and invaded by highly flammable fire-adapted conifers, not only increasing overall landscape flammability but also reducing the capacity of native ecosystems to regenerate due to strong postfire competition and increased fire frequency (Franzese and Raffaele 2017; Taylor et al. 2017; González et al. 2022b, c; Leal-Medina et al. 2024). Sustainable forestry should ban the use of highly flammable and/or serotinous species that self-reinforce themselves by altering fire regimes. In addition, fire managers should deploy resources for the eradication or control of flammable invasive trees and shrubs especially in areas with high potential of ignition. Third, both anthropogenic and natural ignitions in the region are expected to increase due to higher presence of people in the forest as well as increased lightning. Efforts should be made to improve early detection (cameras, sensors, satellite monitoring) and early suppression (fast deployment of crews in helicopters with a priori defined protocols). Fourth, climate change is going to increase the frequency of climatic opportunities for fires to develop extreme events (Kitzberger et al. 2022). Fire spread models that consider fire weather as predictors could help generate spatially explicit fire risk maps upon which to develop more realistic and efficient fire suppression contingency plans and identify ecologically and/or socially fire-vulnerable areas as well as potential fire refugia.

References

Abram N, Mulvaney R, Vimeux F, Phipps SJ, Turner J, England MH (2014) Evolution of the southern annular mode during the past millennium. Nat Clim Chang 4:564–569. https://doi.org/10.1038/NCLIMATE2235

Agee JK (1993) Fire ecology of the Pacific northwest forests. Island Press, Washington, DC IRCA USA

Albornoz FE, Gaxiola A, Seaman BJ, Pugnaire FI, Armesto JJ (2013) Nucleation-driven regeneration promotes post-fire recovery in a Chilean temperate forest. Plant Ecol 214:765–776. https://doi.org/10.1007/s11258-013-0206-x

Aravena JC, LeQuesne C, Jiménez H, Lara A, Armesto JJ (2003) Fire history in Central Chile: tree-ring evidence and modern records. In: Veblen TT, Baker WL, Montenegro G, Swetnam TW (eds) Fire and climatic change in temperate ecosystems of the Western Americas. Ecological studies, vol 160. Springer, New York, NY

Armesto JJ, León-Lobos P, Arroyo MTK (1996) Los bosques templados del sur de Chile y Argentina: una isla biogeográfica. In: Armesto JJ, Villagrán C, MTK A (eds) Ecología de los bosques nativos de Chile, pp 23–28

Assal T, González ME, Sibold J (2018) Burn severity controls on post-fire *araucaria-Nothofagus* regeneration in the Andean cordillera. J Biogeogr 45(11):2483–2494. https://doi.org/10.1111/jbi.13428

Baker W (2009) Fire ecology in Rocky Mountain landscapes. Island Press, Washington DC, p 605

Barberá I, Paritsis J, Ammassari L, Morales JM, Kitzberger T (2023) Microclimate and species composition shape the contribution of fuel moisture to positive fire-vegetation feedbacks. Agric For Meteorol 330:109289. https://doi.org/10.1016/j.agrformet.2022.109289

Barría J (2007) Estructura y regeneración de *Austrocedrus chilensis* (D. Don) Picirca Serm Bizzarri (Ciprés de la Cordillera), en los Corrales de Urriola, Reserva Nacional Río de Los Cipreses (34° 43' S). Tesis Ingeniero Forestal. Valdivia, Chile. Facultad de Ciencias Forestales, Universidad Austral de Chile, p 38

Bianchi LO, Defossé GE (2015) Live fuel moisture content and leaf ignition of forest species in Andean Patagonia, Argentina. International Journal of Wildland Fire 24: 340-348. https://doi.org/10.1071/WF13099

Bianchi LO, Oddi FJ, Muñoz M, Defossé GE (2018) Comparison of leaf moisture content and ignition characteristics among native species and exotic conifers in northwestern Patagonia. Argentina For Sci 65:375–386. https://doi.org/10.1093/forsci/fxy054

Blackhall M, Raffaele E (2019) Flammability of Patagonian invaders and natives: when exotic plant species affect live fine fuel ignitability in wildland-urban interfaces. Landsc Urban Plan 189:1–10. https://doi.org/10.1016/j.landurbplan.2019.04.002

Blackhall M, Veblen TT, Raffaele E (2015) Recent fire and cattle herbivory enhance plant-level fuel flammability in shrublands. J Veg Sci 26:123–133. https://doi.org/10.1111/jvs.12216

Blackhall M, Raffaele E, Paritsis J, Tiribelli F, Morales JM, Kitzberger T, Gowda JH, Veblen TT (2017) Effects of biological legacies and herbivory on fuels and flammability traits: a long-term experimental study of alternative stable states. J Ecol 105:1309–1322. https://doi.org/10.1111/1365-2745.12796

Bowman DMJS, Murphy B, Williamson GJ, Cochrane MA (2014) Pyrogeographic models, feedbacks and the future of global fire regimes. Glob Ecol Biogeogr 23(7):821–824. https://doi.org/10.1111/geb.12180

Bowman DMJS, Moreira-Muñoz A, Kolden CA, Chávez RO, Muñoz AA et al (2019) Human-environmental drivers and impacts of the globally extreme 2017 Chilean fires. Ambio 48:350–362. https://doi.org/10.1007/s13280-018-1084-1

Bruno J, Martín G (1982) Los incendios forestales en los Parques Nacionales. Report Administración de Parques Nacionales, Buenos Aires

Burger N (2021) Efectos del megaincendio "Las Máquinas" del 2017 en bosques fragmentados de *Nothofagus glauca* en la región del Maule. Ms Thesis, Escuela de Graduados Facultad de Ciencias Forestales y Recursos Naturales, Universidad Austral de Chile, p 68

Burns B (1991) Regeneration dynamics of *Araucaria araucana*. PhD. Thesis. Department of Geography, University of Colorado, Boulder, p 211

Bustos-Schindler C, Le Quesne C, González ME, Solari ME 2010 Historia preliminar de incendios y prácticas (multi) culturales en la cuenca media del río Cachapoal (34° S), Chile central. Bosque 31: 17-27

Burns B (1993) Fire-induced dynamics of *Araucaria araucana-Nothofagus Antarctica* forest in the southern Andes. J Biogeogr 20:669–685. https://doi.org/10.2307/2845522

Cavallero L, Raffaele E, Aizen MA (2013) Birds as mediators of passive restoration during early post-fire recovery. Biol Conserv 158:342–350. https://doi.org/10.1016/j.biocon.2012.10.004

Cóbar-Carranza AJ, García RA, Pauchard A, Peña E (2014) Effect of *Pinus contorta* invasion on forest fuel properties and its potential implications on the fire regimes of *Araucaria araucana* and *Nothofagus Antarctica* forests. Biol Invasions 16(11):2273–2291. https://doi.org/10.1007/s10530-014-0663-8

Cóbar-Carranza AJ, García R, Pauchard A (2015) Efecto de la alta temperatura en la germinación y supervivencia de semillas de la especie invasora *Pinus contorta* y dos especies nativas del sur de Chile. Bosque 36(1):53–60. https://doi.org/10.4067/S0717-92002015000100006

CONAF (2017) Corporación Nacional Forestal of Chile Fire dataset. In: Chile CNFo, (ed) Santiago, Chile

Covington WW, Moore MM (1994) Southwestern ponderosa forest structure: changes since euro-American settlement. J For 92:39–47. https://doi.org/10.1093/jof/92.1.39

Cox G (1863) Viajes a las regiones septentrionales de Patagonia 1862–1863. Anales de la Universidad de Chile 23:3–239. and 437–509

Díaz MF, Bigelow S, Armesto JJ (2007) Alteration of the hydrologic cycle due to forest clearing and its consequences for rainforest succession. Forest Ecol Manag 244:32–40. https://doi.org/10.1016/j.foreco.2007.03.030

Díaz-Hormazabal I, González ME (2016) Análisis espacio-temporal de incendios forestales en la región del Maule. Chile Bosque 37(1):147–158. https://doi.org/10.4067/S0717-92002016000100014

Donoso P (1988) Caracterización y proposiciones silviculturales para comunidades de Roble y Raulí en el área de protección "Rodal 7 Tazas" VIII Región. Bosque 9:103–114

Donoso C (1993) Bosques templados de Chile y Argentina Variación, estructura y dinámica. Editorial Universitaria, p 483

Donoso C (1996) Ecology of *Nothofagus* forests in Central Chile. In: Veblen TT, Hill RS, Hill J (eds) The ecology and biogeography of *Nothofagus* forests. Yale University Press, United States, pp 271–291

Donoso C, Lara A (1996) Utilización de los bosques nativos de Chile: pasado, presente y futuro. In: Armesto JJ, Villagrán C, Arroyo MTK (eds.), Ecología de los Bosques Nativos de Chile Editorial Universitaria, Santiago, p 363–388

Dudynski N, Ghermandi L (2013) Fire as a stimulant of shrub recruitment in northwestern Patagonian (Argentina) grasslands. Ecol Res 28:981–990. https://doi.org/10.1007/s11284-013-1080-7

Elizalde R (1958) La sobrevivencia de Chile. Ministerio de Agricultura. Dirección General de Producción Agraria y Pesquera, Santiago de Chile

Enright NJ, Fontaine JB, Bowman DM, Bradstock RA, Williams RJ (2015) Interval squeeze: altered fire regimes and demographic responses interact to threaten woody species persistence as climate changes. Front Ecol Environ 13:265–272. https://doi.org/10.1890/140231

Fonck F (1896) Libro de los Diarios de Fray Francisco Menéndez. CF Niemeyer, Valparaiso, Chile

Franco MG, Mundo IA, Veblen TT (2022) Burn severity in Araucaria araucana forests of northern Patagonia: tree mortality scales up to burn severity at plot scale, mediated by topography and climatic context. Plant Ecol 223, 811–828. https://doi.org/10.1007/s11258-022-01241-w

Franzese J, Raffaele E (2017) Fire as a driver of pine invasions in the southern hemisphere: a review. Biol Invasions 19(8):2237–2246. https://doi.org/10.1007/s10530-017-1435-z

Franzese J, Raffaele E et al (2022) The legacy of pine introduction threatens the fuel traits of Patagonian native forests. Biol Conserv 267:109472. https://doi.org/10.1016/j.biocon.2022.109472

Fuentes-Ramirez A, Arroyo-Vargas P, Del Fierro A, Perez F (2019) Post-fire response of *Araucaria araucana* (Mol.) K Koch: assessment of vegetative resprouting, seed production and germination. Gayana Bot 76:119–122. https://doi.org/10.4067/S0717-66432019000100119

Fuentes-Ramírez A, Salas-Eljatib C, González ME, Urrutia-Estrada J, Arroyo-Vargas P, Santibañez P (2020) Initial response of understorey vegetation and tree regeneration to a mixed-severity fire in old-growth *Araucaria–Nothofagus* forests. Appl Veg Sci 23(2):210–222. https://doi.org/10.1111/avsc.12479

Fuentes-Ramírez A, Almonacid-Muñoz L, Muñoz-Gómez N, Moloney KL (2022) Spatio-temporal variation in soil nutrients and plant recovery across a fire-severity gradient in old-growth *araucaria-Nothofagus* forests of south-Central Chile. Forests 13(3):448. https://doi.org/10.3390/f13030448

Fyfe JC, Saenko OA (2006) Simulated changes in the extratropical southern hemisphere winds and currents. Geophys Res Lett 33(6). https://doi.org/10.1029/2005GL025332

Garreaud RD, Vuille M, Compagnucci R, Marengo J (2009) Present-day south American climate. Palaeogeogr Palaeoclimatol Palaeoecol 281:180–195. https://doi.org/10.1016/j.palaeo.2007.10.032

Goldenberg MG, Oddi FJ, Gowda JH, Garibaldi LA (2020) Effects of firewood harvesting intensity on biodiversity and ecosystem services in shrublands of northern Patagonia. For Ecosyst. https://doi.org/10.1186/s40663-020-00255-y

Gómez-González S, González ME, Paula S, Diaz-Hormazábal I, Lara A, Baquerizo-Delgado M (2019) Temperature and agriculture are largely associated with fire activity in Central Chile across different temporal periods. Forest Ecol Manag 433:535–543. https://doi.org/10.1016/j.foreco.2018.11.041

González ME, Donoso C (1999) Producción de semillas y hojarasca de la bambúcea *Chusquea quila*, posterior a su floración sincrónica. Rev Chil Hist Nat 72(2):169–180

González ME, Lara A (2015) Large fires in the Andean *araucaria* forests: when a natural ecological process becomes a threat. Oryx 49(3):394. https://doi.org/10.1017/S0030605315000599

González ME, Veblen TT (2006) Climatic influences on fire in *Araucaria araucana-Nothofagus* forests in the Andean cordillera of south-Central Chile. Ecoscience 13(3):342–350. https://doi.org/10.2980/i1195-6860-13-3-342.1

González ME, Veblen TT (2007) Incendios en bosques de *Araucaria araucana* y consideraciones ecológicas al madereo de aprovechamiento en áreas recientemente quemadas. Rev Chil Hist Nat 80:243–253. https://doi.org/10.4067/S0716-078X2007000200009

González ME, Veblen TT, Sibold JS (2005) Fire history of *Araucaria–Nothofagus* forests in Villarrica National Park, Chile. J Biogeogr 32:1187–1202. https://doi.org/10.1111/j.1365-2699.2005.01262.x

González ME, Veblen TT, Sibold JS (2010a) Influence of fire severity on stand development of *Araucaria araucana—Nothofagus pumilio* stands in the Andean cordillera of south-Central Chile. Austral Ecol 35:597–615. https://doi.org/10.1111/j.1442-9993.2009.02064.x

González ME, Szejner M, Muñoz A, Silva J (2010b) Incendios catastróficos en bosques andinos de *Araucaria-Nothofagus*: efecto de la severidad y respuesta de la vegetación. Bosque Nativo 46:12–17

González ME, Cortes M, Gallo L, Bekessy S, Echeverría C, Izquierdo F, Montaldo P (2013a) Coníferas chilenas: *Araucaria araucana*. In: Donoso C (ed.), Las especies arbóreas de los bosques Templados de Chile y Argentina. Autoecología, Marisa Cuneo Ediciones (2da edición), Valdivia, Chile, p. 36–53

González ME, Donoso C, Ovalle P, Martínez-Pastur G (2013b) Latifoliadas chilenas: *Nothofagus pumilio*. In: Donoso C (ed.), Las especies arbóreas de los bosques Templados de Chile y Argentina. Autoecología. Marisa Cuneo Ediciones (2da edición), Valdivia, Chile, p. 486–500

González ME, Amoroso M, Lara A, Veblen TT, Donoso C, Kitzberger T et al (2014) Ecología de disturbios y su influencia en los bosques templados de Chile y Argentina. In: Donoso C, González ME, Lara A (eds) Ecología Forestal. Bases para el Manejo Sustentable y Conservación de los Bosques Nativos de Chile, Valdivia Chile, Ediciones UACh, pp 411–502

González ME, Szejner P, Donoso PJ, Salas C (2015) Fire, logging and establishment patterns of second-growth forests in south-Central Chile: implications for their management and restoration. Ciencia e Investigación Agraria 42(3):427–441. https://doi.org/10.4067/S0718-16202015000300011

González ME, Gómez-González S, Lara A, Garreaud R, Díaz-Hormazábal I (2018) The 2010–2015 megadrought and its influence on the fire regime in central and south-Central Chile. Ecosphere 9. https://doi.org/10.1002/ecs2.2300

González ME, Muñoz AA, González-Reyes A, Christie DA, Sibold J (2020) Fire history in Andean *araucaria-Nothofagus* forests: coupled influences of past human land-use and climate on fire regimes in Northwest Patagonia. Int J Wildland Fire 29(8):649–660. https://doi.org/10.1071/WF19174

González ME, Lara A, Urrutia-Jalabert R, Bustos-Salazar A, Ruiz-Gómez C, Aravena JC (2022a) Carbon stocks across different environments, disturbance regimes, and stand age in *Fitzroya cupressoides* forests, the longest-lived species of the southern hemisphere. Front For Glob Change. https://doi.org/10.3389/ffgcirca2022.960429

González ME, Galleguillos M, Lopatin J, Leal C, Becerra-Rodas C, Lara A, San Martín J (2022b) Surviving in a hostile landscape: *Nothofagus alessandrii* remnant forests threatened by megafires and exotic pine invasion in the coastal range of Central Chile. Oryx 57:228–238. https://doi.org/10.1017/S0030605322000102

González ME, Becerra-Rodas C, Lara A, Little C (2022c) La influencia del fuego en la dinámica de los bosques de *Nothofagus alessandrii*. In: San Martin J (ed) Los Bosques relictos de Ruil: Ecología, Biodiversidad, Conservación y Restauración. Editorial El Sur Impresores Ltd, Chile, pp 417–427. (p 528)

Gowda JH, Kitzberger T, Premoli AC (2012) Landscape responses to a century of land use along the northern Patagonian forest-steppe transition. Plant Ecol 213:259–272. https://doi.org/10.1007/s11258-011-9972-5

Gowda JH, Tiribelli F, Mermoz M, Kitzberger T, Morales JM (2019) Fragmentation modulates the response of dichotomous landscapes to fire and seed dispersal. Ecol Model 392:22–30. https://doi.org/10.1016/j.ecolmodel.2018.10.014

Gowda JH, González Musso R, Kitzberger T (2023) Modelos de cambio en cobertura forestal de la cuenca del río Manso inferior ≐Una herramienta para definir estrategias de manejo? Bosque 44:273–284. https://doi.org/10.4067/s0717-92002023000100273

Grigera D, Pavic C (2007) Ensambles de aves en un sitio quemado y en un sitio no alterado en un área forestal del noroeste de la Patagonia Argentina. Hornero 22(1):29–37

Heusser CJ (1994) Paleoindians and fire during the late quaternary in southern South America. Rev Chil Hist Nat 67:435–443

Holz A (2009) Climatic and human influences on Fire regimes and Forest dynamics in temperate rainforests in southern Chile., PhD Thesis. University of Colorado, Boulder, CO

Holz A, Palma RE (2012) Floraciones de bambúes en Chile y Argentina: actual floración masiva del colihue, historia natural y riesgos asociados. Rev Bosque Nativo 50:40–46. https://repositorio.uc.cl/handle/11534/14194

Holz A, Veblen TT (2009) *Pilgerodendron uviferum*: the southernmost tree-ring fire recorder species. Ecoscience 16:322–329. https://doi.org/10.2980/16-3-3262

Holz A, Veblen TT (2011) Variability in the Southern Annular Mode determines wildfire activity in Patagonia, Geophys. Res. Lett., 38, L14710. doi:10.1029/2011GL047674.

Holz A, Veblen TT (2012) Wildfire activity in rainforests in western Patagonia linked to the southern annular mode. Int J Wildland Fire 21:114–126. https://doi.org/10.1071/WF10121

Holz A, Kitzberger T, Paritsis J, Veblen TT (2012) Ecological and climatic controls of modern wildfire activity patterns across southwestern South America. Ecosphere 3(11):1–25. https://doi.org/10.1890/ES12-00234.1

Holz A, Paritsis J, Mundo I, Veblen TT, Kitzberger T (2017) Southern annular mode drives multicentury wildfire activity in southern South America. Proc Natl Acad Sci U S A 114:9552–9557. https://doi.org/10.1073/pnas.1705168114

Ibarra M, Mourgues V (1976) Estudio de las relaciones entre suelos y asociaciones forestales de *Austrocedrus chilensis* (D.Don) Florin et Boutelje y *Nothofagus glauca* (Phil.) en la precordillera de Parral. Thesis, Facirca Cs. Forestales, Universidad de Chile, Santiago

Infante J, Novoa FJ, Ibarra JT, Melnick DJ, Griffin KL, Bonacic C (2021) Altered fire regimes modify lizard communities in globally endangered araucaria forests of the southern Andes. Sci Rep 11. https://doi.org/10.1038/s41598-021-02169-3

IPCC (2013) Annex I: Atlas of Global and Regional Climate Projections. In: van Oldenborgh, G.J., Collins, M., Arblaster, J., Christensen, J.H., Marotzke, J., Power, S.B., Rummukainen, M., Zhou, T. (Eds.), Climate Change 2013: The Physical Science Basis. Contribution of Working Group I to the Fifth Assessment Report of the Intergovernmental Panel on Climate Change [Stocker, T.F., D. Qin, G.-K. Plattner, M. Tignor, S.K. Allen, J. Boschung, A. Nauels, Y. Xia

Keeley JE, Bradstock RA, Pausas JG, Rundel PW (2012) Fire in Mediterranean ecosystems: ecology, evolution and management. Cambridge University Press, p 515

Kitzberger T, Veblen TT (1997) Influences of humans and ENSO on fire history of *Austrocedrus chilensis* woodlands in northern Patagonia, Argentina. Ecoscience 4:508–520. https://doi.org/10.1080/11956860.1997.11682430

Kitzberger T, Veblen TT (1999) Fire-induced changes in northern Patagonian landscapes. Landsc Ecol 14:1–15. https://doi.org/10.1023/A:1008069712826

Kitzberger T, Veblen TT, Villalba R (1997) Climatic influences on fire regimes along a rainforest-to-xeric woodland gradient in northern Patagonia, Argentina. J Biogeogr 23:35–47. https://doi.org/10.1111/j.1365-2699.1997.tb00048.x

Kitzberger T, Raffaele E, Heinemann K, Mazzarino MJ (2005a) Effects of fire severity in a north Patagonian subalpine forest. J Veg Sci 16:5–12. https://doi.org/10.1111/j.1654-1103.2005.tb02333.x

Kitzberger T, Raffaele E, Veblen TT (2005b) Variable community responses to herbivory in fire-altered landscapes of northern Patagonia, Argentina. Afr J Range Forage Sci 22:85–91. https://doi.org/10.2989/10220110509485865

Kitzberger T, Aráoz E, Gowda JH, Mermoz M, Morales JM (2012) Decreases in fire spread probability with forest age promotes alternative community states, reduced resilience to climate variability and large fire regime shifts. Ecosystems 15:97–112. https://doi.org/10.1007/s10021-011-9494-y

Kitzberger T, Grosfeld J (2016) Diagnóstico de la severidad de fuego y propuestas de restauración y manejo a nivel predial para áreas afectadas por el incendio de Cholila de 2015. Unpublished report INIBIOMA. 84pp

Kitzberger T, Perry GLW, Paritsis J, Gowda JH, Tepley AJ, Holz A, Veblen TT (2016) Fire–vegetation feedbacks and alternative states: common mechanisms of temperate forest vulnerability to fire in southern South America and New Zealand. N Z J Bot 54:247–272. https://doi.org/10.1080/0028825X.2016.1151903

Kitzberger T, Tiribelli F, Barberá I, Gowda JH, Morales JM, Zalazar L, Paritsis J (2022) Projections of fire probability and ecosystem vulnerability under 21st century climate across a trans-Andean productivity gradient in Patagonia. Sci Total Environ 839:156303. https://doi.org/10.1016/j.scitotenv.2022.156303

Krawchuk, MA, Moritz, M, Parisien, MA (2009) Global pyrogeography: the current and future distribution of wildfire. PLoS One 4 (4), e5102 http://dx.plos.org/10.1371/journal. pone.0005102

Landesmann JB, Gowda JH, Garibaldi L, Kitzberger T (2015) Survival, growth and vulnerability to drought in fire refuges: implications for the persistence of a fire-sensitive conifer in northern Patagonia. Oecologia 179:1111–1122. https://doi.org/10.1007/s00442-015-3431-2

Landesmann JB, Gowda JH, Kitzberger T (2016) Temporal shifts in the interaction between woody resprouters and an obligate seeder tree during a post-fire succession in Patagonia. J Veg Sci 27:1198–1208. https://doi.org/10.1111/jvs.12430

Landesmann JB, Tiribelli F, Paritsis J, Veblen TT, Kitzberger T (2021) Increased fire severity triggers positive feedbacks of greater vegetation flammability and favors plant community-type conversions. J Veg Sci 32:e12936. https://doi.org/10.1111/jvs.12936

Lara A, Fraver S, Aravena JC and Wolodarsky-Franke A (1999) Fire and the dynamics of Fitzroya cupressoides (alerce) forests of Chile's Cordillera Pelada. Ecoscience. 6, 100–109. https://doi.org/10.1080/11956860.1999.11952199

Le Quesne C (1988) Caracterización de bosques de ciprés de la cordillera (Austrocedrus chilensis (D. Don) Florin et Boutelje), en Radal-7 Tazas, Séptima Región, Chile. Tesis Ingeniero Forestal. Valdivia, Chile. Facultad de Ciencias Forestales, Universidad Austral de Chile. 66 p

Leal-Medina C, Lopatin J, Contreras A, González ME, Galleguillos M (2024) Post-fire Pinus radiata invasion in a threatened biodiversity hotspot forest: a multi-scale remote sensing assessment. For Ecol Manag 561:121861. https://doi.org/10.1016/j.foreco.2024.121861

Letorneau FJ, Andenmatten E, Schlichter T (2004) Effect of climatic conditions and tree size on Austrocedrus chilensis–shrub interactions in northern Patagonia. Forest Ecology and Management 191 (1-3). https://doi.org/10.1016/j.foreco.2003.11.002

Litton CM, Santelices R (2002) Early post-fire succession in a *Nothofagus glauca* forest in the coastal cordillera of south-Central Chile. Int J Wildland Fire 11(2):115–125. https://doi.org/10.1071/WF02039

Marshall GJ (2003) Trends in the southern annular mode from observations and reanalyses. J Clim 16:4134–4414. https://doi.org/10.1175/1520-0442(2003)016<4134:TITSAM>2.0.CO;2

Martel-Cea J, Abarzúa AM, González ME, Jarpa L, Hernández M (2024) Fire-climate-human dynamics over the last 1800 years in the Mesic *araucaria-Nothofagus* forests. J Biogeogr 51:1490–1504. https://doi.org/10.1111/jbi.14839

McWethy D, Pauchard A, García R, Holz A, González ME et al (2018) Landscape drivers of recent fire activity (2001–2017) in south-Central Chile. PLoS ONE 13:e0201195. https://doi.org/10.1371/journal.pone.0205287

Mermoz M, Kitzberger T, Veblen TT (2005) Landscape influences on occurrence and spread of wildfires in Patagonian forests and shrublands. Ecology 86:2705–2715. https://doi.org/10.1890/04-1850

Miranda A, Mentler R, Moletto-Lobos Í, Alfaro G, Aliaga L et al (2022) The landscape fire scars database: mapping historical burned area and fire severity in Chile. Earth Syst Sci Data 14:3599–3613. https://doi.org/10.5194/essd-14-3599-2022

Montaldo P (1974) La bio-ecología de *Araucaria araucana* (Mol.) Koch. Boletín Instituto Forestal Latino-Americano de Investigación y Capacitación 46(48):3–55

Morales JM, Mermoz M, Gowda JH, Kitzberger T (2015) A stochastic fire spread model for North Patagonia based on fire occurrence maps. Ecol Model 300:73–80. https://doi.org/10.1016/j.ecolmodel.2015.01.004

Mundo IA, Kitzberger T, Roig Juñent FA, Villalba R, Barrera MD (2012) Fire history in the *Araucaria araucana* forests of Argentina: human and climate influences. Int J Wildland Fire 22:194–206. https://doi.org/10.1071/WF11164

Mundo IA, Wiegand T, Kanagaraj R, Kitzberger T (2013) Environmental drivers and spatial dependency in wildfire ignition patterns of northwestern Patagonia. J Environ Manag 123:77–87. https://doi.org/10.1016/j.jenvman.2013.03.011

Mundo I, Villalba R, Veblen TT, Kitzberger T, Holz A, Paritsis J, Ripalta A (2017) Fire history in southern Patagonia: human and climate influences on fire activity in *Nothofagus pumilio* forests. Ecosphere 8(9):e01932. https://doi.org/10.1002/ecs2.1932

Muñoz A, González ME, Schneider-Valenzuela I, Klock-Barría K (2023) Multiproxy approach to reconstruct the Fire history of *Araucaria araucana* forests in the Nahuelbuta coastal range. Chile Forests 14:1082. https://doi.org/10.3390/f14061082

Musters GC (1871) At home with Patagonians: a Year's wandering over untrodden ground from the straits of Magellan to the Rio Negro. John Murray, London, UK

Nanavati WP, Whitlock C, Iglesias V, de Porras ME (2019) Postglacial vegetation, fire, and climate history along the eastern Andes, Argentina and Chile (lat. 41–55°S). Quat Sci Rev 207:145–160. https://doi.org/10.1016/j.quascirev.2019.01.014

Nelson K, Thompson DK, Hopkinson C, Petrone RM, Chasmer L (2021) Peatland-fire interactions: a review of wildland fire feedbacks and interactions in Canadian boreal peatlands. Sci Total Environ 769:45212. https://doi.org/10.1016/j.scitotenv.2021.145212

Novoa FJ, Altamirano TA, Bonacic C, Martin K, Ibarra JT (2021) Fire regimes shape biodiversity: responses of avian guilds to burned forests in Andean temperate ecosystems of southern Chile. Avian Conserv Ecol 16(2):22. https://doi.org/10.5751/ACE-01999-160222

Olson DM, Dinerstein E, Wikramanayake ED, Burgess ND, Powell GV et al (2001) Terrestrial ecoregions of the world: a new map of life on earth. Bioscience 51:933–938. https://doi.org/10.1641/0006-3568(2001)051[0933:TEOTWA]2.0.CO;2

Paritsis J, Holz A, Veblen TT, Kitzberger T (2013) Habitat distribution modeling reveals vegetation flammability and land use as drivers of wildfire in SW Patagonia. Ecosphere 4:art53–art20. https://doi.org/10.1890/ES12-00378.1

Paritsis J, Veblen TT, Holz A (2015) Positive fire feedbacks contribute to shifts from *Nothofagus pumilio* forests to fire-prone shrublands in Patagonia. J Veg Sci 26:89–101. https://doi.org/10.1111/jvs.12225

Paritsis J, Landesmann JB, Kitzberger T, Tiribelli F, Sasal Y et al (2018) Pine plantations and invasion alter fuel structure and potential fire behavior in a Patagonian forest-steppe ecotone. Forests 9(3):117. https://doi.org/10.3390/f9030117

Pausas JG (2012) Incendios forestales, una visión desde la ecología. CSIC, Spain, p 113

Pausas JG, Ribeiro E (2013) The global fire-productivity relationship. Glob Ecol Biogeogr 22:728–736. https://doi.org/10.1111/geb.12043

Pausas JG, Bradstock RA, Keith DA, Keeley JE, Fire Network GCTE (2004) Plant functional traits in relation to fire in crown-fire ecosystems. Ecology 85:1085–1100. https://doi.org/10.1890/02-4094

Pérez Rosales V (1886) Recuerdos del pasado (1814–1860). Imprenta Gutenberg, Santiago, Chile, p 432

Pozo RA, Galleguillos M, González ME, Vásquez F, Arriagada R (2022) Assessing the socio-economic and land-cover drivers of wildfire activity and its spatiotemporal distribution in south-Central Chile. Sci Total Environ 810:152002. https://doi.org/10.1016/j.scitotenv.2021.152002

Quinteros P, Hansen N, Kutschker A (2010) Composition and diversity of the understory of ñire (*Nothofagus Antarctica*) forests in relation to forest structure. Ecol Austral 20:225–234

Raffaele E, Veblen TT, Blackhall M, Tercero-Bucardo N (2011) Synergistic influences of introduced herbivores and fire on vegetation change in northern Patagonia, Argentina. J Veg Sci 22:59–71. https://doi.org/10.1111/j.1654-1103.2010.01233.x

Raffaele E, Nuñez MA, Eneström J et al (2016) Fire as mediator of pine invasion: evidence from Patagonia, Argentina. Biol Invasions 18:597–601. https://doi.org/10.1007/s10530-015-1038-5

Riquelme T (2021) Historia de incendios en bosques de Araucaria araucana (Mol.) K. Koch en la Reserva Nacional Malalcahuello. Tesis de Magister, Facultad de Ciencias Forestales y Recursos Naturales, Universidad Austral de Chile. Valdivia. Chile. 36 p

Rothkugel M (1916) Los Bosques Patagónicos. Ministerio de Agricultura, Buenos Aires, Argentina

Rowe JS (1983) Concepts of fire effects on plant individuals and species. In: Wein RW, MacLean DA (eds) The role of Fire in northern circumpolar ecosystems. Wiley, London, UK., pp 135–154

Rozas V, C Le Quesne, M Osorio-Badilla, ME González, A González-Reyes (2018) Coupled human-climate signals on the fire history of upper Cachapoal Valley, Mediterranean Andes of Chile, since AD 1201. Global and Planetary Change 167: 137–147

Sackmann P, Farji-Brener A (2006) Effect of fire on ground beetles and ant assemblages along an environmental gradient in NW Patagonia: does habitat type matter? Ecoscience 13(3):360–371. https://doi.org/10.2980/i1195-6860-13-3-360.1

Sanguinetti J (2015) Cambio Climático, incendios forestales y su interacción con granívoros no-nativos:una amenaza potencial para la conservación de los bosques de *Araucaria* en Argentina. Proyecto Pehuén—Report. Departamento Conservación y Manejo—Parque Nacional Lanín. 25 p

Sasal Y, Raffaele E, Farji-Brener AG (2010) Succession of ground dwelling beetle assemblages after fire in three habitat types in the Andean forest of NW Patagonia, Argentina. J Insect Sci 10:1–17. https://doi.org/10.1673/031.010.3701

Segura AM, Holmgren M, Anabalón JJ, Fuentes ER (1998) The significance of fire intensity in creating local patchiness in the Chilean matorral. Plant Ecol 139:259–264. https://doi.org/10.1023/A:1009715328868

Silla F, Fraver S, Lara A, Allnut T and Newton A (2002) Regeneration and stand dynamics of Fitzroya cupressoides (Cupressaceae) forests of southern Chile's Central Depression. For. Ecol. Manage. 165, 213–224. https://doi.org/10.1016/S0378-1127(01)00619-3

Taylor KT, Maxwell BD, McWethy DB, Pauchard A, Nuñez MA, Whitlock C (2017) *Pinus contorta* invasions increase wildfire fuel loads and may create a positive feedback with fire. Ecology 98:678–687. https://doi.org/10.1002/ecy.1673

Tello F, González ME, Valdivia N, Torres F, Lara A, García-López A (2020) Diversity loss and changes in saproxylic beetle assemblages following a high-severity fire in Araucaria–Nothofagus forests. J Insect Conserv 24:585–601. https://doi.org/10.1007/s10841-020-00223-5

Tello F, González ME, Micó E, Valdivia N, Torres F, Lara A, García-López A (2022) Short-interval, severe wildfires alter saproxylic beetle diversity in andean araucaria forests in northwest chilean Patagonia. Forests 13(3):441. https://doi.org/10.3390/f13030441

Tepley AJ, Thomann E, Veblen TT, Perry GLW, Holz A et al (2018) Influences of fire–vegetation feedbacks and post-fire recovery rates on forest landscape vulnerability to altered fire regimes. J Ecol 106:1925–1940. https://doi.org/10.1111/1365-2745.12950

Tercero-Bucardo N, Kitzberger T, Veblen TT, Raffaele E (2007) A field experiment on climatic and herbivore impacts on post-fire tree regeneration in North-Western Patagonia. J Ecol 95:771–779. https://doi.org/10.1111/j.1365-2745.2007.01249.x

Thompson DWJ, Solomon S (2002) Interpretation of recent southern hemisphere climate change. Science 296:895–899. https://doi.org/10.1126/science.1069270

Tiribelli F, Kitzberger T, Morales JM (2018) Changes in vegetation structure and fuel characteristics along post-fire succession promote alternative stable states and positive fire-vegetation feedbacks. J Veg Sci 29:147–156. https://doi.org/10.1111/jvs.12620

Tiribelli F, Morales JM, Gowda JH, Mermoz M, Kitzberger T (2019) Non-additive effects of alternative stable states on landscape flammability in NW Patagonia: fire history and simulation modeling evidence. Int J Wildland Fire 28:149–159. https://doi.org/10.1071/WF18073

Tiribelli F, Paritsis J, Barberá I, Kitzberger T (2024) Spatial and temporal opportunities for forest resilience promoted by burn severity attenuation across a productivity gradient in north western Patagonia. Int J Wildland Fire 33. https://doi.org/10.1071/WF23098

Tortorelli L (1942) La explotación racional de los bosques de Araucaria de Neuquén, su importancia económica. Escuela de Estudios Argentinos, Buenos Aires, p 74

Urrutia-Estrada J, Fuentes-Ramírez A, Hauenstein E (2018) Differences in floristic composition of *araucaria-Nothofagus* forests affected by distinct fire severities. Gayana Bot 75:625–638

Veblen TT (1982a) Growth patterns of *Chusquea* bamboos in the understory of Chilean *Nothofagus* forests and their influences in forest dynamics. Bullet Torrey Botanic Club 109:474–487. https://doi.org/10.2307/2996488

Veblen TT (1982b) Regeneration patterns in *Araucaria araucana* forests in Chile. J Biogeogr 9:11–28. https://doi.org/10.2307/2844727

Veblen TT (1985) Stand dynamics in Chilean *Nothofagus* forests. In: Pickett STA, White PP (eds) The ecology of natural disturbance and patch dynamics. Academic, New York, pp 35–51

Veblen TT, Ashton DH (1978) Catastrophic influences on the vegetation of the Valdivian Andes. Vegetation 36:149–167. https://doi.org/10.1007/BF02342598

Veblen TT and Ashton DH (1982) The regeneration status of Fitzroya cupressoides in the Cordillera Pelada, Chile. Biol. Conserv. 23, 141–161. https://doi.org/10.1016/0006-3207(82)90036-2

Veblen TT, Lorenz DC (1987) Post-fire stand development of *Austrocedrus-Nothofagus* forests in Patagonia. Vegetatio 71:113–126. https://doi.org/10.1007/BF00044825

Veblen TT, Schlegel FM, Escobar B (1980) Dry matter production of two species of bamboos (*Chusquea culeou* and *Chusquea tenuiflora*) in south-Central Chile. J Ecol 68:397–404. https://doi.org/10.2307/2259412

Veblen TT, Donoso C, Schlegel FM, Escobar R (1981) Forest dynamics in south-Central Chile. J Biogeogr 8:211–247. https://doi.org/10.2307/2844678

Veblen, TT, T Kitzberger, Lara A (1992) Disturbance and vegetation dynamics along a transect from rain forest to Patagonian shrublands. Journal of Vegetation Science 3: 507-520. https://doi.org/10.2307/3235807

Veblen TT, Kitzberger T, Lara A (1992a) Disturbance and forest dynamics along a transect from Andean rainforest to Patagonian shrubland. J Veg Sci 3:507–520. https://doi.org/10.2307/3235807

Veblen TT, Mermoz M, Martin C, Kitzberger T (1992b) Ecological impacts of introduced animals in Nahuel Huapi National Park, Argentina. Cons Biol 6:71–83. https://doi.org/10.1046/j.1523-1739.1992.610071.x

Veblen TT, Burns BR, Kitzberger T, Lara A, Villalba R (1995) The ecology of the conifers of southern South America. In: Enright N, Hill R (eds) Ecology of the southern conifers. Melbourne University Press, Melbourne, Australia, pp 120–155

Veblen TT, Donoso C, Kitzberger T, Rebertus AJ (1996) Ecology of southern Chilean and Argentinean *Nothofagus* forests. In: Veblen TT, Hill RS, Read J (eds) The ecology and biogeography of *Nothofagus* forests. Yale University Press, New Haven, USA, pp 293–353

Veblen TT, Kitzberger T, Villalba R, Donnegan J (1999) Fire history in northern Patagonia: the roles of humans and climatic variation. Ecol Monogr 69:47–67. https://doi.org/10.1890/0012-9615(1999)069[0047:FHINPT]2.0.CO;2

Veblen TT, Kitzberger T, Raffaele E, Lorenz D (2003) Fire history and vegetation changes in northern Patagonia, Argentina. In: Veblen TT, Baker W, Montenegro G, Swetnam TW (eds) Fire and climatic change in temperate ecosystems of the western Americas, ecological studies 160. Springer Verlag, NewYork, pp 265–295

Veblen TT, Kitzberger T, Raffaele E, Mermoz M, González ME, Sibold JS, Holz A (2008) The historical range of variability of fires in the Andean-Patagonian *Nothofagus* forest region. Int J Wildland Fire 17:724–741. https://doi.org/10.1071/WF07152

Whelan RJ (1995) The ecology of fire. Cambridge University Press, p 360

Whitlock C, Moreno P, Bartlein P (2007) Climatic controls of Holocene fire patterns in southern South America. Quat Res 68(1):28–36. https://doi.org/10.1016/j.yqres.2007.01.012

Willis B (1914) El Norte de la Patagonia. Dirección de Parques Nacionales, Buenos Aires

Fire in South American Wetlands

10

Geraldo Alves Damasceno Junior, Pia Parolin,
Maxwell da Rosa Oliveira, Danilo Bandini Ribeiro,
Thiago Silva Teles, Ieda Maria Bortolotto,
Geraldo Wilson Fernandes, and Juan Jose Neiff

Abstract

Fire in wetlands poses a considerable challenge to their biota. This chapter presents data on some important South American wetlands with their different climatic patterns. Some are fire-dependent, others fire-independent. Fire incidence can vary from 9 to 206 fires per km^2, often at the end of the dry season. The seasonally flooded grasslands are the most resistant, and the flooded forests of

G. A. Damasceno Junior (✉)
Laboratório de Ecologia Vegetal, Instituto de Biociências, Universidade Federal de Mato Grosso do Sul, Campo Grande, Brazil

P. Parolin
Max-Planck-Institute of Animal Behaviour, Radolfzell am Bodensee, Germany
e-mail: piaparolin@gmx.de

M. da Rosa Oliveira
Instituto de Ciências Biológicas, Universidade Federal de Minas Gerais, Belo Horizonte, MG, Brazil

Laboratório de Ecologia Vegetal, Instituto de Biociências, Universidade Federal de Mato Grosso do Sul, Campo Grande, Brazil

D. B. Ribeiro · T. S. Teles · I. M. Bortolotto
Instituto de Biociências, Universidade Federal de Mato Grosso do Sul, Campo Grande, Brazil

G. W. Fernandes
Departamento de Genetica, Ecologia & Evolucao, Instituto de Ciências Biológicas, Universidade Federal de Minas Gerais, Belo Horizonte, Brazil
e-mail: gwilson@icb.ufmg.br

J. J. Neiff
Centro de Ecología Aplicada del Litoral (CECOAL), CONICET-UNNE, Corrientes, Argentina
e-mail: jj@neiff.com.ar

A. Fidelis, V. R. Pivello (eds.), *Fire in the South American Ecosystems*, Ecological Studies 250, https://doi.org/10.1007/978-3-031-89372-8_10

the Amazon River are the most sensitive to fire. The most floodable areas can have intense fire events, forming corridors of fire and flooding along rivers. Fire can induce water repellency on wetland soils and produce ashes that modify the environment during the flood. Fire can change, smooth, or trigger vegetation responses along the inundation gradient and promote the monodominance of certain species and vegetation openness. Megafires can produce high mortality in both flora and fauna. Fire is used to promote the abundance of some plants that serve as food, to make handicrafts, or even to improve the quality of native pastures. The regulation of Integrated Fire Management in wetlands is under discussion in South American countries. We understand that the research about fire in tropical and subtropical wetlands has many gaps that can subside the regulation process.

Resumen

Los incendios en humedales presentan un desafío considerable para su biota. Este capítulo ofrece información sobre algunos de los humedales más importantes de Sudamérica, con diferentes patrones climáticos. Algunos dependen del fuego y otros no. Hay una incidencia anual de puntos calientes/km^2 que oscila entre 9 y 206, normalmente al final de la sequía. Los pastizales inundados estacionalmente son los más resistentes al fuego, y los bosques inundados del río Amazonas los más sensibles. Las zonas más inundadas suelen sufrir quemas más intensas, formando corredores de incendios e inundaciones a lo largo de los ríos. El fuego puede inducir repelencia hídrica en los suelos y las cenizas pueden modificar el ambiente durante las inundaciones; alterar, suavizar o desencadenar respuestas de la vegetación a lo largo del gradiente de inundación; promover la monodominancia de algunas plantas y la apertura de la vegetación. Los megaincendios pueden causar elevada mortalidad en la flora y la fauna. El fuego se utiliza para manejar plantas alimenticias o utilizadas en la artesanía y pastos nativos. Los países sudamericanos están debatiendo la regulación de la gestión integrada del fuego en los humedales. La investigación sobre incendios en humedales tropicales y subtropicales para subvencionar el proceso de regulación aún tiene muchas lagunas.

Resumo

O fogo nas áreas úmidas representa um desafio considerável para a sua biota. Este capítulo apresenta dados sobre algumas das áreas úmidas mais importantes da América do Sul, com diferentes padrões climáticos. Algumas são dependentes do fogo e outras independentes. Apresentam incidência anual de focos de calor/km^2 variando de 9 a 206, geralmente no final da estação seca. Os campos sazonalmente inundados são os mais resistentes ao fogo e as florestas inundadas do rio Amazonas, as mais sensíveis. As áreas mais inundáveis podem sofrer queimas mais intensas, formando corredores de fogo e inundações ao longo dos rios. O fogo pode induzir a repelência hídrica nos solos e as cinzas podem modificar o ambiente durante as cheias; alterar, suavizar ou desencadear respostas da vegetação ao longo do gradiente de inundação; promover a monodominância de algumas plantas e a abertura da vegetação. Mega-incêndios podem causar alta mortalidade na flora e fauna. O fogo é utilizado no manejo de plantas alimentícias, utilizadas no artesanato, ou pastagens nativas. A regulamentação do manejo integrado do fogo em áreas úmidas está em discussão nos países sul-americanos. A pesquisa sobre incêndios em áreas úmidas tropicais e subtropicais para subsidiar o processo de regulamentação ainda apresenta muitas lacunas.

10.1 Introduction

Fire in wetlands is a significant and pressing issue, posing a considerable challenge to the biota living in these environments (Damasceno-Junior 2021). In environments subject to seasonal flooding, waterlogging presents challenges due to drastic fluctuations in water levels and root aeration conditions. Only species with adaptations can survive these events (Ferreira et al. 2009; Parolin et al. 2022). Given the high predictability of the flood pulse, nearly all organisms inhabiting flood-prone areas have evolved genetic adaptations, as noted by Junk et al. (1989). On the other hand, fire is also an extreme event. It can consume large amounts of organic matter and raise temperatures to levels some organisms cannot tolerate (Lukac et al. 2011; Silva et al. 2021; Batista et al. 2023). In fire-dependent environments, adaptations have evolved due to the predictability of fire occurrences, albeit less frequent than inundation events. This fire regime has played a significant role in shaping the environment over evolutionary time scales. Adaptations such as the presence of thick bark in trees, the resistance of seeds to fire, and flowering after fire events are frequent in fire-dependent environments (e.g., Fichino et al. 2016; Pivello et al. 2021). Megafires are increasingly prevalent in South American environments, particularly due to significant land-use changes witnessed in recent decades. Additionally, these fires may worsen due to projected climate change effects, as one of the primary impacts of these changes is the escalation of extreme events, as noted by Pivello et al. (2021).

In this context, the South American wetlands are very diverse because there are many flooding regimes in different climates (Junk 2013; Neiff et al. 1994). This chapter deals with fire in South American wetlands, considering their diversity, variety, and peculiarities.

10.2 The Classification System of Wetlands in South America

South America has different wetlands and climatic patterns (Figs. 10.1, 10.2, 10.3, and 10.4). However, due to climatic seasonality, they have wet and dry phases with irregular periodicity and extraordinary droughts lasting for months or more than a year (Fig. 10.2).

The continental area of South America has three major mountain massifs: the Andes, the Guiana, and the Brazilian Shields. The largest areas are tropical and

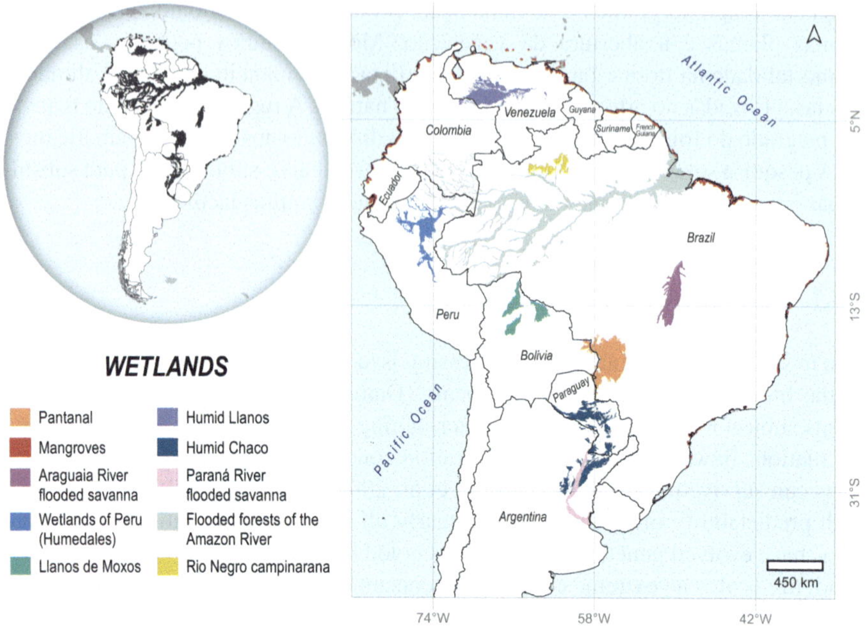

Fig. 10.1 Distribution of the main South American wetlands. Sources: Pantanal—Padovani (2010); Mangroves, Paraná River Flooded Savannas, Wetlands of Peru (Humedales), Rio Negro Campinarana—Olson et al. (2001); Araguaia River Flooded Savanna and Humid Chaco—Lehner and Döll (2004); Flooded Forests of the Amazon River—Olson et al. (2001) and Lehner and Döll (2004); Llanos de Moxos—GeoBolivia (2020); Llanos—Colombia (2021) and Lehner and Döll (2004). Produced by Maxwell Oliveira and Júlia Abrantes Rodrigues

subtropical lowlands. The subtropical lowland regions, which are predominantly not included in this analysis, have the world's most expansive wetlands. This is because these areas' precipitation and mountain runoff surpass the evapotranspiration rate. Wetlands, characterized by their flat or slightly concave terrain, typically have less storage capacity than deeper lakes.

Information about South American wetlands is unevenly distributed, with certain regions being extensively studied while others have limited available data. The most important South American Wetlands, in terms of area, are presented in Fig. 10.1. For some of them, there is information on the floodable area, limits, and size, but in other cases, the mapped area corresponds to the entire basin, including the highlands. The flooded forests of the Amazon River (várzeas) occupy around 98,110 km^2 (Fig. 10.3)**,** and the Araguaia River Flooded Savannas, 58,600 km^2; Brazilian coastal mangroves cover around 13,888 km^2, being 7000 in the Amazon region (Nunes-da-Cunha et al. 2015). The Pantanal is recognized as the largest continuous tropical wetland, with circa 150,000 km^2 in Brazil, Bolivia, and Paraguay with a maximum inundated area of 131,000 km^2 (Hamilton et al. 1996; Merino and Assine 2020) (Fig. 10.4)**.** The Peruvian lowland areas of Ucayali and Marañon rivers (Wetlands of Peru in Fig. 10.1) have a floodable area of 60,000 km^2 (Kvist and Nebel 2001). The

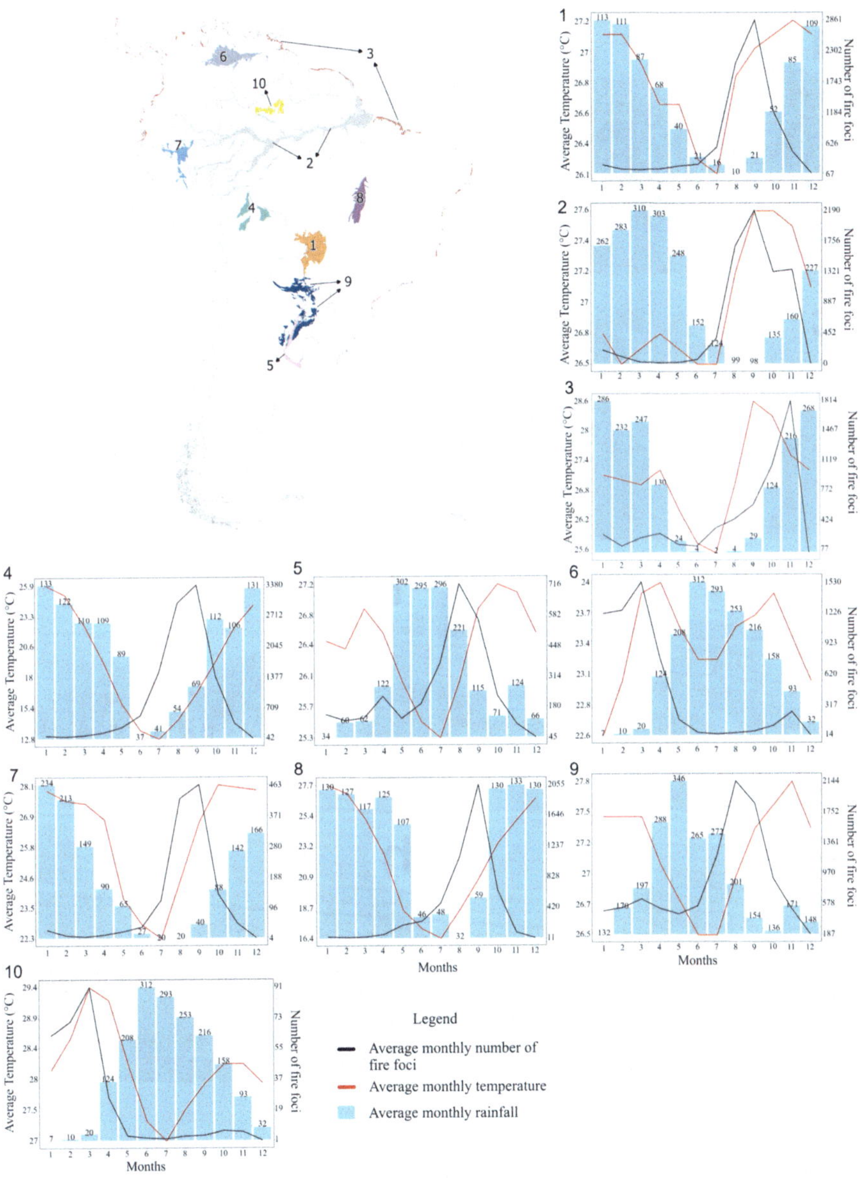

Fig. 10.2 Yearly temperature patterns, rainfall, and fire foci in the main South American Wetlands. Data Sources: fire foci—INPE (2023); climate variables—Cerasoli et al. (2022)

Humid Chaco occurs in Paraguay, Argentina, and a small extent in Brazil south of Pantanal, covering an area of 200,000 km^2 (Ginzburg et al. 2005; Sartori et al. 2018) (Fig. 10.3). The Orinoco Llanos is a lowland area with circa 450,000 km^2 (see Chap. 6), but the seasonally flooded area ranges from 1278 to 105,454 km^2. The Llanos de

Fig. 10.3 (**a**) Riverine human populations along the Amazon river in Caxiuanã National Forest, near (Pará, Brazil) (Photo: Pia Parolin); (**b**) Blackwater flooded forest (igapó) in Caxiuanã NationalFig. 10.3 (continued) Forest (Pará, Brazil) (Photo: Pia Parolin); (**c**) Flooded vegetation on

Moxos is a savanna region of nearly 150,000 km^2 located in the upper Madeira River in Bolivia, and the seasonally flooded area varies from 2069 to 78,460 km^2 (Hamilton et al. 2004). Besides the mapped areas, we have some interesting mentions: the flooded areas of Roraima/Rupununi savannas (not included in Fig. 10.1), cover *circa* 16,500 km^2, and the Rio Grande do Sul wetland areas cover 30,332 km^2 (also not included in Fig. 10.1). These large wetlands have a monomodal, sometimes bimodal, and predictable flood pulse, which has evolutionary consequences for the biota (Junk 2013; Neiff et al. 1994; San Martín et al. 2023).

The Cerrado Region in Central Brazil is full of small swamps locally known as *vereda* or *varjão* (Fig. 10.4). *Vereda* is a seasonally flooded savanna covered by grassy vegetation, clusters of shrubs, and the palm tree *Mauritia flexuosa* without forming a canopy (Brasil 2012). However, this vegetation can also occurs without *Mauritia flexuosa* (Moreira et al. 2014) (Fig. 10.4d). These areas are often underlain by impermeable layers (lateritic) in the soil, which maintain these areas as wetlands (Da Silva Meneses and Da Costa 2012; Furlan et al. 2021). The *veredas* could not be mapped in our maps because of the adopted working scale. There was still no estimate of the total area of the *vereda* in the Cerrado region. However, to give an idea, from 90,545 km^2 of the Triangulo Mineriro Cerrado region in central Brazil, 765.4 km^2 are *veredas* (0,8% of the land cover), corresponding to 6782 mapped areas (Gonçalves et al. 2022a, b). The *veredas* are vital to the Cerrado rivers' water supply and many of Central Brazil's cities.

10.3 Fire Regimes in the South American Wetlands

The main wetlands of South America also show variations in their wildfire regimes in the last 20 years. Fire incidence can vary from 9 fires per km^2 for the lowest incidence wetlands to 206 fires per km^2 for the highest incidence wetlands (Fig. 10.5). Throughout the year, in the months of highest incidence, some wetlands can register more than 3000 fire foci; in others, this value does not exceed 100 (Fig. 10.5). The month with the highest fire foci varies among the wetlands. However, this period tends to occur during dry periods, where the lowest rainfall and high temperatures are recorded at the end of the dry season (Figs. 10.2 and 10.5). This pattern may indicate seasonality as an important factor in the occurrence of wildfires in these areas. Seasonality may be related to temporal continuity in fuel availability. Rainy periods increase plant productivity and biomass growth (Grau and Veblen 2000; Nunes et al. 2012).

Fig. 10.3 (continued) the at the Amazon river margin near the Amazon Tall Tower Observatory-ATTO (Amazonas, Brazil) (Photo: Pia Parolin); (**d**) Flooded grassland on the Solimões river (Amazonas, Brazil) (Photo: Pia Parolin); (**e**) Floodplains of Amazon river (Pará, Brazil) (Photo: Pia Parolin); (**f**) Vegetation on the margin of the Amazon river (Pará, Brazil) (Photo: Pia Parolin); (**g**) Wetlands of Iberá (humid Chaco, Argentina) (Photo: Juan Jose Neiff); (**h**) Monodominant stands of *Copernicia alba* (humid Chaco, Argentina) (Photo: Juan Jose Neiff); (**i**) Pilcomayo river in the dry season (humid Chaco, Argentina) (Photo: Juan Jose Neiff); (**j**) Flight of birds (güiriri duck) (Llanos of the Apure state, Venezuela (Photo: Río Verde: @rioverdeoficial)

Fig. 10.4 (**a**) Riparian forest of Paraguay river after a fire in Pantanal (Brazil)(Photo: Ieda Maria Bortolotto); (**b**) Flooded grasslands in Paraguai Mirim river (Pantanal, Brazil) (Photo: Maxwell R. Oliveira); (**c**) Flooded grassland in Baía do Burro, Pantanal (Brazil) (Photo: Maxwell R. Oliveira); (**d**) *Vereda* in the APA do Guariroba (Campo Grande, Brazil) (Photo: Geraldo A. Damasceno Junior); (**e**) Resprouting of *Ocotea dyospirifolia* after a fire in a Riparian forest of Pantanal (Brazil) (Photo: Ieda Maria Bortolotto); (**f**) Flooded individuals of *Copernicia alba* after a fire (Pantanal, Brazil) (Photo: Ieda Maria Bortolotto); (**g**) *Heterospizias meridionalis* during a fire in the Pantanal (Brazil) (Photo: Geraldo A. Damasceno Junior). (**h**) *Vereda* in the municipality of Bonito de Minas -MG (Brazil) (Photo: Ricardo Solar); (**i**) Burnt *Vereda* in Bonito de Minas—MG (Brazil) (Photo: Ricardo Solar)

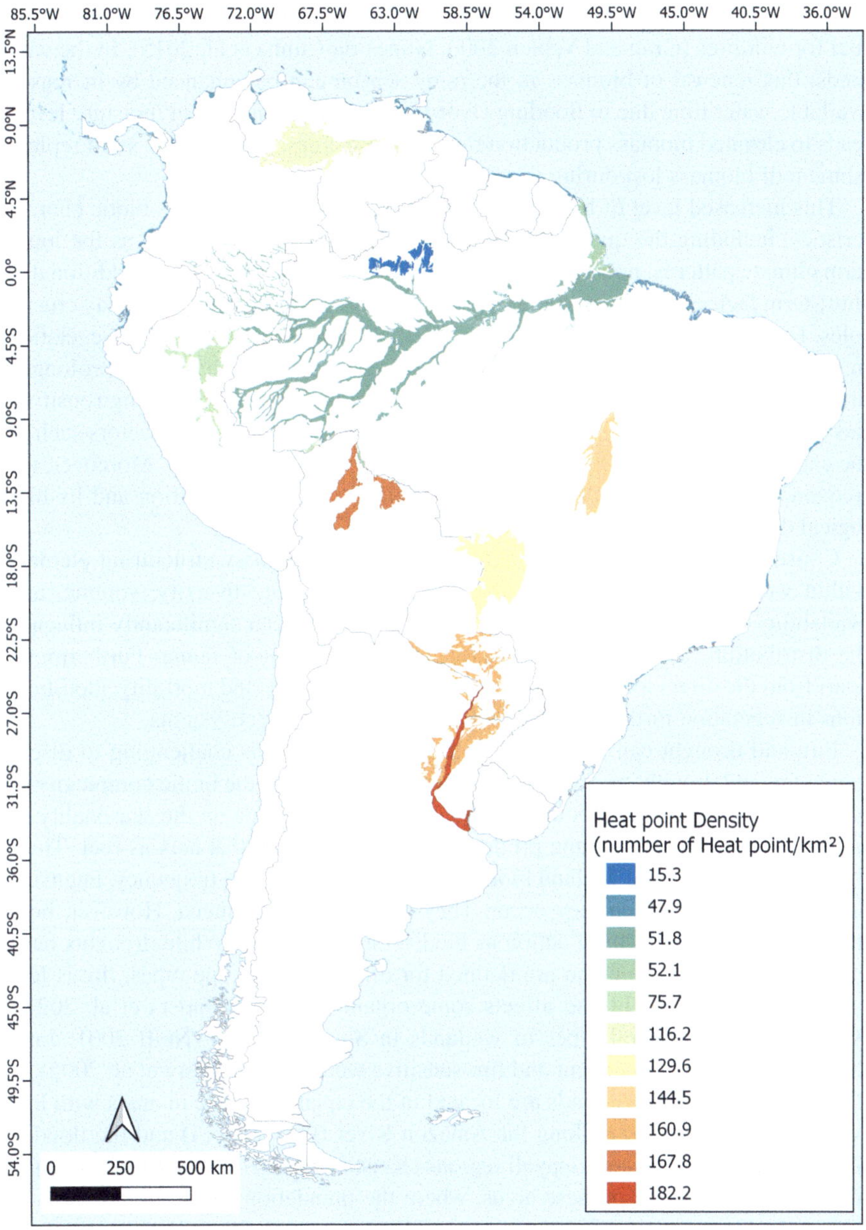

Fig. 10.5 Heat point density/km^2 in the main South American wetlands over two decades (2002–2022). Source: fire foci—INPE (2023); wetlands—see Fig. 10.1

On the other hand, during prolonged drought, the biomass dries up and becomes fuel for wildfires (Grau and Veblen 2000; Nunes-da-Cunha et al. 2015). In the wetlands, this renewal of biomass in the rainy season can be enhanced by increased available water time due to flooding (Ivory et al. 2019). This rise in moisture levels leads to elevated biomass production rates in these areas, facilitating a swift replenishment of biomass lost during drought periods.

This increased level of biomass led to the examination of various biotic characteristics, including the quality and quantity of fuel available, as well as the long-term climate patterns, notably the frequency and duration of droughts. Additionally, short-term factors such as temperature, atmospheric humidity, and wind play crucial roles. Furthermore, it is essential to consider the evolutionary history of vegetation in relation to fire dynamics. Wetlands are naturally vulnerable during prolonged droughts. Different vegetation types and plant life forms exhibit varying sensitivities to wetland fire events. This sensitivity is primarily influenced by factors such as the capacity and duration required for vegetation structure recovery. Moreover, the frequency of fires can significantly alter the landscape's composition and its biological diversity.

Consequently, given that vegetation serves as the primary structuring element within wetland ecosystems, any alterations in its extent, diversity, volume, and availability as a resource for food, shelter, and nesting can significantly influence the distribution, complexity, abundance, and recruitment of fauna. Furthermore, apart from the direct impact of fire on fauna, such as injury and mortality, modifications in vegetation further shape the dynamics of wetland ecosystems.

Fire and drought can be natural or human-induced. It is challenging to disentangle fire and drought as ecological factors that determine the biotic complexity of wetlands, especially since both occur together. This depends on the seasonality of rainfall and the excess organic production from vegetation that acts as fuel. Thus, the effects of fire on the wetland biota are largely related to the frequency, intensity, and seasonality in which they occur. They are pulsed phenomena. However, both differ in their mechanism of action as biodiversity modifiers. While droughts have gradual effects over time and are defined for different landscape types, fire is less predictable than drought and affects some organisms more (Batista et al. 2023). Among the nine general types of wetlands in South America (Neiff 2001; Junk 2013), we have fire-dependent and fire-sensitive wetlands (Hardesty et al. 2005).

Most fire-sensitive wetlands are located in the equatorial zone in areas with low seasonality, such as those along the Amazon River (Fig. 10.6b, i) and the flooded forests in the Marañon and Ucayali regions (Kvist and Nebel 2001). Flooding is the main ecological driver in these areas, where the inundation can reach up to 16 m high and takes more than 270 days a year (Wittmann et al. 2006; De Simone et al. 2002). In the Amazon, the vegetation of the floodplains is mainly composed of trees, except for seasonally flooded grasslands in the lower reaches, which remain underwater most of the year. On the other hand, in the fire-dependent areas (Fig. 10.6c, d, f, g, and h), seasonality allows fire occurrence during the dry season, and the landscape is more fragmented, with grasslands in the low line areas and forests in higher areas subjected to less time of inundation or flooded only in the edge. These areas

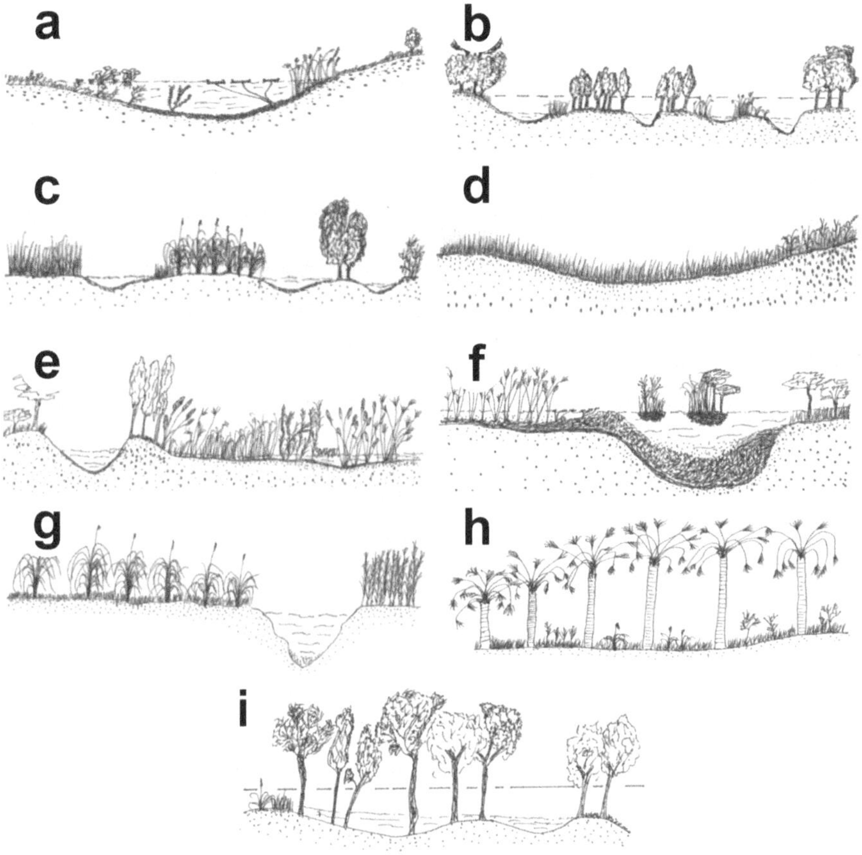

Fig. 10.6 Main types of South American Wetlands (Adapted from Neiff 2001). (**a**) Shalow lakes; (**b**) Floodplains (Várzeas); (**c**) Pantanals; (**d**) Swamps with short grasses; (**e**) Meander Scrolls; (**f**) Peatlands; (**g**) Swamps with high grasses; (**h**) Palm groves (Morichales); (**i**) Seasonally flooded forests (Igapó). (Drawings: JJ Neiff)

include the Pantanal, Llanos de Mojos, Bananal Island (Araguaia region), Ruponini Savannas (Roraima), and the Chaco (humid and dry). The vegetation of this savanna landscape occupies large areas where the fire is millennial. It is important to note here that where we have peatlands (Fig. 10.6f), the sensibility to fire can be very high when we consider climatic changes with less amount of rain every year, as has been occurring in the Cerrado and Pantanal region. If burned constantly, these peatlands will completely lose the Histosol layer, with a great amount of carbon released to the atmosphere or dissolved in water. It can also cause a reduction of biodiversity and exposition of the soil to erosion, and the sediments carried to the peatland can trigger a gradual and perennial decrease in the ability to retain water (Silva et al. 2022). The sensibility of peatlands to fire was also noted in the region of Iberá, Argentina (Poi 2017). Areas with shallow lakes tend to be fire-dependent because, in most cases, they can remain without water during the dry season and have aquatic plants with resistant underground systems (Fig. 10.6a). On the other hand, the meander scrolls are composed by a mix of riparian forests and flooded grasslands in areas with high seasonality, having fire-dependent and fire-sensitive vegetation side by side (Fig. 10.6e).

10.4 Effects of Fire on the Wetland Environment

At the landscape level, in fire-dependent wetlands, areas with higher inundation levels tend to present higher fire frequency. This phenomenon occurs because higher inundation levels lead to greater biomass production and, in many cases, lower biomass consumption by large herbivores. When the dry season comes, if the inundation is not that great, biomass becomes available to be burned. The number of fire foci and intensity will depend on the heat intensity, distance of the people's settlements, air humidity, and other features. This type of phenomenon generally occurs along the most floodable rivers. Because of that, the fire in these cases tends to create a fire corridor where the double filter (fire-inundation) effects are more intense and the selective action is stronger (Damasceno-Junior 2021; Garcia et al. 2021). This phenomenon has been documented along the Paraguay River in Pantanal (Damasceno-Junior et al. 2021), and similar patterns may be observed along the Araguaia River, where higher fire frequencies occur along its course (Andrade et al. 2021; Assis et al. 2021). Moreover, this dynamic is not unique to these regions; it is also observed in other wetlands worldwide, such as the Okavango Delta (Heinl et al. 2007).

Fires in wetlands can occur in different forms: on the surface (surface fire) and beneath the surface (ground fire), which is prevalent in regions with histosols capable of burning substantial organic matter. Additionally, fires can occasionally occur over water, particularly on floating vegetation, as documented in stands of *Cyperus giganteus* in the Pantanal (Rocha et al. 2015). The effects of fire cannot be assessed solely based on the temperature at the fire's height or the fire's duration, as several local factors are involved, such as soil structure, moisture in the profile, and terrain relief. Temperatures above 60° must be considered restrictive for the permanence of the organisms because, at this temperature, the proteins coagulate. However, this temperature is rarely reached at ground level since the most remarkable heat production of fire occurs in the upper third of the flame.

The soils thermal inertia is adequate to prevent critical heating from penetrating beyond a depth of 10 cm. To maintain this condition, the soil's moisture content must fall within 10–15% range. Deviating from this range, either above or below, leads to a significant decrease in thermal conductivity. However, it's worth noting that there is limited available data concerning fires in wetlands, which necessitates further research in this area. In experiments conducted in *vereda* wetlands, the fire burned 60–98% of the fuel with low intensity (240–1083 kW m^{-1}). Late dry-season fires tend to be more intense than early dry-season fires, with lower temperatures (149–442 °C) than dry savannas around (Schmidt et al. 2017). In Iberá wetlands,

Argentina, experiments showed that fire attained higher temperatures (160–180 °C) circa 50 cm above soil with a duration below 5 min. The soil temperature increased <3 °C in the five first centimeters below ground level. The decomposition of organic matter tends to be lower until circa 180 days after fire (J.J Neiff unpublished data).

The amount of organic carbon stored in the soil as organic matter (OM) varies greatly, depending on the geomorphology of the wetland, the duration of water storage in the soil, and the production of the vegetation. It can be <5% in some marshes or more than 70% in peat bogs. The functions of different forms of OM in wetlands are diverse, and their benefits are mainly in terms of moisture and nutrient retention and promoting root activity. OM deposited in the soil protects against the effects of radiation and raindrops, improves aeration conditions, and contributes to soil structure by increasing the ion exchange capacity. However, the accumulation of compounds with a high lignin content on the soil surface, such as those contained in some types of litter, acidifies the pH, blocks phosphorus, and delays the mineralization of organic matter due to the bacteriostatic effect of lignin, as occurs in peat bogs (Neiff et al. 1994).

Fire can have negative impacts due to the export of nutrients from the site and the disturbances it causes to the local biota. It led to a fruitful discussion among researchers as the biological fertility of the soil is not always affected as a favorable biotic response has been recorded, with rapid microbial activity after burning due to pH change, increase in cations, and phosphorus (Fenn et al. 1993). Ashes can also affect wetlands, as after a fire, the ashes are deposited on wetland soils and water. In a burned *vereda*, there was an increase in Ca, Mg, and P levels after 2 months of fire and for K 4 months after the fire (Oliveira-Filho et al. 2018).

Depending on the texture of the soil, fire intensity, and soil temperature, fire can induce water repellency on soils. The repellency is caused by substances produced during soil heating and can penetrate the ground, creating a hydrophobic layer. Fires with temperatures between 175 °C and 200 °C can produce intense water repellency, and destruction of water repellency occurs when soils are heated between 280 and 400 °C (Debano 2000). This phenomenon is known to cause erosion of the first layer of the soil and can also increase the inundation levels after a fire (Debano 2000). In South American wetlands, this still needs to be studied. Some experiments conducted in Pantanal after the 2020 Megafires did not detect changes in the soil water repellency (GA Carvalho, com pess).

10.5 Effects of Fire on Wetland Vegetation

It is not unusual for grasslands, swamps, and forests to exhibit varying degrees of adaptation and recovery following fire disturbances in these natural systems, particularly in the presence of human intervention post-fire. A notable adaptation involves the selection of plants with roots and rhizomes extending deeper than 5 cm into the soil. Additionally, the prevalence of monocots in South American wetlands may be linked to the frequency and duration of fires in these ecosystems.

In most fire-dependent areas, fire is combined with inundation, which excludes the woody species from the lower parts of the flood gradient and promotes the openness of the landscape. In this process, the exclusion of trees and shrubs is promoted by inundation, which kills non-resistant species, mostly all seedlings. On the other hand, most flood-resistant trees are burned during fire events. This double filter makes vegetation more open than its counterparts in non-floodable areas such as the Cerrado (Arruda et al. 2016; Damasceno-Junior et al. 2021; Manrique-Pineda et al. 2021). In addition, some of these areas are used as native pastures by cattle ranchers, and they also promote the cleaning of the small shrubs and trees. Resprouting is another adaptation for many species in riparian forests (Fig. 10.4e). This strategy is more known for Cerrado and other savanna vegetation but poorly known for these wetlands.

Another effect of fire and flood in wetlands is monodominance. While flooding alone is responsible for the monodominance in some wetlands (Gris et al. 2020; Torti et al. 2001), the combination of flooding and fire can also lead to monodominance. This is the case of *Tabebuia aurea* in the Pantanal. This species grows preferably over earth mounds locally known as murundus in Pantanal. These earthmounds can protect the trees from inundation and also from fire. Areas dominated by these plants growing atop earthmounds, subjected to frequent fires and high inundation levels, tend to exhibit lower species diversity and reduced populations of other species compared to *Tabebuia aurea* (Manrique-Pineda et al. 2021).

Fire and flooding favor the monodominance of *Attalea phalerata* in small islands (locally known as capões) in the Pantanal. This species acts as grasses in some forested islands. Its preferred habitat is on the edge of these islands in transition to seasonally flooded grasslands. This palm species possesses a remarkable ability to withstand certain levels of inundation. When a fire occurs, it tends to infiltrate the forest and ignite these palms, which typically accumulate a significant amount of dry leaves on the forest floor. Some leaves remain in contact with the ground, forming a pathway for the fire to ascend towards the crown of the palm rapidly. The protective sheaths left by the fallen leaves shield the palm from the fire's lethal effects, allowing it to survive. Consequently, this palm species eliminates neighboring trees along the border of these islands and establishes dominance, resulting in a monodominant ecosystem (Rivaben et al. 2021).

Another monodominant species that rely on fire and flood is *Copernicia alba*. This species tolerates more than 120 days/y of flooding but is a very flammable palm that burns quickly and benefits from areas with repeated fire events (R. H. Silva, pers. comm.) (Fig. 10.4f). Many aquatic species are also fire tolerant. *Cyperus giganteus* is an aquatic sedge that can grow up floating over the water and form extensive areas of monodominant floating mat. This species can burn over the water and recover the vegetation structure after 2 months (Rocha et al. 2015).

The occurrence of fire in floodable environments can also change the preferable position of some species along the flooding gradient. Species that occur in riparian forests have the preferable belt of occurrence along the flooding gradient according to their flooding tolerances (Damasceno-Junior et al. 2005; Nunes da Cunha 2001), patterns of dispersal (Kubitzki and Ziburski 1994) and other ecological filters. In repeatedly burned riparian forests of the Paraguay River, many species can occur in higher areas along the flooding gradient than expected because of the combined fire and flood effect in the low-line areas of the flooding gradient, burning and drowning the small trees and shrubs (Arruda et al. 2016). This effect reduces species richness and abundance in the more flooded riparian forests and increases the richness and abundance in the less floodable riparian forest areas (Arruda et al. 2016; Oliveira et al. 2014).

Fire can trigger functional responses of trees to the flooding gradient. In the forested islands of the Pantanal, tree species increase the relative bark thickness towards the most floodable areas only if the area was previously exposed to fire. In addition, leaf nitrogen decreases along the flooding gradient without fire but increases if the area was previously burned. The same is true for leaf dry matter content, a proxy for flammability (Ortiz et al. 2023). Leaf flammability decreases along the inundation gradient only if the forest was previously burned; if not, it tends to increase (de Almeida Souza et al. 2019). Even in fire-dependent environments, seasonally flooded forests can lose diversity after fire. In seasonally flooded forests of the Orinoco River, seasonally flooded forests decreased the α, β, and functional diversity after fire both in high and low intensity. Fire also reduced the richness of animal-dispersed species in these forests and increased the proportion of deciduous species (Meza et al. 2023). For forested *impucas* (patches of seasonally flooded forests in the middle of savanna vegetation) in the transition of Amazon and Cerrado, the fire increased the mortality and turnover of tree species along tree years of observation (Maracahipes et al. 2014).

Many tree species in flooded areas are resistant or tolerant to fire. Fire resistance in the floodable forests of Pantanal is linked to bark thickness. However, trees without bark thickness depend on the amount of phenolic compounds found in the bark, less parenchyma, and more sclerenchyma for fire resistance (da Silva et al. 2021). Aquatic species tend to be more sensitive to fire. Nevertheless, many of these species can have underground systems, such as rhizomes found in species of Alismataceae, Nymphaeaceae, and Poaceae, that can tolerate and resprout after fire events. Some examples are *Echinodorus paniculatus* and *Nymphaea amazonum* (Scremin-Dias et al. 2021).

Seed banks in open floodable areas are composed of terrestrial and aquatic species, germinating at different times of the year. They are not significantly altered after fire events in Pantanal except for the top 2 cm layer. One of the most successful tree species after a fire in Pantanal is *Cecropia pachystachya,* which can be present in the seed bank and benefit from the openness of the canopy (Kohagura et al. 2020; Pott et al. 2021). Some traits that are adaptations to waterlogged areas can be useful for fire tolerance. In *Sesbania marginata*, the indehiscent fruit that is full of aerenchyma and allows the diaspore to float after being released from the plant, but when fire comes, this fruit confers protection to the seeds and seed germination is not affected by fire (Soares et al. 2021). In addition, this species can become monodominant after significant fire events (Damasceno-Junior pers. obs). Repeated fires can reduce the availability of ripe fruits in a mosaic of floodable and not floodable vegetation (dos Santos Ferreira et al. 2023).

In *veredas* of Central Brazil, peatlands are typical (Fig. 10.4d, h, i), with many layers of vegetation constituted by sedges, Xyridaceae, grasses, and other herbaceous species (Araújo et al. 2002; Moreira et al. 2011), sometimes with *Mauritia flexuosa* forming a line in the lower part of the wetland sometimes without (Moreira et al. 2017). Fire in these environments can have the same intensity when applied early or late fire and is more intense only if there is more wind during the fire (Schmidt et al. 2017). Environmental recovery after a fire is due mainly to the wetness of the layers below the surface (Schmidt et al. 2017), and we believe that fires in these environments during parched years can cause underground fire, which is difficult to fight and has more destructive potential.

However, low-intensity fires are often set by the local people in Jalapão *veredas* to improve the quality of forage for cattle and to increase the number of flowers of *Syngonanthus nitens* (golden grass), which is used for handicrafts in Minas Gerais and Goiás state in Brazil (Schmidt et al. 2007). The main effects of fire in these environments are penetration in the soils with high temperatures when it burns in very dry conditions, burning the roots and the seed bank. Many invasive species can develop and outcompete in burned areas for an undetermined time (Maillard et al. 2009). In some situations, the destruction of the *vereda* can accelerate the succession, and the vegetation can become a gallery forest (Borges et al. 2016). It can last many days and burn the *Mauritia flexuosa* palms (Maillard et al. 2009), which can take 10 years to recover (Schmidt et al. 2017).

Available data are scarce on fire-independent flooded environments. These environments are primarily found in regions with lower seasonality, where the dry season's duration and intensity are not conducive to the same levels of fire recurrence observed in savanna environments. Consequently, research on these flooded environments remains limited. The Amazon floodplains are estimated at 456,000 km^2 of várzea areas and 302,000 km^2 of *igapó* (Schöngart et al. 2017). White waters dominate the first, and the last by black waters. The central floodplain areas of the Amazon Basin are subjected to unimodal flood pulse and lower seasonality (Melack and Hess 2010). Most of these areas with lower seasonality (Fig. 10.2) are covered by floodplain forests (Melack and Hess 2010). Because of the lower seasonality, fire in these environments occurs mainly in years of appropriate climatic conditions. In a study of an *igapó* in the Amazon, the fire scars are located around human settlements and probably set by humans. These fire scars were only 3.4% of the total area, and more than 90% of fire scars were associated with El Nino events (Carvalho et al. 2021).

The impact of fire in these environments can reach a different magnitude from the fire-dependent ones. In another study, there was no full recovery of canopy cover after 10–30 years of the fire events in Igapó forests (Smith et al. 2023). In some cases, post-fire tree mortality can reach 91% with very low forest recovery (Flores et al. 2014). Nevertheless, the nature of these seasonally flooded forests of the Amazon is very heterogeneous, and it is impossible with the present knowledge to generalize if all these forests react similarly to fire (Schöngart et al. 2017).

10.6 Effects of Fire on Wetland Fauna

Studies that have used animals as the response organisms to fire effects in wetlands are much less frequent compared with plants (T.S. Teles, pers. comm.), which also happens in many South American ecosystems (Giorgis et al. 2021). Generally, the effects of fire on animal biodiversity in wetlands are heterogeneous and contrasting, and a large magnitude of effects have been detected for the same groups. In many cases, the negative effect of fire, like reducing the abundance or richness, is not only a pure fire effect but is influenced by weather conditions or spatial heterogeneity among sample sites. Also, neutral effects of fire on animal diversity in wetlands are commonly reported (e.g., Brennan et al. 2005; Ordano et al. 2008; Gonçalves et al. 2022a, b).

Animals can explore burned environments in several ways. Fires generally have little effect on savanna wetlands and birds, except during the breeding season. Large mammals manage to escape the fire by migrating or burrowing themselves (like many amphibians), and many taxa are adapted to explore burned areas, which can even increase their populations after a fire (Pastro et al. 2011). Most of these tendencies must be confirmed by the South American fauna. In South American wetlands, the most visible benefitted animals are scavenger birds that can eat carcasses of animals killed by fire and can also be viewed following the smokes, like the crested carcara (*Caracara plancus*) or hunting animals escaping from flames, as the savanna hawk (e.g., *Heterospizias meridionalis*—Fig. 10.4g). Fire can also expose some resources, such as seeds, facilitating the foraging of granivores, such as the rufous-collared sparrow (*Zonotrichia capensis*); it also increases the visibility of predators owing to the destruction of biomass, creating a safe space for herbivores. Among these, especially large ungulates will also benefit from the herbs sprouting stimulated by fire. Anjos et al. (unpublished data) have found that the species composition of birds is affected by fire in savannas compared to areas not burned for 3 years, which does not happen in grasslands. Fire also reduces functional diversity in both grasslands and savannas, and for some species, such as birds that nest in the ground, this can be a problem (Anjos et al., unpublished data). Fire also softens the wood, making this resource more available for burrowing insects (Pausas and Parr 2018; Batista et al. 2023). Ribeiro et al. (unpublished data) verified that butterflies were not affected by fire regimes, but invertebrates of South American wetlands respond with an increase in abundance 1–3 months after fire, such as the Arachnida *Solifuga* (Arrua et al. 2023) and other spiders (Arrua et al., unpublished data). Fire also interacts with flooding, changing the species composition of spider communities in flooded areas compared to non-flooded ones in the Pantanal, and it modulates the response of those communities to fire. While fire has a positive response to spider richness in flooded areas, this response is negative in non-flooded ones (Arrua et al., unpublished data).

Nevertheless, this is not the same during megafires, when the intensity and area of fire are abnormally big, hindering fauna from escaping from flames, even those with great displacement capability (Tomas et al. 2021). One year after megafires, most mammals have their abundance significantly reduced, and some species were benefited or not affected by these events. Recent studies show that jaguars increased habitat use 3 years after the megafire events in the Pantanal and puma was not affected, in contrast with ocelot (*Leopardus pardalis*), giant armadillos (*Priodontes maximus*), tapir (*Tapirus terrestris*), red-brocket deer (*Mazama americana*), collared peccary (*Dicotyles tajacu*) and azara agouti (*Dasyprocta azarae*), which reduced their occupancy (Bardales et al. 2024; Maggioli et al. 2024). When looking for fire extension and, the recovery time after burning, the increase in the burnt area led to a reduction in the abundance of rabbits (*Sylvilagus brasiliensis*) and collared peccary (*Dicotyles tajacu*) and did not change the abundance of tapirs (*Tapirus terrestris*) and brown brocket deer (*Mazama gouazoubira*). Still, a significant effect of the proportion of area burned in the 12 months preceding monitoring on the relative abundance of the small species of rabbits (*Sylvilagus brasiliensis*) was detected, exclusively at the smallest spatial extent (within 500-m radius buffers). For collared peccary (*Dicotyles tajacu*), a medium-sized forest specialist, the most significant result was obtained at the largest scale (1000 m), regarding the proportion of land burned twice in the 12 months preceding monitoring, emphasizing that fire affects animals in different scales, according to their biological life traits (Harmange et al. 2024, in press).

Contrarily to the terrestrial ones, wetland bats showed a positive response to fire in the Pantanal burned patches, which presented a high abundance of predator bats right after the fire, including gleaning insectivorous, carnivorous, and sanguivorous bats (Santos et al. 2021). Fishes live in small riparian wetlands, and many migrate to streams when a drought occurs, therefore not being exposed to fire; however, the ashes carried to the water can reduce their fitness in some cases (Gonino et al. 2019). In fire-sensitive environments, fire disrupts the fauna composition, reduces the number of specialists, and increases generalists (Andrade et al. 2017), however, this is usually taxon-dependent. Some studies have shown that ants are more resilient to fire than beetles and birds (Silveira et al. 2016). In some groups (e.g., dung beetles), large-bodied species are replaced by small-bodied ones after wildfires (Andrade et al. 2014). Invertebrates in fire-dependent ecosystems usually survive the fire and rebuild their populations in the rainy season, when small puddles of water are available. Some species of ants and termites build dome-shaped nests (called *tacurúes* in Argentina, *tatucos* in Colombia, and *murundus* in Brazil) up to 1 m high, protecting the population from fire and flood. This also occurs in the drylands of Cerrado (see Chap. 5).

Some ant species are favored in humid areas with annual fire frequency, and many fishes from savanna wetlands are well adapted to drought-induced anoxia. *Oplosternon littorae* has a well-vascularized stomach with abundant blood flow, taking oxygen from the air by osmosis, captured in its mouth. The dipnoan *Lepidosiren paradoxa* ("lola" or "peixe-pulmão"), a species endemic to South American swamps, builds a cave in the mud at the bottom, blocking the cave entrance with mucus and sediment. This regulates humidity and gas intake during long dry periods, permitting the individuals to remain alive in the cave after a fire (Neiff 2001).

Fire which usually occurs when wetlands dry out, does not cause significant damage to soil biota because increases in the soil temperature are very low, by 3–5 °C in just the first 5 cm of soil and only for a few hours, as we have seen in controlled burning experiments (Neiff 2001; 2004).

10.7 Fire Management and Traditional Knowledge in Wetlands

The management of fire in wetlands necessitates a deep understanding of inundation patterns. Prescribed burning must consider the timing, depth, and duration of inundation in a given environment. If the environment is inundated at the end of the rainy season, prescribed fire can be performed in three situations of the terrestrial phase of flood pulsing: just after inundation, in the middle of the dry season, or at the beginning of the new rainy season. It will depend on the needs of the management. For instance, in areas like the Paraguay River flooding zone, where peak inundation occurs 3 months after the rainy season ends (Damasceno-Junior et al. 2021), prescribed fire can only be safely implemented during the rainy season when there is a sufficient dry spell of around 10 days. Alternatively, prescribed burning can also be conducted in years when the river level does not reach the floodplain, exposing accumulated biomass to fire. In these years, the window to burn is greater than in the first situation, and the prescribed fire can occur likewise at the beginning of the dry season, in the middle, or at the end of the dry season, just like in terrestrial grasslands.

Many monodominant plant species with economic potential can be managed using prescribed fire. For example, *Copernicia alba* has edible fruits and benefits from the combination of fire, inundation, and soils with high pH (Damasceno-Junior et al. 2021). Cattle ranching in wetlands, based on native pastures, needs fire as a fundamental management tool. From a conservation perspective, fire regulation in these wetlands must consider these economic possibilities and other aspects, such as carbon balance. Regulation of fire use in floodplains is in its nascent stages, and there's a pressing need for more research programs across various South American wetlands. In Brazil, the parliament recently approved the regulation of Integrated Fire Management (see Chap. 14).

Fire is not a regular tool for the deforestation process in wetlands because most fires are used to improve the quality of pastures or for similar purposes. A good example is the Chaco region. The link between fire and deforestation is more consistent in Dry Chaco than in Wet Chaco. Of 51,000 km^2 of deforestation in Dry Chaco of Argentina, 40% was linked to burned areas (San Martín et al. 2023). One important issue about the fire occurrence in wetlands is the possible decrease of potable water for human consumption. Fortunately, the water supply for human consumption from the South American wetlands is still at low risk with the increasing megafires occurrence (Robinne et al. 2018).

Human populations traditionally use Wetlands (Fig. 10.3a) who manage food, medicinal plants, and several other essential resources for the local economy, using fire as a management tool throughout the seasonal cycle. Traditional fire management in wetlands includes reports of management to reduce the density of unwanted species and the most suitable season for burning in many parts of the world (Russell-Smith et al. 1997). Likewise, there are extensive wetlands in South America that are subject to fire but provide an ideal environment for the growth of important species for the food security of human populations.

Hilbert et al. (2017) provide evidence for wild rice domestication (*Oryza* spp.) in wetlands of the Amazon region and suggest that in the past, likewise in India nowadays, after the rice grains fell into the soil, people may have burned patches of these rice grains embedded in the soil during the dry season to remove competing vegetation. This practice may have been used in other wetlands, such as the Pantanal, where three species of *Oryza* occur in seasonally flooded areas. These species are important for biological conservation, local culture and also bio-economy (Bortolotto et al. 2021). Another species widely distributed in South America, in fire-dependent environments and important as food (especially as a protein and vitamin supplement), is *Typha domingensis*. Arenas and Scarpa (2003) suggest that the management necessary for its exploitation in the humid Chaco should consist of protection against cattle and fires. On the other hand, in Central Mexico, *Typha domingensis* expands after fires to colonize areas formerly dominated by other natives (López-Arcos et al. 2012). According to this study, this invasion responds to increased phosphate availability from the ash of other species. With more phosphate, this species grows and allocates more into storage tissues.

In addition to aquatic species, which are subject to fire in the dry season, domesticated species cultivated in humid areas in the dry season can also be managed with fires. This is the case of cassava (*Manihot esculenta*), cultivated in traditional agricultural systems called "roças de esgoto" (drained peat swamp swidden field) in Jalapão, Tocantins, Brazil (Borges et al. 2016). In this region, farmers often practice draining, cutting and burning *vereda* areas. This agricultural practice helps clear vegetation and manage land for cultivation or other purposes in flood-prone regions.

Welch (2015) documented various hunting strategies employed by the Xavante people in Brazil (see Chap. 3), which include avoiding burning grasslands in seasonally flooded areas towards the end of the dry season to mitigate the adverse effects of dryness and increased wind. Additionally, in Jalapão, Tocantins, Brazil, the local population frequently uses low-intensity fires to enhance the quality of forage for livestock in humid grasslands or *vereda* and to promote the proliferation of *Syngonanthus nitens* (golden grass, a plant typical of wet fields), which is valuable for handicrafts (Schmidt et al. 2007). Moreover, in Tocantins, Brazil, fire is utilized in paths associated with drained peat swamp-swidden fields to help maintain tree cover during fallow periods, potentially expediting natural succession processes (Borges et al. 2016). Fire is prohibited in the *veredas*, which are considered permanent preservation areas. Still, the examples above show that its use can be associated with traditional forms of management that benefit the environment and the local economy (see Chap. 14).

Wetlands management is historically based on traditional ecological knowledge, representing a biocultural heritage of great relevance and increasing recognition of its role in contemporary science. In Venezuela, for example, there is growing recognition that Indigenous knowledge can contribute to more sustainable management of protected areas in the Gran Sabana (Sletto and Rodriguez 2013). On the other hand, these authors noted marked differences in how young and old people see fire and its impact, just as there are differences in the perception of fire between communities close to highways and those in remote areas.

San Martin et al. (2023) point to the need for studies combining the environmental sciences with the social sciences to understand the role of human action associated with fire in the Gran Chaco. Schulz et al. (2019) also suggested this strategy for the Pantanal. Based on a literature review, these authors suggest that it is necessary to integrate insights from various disciplines and that environmental management can be improved through a better grounding in traditional practices and local perspectives.

Ethnobiological studies conducted in humid areas prone to fire in South America represent a crucial interdisciplinary research field. They facilitate the exploration of traditional knowledge and enable its integration into broader scientific frameworks. However, there are numerous gaps in research, posing challenges to developing policies focused on Integrated Fire Management informed by traditional ecological knowledge.

Moreover, documentation of ecological knowledge regarding fire usage before European colonization is lacking in the Americas. Many Indigenous Peoples have faced decimation, and traditional cultural practices among the remaining communities are rapidly eroding.

The knowledge of when and how to burn in each location, considering climatic conditions and the type of vegetation, are questions that have guided scientific studies contemplating these characteristics. For example, Díaz et al. (2023) noted that the knowledge of the use of fire by the Kichwa Saraguro Indigenous People from southern Ecuador is transmitted to new generations, and they know how and where to carry out traditional fires.

As wildfires continue to escalate in South America, there is an urgent need for ethnobiological studies that encompass a broader range of stakeholders beyond just Indigenous Peoples and Traditional and Local Communities. In regions like Acre, Brazilian Amazon, between 2016 and 2019, a small percentage of fires, ranging from 1 to 2%, occurred in Indigenous Lands, while a larger proportion, between 11% to 14%, took place in conservation units, including extractive reserves inhabited by Traditional Communities. In contrast, a significant portion, between 32–35%, of fires were recorded on public lands, which are often targeted by large land grabbers (Silva et al. 2021). Moreover, from 2002 to 2020, approximately 95% of wildfires in the Pantanal were in natural vegetation, with 28% occurring in areas classified as wetlands, likely exacerbated by drought conditions (Mataveli et al. 2021). Studies focusing on human interactions and wetland fire management hold the potential to offer valuable insights and contribute to the development of sustainable solutions. Consequently, there is an urgent need to prioritize research to preserve traditional ecological knowledge, given its invaluable insights into sustainable fire management practices.

10.8 Megafires in South American Wetlands

From mid-2019 to the first quarter of 2023, an extraordinary La Niña event of exceptional duration occurred, producing intense drought stress in many South American ecosystems, especially in wetlands below the equator line. Numerous wildfires were recorded during these periods, the consequences of which are still poorly understood, particularly regarding ecosystem resilience and changes in organisms' distribution and abundance patterns. The fire associated with this drought is the "trigger effect" of numerous biotic transformations that constitute an abrupt "reset process" in the dynamics of wetland pulses. Certain floodable environments pose a higher risk of mega-fires, particularly those inundated during years of high biomass production. The more dangerous are the more floodable ones. These environments can become tinderboxes in exceptionally dry years, generally with hot waves, leading to catastrophic fires. Effective prevention of such fires relies on accurate forecasting of dry periods.

Like those in the Pantanal, riparian forests in fire-dependent floodplains exhibit some resilience to low-intensity fires and harbor many fire-resistant species (see Arruda et al. 2016; Oliveira et al. 2014). These areas can be managed using controlled burns of low intensity. However, even these forests are vulnerable to megafires, and there is limited data on their effects on such ecosystems.

Our knowledge about the effects, consequences, and recovery of megafires in South American wetlands is very limited and usually restricted to the area burnt, carbon emissions, or effects in some taxa, as, for example, vertebrates (Tomas et al. 2021). Owing to the great area affected and usually its great intensity, it is expected that those events will act as a strong environmental filter, affecting even organisms that have some degree of adaptations to resist fire events. For instance, fire-adapted trees with thick bark can lose their aerial parts in these events, and animals with great mobility and capacity for fire detection can be killed (Garcia et al. 2021).

10.9 Research Gaps on Fire in Wetlands

Despite some progress, our understanding of the effects of fire on fauna in South American wetlands remains limited. Additionally, there are significant knowledge gaps regarding the biological impacts of megafires and the resilience of ecosystems, communities, and populations to these extreme events. Another area requiring further investigation is how fire alters the direction or intensity of responses of species, communities, and landscapes to natural flooding cycles. While some research has been conducted in these areas (Neiff et al. 1985, 2005), more data are needed to understand these dynamics comprehensively. Another aspect is managing economically important species that benefit from fire and inundation. If the management of these species relies on fire, it is important to know the frequency, time of year, and intensity of the areas that can be burned. It is also important to know how these things can vary over the years with different levels and intensities of flooding.

Furthermore, the role of fire associated with inundation in either preventing or promoting biological invasions in wetlands represents another critical gap in our knowledge. Investigating how fire interacts with inundation patterns to influence the spread of invasive species is essential for developing effective management strategies for wetland ecosystems. Closing these gaps through continued research efforts will be crucial for advancing our understanding of fire ecology in South American wetlands and informing conservation and management practices.

The scarcity of studies precludes us from forecasting the time necessary to recover fauna and flora from megafires events and determining which taxa will be the winners and losers in the long run.

10.10 Conclusion

Our analysis of data and studies from the whole South American continent indicate that fire occurrence in wetlands is strongly related to seasonality. For areas with low seasonality, here termed as fire-independent wetlands, the effects of fire are still poorly understood but can persist for extended periods, sometimes exceeding 30 years. Fire is more prevalent in wetlands characterized by pronounced seasonality, where there is a higher likelihood of occurrence. We found that these areas, which we termed fire-dependent wetlands, exhibit several distinctive features:

1. Intensified landscape openness: Areas with frequent fires and high inundation levels become open grasslands because both fire and flood can kill trees and bushes.
2. Occurrence of monodominant plant species: Some of these monodominant plants are resistant to flood and fire and are common in fire-dependent wetlands.
3. Changes in biota responses across the flooding gradient: In seasonally flooded gradients the biota tends to be distributed or respond according to levels of inundation to which they are subjected. The fire occurrence can smooth, change these responses, or even trigger inundation responses when they do not appear.
4. High probability of ground fires: Such fires are more likely to occur in humid environments primarily due to the presence of histosols.
5. Reduction in species richness and abundance: Fire, in combination with inundation, can reduce species richness and abundance changes due to its double filtering effects.
6. Changes in species composition: Fire, combined or not with inundation, can lead to changes in species composition within fire-dependent wetlands. The time and levels of inundation mediate these changes.

Most of these effects are observed under low-intensity fires and may differ under conditions of mega-fires. Effective fire management in fire-dependent wetlands is crucial, and the conservation of these environments often relies on the use of fire. Traditional knowledge is fundamental to achieving successful outcomes in fire management practices.

References

Andrade ASR, Ramos RM, Sano EE, Libonati R, Santos FLM et al (2021) Implementation of fire policies in Brazil: an assessment of fire dynamics in Brazilian Savana. Sustainability (Switzerland) 13:11532. https://doi.org/10.3390/su132011532

Araújo GM, Barbosa AAA, Arantes AA, Amaral AF (2002) Composição florística de veredas no Município de Uberlândia, MG. Braz J Bot 25:475–493. https://doi.org/10.1590/S0100-84042002012000012

Arenas P, Scarpa GF (2003) The consumption of Typha Domingensis Pers. (Typhaceae) pollen among the ethnic groups of the Gran Chaco, South America. Econ Bot 57:181–188. https://doi.org/10.1663/0013-0001(2003)057[0181:TCOTDP]2.0.CO;2

Arruda WDS, Oldeland J, Paranhos Filho AC, Pott A, Cunha NL et al (2016) Inundation and fire shape the structure of riparian forests in the Pantanal, Brazil. PLoS ONE 11:e0156825. https://doi.org/10.1371/journal.pone.0156825

Arrua BA, Carvalho LS, Teles TS, Oliveira MR, Ribeiro DB (2023) Fire Has a Positive Effect on the Abundance of Sun Spiders (Arachnida: Solifugae) in the Cerrado-Pantanal Ecotone Fire 6(2) 69-10.3390/fire6020069

Assis PC, Faria KMS, Bayer M (2021) Unidades de Conservação e sua efetividade na proteção dos recursos hídricos na Bacia do Rio Araguaia. Soc Nat 34:e60335. https://doi.org/10.14393/sn-v34-2022-60335

Bardales R, Boron V, Passos Viana DF, Sousa LL, Dröge E et al (2024) Neotropical mammal responses to megafires in the Brazilian Pantanal. Glob Change Biol 30:e17278. https://doi.org/10.1111/gcb.17278

Batista EKL, Figueira JEC, Solar RRC, de Azevedo CS, Beirã MV et al (2023) In case of fire, escape or die: a trait-based approach for identifying animal species threatened by fire. Fire 6:242. https://doi.org/10.3390/fire6060242

Borges SL, Eloy L, Schimidt IB, Barradas ACS, Santos IA (2016) Fire Management in Veredas (palm swamps): new perspectives on traditional farming Systems in Jalapão, Brazil. Ambiente Soc 19:269–294. https://doi.org/10.1590/1809-4422ASOC20150020R1V1932016

Bortolotto IM, de Souza PR, Pott A, Damasceno-Junior GA (2021) Wild food plants of the Pantanal: past, present, and future. In: Damasceno-Junior GA, Pott A (eds) Flora and vegetation of the Pantanal wetland, 1st edn. Springer, Cham, pp 689–738. https://doi.org/10.1007/978-3-030-83375-6_19

Brasil (2012) Lei N° 12.651, de 25 de Maio de 2021—Código Florestal. https://www.planalto.gov.br/ccivil_03/_ato2011-2014/2012/lei/l12651.htm. Accessed July of 2024

Brennan EK, Smith LM, Haukos DA, Lagrange TG (2005) Short-term response of wetland birds to prescribed burning in rainwater basin wetlands. Wetlands 25:667–674. https://doi.org/10.1672/0277-5212(2005)025[0667:SROWBT]2.0.CO;2

Carvalho TC, Wittmann F, Piedade MTF, Resende AFS, Silva TSF, Schöngart J (2021) Fires in Amazonian Blackwater Floodplain Forests: Causes Human Dimension and Implications for Conservation. Frontiers in Forests and Global Change 410.3389/ffgc.2021.755441

Cerasoli F, D'Alessandro P, Biondi M (2022) Worldclim 2.1 versus Worldclim 1.4: climatic niche and grid resolution affect between-version mismatches in habitat suitability models predictions across Europe. Ecol Evol 12:e8430. https://doi.org/10.1002/ece3.8430

Colombia (2021) Catálogo de Metadatos del Ministerio de Ambiente y Desarrollo Sostenible. https://Geonetwork.Minambiente.Gov.Co/Geonetwork/Srv/Por/Catalog.Search#/Home. Accessed July of 2024

Da Silva Meneses MEN, Da Costa ML (2012) Caracterização mineralógica e química dos regolitos de uma área de transição savana-floresta em Roraima: uma análise da evolução da paisagem. Rev Bras Geosci 42(Suppl. 1):42–56. https://doi.org/10.5327/Z0375-75362012000500005

Da Silva JR, da Silva MD, Boaretto AG, Carollo CA, Silva de Oliveira LC et al (2021) Fire tolerance strategies in woody species from Pantanal riparian forests involve phenolic compounds and structure of the inner bark. Flora 281:151865. https://doi.org/10.1016/j.flora.2021.151865

Damasceno-Junior GA, Semir J, Dos Santos FAM, de Freitas Leitão-Filho H (2005) Structure, distribution of species and inundation in a riparian forest of Rio Paraguai, Pantanal, Brazil. Flora 200:119–135. https://doi.org/10.1016/j.flora.2004.09.002

Damasceno-Junior GA, Pereira Ade MM, Oldeland J, Parolin P, Pott A (2021) Fire, flood and Pantanal vegetation. In: Damasceno-Junior GA, Pott A (eds) Flora and vegetation of the Pantanal wetland, 1st edn. Springer, Cham, pp 661–688. https://doi.org/10.1007/978-3-030-83375-6_18

de Almeida Souza AH, Batalha MA, Casagrande JC, Rivaben RC, Assunção VA et al (2019) Fire can weaken or trigger functional responses of trees to flooding in wetland forest patches. J Veg Sci 30:521–532. https://doi.org/10.1111/jvs.12719

Andrade RB, Barlow J, Louzada J, Vaz-de-Mello FZ, Silveira JM, Cochrane MA (2014) Tropical forest fires and biodiversity: dung beetle community and biomass responses in a northern Brazilian Amazon forest. J Insect Conserv 18:1097–1104. https://doi.org/10.1007/s10841-014-9719-4

Andrade RB, Balch JK, Carreira JYO, Brando PM, Freitas AVL (2017) The impacts of recurrent fires on diversity of fruit-feeding butterflies in a South-Eastern Amazon forest. J Trop Ecol 33:22–32. https://doi.org/10.1017/S0266467416000559

de Oliveira MT, Damasceno-Junior GA, Pott A, Paranhos Filho AC, Suarez YR, Parolin P (2014) Regeneration of riparian forests of the Brazilian Pantanal under flood and fire influence. For Ecol Manag 331:256–263. https://doi.org/10.1016/j.foreco.2014.08.011

De Simone O, Müller E, Junk WJ, Schmidt W (2002) Adaptations of Central Amazon Tree Species to Prolonged Flooding: Root Morphology and Leaf Longevity Plant Biology 4(4):515–522. https://doi.org/10.1055/s-2002-34127

Debano LF (2000) The role of fire and soil heating on water repellency in wildland environments: a review. J Hydrol 231–232:195–206. https://doi.org/10.1016/S0022-1694(00)00194-3

Díaz SC, Quezada LC, Álvarez LJ, Loján-Córdova J, Carrión-Paladines V (2023) Indigenous use of fire in the Paramo ecosystem of southern Ecuador: a case study using remote sensing methods and ancestral knowledge of the Kichwa Saraguro people. Fire Ecol 19:1–18. https://doi.org/10.1186/s42408-022-00164-1

dos Santos Ferreira BH, da Rosa Oliveira M, Fernandes RA, Nacagava VA, Arguelho BA et al (2023) Flowering and fruiting show phenological complementarity in both trees and non-trees in mosaic-burnt floodable savanna. J Environ Manag 337:117665. https://doi.org/10.1016/j.jenvman.2023.117665

Fenn ME, Poth MA, Dunn PH, Barro SC (1993) Microbial nitrogen and biomass respiration and nitrogen mineralization in soil beneath two chaparral species along a fire-induced age gradient. Soil Biol. Biochemical 25 (4):457–466

Ferreira CS, Piedade MTF, Tiné MA, Rossatto DR, Parolin P, Buckeridge MS (2009) The role of carbohydrates in seed germination and seedling establishment of Himatanthus sucuuba, an Amazonian tree with populations adapted to flooded and non-flooded conditions. Ann Bot 104:1111–1119

Fichino BS, Dombroski JRG, Pivello VR, Fidelis A (2016) Does fire trigger seed germination in the Neotropical savannas? Experimental tests with six Cerrado species. Biotropica 48:181–187. https://doi.org/10.1111/btp.12276

Flores BM, Piedade MTF, Nelson BW (2014) Fire disturbance in Amazonian Blackwater floodplain forests. Plant Ecol Divers 7:319–327. https://doi.org/10.1080/17550874.2012.716086

Furlan LM, Rosolen V, Moreira CA, Bueno GT, Ferreira ME (2021) The interactive pedological-hydrological processes and environmental sensitivity of a tropical isolated wetland in the Brazilian Cerrado. SN Appl Sci 3(2):144. https://doi.org/10.1007/s42452-021-04174-7

Garcia LC, Szabo JK, Roque FO, Pereira de AMM, da Cunha CN, Damasceno-Júnior GA, Morato RG, Tomas WM, Libonati R, Ribeiro DB (2021) Record-breaking wildfires in the world's largest continuous tropical wetland: integrative fire management is urgently needed for both biodiversity and humans. J Environ Manag 293:112870. https://doi.org/10.1016/J.JENVMAN.2021.112870

GeoBolivia (2020) Localizacíon de los siteos Ransar. Vicepresidencia del Estado Plurinacional de Bolivia. https://geo.gob.bo/catalogue/#/dataset/2340. Accessed July of 2024

Ginzburg R, Adámoli J, Herrera P, Torrella S (2005) Los Humedales del Chaco: Clasificación, Inventario y Mapeo a Escala Regional. INSUGEO, Miscelánea 14:121–138

Giorgis MA, Zeballos SR, Carbone L, Zimmermann H, von Wehrden H et al (2021) A review of fire effects across south American ecosystems: the role of climate and time since fire. Fire Ecol 17:1–20. https://doi.org/10.1186/s42408-021-00100-9

Gonçalves RVS, Cardoso JCF, Oliveira PE, Raymundo D, de Oliveira DC (2022a) The role of topography, climate, soil and the surrounding matrix in the distribution of Veredas wetlands in Central Brazil. Wetlands Ecol Manag 30:1261–1279. https://doi.org/10.1007/s11273-022-09895-z

Gonçalves TF, Correa CMA, Audino LD, Vaz-de-Mello FZ, Fontoura FM, Guedes NMR (2022b) Quantifying the post-fire recovery of taxonomic and functional diversity of dung beetles in the Brazilian Pantanal. Ecol Entomol 47:601–612. https://doi.org/10.1111/een.13144

Gonino GMR, Figueiredo BRS, Manetta GI, Zaia Alves GH, Benedito E (2019) Fire increases the productivity of sugarcane, but it also generates ashes that negatively affect native fish species in aquatic systems. Sci Total Environ 664:215–221. https://doi.org/10.1016/j.scitotenv.2019.02.022

Grau HR, Veblen TT (2000) Rainfall variability, fire and vegetation dynamics in neotropical montane ecosystems in North-Western Argentina. J Biogeogr 27:1107–1121

Gris D, Paixão E, Arruda RCO, Ishii IH, Marques MR, Damasceno-Junior GA (2020) Growth and establishment of monodominant stands affected by ENSO and flooding in the Pantanal. Sci Rep 10:1–13. https://doi.org/10.1038/s41598-020-60402-x

Hamilton SK, Sippel SJ, Melack JM (1996) Inundation patterns in the Pantanal wetland of South America determined from passive microwave remote sensing. Arch Hydrobiol 137:1–23

Hamilton SK, Sippel SJ, Melack JM (2004) Seasonal inundation patterns in two large savanna floodplains of South America: the llanos de Moxos (Bolivia) and the llanos del Orinoco (Venezuela and Colombia). Hydrol Process 18:2103–2116. https://doi.org/10.1002/hyp.5559

Hardesty J, Myers R, Fulks W (2005) Fire, ecosystems and people: a preliminary assessment of fire as a global conservation issue. George Wright Forum 22:78–87

Harmange C, Teles TS, Ribeiro DB, Costa AM, Godoi MN, Roque FO, Souza FL, Pays O (2024) Fire shapes mammal abundance at the Cerrado-Pantanal ecotone: Scale of effect, species traits and land-cover interaction. J Nat Conserv 82:126728. https://doi.org/10.1016/j.jnc.2024.126728.

Heinl M, Frost P, Vanderpost C, Sliva J (2007) Fire activity on drylands and floodplains in the southern Okavango Delta, Botswana. J Arid Environ 68:77–87. https://doi.org/10.1016/j.jaridenv.2005.10.023

Hilbert L, Neves EG, Pugliese F, Whitney BS, Shock M et al (2017) Evidence for mid-Holocene Rice domestication in the Americas. Nat Ecol Evol 1:1693–1698. https://doi.org/10.1038/s41559-017-0322-4

INPE Programa Queimadas (2023) BDQueimadas—Base de dados de focos de calor. Available in: http://terrabrasilis.dpi.inpe.br/queimadas/portal/faq/index.htm. Accessed 16 Oct 2023

Ivory SJ, McGlue MM, Spera S, Silva A, Bergier I (2019) Vegetation, rainfall, and pulsing hydrology in the Pantanal, the world's largest tropical wetland. Environ Res Lett 14(12):124017. https://doi.org/10.1088/1748-9326/ab4ffe

Junk WJ (2013) Current state of knowledge regarding South America wetlands and their future under global climate change. Aquat Sci 75:113–131. https://doi.org/10.1007/s00027-012-0253-8

Junk WJ, Bayley PB, Sparks RE (1989) The flood-pulse concept in river-floodplain systems. In: Dodge DP (ed) Proceedings of the international large river symposium, vol 106, pp 110–127

Kohagura C, De Souza EB, Bao F, Ferreira FA, Pott A (2020) Flood and fire affect the soil seed bank of riparian forest in the Pantanal wetland. Rodriguésia 71:e00052018. https://doi.org/10.1590/2175-7860202071013

Kubitzki K, Ziburski A (1994) Seed dispersal in flood plain forests of Amazonia. Biotropica 26:30–43. https://doi.org/10.2307/2389108

Kvist LP, Nebel G (2001) A review of Peruvian flood plain forests: ecosystems, inhabitants and resource use. For Ecol Manag 150:3–26. https://doi.org/10.1016/S0378-1127(00)00679-4

Lehner B, Döll P (2004) Development and validation of a global database of lakes, reservoirs and wetlands. J Hydrol 296:1–22

López-Arcos D, Gómez-Romero M, Lindig-Cisneros R, Zedler PH (2012) Fire-mobilized nutrients from hydrophyte leaves favor differentially Typha Domingensis seedling growth. Environ Exp Bot 78:33–38. https://doi.org/10.1016/j.envexpbot.2011.12.023

Lukac M, Pensa M, Shiller G (2011) Tree species' tolerance to water stress, salinity and fire. In: Bredemeier M, Cohen S, Godbold D, Lode E, Pichler V, Schleppi P (eds) Forest management and the water cycle. Ecological studies, vol 212. Springer, Dordrecht, pp 453–469. https://doi.org/10.1007/978-90-481-9834-4

Magioli M, Lima LHA, Villela PMS, Sampaio R, Bonjorne L, Ribeiro RLA et al (2024) Forest type modulates mammalian responses to megafires. Sci Rep 14:13538. https://doi.org/10.1038/s41598-024-64460-3

Maillard P, Pereira DB, De SCG (2009) Incêndios florestais em veredas: conceitos e estudo de Caso no Peruaçu wildfires in palm swamps: concepts and case study in the Peruaçu. Rev Bras Cartogr 61:321–330. https://doi.org/10.14393/rbcv61n4-43653

Manrique-Pineda DA, de Souza EB, Paranhos Filho AC, Encina CCC, Damasceno-Junior GA (2021) Fire, flood and monodominance of Tabebuia aurea in Pantanal. For Ecol Manag 479:118599. https://doi.org/10.1016/j.foreco.2020.118599

Maracahipes L, Marimon BS, Lenza E, Marimon-Junior BH, De Oliveira EA, Mews HA, Gomes L, Feldpausch TR (2014) Post-fire dynamics of woody vegetation in seasonally flooded forests (impucas) in the Cerrado-Amazonian Forest transition zone. Flora 209:260–270. https://doi.org/10.1016/j.flora.2014.02.008

Mataveli GAV, Pereira G, de Oliveira G, Seixas HT, Cardozo Fda S, Shimabukuro YE, Kawakubo FS, Brunsell NA (2021) 2020 Pantanal's widespread fire: short- and long-term implications for biodiversity and conservation. Biodivers Conserv 30:3299–3303. https://doi.org/10.1007/s10531-021-02243-2

Melack JM, Hess LL (2010) Remote sensing of the distribution and extent of wetlands in the Amazon Basin. In: Junk WJ, Piedade MTF, Wittmann F, Schöngart J, Parolin P (eds) . Amazonian floodplain forests: ecophysiology, biodiversity and sustainable management, Springer, Cham, pp 43–59. https://doi.org/10.1007/978-90-481-8725-6_3

Merino ER, Assine ML (2020) Hidden in plain sight: how finding a lake in the Brazilian Pantanal improves understanding of wetland hydrogeomorphology. Earth Surf Process Landf 45:440–458. https://doi.org/10.1002/esp.4745

Meza MC, Espelta JM, González TM, Armenteras D (2023) Fire reduces taxonomic and functional diversity in Neotropical moist seasonally flooded forests. Perspect Ecol Conserv 21:101–111. https://doi.org/10.1016/j.pecon.2023.04.003

Moreira SN, Pott A, Pott VJ, Damasceno-Junior GA (2011) Structure of pond vegetation of a Vereda in the Brazilian Cerrado. Rodriguésia 62:721–729. https://doi.org/10.1590/S2175-78602011000400002

Moreira SN, Eisenlohr PV, Pott A, Pott VJ, Oliveira-Filho AT (2014) Similar vegetation structure in protected and non-protected wetlands in Central Brazil: conservation significance. Environ Conserv 42:356–362. https://doi.org/10.1017/S0376892915000107

Moreira SN, Pott VJ, de Souza EB, Pott A (2017) Are *Mauritia flexuosa* L. f. Palm swamps in the Brazilian Pantanal true veredas? A floristic appraisal. Bol Mus Para Emílio Goeldi Cienc Nat 12:221–238. https://doi.org/10.46357/bcnaturais.v12i2.386

Neiff JJ (2001) Diversity in some tropical wetland systems of South America. In: Gopal B, Junk WJ, Davis J (eds) Biodiversity in wetlands: assessment, function and conservation, vol II. Backhuys Publishers, Leiden, pp 157–186

Neiff JJ, Reboratti HJ, Gorleri MC, Basualdo M (1985) Impacto de las crecientes extraordinarias sobre los bosques fluviales del Bajo Paraguay. Bol Com Especial Río Bermejo 4:13–30

Neiff JJ, Iriondo MH, Carignan R (1994) Large tropical south American wetlands: an overview. In: Proceedings of the international workshop on the ecology and management of aquatic-terrestrial ecotones, pp 156–165

Neiff JJ, Patiño CAE, Casco SL (2005) Atenuación de las crecidas por los humedales del Bajo Paraguay. In: Petean J, Cappato J (eds) Humedales fluviales de América del Sur, Hacia un manejo sustentable. Fundación Proteger, Buenos Aires, pp 261–276

Nunes da Cunha C (2001) Distribution of Woody Plant communities along the flood gradient in the Pantanal of Poconé, Mato Grosso, Brazil. Int J Ecol Environ Sci 27:63–70

Nunes-da-Cunha C, Piedade MTF, Junk WJ (eds) (2015) Classificação e Delineamento das Áreas Úmidas Brasileiras e de seus Macrohabitats, vol I. EdUFMT Publishers, Cuiabá. 166 p

Nunes EL, Costa MH, Malhado AC, Dias LCP, Vieira SA, Pinto LB, Ladle RJ (2012) Monitoring carbon assimilation in South America's tropical forests: Model specification and application to the Amazonian droughts of 2005 and 2010 Remote Sensing of Environment 117449-463 https://doi.org/10.1016/j.rse.2011.10.022

Oliveira-Filho EC, Brito DQ, Dias ZMB, Guarieiro MS, Carvalho EL, Fascineli ML, Niva C C, Grisolia CK (2018) Effects of ashes from a Brazilian savanna wildfire on water soil and biota: An ecotoxicological approach. Science of The Total Environment 618101–111. https://doi.org/10.1016/j.scitotenv.2017.11.051

Olson DM, Dinerstein E, Wikramanayake ED, Burgess ND, Powell GVN et al (2001) Terrestrial ecoregions of the world: a new map of life on earth. Bioscience 51:933–938. https://doi.org/10.1641/0006-3568(2001)051[0933:TEOTWA]2.0.CO;2

Ordano M, Bortoluzzi A, Chatellenaz M, Biancucci L (2008) Respuesta a corto plazo de las aves a la quema y al pastoreo en pastizales del Parque Nacional Mburucuyá, Argentina. INSUGEO, Miscelánea 17:425–446

Ortiz A, Marinace HM, Fisk JJ, Hankenson L, Pacheco AS, Goodman HD, Potts E, Laurance SGW, Bischoff EA, Holman VF, Love SM, Tng DYP, Apgaua DMG (2023) Shoot flammability differences between forest and savanna trees are driven by leaf dry matter content. Flora 307:152370. https://doi.org/10.1016/j.flora.2023.152370

Padovani CR (2010) Dinâmica espaço-temporal das inundações do Pantanal. PhD Thesis. Universidade de São Paulo

Parolin P, Wittmann F, Schöngart J, Nunes-da-Cunha C, Junk WJ, Piedade MTF (2022) Wetland plants: adaptations, classification, ecology and distribution. In: Mehner T, Tockner K (eds) Encyclopedia of inland waters, vol 2, 2nd edn. Elsevier, Oxford, pp 214–230

Pastro LA, Dickman CR, Letnic M (2011) Burning for biodiversity or burning biodiversity? Prescribed burn vs. wildfire impacts on plants, lizards, and mammals. Ecol Appl 21:3238–3253. https://doi.org/10.1890/10-2351.1

Pausas JG, Parr CL (2018) Towards an understanding of the evolutionary role of fire in animals. Evol Ecol 32:113–125. https://doi.org/10.1007/s10682-018-9927-6

Pivello VR, Vieira I, Christianini AV, Ribeiro DB, da Silva ML et al (2021) Understanding Brazil's catastrophic fires: causes, consequences and policy needed to prevent future tragedies. Perspect Ecol Conserv 19:233–255. https://doi.org/10.1016/j.pecon.2021.06.005

Poi ASG (2017) Biodiversidad en las aguas del Iberá, vol I. Editorial de la Universidad del Nordeste (EUDENE), Argentina, p 132

Pott A, Bao F, de Souza EB (2021) Soil seed banks in the Pantanal wetland. In: Damasceno-Junior GA, Pott A (eds) Flora and vegetation of the Pantanal wetland, 1st edn. Springer, Cham, pp 579–607. https://doi.org/10.1007/978-3-030-83375-6_15

Rivaben RC, Pott A, Bueno ML, Parolin P, Cordova MO, Oldeland J, da Silva RH, Damasceno-Junior GA (2021) Do fire and flood interact to determine forest islet structure and diversity in a Neotropical wetland? Flora 281:151874. https://doi.org/10.1016/j.flora.2021.151874

Robinne FN, Bladon KD, Miller C, Parisien MA, Mathieu J, Flannigan MD (2018) A spatial evaluation of global wildfire-water risks to human and natural systems. Sci Total Environ 610–611:1193–1206. https://doi.org/10.1016/j.scitotenv.2017.08.112

Rocha M, Santos Júnior CC, Damasceno-Júnior GA, Pott VJ, Pott A (2015) Effect of fire on a monodominant floating mat of *Cyperus giganteus* Vahl in a neotropical wetland. Braz J Biol 75:114–124. https://doi.org/10.1590/1519-6984.08613

Russell-Smith J, Lucas D, Gapindi M, Gunbunuka B, Kapirigi N, Namingum G, Lucas K, Giuliani P, Chaloupka G (1997) Aboriginal Resource Utilization and Fire Management Practice in Western Arnherm Land, Monsoonal Northern Australia: Notes for Prehistory, Lessons for the Future1. Human Ecology, 25(2): 159-195

San Martín R, Ottlé C, Sörensson A (2023) Fires in the south American Chaco, from dry forests to wetlands: response to climate depends on land cover. Fire Ecol 19(1):57. https://doi.org/10.1186/s42408-023-00212-4

Santos CF, Teixeira RC, Raizer J, Fischer E (2021) Post-fire phyllostomid assemblages in forest patches of the Pantanal wetland. Mammalia 85:155–158. https://doi.org/10.1515/mammalia-2020-0037

Sartori ALB, Pott VJ, Pott A, De Carvalho FS (2018) Checklist of angiosperm from the Chaco of Mato Grosso do Sul. Iheringia Serie Botanica 73:22–33. https://doi.org/10.21826/2446-8231201873s22

Schmidt IB, Figueiredo IB, Scariot A (2007) Ethnobotany and effects of harvesting on the population ecology of Syngonanthus nitens (bong.) Ruhland (Eriocaulaceae), a NTFP from Jalapão region, Central Brazil 1 ethnobotany and effects of harvesting on the population ecology of *Syngonanthus nitens*. Econ Bot 61:73–85. https://doi.org/10.1663/0013-0001(2007)61[73:EAEOHO]2.0.CO;2

Schmidt IB, Fidelis A, Miranda HS, Ticktin T (2017) How do the wets burn? Fire behavior and intensity in wet grasslands in the Brazilian savanna. Braz J Bot 40:167–175. https://doi.org/10.1007/s40415-016-0330-7

Schöngart J, Wittmann F, Junk WJ, Piedade MTF (2017) Vulnerability of Amazonian floodplains to wildfires differs according to their typologies impeding generalizations Proceedings of the National Academy of Sciences 114(41). https://doi.org/10.1073/pnas.1713734114

Schulz C, Whitney BS, Rossetto OC, Neves DM, Crabb L, de Oliveira EC et al (2019) Physical, ecological and human dimensions of environmental change in Brazil's Pantanal wetland: synthesis and research agenda. Sci Total Environ 687:1011–1027. https://doi.org/10.1016/j.scitotenv.2019.06.023

Scremin-Dias E, da Silva JR, Catian G, Fabiano VS, do Carmo de Oliveira Arruda R (2021) Plant morphoanatomical adaptations to environmental conditions of the Pantanal wetland. In: Damasceno-Junior GA, Pott A (eds) Flora and vegetation of the Pantanal wetland, 1st edn. Springer, Cham, pp 609–636. https://doi.org/10.1007/978-3-030-83375-6_16

Silva SS da, Oliveira I, Morello TF, Anderson LO, Karlokoski A, Brando PM, Melo AWF de, Costa JG da, Souza FSC de, Silva IS da, Nascimento E de S, Pereira MP, Almeida M RN de, Alencar A, Aragão LEOC de, Brown IF, Graça PML de A, Fearnside PM (2021) Burning in southwestern Brazilian Amazonia 2016–2019 J Environ Manag 286112189-10.1016/j.jenvman.2021.112189

Silva AC, Rech AR, Tassinari D (2022) Peatlands of southern Espinhaço Mountain range. Brazil Ecosystem Services, Biotic Interactions and Paleoenvironments. Appris Editora. https://www.editoraappris.com.br

Silveira JM, Louzada J, Barlow J, Andrade R, Mestre L, Solar R, Lacau S, Cochrane MA (2016) A multi-taxa assessment of biodiversity change after single and recurrent wildfires in a Brazilian Amazon Forest. Biotropica 48:170–180. https://doi.org/10.2307/48574841

Sletto B, Rodriguez I (2013) Burning, fire prevention and landscape productions among the Pemon, gran Sabana, Venezuela: toward an intercultural approach to wildland fire Management in Neotropical Savannas. J Environ Manag 115:155–166. https://doi.org/10.1016/j.jenvman.2012.10.041

Smith MN, Stark SC, Taylor TC, Schietti J, de Almeida DRA et al (2023) Diverse anthropogenic disturbances shift Amazon forests along a structural spectrum. Front Ecol Environ 21:24–32. https://doi.org/10.1002/fee.2590

Soares VC, Scremin-Dias E, Daibes LF, Damasceno-Junior GA, Pott A, de Lima LB (2021) Fire has little to no effect on the enhancement of germination, but buried seeds may survive in a Neotropical wetland. Flora 278:151801. https://doi.org/10.1016/j.flora.2021.151801

Tomas WM, Berlinck CN, Chiaravalloti RM, Faggioni GP, Strüssmann C et al (2021) Distance sampling surveys reveal 17 million vertebrates directly killed by the 2020's wildfires in the Pantanal. Brazil Sci Rep 11. https://doi.org/10.1038/s41598-021-02844-5

Torti SD, Coley PD, Kursar TA (2001) Causes and consequences of monodominance in tropical lowland forests. Am Nat 157:141–153. https://doi.org/10.1086/318629

Welch JR (2015) Learning to Hunt by tending the fire: Xavante youth, Ethnoecology, and ceremony in Central Brazil. J Ethnobiol 35:183–208. https://doi.org/10.2993/0278-0771-35.1.183

Wittmann F, Schöngart J, Montero JC, Motzer T, Junk WJ, Piedade MTF, Queiroz HL, Worbes M (2006) Tree species composition and diversity gradients in white-water forests across the Amazon Basin. Journal of Biogeography 33(8):1334–1347. https://doi.org/10.1111/jbi.2006.33.issue-8; https://doi.org/10.1111/j.1365–2699.2006.01495.x

Fire in Mediterranean Ecosystems of South America

11

Susana Gómez-González, Susana Paula, Alejandro Miranda, Mabel Ortega, and Korina Ocampo-Zuleta

Abstract

The South American Mediterranean region is located in central Chile (23–39°S) and has a climate with hot dry summers and mild rainy winters. There, a typical sclerophyllous vegetation develops, called *matorral*. Although central Chile has not had a marked natural fire regime during the Quaternary, anthropogenic fires have been highly prevalent in the last two centuries associated with agricultural and livestock expansion. More recently, extensive and homogeneous forest plantations have contributed to the occurrence of large fires in a context of extended

S. Gómez-González (✉)
Departamento de Biología-IVAGRO, Universidad de Cádiz, Puerto Real, Spain

Centro de Ciencia del Clima y la Resiliencia, (CR)2, Santiago, Chile
e-mail: susana.gomez@uca.es

S. Paula
Facultad de Ciencias, Instituto de Ciencias Ambientales y Evolutivas, Universidad Austral de Chile, Valdivia, Chile

Instituto de Ecología y Biodiversidad (IEB), Concepción, Chile
e-mail: spaula.julia@uach.cl

A. Miranda
Centro de Ciencia del Clima y la Resiliencia, (CR)2, Santiago, Chile

Laboratorio de Ecología del Paisaje y Conservación, Departamento de Ciencias Forestales, Universidad de La Frontera, Temuco, Chile
e-mail: alejandro.miranda@ufrontera.cl

M. Ortega
Departamento Desarrollo e Investigación, Gerencia Protección contra Incendios Forestales, Corporación Nacional Forestal, Santiago, Chile
e-mail: mabel.ortega@conaf.cl

K. Ocampo-Zuleta
Instituto de Ecología y Biodiversidad (IEB), Concepción, Chile
e-mail: korinaocampo@udec.cl

A. Fidelis, V. R. Pivello (eds.), *Fire in the South American Ecosystems*, Ecological Studies 250, https://doi.org/10.1007/978-3-031-89372-8_11

droughts. These anthropogenic fires put biodiversity at risk, as most native species are not fire-adapted. Frequent fires increase the mortality of the woody layer, creating open spaces that are colonized by annual species, mostly exotic. The matorral require active restoration measures in the face of recurrent fires, and more research is needed to determine the conditions that favor their recovery. The large fires that have recently affected the Mediterranean region of Chile offer an opportunity for restoration and the design of more resilient landscapes that integrate biodiversity conservation, ecosystem service provision, and fire protection in a context of climate change.

Resumen

La región mediterránea de Sudamérica se encuentra en Chile central (23–39°S) y presenta un clima con veranos secos y cálidos e inviernos templados y lluviosos. Allí se desarrolla una vegetación típica esclerófila llamada *matorral*. Aunque Chile central no ha tenido un régimen natural de incendios marcado durante el Cuarternario, los incendios antrópicos han tenido una gran prevalencia en los últimos dos siglos asociado a la expansión agrícola y ganadera. Más recientemente, las plantaciones forestales extensas y homogéneas han contribuido a la ocurrencia de grandes incendios en un contexto de sequía prolongada. Estos incendios antrópicos ponen en riesgo la biodiversidad, ya que la mayoría de especies nativas no están adaptadas al fuego. Los incendios frecuentes incrementan la mortalidad del estrato arbóreo y arbustivo, generando espacios abiertos que se colonizan por especies anuales, mayoritariamente exóticas. El matorral requiere medidas de restauración activas cuando los incendios son recurrentes, pero se necesita mayor investigación para determinar las condiciones que favorezcan su recuperación. Los grandes incendios que han afectado recientemente la región mediterránea de Chile ofrece una oportunidad para la restauración y el diseño de paisajes más resilientes, que integren la conservación de la biodiversidad, la provisión de servicios ecosistémicos y la protección de incendios en un contexto de cambio climático.

Resumo

A região mediterrânea da América do Sul está localizada no centro do Chile (23–39°S) e tem um clima com verões quentes e secos, e invernos amenos e chuvosos. Ali se desenvolve uma típica vegetação esclerófila, chamada *matorral*. Embora a região central do Chile não tenha tido um regime de incêndios naturais marcantes durante o Quarentenário, os incêndios antropogênicos foram altamente prevalentes nos últimos dois séculos, associados à expansão da agricultura e da pecuária. Mais recentemente, plantações florestais extensas e homogêneas contribuíram para a ocorrência de grandes incêndios, num contexto de seca prolongada. Esses incêndios antropogênicos colocam a biodiversidade em risco, pois a maioria das espécies nativas não está adaptada ao fogo. Os incêndios frequentes aumentam a mortalidade do estrato arbóreo e arbustivo, gerando espaços abertos que são colonizados por espécies anuais, em sua maioria exóticas. O *matorral* exige medidas ativas de restauração quando os incêndios são recorrentes, sendo necessárias mais pesquisas para determinar as condições que favorecem sua recuperação. Os grandes incêndios recentes na região mediterrânea do Chile oferecem uma oportunidade para a restauração e o planejamento de paisagens mais resilientes, que integrem a conservação da biodiversidade, o fornecimento de serviços ecossistêmicos e a proteção contra incêndios num contexto de mudança climática.

11.1 The Mediterranean Region of South America

The Mediterranean regions of the world are characterized by presenting cold, wet winters that alternate with hot, dry summers. During the rainy months, the productivity of the vegetation increases, which, when it dries in the summer, becomes flammable, making these ecosystems prone to summer fires (Keeley et al. 2012). Mediterranean ecosystems are represented in five regions of the world; the Mediterranean Basin, California, southwest Australia, the Cape region in South Africa, and central Chile, and occupy approximately 1.2% of the Earth's surface (Olson et al. 2001). According to Luebert and Pliscoff (2006), the Mediterranean region of Chile is distributed between 23° and 39°S; it is restricted to the coastal strip in its southernmost part, but from 25°S it progressively extends inland, comprising both mountain ranges (the Coast and the Andes) and the central valley that separates them between 31 and ca. 37°S; towards the northern end, the distribution narrows progressively, being mainly confined to the central valley (Fig. 11.1).

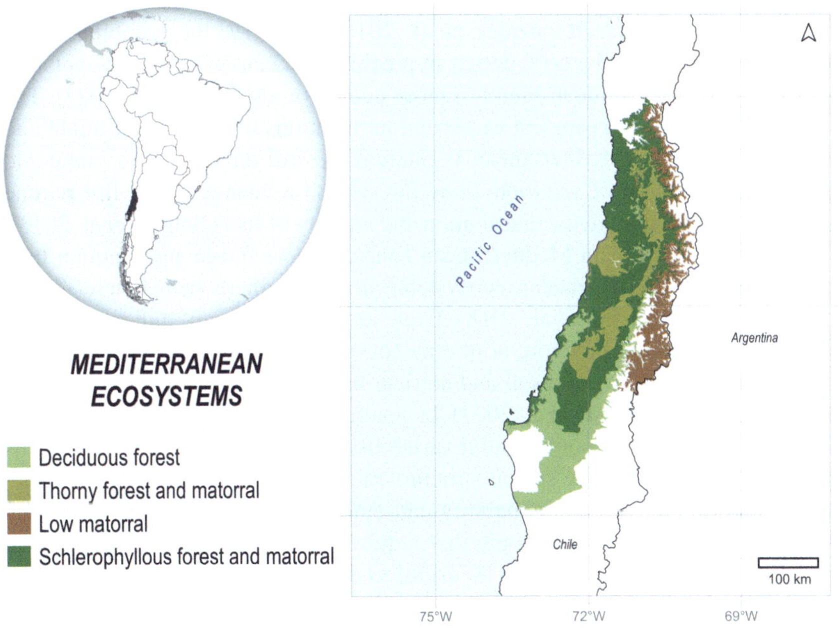

Fig. 11.1 Potential distribution of the main Mediterranean vegetation formations in South America, according to Luebert and Pliscoff (2006). Produced by Júlia Abrantes Rodrigues

However, it is important to consider that other authors limit the Mediterranean ecoregion of Chile to latitudes that range from 30°S to 37°S; that is, excluding the more arid extremes of the north and the more humid ones of the south (e.g., Di Castri 1973; Rundel 1981; Arroyo et al. 1995; Armesto et al. 2007).

In central Chile, annual precipitation varies between 250 and 700 mm and is concentrated mainly in the months of the austral autumn and winter (i.e., from May to August). The topographic heterogeneity of the region, with two parallel mountain ranges separated by a central valley, also plays an important role in the distribution of precipitation, as higher areas can receive more rain than low areas (Arroyo et al. 1995). During the summer, the region experiences a prolonged drought, with maximum temperatures ranging between 25 and 30 °C (exceptionally 35–40 °C). In contrast, in winter, the temperature drops considerably, with maximums around 10–15 °C, but without prolonged frosts. In terms of spatial variability, coastal regions have more moderate temperatures due to the influence of the Pacific Ocean, while temperatures tend to be more extreme in the central valley and in the Andean mountain range.

In the last decade, the central area of Chile has experienced a significant climate change, which has been partly attributed (~25%) to anthropogenic activities (Boisier et al. 2016; Garreaud et al. 2017). Since 2010, annual precipitation has been 30% lower compared to the historical record average, a deficit that has been maintained until 2022 without occurrence of wet years (Miranda et al. 2023). Given its severity and duration, this dry event is considered as a mega-drought (Garreaud et al. 2017). In addition, the mega-drought has been accompanied by an anomalous increase in average temperatures and heatwaves, enhancing water stress and desiccation of vegetation (Boisier et al. 2016; Garreaud et al. 2017) which, in turn, has favored the occurrence of mega-fires (González et al. 2018). In 2019, the combination of extreme temperatures and a 60% deficit in precipitation caused massive vegetation die-off in much of the Chilean Mediterranean ecoregion (Miranda et al. 2023). The medium and long-term ecological and social implications of the mega-drought and the massive die-off of Mediterranean vegetation are still unknown, but one could expect a *shrubification* of sclerophyllous forests and a change in the fire regime associated with variation in the distribution and identity of fuel (Batllori et al. 2013).

The soils of the Chilean Mediterranean zone are more fertile and younger than the soils of other Mediterranean ecosystems in the southern hemisphere (South Africa and Australia, Esler et al. 2018). Even so, they are poorer, thinner, and less deep soils, compared to the more northern ecosystems of Chile. Thus, erosion and torrential rains can cause rapid soil and nutrient loss in areas with steep slopes, particularly after fires (Armesto et al. 2007). In general, these soils have low acidity, but within the region there is some spatial variability in soil types (Sierra 1982). This variability can determine the spatial structure of vegetation at the landscape scale, have implications for plant flammability and potentially in the fire regime (Ojeda et al. 2010). However, this is a topic that requires further research in the Chilean Mediterranean zone and that could be useful to determine areas of higher fire risk and that require specific management strategies.

The vegetation of Mediterranean Chile is called the *matorral* in the broad sense, but different types of vegetation can be found, whose distribution and composition can vary depending on factors such as climate, altitude, and soil type. The vegetation formations described by Rundel (1981) and Arroyo et al. (1995) consider the following main groups: the sclerophyllous matorral (32–36°S), which extends along both mountain ranges (Coastal and Andes); the savanna of *Vachellia caven* (30–36°S) along the Central Valley; the coastal matorral, which is found on the coastal edges between 30–36°S; the Andean montane forest (32–36°S) on the medium altitude slopes of the Andes; and the forest of *Nothofagus* spp. (33–36°S) in the southern transition Mediterranean-temperate zone. Some more unique types of vegetation within the Mediterranean region are noteworthy, such as the relict palm forests (*Jubaea chilensis*) found on the slopes and deep canyons in the Valparaíso Region (33–34°S) and the forests of the endemic and monotypic species *Aextoxicon punctatum,* distributed along the Coastal Range from 30°S (Armesto et al. 2007).

On the other hand, the approach of Luebert and Pliscoff (2006) takes into account the bioclimatic floors to determine the units of potential vegetation (Fig. 11.1). This approach establishes numerous very specific types of vegetation, which we have grouped into more general formations, and we will refer to them throughout this chapter (Fig. 11.1). Under this scheme, the Mediterranean zone of Chile would present low shrubs in the elevated areas, mainly along the Andean mountain range, between 31–36°S; sclerophyllous forests and shrubs between 31–37°S, distributed along the coast, on the inner slopes of the Coastal mountain range and the Andean foothills, dominated by species such as *Lithraea caustica, Quillaja saponaria, Vachellia caven, Cryptocarya alba, Peumus boldus*; thorny forests and shrubs in the central valley between 32–36°S and in the coastal zone that joins the regions of O'Higgins and Valparaíso; and finally, the deciduous forests of trees of the genus *Nothofagus* spp. in the Coastal mountain range and the Andean foothills between 35–37°S, extending through the central valley up to 39°S. Although Luebert and Pliscoff (2006) consider within the Mediterranean region various types of arid shrubs that occupy the coastal zone between 23° and 25°S, as well as some types of Andean grasslands, for practical purposes, we will not consider them in this chapter (Fig. 11.1), as there is no information on fire ecology in these areas and fires are anecdotal.

The Mediterranean region of Chile is considered a global biodiversity hotspot (Myers et al. 2000; Trew and Maclean 2021), both for the uniqueness and diversity of its flora (50% of its vascular plants are endemic; Arroyo et al. 2006), as well as its degree of threat, given that this area concentrates most of the country's population (approx. 80%) and is where most of the economic activities take place. This landscape harbors <30% of the country's native forests, but it is the area with the highest biodiversity. The plants have special adaptations to survive in conditions of water and nutrient scarcity, such as sclerophyllous or thorny leaves, deep roots, and water storage systems in stems and leaves (Montenegro et al. 2003; Armesto et al. 2007). High species diversity, high endemism and prevalence of traits adaptive to drought are characteristics of the vegetation shared with the other Mediterranean regions of the world (Cowling et al. 1996). However, the Mediterranean zone of Chile presents distinctive relevant features. One of them is the low frequency of natural fires during the Quaternary, associated with the blocking of thunderstorms by the Andes (see Sect. 11.3). This has determined that the adaptive responses to fire are less frequent compared to other Mediterranean ecosystems, where fire has indeed been a selective key factor in shaping the traits of the flora (see Sect. 11.4). Another difference is the relictual character and the lower diversity of its flora compared to other Mediterranean regions of the planet. It has been proposed that this could correspond to higher extinction rates in the pre-Pliocene lineages due to lower climatic and topographic stability (Cowling et al. 2015). Finally, another peculiarity of the Chilean Mediterranean flora is that it is relatively poor in annual species (Arroyo et al. 1995). The combination of these factors (flora poorly adapted to fire and poorly represented annual flora), could explain why the Mediterranean matorral is a very vulnerable ecosystem to the invasion of exotic annual species after fire (see Sect. 11.7).

The natural vegetation has been transformed by human activities (mainly agricultural, livestock and forestry), changing land use and landscape configuration rapidly over the last 100 years. The agricultural use, distributed along the central valley, has replaced a large part of the shrubs and thorny forests, while the forest plantations of *Pinus* and *Eucalyptus* spp. have been established in areas previously occupied by deciduous Mediterranean forests and coastal sclerophyllous forests (Miranda et al. 2017; Heilmayr et al. 2016; Fig. 11.2a). Changes in land use have been accompanied by the invasion of exotic species, which today dominate the herbaceous layer of most shrubs in Chile, as well as the edges of roads and forest areas (Figueroa et al. 2004; Pauchard and Alaback 2004). Such alterations in the landscape have changed the amount, quality, and configuration of fuel, favoring the occurrence of large forest fires, particularly in a context of drought derived from anthropogenic climate change (see Sects. 11.5 and 11.7).

11.2 Fire Regime

The administrative regions that occupy the Mediterranean zone in Chile (from Valparaíso to Araucanía) concentrate 90% of all fires that happened in the country and 86% of the total area burned between 1977–2022 (CONAF 2023; Fig. 11.2b). Therefore, the fire statistics calculated at the country level strongly correspond with what is observed for this region. According to these statistics, 99.7% of the forest fires that occur in Chile are of anthropogenic origin (intentional or accidental) and only 0.3% have natural causes. In the last 20 years, there have been 464 fires due to natural causes throughout the country, mainly concentrated between the regions of Maule (34°S) and La Araucanía (39°S). Most of these natural fires are caused by lightning (81%), with fires started by volcanic activity being episodic (0.4%). Due to lack of information, 18% of the natural fires are not classified. Therefore, Chile's fire regime is characterized by having a strong relationship with human activities and an almost negligible natural component (see Sect. 11.3). The severity of the

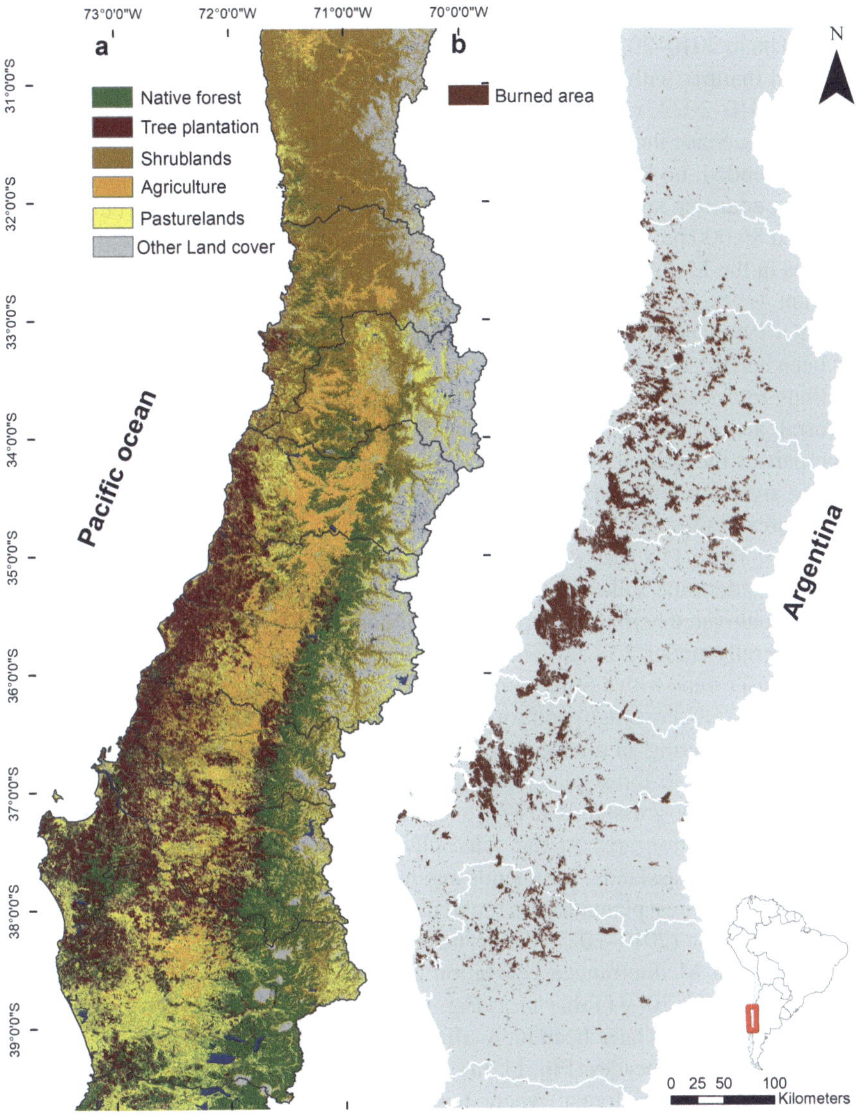

Fig. 11.2 (**a**) Land cover for the year 2014 (based on Zhao et al. 2016). (**b**) Fire scars for the period 1985–2018; these data represent about 60% of the fires that occurred in the period, since the record is incomplete between the years 1985 and 2000 (Based on Miranda et al. 2022)

fires depends on the type of vegetation affected. Generally, in shrub and thorn areas, there are surface fires with low severity, while sclerophyllous forests and forest plantations can suffer quite severe crown fires (CONAF 2017).

The historical average of fire occurrences is 5611 each year, with a burned area of 68,060 hectares annually. In the last decade, the burned area has practically doubled in relation to the 1970s and 1980s (53,600 ha average in 1977–1990 vs. 115,350 ha in 2010–2022). The occurrence of fires has also increased over time in a sustained manner with 4509 average fires in 1977–1990, 5846 in 1990–2010 and 6502 in 2010–2022, reaching its peak in the 2019–2020 seasons with more than 8000 events. Since the mega-drought began in 2010, the fire season has also significantly extended, no longer being exclusive to the summer months (González et al. 2018). In addition, since 2010 there have been nine seasons where the burned area exceeded 80,000 ha; an unprecedented event since records have been kept.

Fires in the Mediterranean zone of Chile are not distributed equitably along the different types of potential vegetation (Luebert and Pliscoff 2006; Fig. 11.1). Between 1985 and 2022, within the study area (Fig. 11.1) a total of 1.9 million ha has been burned, of which 50% is located within the potential distribution of sclerophyllous forest and matorral, 28% in deciduous forest, 22% in thorny forest and matorral, and only 0.1% in low matorral (Table 11.1). Considering that currently a large part of the natural vegetation has been replaced by other land uses, such as grazing areas, agricultural or forestry crops, we have explored the contribution in burned area of these uses over time, within each area of potential natural vegetation (Fig. 11.3; Table 11.1). To do this, we used the land use classification of CONAF, which considers the following categories: grasslands (areas dominated by native, exotic or cultivated herbaceous species); scrublands (areas dominated by shrub species, generally under 2 m in height, considering both native and exotic species); native forest (areas with a minimum of 0.5 ha, dominated by native trees and with a minimum of 10% coverage) and forest plantations.

When considering all potential vegetation areas, the land cover most affected by fires corresponds to forest plantations, representing 37% of the total, followed by scrublands (27%), grasslands (21%) and native forest (15%) (Table 11.1). In the potential distribution area of the sclerophyllous forest and matorral, the largest affected area of grasslands, scrublands and native forests is found, while the largest affected area of forest plantations is found in the area of deciduous forest (Table 11.1). In the last decade (2010–2022) the burned area has increased in all potential distribution areas of Mediterranean vegetation and for all land covers compared to the period from 1985–2009 (Table 11.1; Fig. 11.3a–d). Except for some specific years, the forest plantations have been more affected by fire than native forests, especially during the last two decades (Fig. 11.3a). This difference is particularly noticeable in the area of deciduous forest, where forest plantations are the most consistently affected land use over time (Fig. 11.3d). Here, it is important to consider that forest plantations have a smaller area than natural forests at the national level (3 Mha vs. 14 Mha, respectively), so the proportion of affected area of plantations is higher (see González et al. 2020). In the areas of sclerophyllous forest and matorral, and thorny forest and matorral, the most affected uses during the first 25 years of the study period are scrublands and grasslands, but in the last decade, forest plantations and native forest have gained greater relevance, largely due to the effect of the megafires of 2017 (Fig. 11.3b, c). The largest net increase in burned area over time has been observed in the area of sclerophyllous forest and matorral, even when the largest proportional increase has been in the formation of the thorny forest and matorral

Table 11.1 Burned area of each land use in each type of potential vegetation form (Luebert and Pliscoff 2006). Total values and annual averages are shown in different time periods

	Total burned area (ha) (1985–2022)	Mean annual burned area (ha) (1985–2022)	Mean annual burned area (ha) (1985–2009)	Mean annual burned area (ha) (2010–2022)
Low matorral				
Forest plantation	20	1	0	2
Scrublands	984	26	28	22
Grasslands	925	24	18	36
Native forest	517	14	7	25
Thorny forest and matorral				
Forest plantation	149,838	3842	522	10,523
Scrublands	91,397	2405	1493	4160
Grasslands	117,630	3096	2308	4611
Native forest	53,464	1407	432	3281
Schlerophyllous forest and matorral				
Forest plantation	268,313	7061	3125	14,629
Scrublands	294,303	7745	6395	10,340
Grasslands	206,407	5432	5240	5800
Native forest	167,015	4395	3185	6722
Deciduous forest				
Forest plantation	273,427	7195	4362	12,645
Scrublands	129,260	3402	1610	6847
Grasslands	65,023	1711	1117	2854
Native forest	64,434	1696	774	3468
Total				
Forest plantation	691,599	18,099	8008	37,799
Scrublands	515,944	13,577	9526	21,370
Grasslands	389,985	10,263	8683	13,300
Native forest	285,429	7511	4399	13,496

(52% vs. 79% increase, respectively; Table 11.1; Fig. 11.3b, c). These patterns suggest that management measures aimed at preventing and fighting forest fires should be focused differently depending on the potential vegetation areas that exist within the Mediterranean region of Chile.

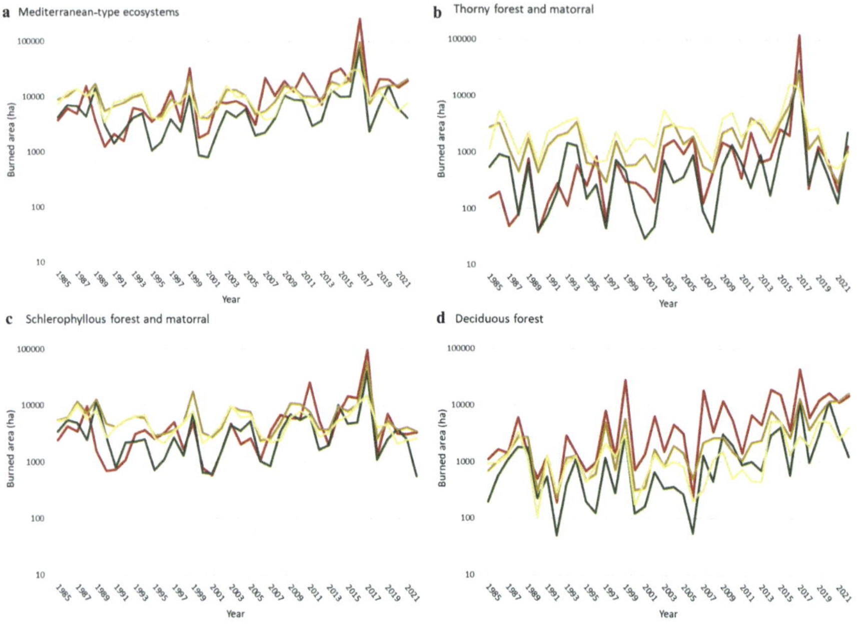

Fig. 11.3 Burned area (represented on a logarithmic scale) by type of cover in the study area (**a**) and in each of the different potential distributions of Mediterranean vegetation forms (**b**–**d**). Red line: forest plantations. Brown line: scrublands. Yellow line: grasslands. Green line: native forest. Data extracted from the public database of the National Forestry Corporation (CONAF 2023)

If we consider the climate projections for the Mediterranean region of Chile, an increase in heatwaves and droughts is expected in the coming decades (González et al. 2020), which could exacerbate the growing trend of fires that has been observed in recent years (Fig. 11.3a). Various studies have shown a direct relationship between temperatures (average and maximum), droughts, and fires in central Chile (e.g., González et al. 2018; Gómez-González et al. 2019). Given this known interaction of fires with climatic conditions, it is possible to project the future fire risk for different areas of the country. In the context of the ARCLIM project, commissioned by the Ministry of the Environment, the change in climate risk was estimated considering the changes in climate variables projected for the year 2050. According to this analysis, an increase in fire risk is expected in much of the Mediterranean area (Fig. 11.4), especially along the central valley and in the Bío-Bío region (Fig. 11.4). In the southern part of the Mediterranean region, plant productivity is higher due to the greater abundance of precipitation, so this increase in future risk could lead to the occurrence of large-scale fires. A possible long-term scenario (2070–2090) is that, under conditions of continued droughts and increasingly higher temperatures, Mediterranean vegetation could change to give way to scattered semi-arid shrubs that would be less prone to fires due to more limited fuel (Batllori et al. 2013). However, it is still necessary to generate more robust databases and models that incorporate climate uncertainty in predicting future fire risk.

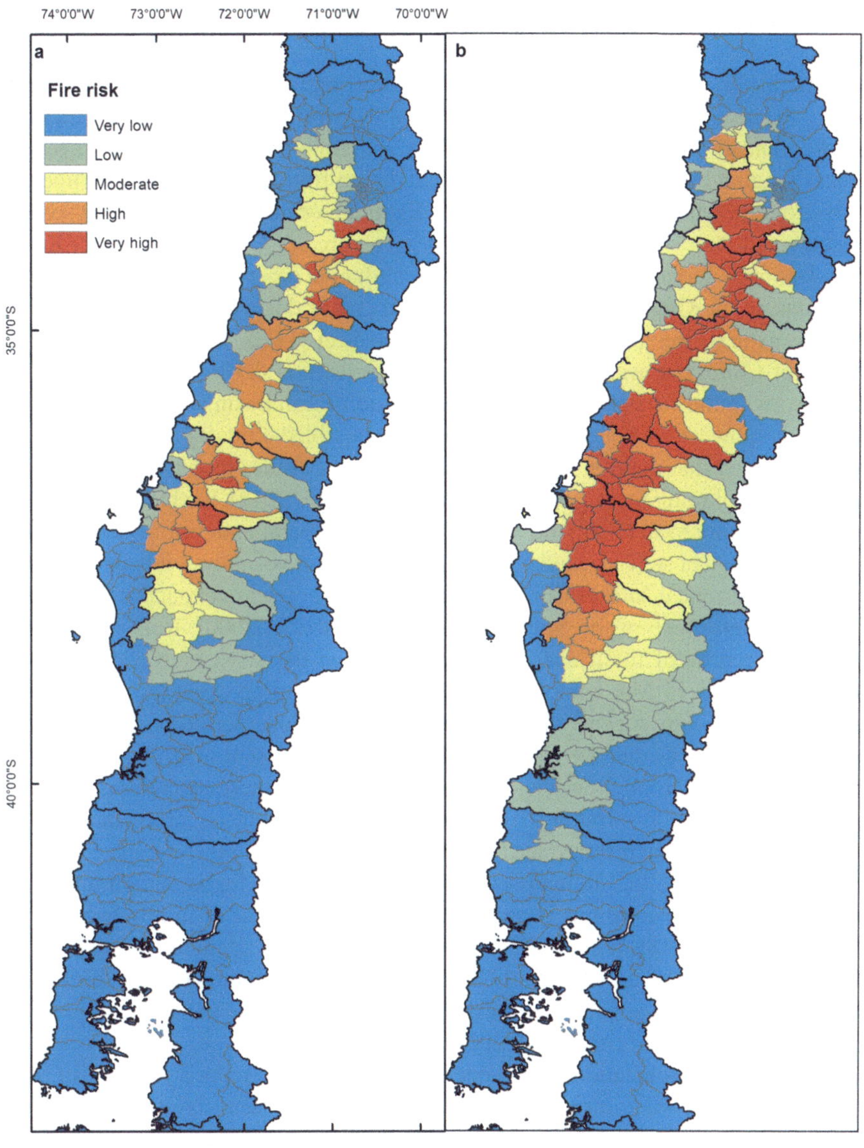

Fig. 11.4 Estimation of present (**a**) and future (**b**) fire risk in the central-southern region of Chile (based on climate risk models; ARCLIM; https://arclim.mma.gob.cl/). The risk of forest fires in the future climate (2035–2065 under the RCP 8.5 scenario) is quantified as the multiplication of the climate threat, the estimated sensitivity in relation to the probability of ignition, and the exposure of forest ecosystems in the various communes of the country

11.3 Fire History

The fire history of the Neotropical Mediterranean region differs significantly from that of the rest of the Mediterranean regions of the world. Except in South America, Mediterranean regions are characterized by a regime of natural fires with recurrences ranging between 10 and 60 years and dating from (at least) the late Pliocene (3.6–2.6 Ma), when high temperatures and lightning occurred during the dry season (Keeley and Pausas 2022). In the Mediterranean area of Chile, on the other hand, thunderstorms seem to have been less frequent than in other Mediterranean ecosystems during the Quaternary (Keeley et al. 2012) and this pattern continues today, making natural fires practically anecdotal (see Sect. 11.2). In the Mediterranean region of Chile, the average annual lightning density is 0.5 lightning/km^2, which contrasts with the 4.5 lightning/km^2 of the Mediterranean Basin or the 4.9 lightning/km^2 of southwestern South Africa (S. Paula, unpublished data). Although it has been proposed that the rain shadow generated by the Andes mountain range could explain the lightning regime in the region (Keeley et al. 2012), the blocking of eastern low-pressure systems by the South Pacific subtropical anticyclone seems to be the most likely cause (Garreaud et al. 2010). Regarding volcanism, it is quite scarce in the Mediterranean area of Chile since ca. 10 Ma, but it was very active during the Oligocene and much of the Miocene (Gregory-Wodzicki 2000). In fact, there are abundant records of fossil charcoal and pumice on the coast of Navidad (33°S) between 17 and 15 Ma, coinciding with a global warming event (Abarzúa et al. 2016). These results suggest the existence of (at least) one period of frequent fires associated with volcanic activity during the Neogene in central Chile. It is worth noting that the sclerophyllous element of the Chilean Mediterranean flora, in which species capable of resprouting from lignotubers predominate, originates at this moment in geological history (see Sect. 11.4).

The study of the first human settlements in the Tagua-Tagua lagoon and later settlements in other localities in central Chile reveals the use of fire by Indigenous Peoples (IP) in domestic and minor productive activities (Heusser 1994; Bustos-Schindler et al. 2010; Jackson et al. 2012). However, it is estimated that this use of fire was minor, or its impact was not significant, probably due to the sparse population in the territory (Aravena et al. 2003).

The Spanish colonization (1536) deeply marks the contemporary history of fires in central Chile, which increase exponentially to the present day (Aravena et al. 2003; Villa-Martínez et al. 2003; Hernández-Riquelme 2014). In the context of the constant hostilities that existed between the Indigenous Peoples and the Spanish, the burning of the Mediterranean matorral became an important part of the military tactics used by the native peoples (Del Pozo 2013). Later, during the colony, the use of fire was quite common for forest clearings. Father Felipe Gómez de Vidaurre (cited by Elizalde 1970), in the "History and Natural and Civil Geography of the Kingdom of Chile" of 1748, points out: "The heat comes from the very bad practice of setting fire to the forests in order to save the trouble of cutting them down to have new lands".

It is important to note, in addition, that mining activity developed in the northern area, intensifying in the eighteenth century, may also have given rise to burns or clearings to extract timber. The legislation of the colony allowed the free exploitation of the forest for the benefit of the mine (Camus 2004). This right was known as the Denunciation of Forests, and it ruled until 1873 (Gallardo 2013). Therefore, the colony was a time of agricultural and mining expansion, as well as of fires in the central area of Chile.

After independence (1810), Chilean agricultural exports were inserted into international trade circuits (California and Australia). In this context, extensive clearing involved the burning of thousands of hectares of native forests and shrubs to allow for the massive cultivation of wheat (Otero 2006). The splendor of Chile's granary lasted about 30 years, and it is estimated that by 1915 about 13 million hectares of forests had been burned (Otero 2006). The consequence of repeated fires was the erosion of 2.85 million hectares in the Coastal Mountain Range (IREN, 1965). The pressure on the sclerophyllous forests during this period was very high and probably fire played a fundamental role in the patchy structure they present today.

In response to the widespread degradation of the forests, in 1925 the Forest Law was enacted (modified in 1931) which, among other issues, prohibits "burning by clearing" and establishes the need to create a regulation for its use. In 1980, Supreme Decree 276 was enacted which delimits and orders the use of fire in the form of "controlled burning" and for agricultural, livestock or preferably forestry lands.

In conclusion, although the Mediterranean area of South America differs from other Mediterranean regions of the world due to the lesser presence of fire on an evolutionary time scale, the existence of a regime of frequent and severe natural fires during the Miocene is quite plausible. The occurrence of natural fires during the Quaternary is also quite probable, although this era is marked by the appearance and expansion of anthropogenic fires that have significantly shaped the structure of the landscape and vegetation.

11.4 Fire Ecology

The development of fire ecology studies in Chile began between the late 1970s and early 1990s, and it was marked by the works of Juan Armesto, Julio Gutiérrez, Eduardo Fuentes, Gloria Montenegro, Rodrigo Villaseñor and their collaborators (e.g., Armesto and Gutiérrez 1978; Montenegro et al. 1983; Fuentes and Espinosa 1986; Villaseñor and Sáiz 1990). Their studies showed the first evidence indicating that the Mediterranean matorral of South America had some resistance to fires, mainly through the regrowth of woody plants and geophytes. Despite this, field observations showed it would not be a type of community typically adapted to fire, as the recruitment of tree and shrub seedlings was quite scarce. This implies a key ecological difference compared to other Mediterranean ecosystems in the world, due to the lower incidence of natural fires during the Quaternary in the central region of Chile (see Sect. 11.3). More recent studies have confirmed the negative effects of fire on the native seed bank, but they also show the existence of certain adaptive responses to fire previously unreported for the Chilean Mediterranean flora such as, for example, germination stimulated by smoke or heat (Gómez-González et al. 2008, 2017; Figueroa et al. 2009). This information has not yet been considered in the ecological theory of Mediterranean ecosystems, so it is still assumed that in the matorral there are no germinating species (e.g., Esler et al. 2018). Incorporating this new knowledge about the Chilean matorral into the ecological theory of fire is key to understanding its role in the evolution of Mediterranean floras and to generate more appropriate management proposals. That is why in this section we will compile all the evidence we have found about adaptive (and negative) responses to fire in the Mediterranean region of South America. Whether these adaptive responses have been selected by fire or by other selective factors is a question that still requires further research (but see Gómez-González et al. 2011a, 2016).

11.4.1 Seed Germination

In the Mediterranean region of Chile, the frequency of species with fire-stimulated germination is relatively low (Gómez-González et al. 2017). When compiling the updated information from studies that have evaluated the effect of smoke and/or heat on the germination of the woody species of the matorral, only 16% of 38 evaluated native species have been found to have germination stimulated by any of these fire signals. This value contrasts with what is observed in other Mediterranean regions of the world, where the frequency of species stimulated by fire can vary between 40% and 77% (see Gómez-González 2018 and cited references). In any case, these percentages should be taken with caution, since the evaluated species represent a small part of the Mediterranean floras, included among the most diverse in the world (Cowling et al. 1996; Grobler and Cowling 2022).

Among the native woody species with germination stimulated by the exposure to smoke are *Vachellia caven* (Fabaceae), *Baccharis vernalis* (Asteraceae) and *Trevoa quinquenervia* (Rhamnaceae) (Gómez-González et al. 2008). *Vachellia caven* can be observed germinating profusely after mild fires in the matorral (Fig. 11.5a). This species dominates in the savannas that are located along the central valley, in areas that have high pressure from fires and herbivores. There are also woody species whose germination is stimulated by heat (100 °C, 5 min), such as *Lithraea caustica* (Anacardiaceae), *Muehlenbeckia hastulata* (Polygonaceae), *Trevoa trinervia* (Rhamnaceae), and the legumes *Otholobium glandulosum*, *Senna candolleana* and *Sophora macrocarpa* (Gómez-González et al. 2017). Just like *Vachellia caven*, seedlings of *Lithraea caustica*, *Muehlenbeckia hastulata* and *Trevoa trinervia* have been observed recruiting after low severity fires (Keeley and Johnson 1977; Muñoz and Fuentes 1989; Segura et al. 1998). However, there are no observations of post-fire recruitment in the field for the case of the three legume species. This suggests that there could be other post-germination factors that are limiting their recruitment in burned areas or that the abundance of seeds in the soil is very scarce. In fact, the seed bank of woody species is very scant in the Mediterranean region of Chile (Figueroa et al. 2004). Recruitment and resprouting after fire has been observed for

Fig. 11.5 Examples of adaptive responses to fire in the Mediterranean flora of Chile. (**a**) Seedlings of *Vachellia caven* after a low severity fire (Talca, Maule Region). (**b**) Seedling of *Dichondra sericea* in a patch burned with very high severity (Camino Lo Orozco, Valparaíso Region). (**c**) Dense patch of *Loasa tricolor* (native) and *Fumaria* sp. (exotic) after a high severity fire (Quintero, Valparaíso Region). (**d**) Fire-stimulated flowering in *Zephyrantes advena* (Camino Lo Orozco, Valparaíso Region). (Photos: S. Gómez-González)

the legume *Adesmia bijuga*. This species is in critical danger and more information about its ecology and relationship with fire is needed for its conservation (Gómez et al. 2021). On the other hand, there are many native woody species (45% of the species evaluated; Gómez-González 2018) that, while their germination is not stimulated by fire, they are resistant to heat and/or smoke (e.g., *Baccharis linearis* or *Colliguaja odorifera*) and, therefore, can also recruit in burned areas (Keeley and Johnson 1977; Segura et al. 1998; Becerra et al. 2022).

Regarding native herbaceous species, there is very little knowledge on the effect of fire on their seeds survival and germination. Much of the evidence comes from seed bank studies, approached with very diverse methodologies, such as the experimental burning of litter on soil collected in the field (Gómez-González and Cavieres 2009), the application of smoke and heat to soil samples (Figueroa et al. 2009; Figueroa and Cavieres 2012), or the collection of soils in burned and unburned areas (Gómez-González 2008; Gómez-González et al. 2011b). In these studies, the increase in the emergence of seedlings of at least 14 native herbs has been observed. Some of them form dense patches in burned areas, such as for example, *Helenium aromaticum* (Asteraceae), *Bromus berteroanus* (Poaceae), *Plagiobothrys fulvus* (Borraginaceae), *Dichondra sericea* (Convolvulaceae; Fig. 11.5b) and *Loasa tricolor* (Loasaceae; Fig. 11.5c) (S. Gómez-González, personal observation). In the case of *Helenium aromaticum,* seeds resist well to high temperatures and smoke, but these signals do not increase their germination (Gómez-González et al. 2011c). Despite this, it is found monopolizing the herbaceous layer after fire in some shrub communities, so it is possible that it presents some type of allelopathy that allows it to dominate the burned patches. It has been shown that anthropogenic fires have generated evolutionary changes in the characteristics of this species seeds (e.g. greater pubescence), giving them greater resistance to high temperatures (Gómez-González et al. 2011a, 2016). It is still unknown if this type of process has occurred in other native species of the matorral.

Unlike what happens in other Mediterranean ecosystems, in Central Chile, the presence of fire-dependent seeders, which are species that only germinate after fire, has not been documented so far (Keeley et al. 2012). This fact, along with the low frequency of species with fire-stimulated germination, support the idea that the Mediterranean vegetation of South America would not be adapted to fire. In this sense, it is worth noting that most species with heat-stimulated germination have hard seeds, so it is very likely that their ability to germinate post-fire is an exaptation to fire, that is, a positive response to fire that could have been selected by another environmental pressure such as, for example, endozoochory (dispersion through animals that ingest the fruits), as has been proposed for the species of the Argentine Chaco (Jaureguiberry and Díaz 2015).

Finally, it is important to consider that most of the species with stimulated and resistant seeds tend to be pioneer herbs or woody plants (many with hard seeds), while the inhibited species are generally endemic or with a more restricted distribution, of high value for conservation (e.g., *Cryptocarya alba, Podanthus mitiqui, Kageneckia* spp. and *Azara petiolaris)* (Gómez-González et al. 2017; Ocampo-Zuleta et al. 2022). Frequent fires in the Mediterranean area of Chile threaten its biodiversity and, in this sense, should be avoided.

11.4.2 Fire-Stimulated Flowering

Fire-stimulated flowering is a trait more common in the Mediterranean ecosystems worldwide than in other types of ecosystems (71% of 386 species observed). This phenomenon occurs predominantly in geophyte plants and is considered an adaptation to fire particularly in those species that rely on fire to bloom (Lamont and Downes 2011). It is considered an adaptive trait as it offers reproductive advantages such as, for example, greater attraction of pollinators and the possibility of dispersing the seeds in a post-fire environment rich in nutrients. In the Mediterranean area of Chile this trait is little known and even has been considered absent (Montenegro et al. 2004; Armesto et al. 2009). However, there is a geophyte species that has been observed blooming right after the fire in several locations in the Valparaíso and Metropolitan regions: *Zephyranthes advena* (Amaryllidaceae) (S. Gómez-González, personal observation; Fig. 11.5d). This species is not exclusive to places subjected to recurrent fires, being very abundant in the coastal sands of the Coquimbo and Valparaíso regions. It would be interesting to analyze the genetic differences between these coastal populations and the populations that appear in burned areas inland, to investigate the possible existence of ecotypes adapted to fire. On the other hand, post-fire flowering has been observed in *Puya chilensis* (Bromeliaceae; Fig. 11.6a), a fire-resistant species because of the tight arrangement of its leaves that protects the apical meristems (Villaseñor and Sáiz 1990; Ramírez et al. 2014). In this species, it would be necessary to evaluate whether this response increases its performance in relation to when it blooms in non-burned areas. To what extent there are more examples of fire-stimulated flowering in the Chilean Mediterranean is an open question that deserves more research.

11.4.3 Resprouting

The post-fire regeneration pattern of South American Mediterranean vegetation is largely explained by the resprouting ability of many of the plant species that compose it. Of the 243 species that have been evaluated, 73% are capable of resprouting after fire or another disturbance that eliminates their aerial biomass; the percentage increases to 83% if the 38 evaluated therophytes are excluded (S. Paula, unpublished data). Some of these species resprout from the lignotuber, a thickening between the stem and the root that houses numerous buds (even thousands), many of them protected from fire several centimeters under the ground (Paula and Pausas 2006). Specifically, the lignotuber has been verified in species such as *Cryptocarya alba* (Lauraceae), *Lithrea caustica* (Anacardiaceae), *Quillaja saponaria* (Quillajaceae), *Azara petiolaris* (Salicaceae), *Podanthus mitiqui* (Asteraceae), *Peumus boldus* (Monimiaceae; Fig. 11.6b) and *Schinus latifolius* (Anacardiaceae; Fig. 11.6c). Most of these species are typical of coastal sclerophyllous forests, survivors of ancient lineages that inhabited central Chile during the early to middle Miocene (Hinojosa et al. 2006), a period of time in which fires were frequent in the territory, according to fossil records (Abarzúa et al. 2016). However, other dominant species such as *Vachellia caven* or *Baccharis* spp. simply resprout from the root collar (Fig. 11.6d, e), which houses a smaller number of buds compared to the lignotuber, and are poorly protected from the high temperatures reached during the fire (Pausas et al. 2018).

The resprouting ability of herbaceous species has been scarcely described, except in the case of geophyte monocots (Sabadin et al. 2015). These species can resprout from bulbs (in species of the families Amaryllidiaceae and Asparagaceae), tubers (in species of Alstroemeriaceae, Orchidaceae, and Liliaceae, among others), corms (Tecophilaeaceae and species of the genus *Conanthera*) and rhizomes (in species of Iridaceae and Poaceae, highlighting among these last the bamboos of the genus *Quila*). Also noteworthy is the profuse vegetative regeneration of *Dioscorea humifusa* (Dioscoreaceae), which resprouts from a tuberized rhizophore (S. Paula, personal observation; Fig. 11.6f); a non-woody rhizome that lacks a primary root system and that (in this case) has several tubers attached (Pausas et al. 2018).

Fig. 11.6 (**a**) Apical resprouting and fire-stimulated flowering in *Puya chilensis* observed by Byron Lamont (Valparaíso Region). (**b**) Burned lignotuber of *Peumus boldus* (O'Higgins region). (**c**) Lignotuber of *Schinus latifolius* (Valparaíso region). (**d**) Root crown of *Vachellia caven* (Valparaíso region). (**e**) Post-fire resprouting from root crown of *Baccharis linearis* (O'Higgins region). (**f**) Post-fire resprouting of *Dioscorea humifusa* from a rhizophore (O'Higgins region). (**g**) Dense and monospecific plantation of *Pinus radiata* with a high load of dry fuel (Maule Region). (Photos: (**a**) F. Ojeda; (**b**, **e**, **f**) J.G. Pausas; (**c**, **d**) S. Paula; (**g**) S. Gómez-González)

The resprouting that occurs from aerial organs is less frequent in the Mediterranean region of Chile. Epicormic resprouting, that is, from aerial meristematic tissues that are protected by the bark or the wood itself, has been considered absent in this region (Montenegro et al. 2004). However, it has been observed after low severity fires or after drought-induced defoliation in *Vachellia caven*, *Lithrea caustica*, *Quillaja saponaria* or *Schinus latifolus* (S. Gómez-González, personal observation). On the other hand, monocotyledonous species as relevant for their uniqueness and conservation status, as the Chilean palm *Jubaea chilensis* and different species of the genus *Puya* are capable of apically resprouting thanks to the protection that the bases of the leaves confer to the meristematic tissues at the top of the stem.

Although in recent years there has been quite a bit of evidence collected about the occurrence of resprouting in the species of the Chilean Mediterranean, little is still known about the ecophysiology of this process; that is, in which tissues the meristematic buds are stored, how many carbohydrate reserves are consumed when resprouting, and how long the plants take to restore previous reserve levels. It is also unknown whether there is variability of this trait among individuals of the same species, which could determine their evolutionary responses to frequent anthropogenic fires. On the other hand, interaction with herbivores could also determine differences in the survival of resprouting species after fire, something that has been observed, but not quantitatively analyzed (e.g., the resprouts of *Lithraea caustica* seem to be more resistant to browsing than those of other woody species; Villaseñor and Sáiz 1990). Future studies should also analyze the evolutionary origin of resprouting in Chilean species, with the aim of knowing the resilience of these species to recurrent fires and to better understand the role of fire in the selection of this character in Mediterranean ecosystems (Paula and Ojeda 2009; Pausas et al. 2018).

11.4.4 Post-Fire Vegetation Recovery

The recovery of post-fire vegetation depends on the species' ability to resist fire *in situ* (through resprouting, post-fire flowering or persistent seed bank) or to colonize burned areas from adjacent unburned areas. Although most of the dominant woody species in the matorral are able to resprout after low to moderate severity fires, the emergence of their seedlings is quite scarce and mostly limited to annual species or pioneer woody species with seed banks that can withstand low-severity fires (see previous sections). Severe fires strongly condition recruitment, as the seed bank is practically destroyed (Segura et al. 1998; Gómez-González and Cavieres 2009). The largest reservoir of woody species seeds is found under the tree canopy, as many species have fleshy fruits that are dispersed by birds, which use trees as perches. It is in this microhabitat where the appropriate environmental conditions for their germination (humidity, shade) are found (Fuentes et al. 1984, 1986), but it is also the place where the fire acquires its greatest severity, due to the high fuel load (Segura et al. 1998; Gómez-González and Cavieres 2009). On the other hand, the dispersion of woody plant seeds from unburned areas is limited (Jiménez and Armesto 1992) and may have little recruitment success due to the combination of abiotic stress, competition with grasses and herbivory (Fuentes et al. 1984, 1986; Holmgren 2002). However, in some locations, colonization and effective recruitment of anemochorous species (e.g., *Baccharis* spp., *Podanthus mitiqui*) and bird-dispersed species in burned areas have been observed (Villaseñor and Sáiz 1990; Becerra et al. 2022).

The sclerophyllous matorral needs more than 10 years to recover its pre-fire vegetation cover (Becerra et al. 2018). This period depends on the fire history and vegetation density. Thus, between 20 and 30 years are required for 70–80% of the vegetation cover to recover after one or two recurrent fires, but this time is insufficient for significant recovery when the place is affected by 3 fires in 30 years (Smith-Ramírez et al. 2021). If fires are frequent, the plants' ability to resprout decreases due to the depletion of non-structural carbohydrate reserves used to supply post-fire growth (Piper and Paula 2020). This could explain the opening of the canopy after recurrent fires and, with it, the expansion of grasslands dominated by exotic species (see Sect. 11.7).

The vegetation response to fire also changes depending on the type of community we consider within the Mediterranean region of Chile. Thus, coastal sclerophyllous forests dominated by species such as *Cryptocarya alba*, *Peumus boldus*, *Quillaja saponaria* and *Schinus latifolius*, would be more sensitive to fire than the shrubs of *Baccharis* spp., *Puya chilensis, Muehlenbeckia hastulata* and *Trevoa trinervia* (Villaseñor and Sáiz 1990; Quintanilla 2000), despite the fact that the dominant species in both types of vegetation have the ability to regenerate vegetatively. Chilean palm forests (*Jubaea chilensis*) seem to withstand fires well, except if they are frequent and severe, as the opening of the understory accelerates soil erosion and the degradation of this unique ecosystem (Quintanilla and Constanzo 2018).

11.4.5 Fire Effects on Non-Plant Organisms

Compared to plants, the ecological effects and evolutionary implications of fire on animals, fungi and bacteria are less known (Pausas 2019). In the Mediterranean region of South America, this knowledge has mainly derived from the monitoring carried out after the La Campana fire (Valparaíso Region), which occurred in 1984, the fire at the China Muerta National Reserve of 2015 (Mediterranean-temperate transition zone, La Araucanía Region), and the mega-fires of 2017 that affected several regions of Chile, mainly between Valparaíso and El Maule. These studies suggest that fire causes significant changes in the structure of bacterial communities (Aponte et al. 2022), microfauna (Saíz and Bascuñan 1990) and mesofauna (Castro-Huerta et al. 2021) of the soil, epigeal arthropods (Saíz et al. 1990; Campodonico et al. 2020), as well as birds (Perfetti-Bolaño et al. 2013; Traeger 2021) and mammals (Zúñiga et al. 2021), with the severity of the fire being a key factor in determining the impacts and recovery capacity of many of these species in the burned areas.

It has been observed that forest fires in the sclerophyllous forests of central Chile affect different functional groups of soil prokaryotes in different ways, which could alter the nutrient cycles of the ecosystem (Aponte et al. 2022). There are studies that propose the application of organic treatments (e.g. pig manure) to improve soil fertility and activate microbial activity after fire in these forests, although this requires further research regarding the ecological impacts that this type of intervention could cause, and its application would depend on the long-term restoration objectives that are required (García-Carmona et al. 2021). In general, the impact of fire on microorganisms is little known in the Mediterranean Chile and, in this sense, the use of new technologies, as the sequencing of environmental DNA, could be useful to explore the microorganisms and soil fungi that emerge after a fire.

Similarly, in these ecosystems it has been observed that fire modifies the structure of the mite community, affecting more the microphytophagous and detritivores than the predator ones (Saíz and Bascuñan 1990). This could be explained by the loss of organic matter in burned soils. On the other hand, the loss of plant biomass is probably related to the reduction of epigeal phytophagous arthropods (Curculionidae, Cicadellidae) and saprophagous (Hybosoridae) that has been observed in some severe fires in shrubs of central Chile. Formicids, on the other hand, are favored by the open spaces and without vegetation that are found in the burned areas (Sáiz and Bascuñan 1990; Campodonico et al. 2020). Beetles are strongly affected by fire and have low recolonization capacity, although there are some groups that seem to be favored, like the species of the genus *Mastogenius* (Saíz et al. 1990). Other arthropods that increase with fire are those of the order Archeognatha and Homoptera (Sáiz and Bascuñan 1990).

As occurs in invertebrate species, the mortality of vertebrate animals due to fires is associated with the fire characteristics such as intensity, magnitude and duration, and the particularities of the species such as mobility, bird access, among others (Traeger 2021). Most of the studies that analyze the effect of fire on vertebrates have been carried out in the forests of *Nothofagus* of the China Muerta National Reserve, which is located in the transition zone between the Mediterranean region and the temperate region, in the foothills of the Andes. From these studies it is concluded that severe fires impoverish the assembly of rodents, with herbivorous rodents being the most affected by the scarcity of food (*Oligoryzomys longicaudatus* and *Irenomys tarsalis*) (Zúñiga et al. 2021). As a consequence, fire can affect the diet of carnivorous vertebrates such as the fox or the puma, by modifying the assembly of their prey, among which are small rodents (Zúñiga et al. 2020). The severity of the fire affects different species of reptiles in different ways; for example, *Liolaemus tenuis* requires places that are not too damaged by fire, while *Liolaemus lemniscatus* prefers severely burned places (Zúñiga 2020). This agrees with studies that show how *Liolaemus lemniscatus* has an easier time feeding in burned shrubs with ashes, as in these places they have the ability to follow the tracks of insects (Mora and Labra 2017). Studies evaluating the impact of fire on birds of the Mediterranean region of Chile are very scarce, but some preliminary studies carried out in the mega-fires of 2017 indicate that the passerine and insectivorous birds are the most affected. This could be due to the decrease in herbaceous cover that provides better nesting and feeding sites. This decrease in insectivorous birds is of great relevance, as it could reduce the ecosystem service of pest control that these species offer in the agricultural and forestry areas (Traeger 2021). However, a study conducted in Cerro

Cayumanque (Bío-Bío Region) showed that certain birds, like *Zenaida auriculata* and *Diuca diuca*, show preference for post-fire sites dominated by herbaceous plants because they provide seeds and shoots, being these species the ones with the greatest ability to exploit these resources (Perfetti-Bolaño et al. 2013).

Adaptive responses to fire have been less studied in animals than in plants, but there is also evidence in animals of specific mechanisms to respond to fires in other ecosystems, such as, for example, the ability to detect smoke (Pausas 2019). Evolutionary responses to fire in animals or microorganisms have not been investigated in the Mediterranean region of South America and require more attention. On the other hand, long-term studies are needed that evaluate the impact of fire on plant-soil and plant-animal interactions, as this would help to better understand the ecosystem's recovery capacity and to establish more appropriate post-fire management measures in the region.

11.5 Fire Drivers

The likelihood of a fire occurring depends on multiple factors related to weather conditions, fuel, and ignition sources, whether natural or anthropogenic. On the other hand, fire spread is largely modulated by the flammability, structure and the composition of vegetation, its continuity across the landscape, in addition to topography and meteorological conditions at the time of the fire (Aldersley et al. 2011). All these fire drivers are related to each other, so it is difficult to establish their relative importance (Gómez-González et al. 2019). However, there is strong consensus on the idea that human activities are the main source of ignition of forest fires in the Mediterranean area of Chile. Thus, the occurrence of fires is concentrated in highly populated areas (McWethy et al. 2018; Miranda et al. 2020), in the vicinity of roads (Díaz-Hormazábal and González 2016) and in productive areas, such as agricultural and forestry (Gómez-González et al. 2019; McWethy et al. 2018).

In central Chile, large fires are related to rainy years before the fire season, associated with the effect of El Niño, and dry springs or summers during the fire season (Holz et al. 2017; Urrutia-Jalabert et al. 2018). The occurrence of heat waves is also strongly related to fires of greater magnitudes (ARCLIM 2020). These favorable conditions have been more recurrent in recent years, with an increase in temperatures since the 1980s and a general reduction of precipitation in the central area of the country in the last decade, in addition to a greater frequency of consecutive dry years between 2010 and 2022 (the mega-drought) (Boisier et al. 2016; Garreaud et al. 2017). Additionally, the return rate of hyper-dry years (<30% of historical precipitation) has been increasing, observed in 1924, 1968, 1998, 2019, and 2021, which has generated adverse conditions by reducing the humidity of forest fuels. These climate changes have been associated with an increase in the magnitude, simultaneity and duration of fires, mainly due to the increased fuel flammability driven by drought and to the occurrence of prolonged periods of extreme temperatures and winds (González et al. 2020).

Land use change is also a key factor in determining changes in fire regimes by modifying the structure and spatial distribution of fuel at a regional scale (Pausas and Keeley 2014). In recent decades, the expansion of forest crops and the degradation of forests into shrubs have been the main driver of land use change in central Chile (Heilmayr et al. 2016; Miranda et al. 2017). There has been also a rapid and extensive process of deforestation-reforestation, favored through state subsidies to forestry established in the 70 s and which have remained in force until 2012. For example, between 1986 and 2011, an average of 0.34 Mha, 0.51 Mha, and 0.71 Mha of forests, shrubs and grasslands, respectively, were replaced by forest plantations in the central-south area of the country (Heilmayr et al. 2020). Currently, in the Mediterranean area of Chile, forest plantations dominate the landscape of the Coastal Range between the regions of O'Higgins and La Araucanía (Fig. 11.1b). This large-scale transformation of the territory has generated strong socio-ecological impacts, such as the increase in poverty of the Indigenous Population (Hofflinger et al. 2021), the decrease in water provision (Álvarez-Garretón et al. 2019) and the loss of biodiversity (Miranda et al. 2017), but it has also been reflected in the increase of fire risk (Carrasco et al. 2021).

Currently, there are numerous studies that have linked forest plantations with fires in central Chile (Carmona et al. 2012; Díaz-Hormazábal and González 2016; de la Barrera et al. 2018; González et al. 2011, 2018, 2020; McWethy et al. 2018; Bowman et al. 2019; Gómez-González et al. 2018, 2019, 2023; Miranda et al. 2020; Carrasco et al. 2021; Hoyos-Santillán et al. 2021). In this region, the burned area of plantations has shown an increasing trend (Hoyos-Santillán et al. 2021; Gómez-González et al. 2023) and the risk of fires is maximized in areas close to cities with greater coverage of forest plantations (Carrasco et al. 2021). In addition, Carmona et al. (2012) found that 33.2% of the high fire danger areas in the Maule region were in forest plantations.

On the other hand, climate interacts with land uses in explaining large fires. Thus, forest plantations are the only land use where fires greater than 200 ha have increased during the mega-drought (González et al. 2018). In fact, since this dry period began in 2010, large fires in areas dominated by plantations have become increasingly frequent in the Mediterranean area of Chile (e.g., 2012 Quillón and 2014 Valparaíso wildfires; 2017 and 2023 mega-fires). According to official reports from the National Forestry Corporation (CONAF), in the mega-fires of 2017 (0.57 Mha) and 2023 (0.44 Mha), forest plantations were the land use most affected (50% and 58% of the total burned area, respectively). In 2017, about 100 MtCO2eq were released, an amount that represents almost 90% of the total national emitted for the year 2016 (González et al. 2020). The regions of Maule, Ñuble and Bío-Bío have been the most affected in the mega-fires of 2017 and 2023. The Maule and Bío-Bío regions account for 48% of the country forest plantations and 49.5% of all fires (CONAF 2023), thus requiring special attention in terms of fire prevention and land management.

Although forest plantations are associated with large fires (González et al. 2018; Bowman et al. 2019; McWethy et al. 2018), when analyzing the contribution of different land uses to fires (of all sizes) in relation to other relevant variables (climatic, topographic, and related to human pressure), agricultural areas are revealed as the use that contributes most to their occurrence and the burned area in the Mediterranean zone (followed by forest plantations; Gómez-González et al. 2019). The burning of grasslands and crop stubble can be responsible for numerous and large fires, particularly in the Maule, Ñuble, and Bío-Bío regions (McWethy et al. 2018). For example, 10% of all fires that occurred between the O'Higgins and Ñuble regions from 2003 to 2022 were caused by agricultural and livestock activities (CONAF 2023). In addition, these areas of pastures and annual crops are distributed in relatively flat areas with a high average annual temperature and strong human impact, variables also relevant in explaining the frequency and size of fires in central Chile (Gómez-González et al. 2019). Therefore, the interaction of climatic and human variables with land use must be considered in landscape planning for fire prevention.

Box 11.1 Why Are Forest Plantations Associated with Large Fires?

The mechanisms by which forest plantations increase fires are related to the species identity (and their evolutionary history), their spatial distribution, and forest management. Forest plantations in central Chile are usually established as intensively managed (short rotation crops) monocultures of exotic species, mainly of *Pinus radiata* and, to a lesser extent, of *Eucalyptus* spp. (Bowman et al. 2019). Pines and eucalyptus are pyrophytic species, highly flammable, that have evolved with fire as a natural disturbance factor in their ecosystems of origin. In fact, they require fire for their regeneration, as they are serotinous species, that is, they need fire to open their cones and release their seeds (Lamont et al. 2020). For example, *Pinus radiata* retains dead branches during its growth, so that when a fire occurs, it can spread to the highest parts of the crown, where the cones are located and accumulated. Therefore, in the absence of forest management (which removes dead branches), the risk of a high severity crown fire is very high in these types of monospecific plantations (Fig. 11.6g). In addition, in the southernmost areas of the Mediterranean region, forest plantations are often accompanied by other exotic pyrophytic species that establish under the canopy (e.g. *Genista monspessulana*), creating greater vertical and horizontal continuity of fuel and increasing the speed and intensity of fires (Pauchard et al. 2008).

Unlike native vegetation, which is more diverse in terms of structure and fuel types, forest plantations are established very homogeneously and continuously across the landscape. This homogeneity contributes to the propagation of fire and with it, the risk of large fires (Fernandes et al. 2016), particularly in the context of prolonged drought that central Chile is experiencing (González et al. 2018). This was observed in the mega-fires of 2017, when forest plantations burned in greater proportion than native vegetation, relative to what was expected from their availability in the landscape (Bowman et al. 2019; McWethy et al. 2018). This "preference" of fire over forest plantations is more noticeable when compared to the deciduous forests of the southern Mediterranean area (McWethy et al. 2018). The soil under these plantations is very dry compared to the soil of the deciduous forests (Huber et al. 2010), which could also increase the probability of ignition and the spread of fire.

11.6 Fire and Plant Invasions

Currently, Chile has about 816 exotic species, of which 70% are concentrated in the central region of the country, with a Mediterranean type climate (Rodríguez et al. 2018). These exotic species come mostly from the Mediterranean Basin, with a strong representation of herbaceous species (Arroyo et al. 2000; Figueroa et al. 2004). Currently, the Mediterranean matorral is characterized by forming a mosaic of tree and shrub patches separated by a herbaceous matrix that is dominated by a diverse community of naturalized exotic species (Figueroa et al. 2004; Gómez-González et al. 2011b). In fact, in some matorral communities, exotic species constitute almost 80% of the total plant abundance in the herbaceous layer (Gómez-González et al. 2011b). This invasion success seems to be due, in part, to the fact that exotic herbs are better adapted to human disturbances (e.g., grazing, agriculture) and fires, since they have been subjected to these selective factors for longer evolutionary time in their native ranges (Holmgren et al. 2000a; Holmgren 2002).

The absence of a historical regime of recurrent fires in the Chilean Mediterranean region could explain the vulnerability of the matorral to plant invasion after fire (Holmgren et al. 2000a; Contreras et al. 2011). As fire is a relatively new disturbance in evolutionary terms, adaptive responses are rare in the matorral flora (see Sect. 11.4). Seed recruitment after fire is very scarce and the mortality of trees and shrubs can be great after high severity fires. These conditions create an opportunity for exotic plants (mainly annual species) to colonize vacant ecological niches and establish successfully in the burned areas of the matorral (Contreras et al. 2011). This is also favored by the fact that the annual life cycle is poorly represented in the native flora of the matorral compared to other Mediterranean ecosystems (e.g., California chaparral, Arroyo et al. 1995) and, therefore, there are few native herb species able to quickly colonize the open spaces that are generated after fires (Gómez-González and Cavieres 2009; Gómez-González et al. 2011b).

On the other hand, since many of the invasive species come from the Mediterranean Basin, they could have more resistant seed banks or germination stimulated by fire, which would allow them to colonize the burned areas more quickly than native species. However, the evidence in this regard is contradictory. Some studies focused on the analysis of seed banks have shown an increase in the richness of exotic species by fire (Gómez-González and Cavieres 2009), while others have documented a reduction (Figueroa et al. 2009) or non-significant changes (Figueroa and Cavieres 2012). This variability in the responses to fire may be due to differences in species composition among different sites, in the severity of the fire, among other possible factors.

A better way to determine if exotic species are better adapted to fire than natives is by conducting germination experiments where seeds are directly subjected to fire signals. However, studies conducted under this approach have been mainly focused on native woody species and there is very little knowledge about both native and exotic herbaceous species. Recently, Gómez-González (2018) compiled bibliographic information and data derived from her own germination tests (unpublished).

With this compilation, the distribution of germination responses (stimulation, inhibition and tolerance) to fire signals (smoke and/or heat) in 38 native species and 14 exotic species of the matorral was analyzed. It was found that inhibition responses were much more frequent in native species (40%) than in exotic ones (7%), while stimulation responses were not so different (16% in native and 21% in exotic species). Among the exotic herbs that have germination stimulated by fire signals are *Anthriscus caucalis* (Apiaceae), stimulated by smoke, and *Aira caryophyllea* (Poaceae), stimulated by heat (S. Gómez-González, unpublished data). Both species can often be found colonizing the matorral burned areas. However, the main response of the exotics was tolerance (neither stimulated nor inhibited by fire signals), with 72% of tolerant species. Therefore, it is the inhibition of the natives and the high tolerance of the exotics that would be favoring post-fire invasion in the Chilean matorral.

Fire frequency is a key factor in determining the invasion process in the matorral. When fires are frequent, the canopy gradually opens up and the exotic herbaceous layer increases its extension (Sax 2002; Contreras et al. 2011; Smith-Ramírez et al. 2021). This modifies the spatial fuel structure of the community, which, in turn, determines the spatial pattern of fire severity and subsequent recruitment (Segura et al. 1998; Gómez-González and Cavieres 2009). In the grasslands, fire is very mild and practically does not affect the native and exotic seed bank (Figueroa and Cavieres 2012), although in some sites the richness of exotic species can even increase (Ávila et al. 1981; Gómez-González and Cavieres 2009) and modify the composition and distribution of plant communities (Contreras et al. 2011). Therefore, although native herbs are practically unaffected, the dominance of exotics is maintained after fire (Holmgren et al. 2000a; Gómez-González et al. 2011b). This dominance represents a competitive barrier that limits the recruitment of woody species and canopy closure after fire (Holmgren et al. 2000b).

On the other hand, under the canopy of trees and shrubs, fire becomes more severe due to having a higher fuel load. In this microhabitat, the high temperatures reached on the ground destroy the native seed bank and only some species of hard seeds are capable of surviving and recruiting (Segura et al. 1998; Gómez-González and Cavieres 2009). These spaces under the canopy usually are subsequently colonized by exotic herbs in the second post-fire year. This process corresponds to the *fire-grass cycle* that has been documented in other ecosystems (D'Antonio and Vitousek 1992), where the fire creates open spaces that are colonized by more flammable exotic herbs that, in turn, promote the occurrence of low severity fires and are favored by them. This process has also been observed in fire-adapted Mediterranean ecosystems like the California chaparral (Keeley et al. 2003), when the recurrence of fires is so high that woody species do not manage to grow enough to produce seeds. Given that the woody species of the Chilean matorral have a lower capacity to regenerate after fire, it is expected that the conversion to a vegetation dominated by exotic species will occur more rapidly than in the Californian chaparral. Therefore, the protection of the matorral from fires is necessary to prevent the invasion process and ensure its long-term conservation.

It is important to consider that the recruitment pattern of these annual exotic species after fire not only depends on their survival in the seed bank, or their ability to disperse from unburned adjacent areas, but they must also overcome other selective post-fire pressures, such as drought and herbivory. (Gómez-González et al. 2011b). The pressure from exotic herbivores (cattle, rabbits, and European hares) is very high after fire in the matorral, so, probably, the invasion success will depend on the ability of these species to overcome both disturbances simultaneously. However, the evidence on the interaction between fire and other factors in the success of Mediterranean matorral invasion is still very scarce. Holmgren et al. (2000a) analyzed the effect of fire and herbivory on the abundance and composition of native and invasive herbs in a matorral of the Metropolitan Region's foothills (33°27'S) and found no effects of fire under any condition, although, in general, exotic herbs were more resistant to herbivores than native species. In contrast, other studies have found the opposite pattern (e.g., Torres-Díaz et al. 2012), which may be due to differences between sites in the relative abundances of native and exotic species. Herbivores could also alter the spatial distribution of post-fire recruitment of native and invasive species, as they prefer to browse in open spaces, where exotics are more abundant (Jaksic and Fuentes 1980; Fuentes et al. 1984).

Instead, the presence of native rodents has also been related to plant invasion in the matorral. The cururo (*Spalacopus cyanus*, Octodontidae), is a native fossorial mammal that by building tunnels generates microperturbations (mounds) that can be very abundant and noticeable at the landscape scale. The invasive species *Fumaria capreolata* L. (Papaveraceae) establishes itself on these mounds (Torres-Díaz et al. 2012). It has been observed that, in the absence of disturbances, *F. capreolata* only establishes itself under the canopy of trees as it cannot compete with the grasses that dominate open spaces (Torres-Díaz et al. 2012). However, in the presence of fire or cururo disturbances, *Fumaria* can colonize open spaces and advance in the invasion process (Fig. 11.5c; Torres-Díaz et al. 2012). This suggests that, in addition to herbivores, the activity of native cururos could interact with fire to modulate the invasion of exotic species. Therefore, complex studies are needed to evaluate the interaction between fire, fauna, and vegetation structure (microhabitats) in different matorral sites. This would be very useful to understand in a more integrative way the invasion success of exotic species in the Chilean matorral and for proposing appropriate management measures to restore burned areas of the matorral (Armesto et al. 2009). In this sense, active management, such as the manual removal of exotic plants and protection against herbivores could be effective in controlling the spread of invasive species and promoting the regeneration of the native ones after a fire (Contreras et al. 2011).

11.7 Concluding Remarks and Future Directions

Knowledge of fire regime history and fire ecology is essential for taking appropriate measures for the management and restoration of natural areas. In general, the South American Mediterranean-type vegetation is not typically adapted to fire, so frequent and severe anthropogenic fires pose a serious risk to its biodiversity. However, more studies are needed that analyze species responses to fire (i.e., resprouting, germination) at the community level, to more accurately determine the vulnerability and recovery needs of the species most sensitive to this disturbance. Differences in the regeneration capacity of different types of vegetation within the Mediterranean area suggest that management and restoration strategies after fire could be context-dependent. In sclerophyllous forest and matorral, passive restoration probably would be sufficient to recover the vegetation if fires are not very recurrent (Smith-Ramírez et al. 2021; Becerra et al. 2022). However, in areas of high fire recurrence, active measures may be required, such as replanting of fire-sensitive species and those of high conservation value. For this, it is important to identify the appropriate microhabitats for their establishment and facilitating species that could accelerate this process (Armesto et al. 2009). In all cases, it would be necessary to reduce herbivory pressure and eliminate exotic species (see Sect. 11.6). Finally, all environmental education measures and fire use regulations that imply reducing the frequency of anthropogenic fires will always be favorable for maintaining the biodiversity of the Mediterranean area of Chile.

Central Chile has undergone a strong modification of land use in the last 200 years, which has transformed much of the native vegetation into crop and grazing areas or into forest plantations. This has generated a change in the quality and spatial distribution of fuel in the landscape. Large fires that have occurred in homogeneous landscapes dominated by forest plantations suggest the need to redesign the landscape. In fact, these fires could offer a window of opportunity to build more heterogeneous landscapes, which maximize ecosystem services to society (Leverkus et al. 2019). In this sense, it is necessary to advance in the search for integrative territorial and forestry management strategies aimed at maximizing economic development while maintaining biodiversity, ecosystem services, and fire protection. These strategies should consider the ecological intensification of forest plantations (see Gómez-González et al. 2022), the diversification of land uses and regional productive systems, the restoration and conservation of native forests, the control of invasive species, and the establishment of water resource protection zones. The design of these landscapes will be crucial to move towards resilience and mitigate the impacts of the expected increase in droughts and forest fires in the near future.

References

Abarzúa AM, Vargas C, Jarpa L, Gutiérrez NM, Hinojosa LF, Paula S (2016) Evidence of Neogene wildfires in Central Chile: Charcoal records from the Navidad formation. Palaeogeogr Palaeoclimatol Palaeoecol 459:76–85. https://doi.org/10.1016/j.palaeo.2016.06.036

Aldersley A, Murray SJ, Cornell SE (2011) Global and regional analysis of climate and human drivers of wildfire. Sci Total Environ 409:3472–3481. https://doi.org/10.1016/j.scitotenv.2011.05.032

Alvarez-Garreton C, Lara A, Boisier JP, Galleguillos M (2019) The impacts of native forests and forest plantations on water supply in Chile. Forests 10:473. https://doi.org/10.3390/f10060473

Aponte H, Galindo-Castañeda T, Yáñez C, Hartmann M, Rojas C (2022) Microbial community-level physiological profiles and genetic prokaryotic structure of burned soils under mediterranean Sclerophyll forests in Central Chile. Front Microbiol 13:824813. https://doi.org/10.3389/fmicb.2022.824813

Aravena JC, LeQuesne C, Jiménez H, Lara A, Armesto JJ (2003) Fire history in central Chile: Tree-ring evidence and modern records. In: Veblen TT, Baker WL, Montenegro G, Swetnam TW (eds) Fire and climatic change in temperate ecosystems of the Western Americas, 1st edn. Springer Science & Business Media, New York, pp 343–356

ARCLIM (2020) Atlas de riesgos climáticos. Ministerio del Medio Ambiente. Gobierno de Chile. https://arclim.mma.gob.cl/. Accesed 01 April 2025

Armesto JJ, Arroyo MTK, Hinojosa LF (2007) The Mediterranean environment of central Chile. In: Veblen TT, Young KR, Orme AR (eds) The physical geography of South America, 1st edn. Oxford University Press, New York, pp 184–199

Armesto JJ, Bustamante-Sánchez MA, Díaz MF, González ME, Holz A et al (2009) Fire disturbance regimes, ecosystem recovery and restoration strategies in Mediterranean and temperate regions of Chile. In: Cerdà A, Robichaud PR (eds) Fire effects on soils and restoration strategies, 1st edn. CRC Press, Boca Raton, pp 553–584

Armesto JJ, Gutiérrez JR (1978) El efecto del fuego en la estructura de la vegetación de Chile central. An Museo Hist Nat Valparaíso 11:43–48

Arroyo M, Marticorena C, Matthei O, Cavieres L (2000) Plant invasions in Chile: present patterns and future predictions. In: Mooney HA, Hobbs RJ (eds) Invasive species in a changing world, 1st edn. Island Press, Covelo, pp 385–421

Arroyo MTK, Cavieres L, Marticorena C, Muñoz-Schick M (1995) Convergence in the Mediterranean floras in central Chile and California: insights from comparative biogeography. In: Arroyo MTK, Zedler PH, Fox MD (eds) Ecology and biogeography of Mediterranean ecosystems in Chile, California, and Australia, 1st edn. Springer, New York, pp 43–88

Arroyo MTK, Marquet PA, Marticorena C, Cavieres LA, Squeo FA, et al. (2006) El hotspot chileno, prioridad mundial para la conservación. Diversidad de ecosistemas, ecosistemas terrestres. In: Comisión Nacional del Medio Ambiente (ed) Biodiversidad de Chile. Patrimonios y Desafíos, 1st ed. Ministerio de Medio Ambiente, Gobierno de Chile, Santiago, pp. 90–95

Ávila G, Aljaro M, Silva B (1981) Observaciones en el estrato herbáceo del matorral después del fuego. An Museo Hist Nat Valparaíso 14:99–105

Batllori E, Parisien MA, Krawchuk MA, Moritz MA (2013) Climate change-induced shifts in fire for Mediterranean ecosystems. Glob Ecol Biogeogr 22:1118–1129. https://doi.org/10.1111/geb.12065

Becerra PI, Figueroa C, Meza A (2022) Dinámica post-incendio de la vegetación en la localidad de Rastrojos, Chile central. Gayana Bot 79:10–26. https://doi.org/10.4067/S0717-66432022000100010

Becerra PI, Smith-Ramírez C, Arellano E (2018) Evaluación de técnicas pasivas y activas para la recuperación del bosque esclerófilo de Chile central. Corporación Nacional Forestal, Santiago

Boisier JP, Rondanelli R, Garreaud RD, Muñoz F (2016) Anthropogenic and natural contributions to the Southeast Pacific precipitation decline and recent megadrought in central Chile. Geophys Res Lett 43:413–421. https://doi.org/10.1002/2015gl067265

Bowman DM, Moreira-Muñoz A, Kolden CA, Chávez RO, Muñoz AA, Salinas F et al (2019) Human–environmental drivers and impacts of the globally extreme 2017 Chilean fires. Ambio 48:350–362. https://doi.org/10.1007/s13280-018-1084-1

Bustos-Schindler C, Quesne C, González M, Solari M (2010) Historia preliminar de incendios y prácticas (multi) culturales en la cuenca media del río Cachapoal (34° S), Chile central. Bosque 31:17–27. https://doi.org/10.4067/S0717-92002010000100003

Campodonico JF, Paula S, Guerrero B, Escárez F, Del Campo ME, et al. (2020) Impacto de la severidad de incendio en el ensamble de artrópodos epigeos mediterráneos: el caso de Quebrada de la Plata, Chile Central. In: Abstracts of the XXVI Reunión Anual de la Sociedad de Ecología de Chile (SOCECOL), Valdivia, 20-22 November 2019

Camus P (2004) Los Bosques y la Minería del Norte Chico, s. XIX. Un mito en la representación del paisaje chileno. Historia 37:289–310. https://doi.org/10.4067/S0717-71942004000200001

Carmona A, González M, Nahuelhual L, Silva J (2012) Spatio-temporal effects of human drivers on fire danger in Mediterranean Chile. Bosque 33:321–328. https://doi.org/10.4067/S0717-92002012000300016

Carrasco J, Acuna M, Miranda A, Alfaro G, Pais C, Weintraub A (2021) Exploring the multidimensional effects of human activity and land cover on fire occurrence for territorial planning. J Environ Manag 297:113428. https://doi.org/10.1016/j.jenvman.2021.113428

Castro-Huerta R, Morales C, Gajardo J, Mundaca EA, Yáñez M (2021) Soil mesofauna responses to fire severity in a sclerophyllous forest in central Chile. Forests 12:1444. https://doi.org/10.3390/f12111444

CONAF (2017) Análisis de la afectación y severidad de los incendios forestales ocurridos en enero y febrero de 2017 sobre los usos de suelo y los ecosistemas naturales presentes entre las regiones de Coquimbo y Los Ríos de Chile. Informe Técnico, Corporación Nacional Forestal, Santiago

CONAF (2023) Incendios forestales en Chile. Estadísticas históricas. https://wwwconafcl/incendios-forestales/incendios-forestales-en-chile/estadisticas-historicas/. Accessed 25 May 2023

Contreras TE, Figueroa JA, Abarca L, Castro SA (2011) Fire regimen and spread of plants naturalized in central Chile. Rev Chil Hist Nat 84:307–323. https://www.redalyc.org/articulo.oa?id=369944299001

Cowling RM, Potts AJ, Bradshaw PL, Colville J, Arianoutsou M et al (2015) Variation in plant diversity in mediterranean-climate ecosystems: The role of climatic and topographical stability. J Biogeogr 42:552–564. https://doi.org/10.1111/jbi.12429

Cowling RM, Rundel PW, Lamont BB, Arroyo MK, Arianoutsou M (1996) Plant diversity in Mediterranean-climate regions. Trends Ecol Evol 11:362–366. https://doi.org/10.1016/0169-5347(96)10044-6

D'Antonio CM, Vitousek PM (1992) Biological Invasions by exotic grasses, the grass/fire cycle, and global change. Annu Rev Ecol Evol Syst 23:63–87. https://doi.org/10.1146/annurev.es.23.110192.000431

De la Barrera F, Barraza F, Favier P, Ruiz V, Quense J (2018) Megafires in Chile 2017: Monitoring multiscale environmental impacts of burned ecosystems. Sci Total Environ 637:1526–1536. https://doi.org/10.1016/j.scitotenv.2018.05.119

Del Pozo S (2013) Historia de los Bosques Naturales y Forestal de Chile. Memoria Chilena. http://www.memoriachilena.gob.cl/602/w3-article-3373.html. Accessed 15 May 2023

Di Castri F (1973) Distribution of the Mediterranean vegetation in Chile. Bol Soc Biol Concepción 47:87–107

Díaz-Hormazábal I, González ME (2016) Análisis espacio-temporal de incendios forestales en la región del Maule, Chile. Bosque 37:147–158. https://doi.org/10.4067/S0717-92002016000100014

Elizalde R (1970) La sobrevivencia de Chile. La conservación de sus recursos naturales renovables, Ministerio de Agricultura, Santiago

Esler KJ, Jacobsen AL, Pratt RB (eds) (2018) The biology of Mediterranean-type ecosystems. Oxford University Press, Oxford

Fernandes PM, Monteiro-Henriques T, Guiomar N, Loureiro C, Barros AM (2016) Bottom-up variables govern large-fire size in Portugal. Ecosystems 19:1362–1375. https://doi.org/10.1007/s10021-016-0010-2

Figueroa JA, Cavieres LA (2012) The effect of heat and smoke on the emergence of exotic and native seedlings in a Mediterranean fire-free matorral of central Chile. Rev Chil Hist Nat 85:101–111. https://doi.org/10.4067/S0716-078X2012000100008

Figueroa JA, Cavieres LA, Gómez-González S, Montenegro MM, Jaksic FM (2009) Do heat and smoke increase emergence of exotic and native plants in the matorral of central Chile? Acta Oecol 35:335–340. https://doi.org/10.1016/j.actao.2008.12.004

Figueroa JA, Teillier S, Jaksic FM (2004) Composition, size, and dynamics of the seed bank in a mediterranean shrubland of Chile. Austral Ecol 29:574–584. https://doi.org/10.1111/j.1442-9993.2004.01392.x

Fuentes ER, Hoffmann AJ, Poiani A, Alliende MC (1986) Vegetation change in large clearings: patterns in the Chilean matorral. Oecologia 68:358–366. https://doi.org/10.1007/BF01036739

Fuentes ER, Otaiza RD, Alliende MC, Hoffmann A, Poiani A (1984) Shrub clumps of the Chilean matorral vegetation: structure and possible maintenance mechanisms. Oecologia 62:405–411. https://doi.org/10.1007/BF00384275

Fuentes ER, Espinosa G (1986) Resilience of shrublands in central Chile: a volcanism-related hypothesis. Interciencia 11:164–165

Gallardo E (2013) Manual de derecho forestal. Corporación Nacional Forestal, Santiago

García-Carmona M, Marin C, Garcia-Orenes F, Rojas C (2021) Contrasting organic amendments induce different short-term responses in soil abiotic and biotic properties in a fire-affected native Mediterranean forest in Chile. J Soil Sci Plant Nutr 21:2105–2114. https://doi.org/10.1007/s42729-021-00506-z

Garreaud RD, Alvarez-Garreton C, Barichivich J, Boisier JP et al (2017) The 2010–2015 megadrought in central Chile: impacts on regional hydroclimate and vegetation. HESS 21:6307–6327. https://doi.org/10.5194/hess-21-6307-2017

Garreaud RD, Molina A, Farias M (2010) Andean uplift, ocean cooling and Atacama hyperaridity: a climate modeling perspective. Earth Planet Sci Lett 292:39–50. https://doi.org/10.1016/j.epsl.2010.01.017

Gómez P, Garrido P, Espinoza S, San Martín J (2021) Primer registro de regeneración posincendio de *Adesmia bijuga* (Fabaceae), una especie en peligro crítico de Chile central. Coll Bot 40:e008. https://doi.org/10.3989/collectbot.2021.v40.008

Gómez-González S (2008) Patrones y mecanismos de regeneración por semillas tras el fuego en el matorral mediterráneo de Chile central: Consecuencias en la invasión de especies alóctonas., Dissertation,. Universidad de Concepción

Gómez-González S (2018) Vulnerabilidad del matorral mediterráneo de Chile a los incendios forestales: apuntes para su recuperación y conservación. In: Novo FG, Pausas JG (eds) Ecología de la regeneración de zonas incendiadas, 1st edn. Academia de Ciencias Sociales y del Medio Ambiente de Andalucía, Sevilla, pp 163–174

Gómez-González S, Cavieres LA (2009) Litter burning does not equally affect seedling emergence of native and alien species of the Mediterranean-type Chilean matorral. Int J Wildland Fire 18:213–221. https://doi.org/10.1071/WF07074

Gómez-González S, González ME, Paula S, Díaz-Hormazábal I, Lara A, Delgado-Baquerizo M (2019) Temperature and agriculture are largely associated with fire activity in Central Chile across different temporal periods. For Ecol Manag 433:535–543. https://doi.org/10.1016/j.foreco.2018.11.041

Gómez-González S, Miranda A, Hoyos-Santillan J, Lara A, Moraga P, Pausas JG (2023) Afforestation and climate mitigation: lessons from Chile. Trends Ecol Evol 39:5–8. https://doi.org/10.1016/j.tree.2023.09.014

Gómez-González S, Ojeda F, Fernandes PM (2018) Portugal and Chile: Longing for sustainable forestry while rising from the ashes. Environ Sci Pol 81:104–107. https://doi.org/10.1016/j.envsci.2017.11.006

Gomez-Gonzalez S, Ojeda F, Torres-Morales P, Palma JE (2016) Seed pubescence and shape modulate adaptive responses to fire cues. PLoS One 11:e0159655. https://doi.org/10.1073/pnas.1108863108

Gómez-González S, Paniw M, Blanco-Pastor JL, García-Cervigón AI, Godoy O, Herrera JM et al (2022) Moving towards the ecological intensification of tree plantations. Trends Plant Sci 27:637–645. https://doi.org/10.1016/j.tplants.2021.12.009

Gómez-González S, Paula S, Cavieres LA, Pausas JG (2017) Postfire responses of the woody flora of Central Chile: Insights from a germination experiment. PLoS One 12:e0180661. https://doi.org/10.1371/journal.pone.0180661

Gómez-González S, Sierra-Almeida A, Cavieres LA (2008) Does plant-derived smoke affect seed germination in dominant woody species of the Mediterranean matorral of central Chile? For Ecol Manag 255:1510–1515. https://doi.org/10.1016/j.foreco.2007.11.006

Gómez-González S, Torres-Díaz C, Bustos-Schindler C, Gianoli E (2011a) Anthropogenic fire drives the evolution of seed traits. P Natl Acad Sci U S A 108:18743–18747. https://doi.org/10.1073/pnas.1108863108

Gómez-González S, Torres-Diaz C, Gianoli E (2011c) The effects of fire-related cues on seed germination and viability of *Helenium aromaticum* (Hook.) HL Bailey (Asteraceae). Gayana Bot 68:86–88. https://doi.org/10.4067/S0717-66432011000100007

Gómez-González S, Torres-Díaz C, Valencia G, Torres-Morales P, Cavieres LA, Pausas JG (2011b) Anthropogenic fires increase alien and native annual species in the Chilean coastal matorral. Divers Distrib 17:58–67. https://doi.org/10.1111/j.1472-4642.2010.00728.x

González ME, Gómez-González S, Lara A, Garreaud R, Díaz-Hormazábal I (2018) The 2010–2015 Megadrought and its influence on the fire regime in central and south-central Chile. Ecosphere 9:e02300. https://doi.org/10.1002/ecs2.2300

González ME, Lara A, Urrutia R, Bosnich J (2011) Cambio climático y su impacto potencial en la ocurrencia de incendios forestales en la zona centro-sur de Chile (33°–42° S). Bosque 32:215–219. https://doi.org/10.4067/S0717-92002011000300002

González ME, Sapiains R, Gómez-González S, Garreaud R, Miranda A, Galleguillos M, et al. (2020) Incendios forestales en Chile: causas, impactos y resiliencia. Centro de Ciencia del Clima y la Resiliencia (CR)2. https://www.cr2.cl/incendios/. Accessed 15 May 2023

Gregory-Wodzicki KM (2000) Uplift history of the Central and Northern Andes: a review. GSA Bull 112:1091–1105. https://doi.org/10.1130/0016-7606(2000)112<1091:UHOTCA>2.0.CO;2

Grobler BA, Cowling RM (2022) Which is the richest of them all? Comparing area-adjusted plant diversities of Mediterranean-and tropical-climate regions. Front Biogeogr 14(3):e56241. https://doi.org/10.21425/F5FBG56241

Heilmayr R, Echeverría C, Fuentes R, Lambin EF (2016) A plantation-dominated forest transition in Chile. Appl Geogr 75:71–82. https://doi.org/10.1016/j.apgeog.2016.07.014

Heilmayr R, Echeverría C, Lambin EF (2020) Impacts of Chilean forest subsidies on forest cover, carbon and biodiversity. Nat Sustain 3:701–709. https://doi.org/10.1038/s41893-020-0547-0

Hernández-Riquelme M (2014) Registro de fuego en sedimentos durante el Holoceno en el sector de Chada Región Metropolitana. Universidad Austral de, Chile

Heusser CJ (1994) Paleoindians and fire during the late Quaternary in southern South America. Rev Chil Hist Nat 67:435–443

Hinojosa LF, Armesto JJ, Villagrán C (2006) Are Chilean coastal forests pre-Pleistocene relicts? Evidence from foliar physiognomy, palaeoclimate, and phytogeography. J Biogeogr 33:331–341. https://doi.org/10.1111/j.1365-2699.2005.01350.x

Hofflinger A, Nahuelpan H, Boso Á, Millalen P (2021) Do large-scale forestry companies generate prosperity in indigenous communities? The socioeconomic impacts of tree plantations in southern Chile. Hum Ecol 49:619–630. https://doi.org/10.1007/s10745-020-00204-x

Holmgren M (2002) Exotic herbivores as drivers of plant invasion and switch to ecosystem alternative states. Biol Invasions 4:25–33. https://doi.org/10.1023/A:1020535628776

Holmgren M, Avilés R, Sierralta L, Segura AM, Fuentes ER (2000a) Why have European herbs so successfully invaded the Chilean matorral? Effects of herbivory, soil nutrients, and fire. J Arid Environ 44(2):197–211. https://doi.org/10.1006/jare.1999.0589

Holmgren M, Segura AM, Fuentes ER (2000b) Limiting mechanisms in the regeneration of the Chilean matorral. Plant Ecol 147:49–57. https://doi.org/10.1023/a:1009804726307

Holz A, Paritsis J, Mundo IA, Veblen TT, Kitzberger T et al (2017) Southern Annular Mode drives multicentury wildfire activity in southern South America. P Natl Acad Sci USA 114:9552-9557. https://doi.org/10.1073/pnas.1705168114

Hoyos-Santillan J, Miranda A, Lara A, Sepulveda-Jauregui A, Zamorano-Elgueta C, Gómez-González S et al (2021) Diversifying Chile's climate action away from industrial plantations. Environ Sci Pol 124:85–89. https://doi.org/10.1016/j.envsci.2021.06.013

Huber A, Iroumé A, Mohr C, Frêne C (2010) Efecto de plantaciones de *Pinus radiata* y *Eucalyptus globulus* sobre el recurso agua en la Cordillera de la Costa de la región del Biobío, Chile. Bosque 31:219–230. https://doi.org/10.4067/S0717-92002010000300006

IREN (1965) Evaluación de la erosión (Cordillera de la Costa entre Valparaíso y Cautín). Instituto Nacional de Investigación de Recursos Naturales (Chile). https://bibliotecadigital.ciren.cl/handle/20.500.13082/14283. Accessed 15 May 2023

Jackson D, Aspillaga E, Rodríguez XP, Jackson D, Santana F, Méndez C (2012) Las ocupaciones humanas del sitio arqueológico de Santa Inés, Laguna de Tagua Tagua, Chile central. Rev Chil Antropol 26:151–168

Jaksic FM, Fuentes ER (1980) Correlates of tail losses in twelve species of *Liolaemus lizards*. J Herpetol 14:137–141. https://doi.org/10.2307/1563844

Jaureguiberry P, Díaz S (2015) Post-burning regeneration of the Chaco seasonally dry forest: germination response of dominant species to experimental heat shock. Oecologia 177:689–699. https://doi.org/10.1007/s00442-014-3161-x

Jiménez HE, Armesto JJ (1992) Importance of the soil seed bank of disturbed sites in Chilean matorral in early secondary succession. J Veg Sci 3:579–586. https://doi.org/10.2307/3235824

Keeley SC, Johnson AW (1977) A comparison of the pattern of herb and shrub growth in comparable sites in Chile and California. Am Midl Nat:120–132. https://doi.org/10.2307/2424690

Keeley JE, Bond WJ, Bradstock RA, Pausas JG, Rundel PW (2012) Fire in Mediterranean ecosystems: ecology, evolution and management. Cambridge University Press, Cambridge

Keeley JE, Lubin D, Fotheringham CJ (2003) Fire and grazing impacts on plant diversity and alien plant invasions in the southern Sierra Nevada. Ecol Appl 13:1355–1374. https://doi.org/10.1890/02-5002

Keeley JE, Pausas JG (2022) Evolutionary ecology of fire. Annu Rev Ecol Evol Syst 53:203–225. https://doi.org/10.1146/annurev-ecolsys-102320-095612

Lamont BB, Downes KS (2011) Fire-stimulated flowering among resprouters and geophytes in Australia and South Africa. Plant Ecol 212:2111–2125. https://doi.org/10.1007/s11258-011-9987-y

Lamont BB, Pausas JG, He T, Witkowski ET, Hanley ME (2020) Fire as a selective agent for both serotiny and nonserotiny over space and time. Crit Rev Plant Sci 39:140–172. https://doi.org/10.1080/07352689.2020.1768465

Leverkus AB, Murillo PG, Doña VJ, Pausas JG (2019) Wildfires: Opportunity for restoration? Science 363:134–135. https://doi.org/10.1126/science.aaw2134

Luebert F, Pliscoff P (2006) Sinopsis bioclimática y vegetacional de Chile. Editorial Universitaria, Santiago

McWethy DB, Pauchard A, García RA, Holz A, González ME, Veblen TT et al (2018) Landscape drivers of recent fire activity (2001–2017) in south-central Chile. PLoS One 13:e0201195. https://doi.org/10.1371/journal.pone.0201195

Miranda A, Altamirano A, Cayuela L, Lara A, González M (2017) Native forest loss in the Chilean biodiversity hotspot: revealing the evidence. Reg Environ Chang 17:285–297. https://doi.org/10.1007/s10113-016-1010-7

Miranda A, Carrasco J, González M, Pais C, Lara A, Altamirano A et al (2020) Evidence-based mapping of the wildland-urban interface to better identify human communities threatened by wildfires. Environ Res Lett 15:094069. https://doi.org/10.1088/1748-9326/ab9be5

Miranda A, Mentler R, Moletto-Lobos Í, Alfaro G et al (2022) The Landscape Fire Scars Database: mapping historical burned area and fire severity in Chile. ESSD 14:3599–3613. https://doi.org/10.5194/essd-2021-467

Miranda A, Syphard AD, Berdugo M, Carrasco J, Gómez-González S et al (2023) Widespread synchronous decline of Mediterranean-type forest driven by accelerated aridity. Nat Plants 9:1810–1817. https://doi.org/10.1038/s41477-023-01541-7

Montenegro G, Avila G, Schatte P (1983) Presence and development of lignotubers in shrubs of the Chilean matorral. Can J Bot 61:1804–1808. https://doi.org/10.1139/b83-189

Montenegro G, Ginocchio R, Segura A, Keely JE, Gómez M (2004) Fire regimes and vegetation responses in two Mediterranean-climate regions. Rev Chil Hist Nat 77:455–464. https://doi.org/10.4067/s0716-078x2004000300005

Montenegro G, Gómez M, Díaz F, Ginocchio R (2003) Regeneration potential of Chilean matorral after fire: an updated view. In: Veblen TT, Baker WL, Montenegro G, Swetnam TW (eds) Fire and climatic change in temperate ecosystems of the western Americas, 1st edn. Springer, New York, pp 381–409

Mora M, Labra A (2017) The response of two *Liolaemus* lizard species to ash from fire and volcanism. J Herpetol 51:388–395. https://doi.org/10.1670/16-105

Muñoz MR, Fuentes ER (1989) Does fire induce shrub germination in the Chilean matorral? Oikos 56:177–181. https://doi.org/10.2307/3565333

Myers N, Mittermeier RA, Mittermeier CG, da Fonseca GAB, Kent J (2000) Biodiversity hotspots for conservation priorities. Nature 403:853–858. https://doi.org/10.1038/35002501

Ocampo-Zuleta K, Gómez-González S, Paula S (2022) Generalised seed mortality driven by heat shock in woody plants from Mediterranean Chile. Int J Wildland Fire 31:1080–1088. https://doi.org/10.1071/WF22027

Ojeda F, Pausas JG, Verdú M (2010) Soil shapes community structure through fire. Oecologia 163:729–735. https://doi.org/10.1007/s00442-009-1550-3

Olson DM, Dinerstein E, Wikramanayake ED, Burgess ND, Powell GV et al (2001) Terrestrial Ecoregions of the World: a new map of life on Earth. Bioscience 51:933–938. https://doi.org/10.1641/0006-3568(2001)051[0933:TEOTWA]2.0.CO;2

Otero L (2006) La huella del fuego: historia de los bosques nativos: poblamiento y cambios en el paisaje del sur de Chile. Pehuén editores, Santiago

Pauchard A, Alaback PB (2004) Influence of elevation, land use, and landscape context on patterns of alien plant invasions along roadsides in protected areas of South-Central Chile. Conserv Biol 18:238–248. https://doi.org/10.1111/j.1523-1739.2004.00300.x

Pauchard A, García RA, Pena E, González C, Cavieres LA, Bustamante RO (2008) Positive feedbacks between plant invasions and fire regimes: *Teline monspessulana* (L.) K. Koch (Fabaceae) in central Chile. Biol Invasions 10:547–553. https://doi.org/10.1007/s10530-007-9151-8

Paula S, Ojeda F (2009) Belowground starch consumption after recurrent severe disturbance in three resprouter species of the genus *Erica*. Botany 87:253–259. https://doi.org/10.1139/B08-134

Paula S, Pausas JG (2006) Leaf traits and resprouting ability in the Mediterranean basin. Funct Ecol 20:941–947. https://doi.org/10.1111/j.1365-2435.2006.01185.x

Pausas JG (2019) Generalized fire response strategies in plants and animals. Oikos 128:147–153. https://doi.org/10.1111/oik.05907

Pausas JG, Keeley JE (2014) Abrupt climate-independent fire regime changes. Ecosystems 17:1109–1120. https://doi.org/10.1007/s10021-014-9773-5

Pausas JG, Lamont BB, Paula S, Appezzato-da-Glória B, Fidelis A (2018) Unearthing belowground bud banks in fire-prone ecosystems. New Phytol 217:1435–1448. https://doi.org/10.1111/nph.14982

Perfetti-Bolaño A, González-Acuña D, Barrientos C, Moreno L (2013) Efectos del fuego sobre la avifauna del Cerro Cayumanque, región del Bío-bío, Chile. Bol Chil Ornitol 19:1–11

Piper F, Paula S (2020) The role of nonstructural carbohydrates storage in forest resilience under climate change. Curr Forestry Rep 6:1–13. https://doi.org/10.1007/s40725-019-00109-z

Quintanilla V (2000) Influencia del fuego en el desequilibrio ecológico de la vegetación en la zona mediterránea de Chile: casos de estudio. Investigaciones Geográficas 34:3–15. https://doi.org/10.5354/0719-5370.2000.27705

Quintanilla V, Constanzo MM (2018) Antecedentes sobre los impactos ecológicos de los fuegos y de otros factores antrópicos en los bosques de *Jubaea chilensis* (Moll) Baillon: caso de estudio: microcuencas periurbanas de las ciudades de Valparaíso y Viña del Mar, Chile. Territorium 25:75–88. https://doi.org/10.14195/1647-7723_25-1_6

Ramírez P, Villaseñor R, Eyzaguirre A, Morales N, Muñoz A (2014) Anatomía foliar de *Puya chilensis* Molina (Bromeliaceae), una planta que resiste al fuego. Chloris Chilensis 17:1

Rodríguez R, Marticorena C, Alarcón D, Baeza C, Cavieres L et al (2018) Catálogo de las plantas vasculares de Chile. Gayana Bot 75:1–430. https://doi.org/10.4067/S0717-66432018000100001

Rundel PW (1981) The matorral zone of central Chile. In: di Castri F, Goodall DW, Specht RL (eds) Mediterranean-type Shrublands, 1st edn. Elsevier, Amsterdam, pp 175–201

Sabadin P, Gómez MC, Ginocchio R, Peña I, Mujica AM, Montenegro G (2015) Effect of fire on herbaceous matorral vegetation of Central Chile. Cien Inv Agr 42:415–425. https://doi.org/10.4067/S0718-16202015000300010

Sáiz F, Bascuñán J (1990) Incendios forestales en el Parque Nacional La Campana, sector Ocoa, V Región, Chile. VII. Artrópodos asociados a hojarasca, excepto Coleoptera, Blattodea, Formicidae y Mutillidae. Anales Mus Hist Nat Valparaíso 21:51–61

Sáiz F, Solervicens J, Vivar C (1990) Incendios forestales en el Parque Nacional La Campana, sector Ocoa, V Región, Chile. VI. Coleópteros epígeos. Impacto y recuperación. Anales Mus Hist Nat Valparaíso 21:131–159

Sax DF (2002) Native and naturalized plant diversity are positively correlated in scrub communities of California and Chile. Divers Distrib 8:193–210. https://doi.org/10.1046/j.1472-4642.2002.00147.x

Segura AM, Holmgren M, Anabalón JJ, Fuentes ER (1998) The significance of fire intensity in creating local patchiness in the Chilean matorral. Plant Ecol 139:259–264. https://doi.org/10.1023/A:1009715328868

Sierra BC (1982) La acidez y alcalinidad de los suelos (ph). Instituto de Investigaciones Agropecuarias. https://hdl.handle.net/20.500.14001/30120. Accessed 16 May 2023

Smith-Ramírez C, Castillo-Mandujano J, Becerra P, Sandoval N, Allende R, Fuentes R (2021) Recovery of Chilean Mediterranean vegetation after different frequencies of fires. For Ecol Manag 485:118922. https://doi.org/10.1016/j.foreco.2021.118922

Torres-Díaz C, Gómez-González S, Torres-Morales P, Gianoli E (2012) Soil disturbance by a native rodent drives microhabitat expansion of an alien plant. Biol Invasions 14:1211–1220. https://doi.org/10.1007/s10530-011-0150-4

Traeger C (2021) Determinación de la riqueza de aves afectadas por incendios forestales en la región mediterránea de Chile durante eventos extremos en los años 2017 y 2019. Dissertation, Pontificia Universidad Católica de Chile. https://doi.org/10.7764/tesisUC/AGR/63116

Trew BT, Maclean IMD (2021) Vulnerability of global biodiversity hotspots to climate change. Glob Ecol Biogeogr 30:768–783. https://doi.org/10.1111/geb.13272

Urrutia-Jalabert R, González ME, González-Reyes Á, Lara A, Garreaud R (2018) Climate variability and forest fires in central and south-central Chile. Ecosphere 9:e02171. https://doi.org/10.1002/ecs2.2171

Villa-Martínez R, Villagrán C, Jenny B (2003) The last 7500 cal yr BP of westerly rainfall in Central Chile inferred from a high-resolution pollen record from Laguna Aculeo (34° S). Quat Res 60:284–293. https://doi.org/10.1016/j.yqres.2003.07.007

Villaseñor R, Sáiz F (1990) Incendios forestales en el Parque Nacional La Campana, Sector Ocoa, V Región, Chile: efecto sobre el estrato arbustivo-arbóreo. Anales Mus Hist Nat Valparaiso 21:15–26

Zhao Y, Feng D, Yu L et al (2016) Detailed dynamic land cover mapping of Chile: Accuracy improvement by integrating multi-temporal data. Remote Sens Environ 183:170–185. https://doi.org/10.1016/j.rse.2016.05.016

Zúñiga AH (2020) Changes in the structure of assemblages of three *Liolaemus* lizards (Iguania, liolaemidae) in a protected area of south-central Chile affected by a mixed-severity wildfire. Zoodiversity 54:265–274. https://doi.org/10.15407/zoo2020.04.265

Zúñiga AH, Rau JR, Fuenzalida V, Fuentes-Ramírez A (2020) Temporal changes in the diet of two sympatric carnivorous mammals in a protected area of south–central Chile affected by a mixed–severity forest fire. Anim Biodivers Conserv 43.177–186. https://doi.org/10.32800/abc.2020.43.0177

Zúñiga AH, Rau JR, Jaksic FM, Vergara PM, Encina-Montoya F, Fuentes-Ramírez A (2021) Rodent assemblage composition as indicator of fire severity in a protected area of south-central Chile. Austral Ecol 46:249–260. https://doi.org/10.1111/aec.12975

Fire in South American Tropical Forests: Examining the Past to Understand the Present and Predict the Future

12

Divino Vicente Silverio, Eddie Lenza, Letícia Gomes, Lucas Paolucci, Paulo Brando, and Marcia Macedo

Abstract

The tropical forests of South America encompass the vast Amazon and Atlantic Forest regions and are recognized for their remarkable biodiversity and carbon sequestration capacity. However, these valuable ecosystems are becoming increasingly threatened by climate change and wildfires. In this chapter, we analyze important aspects of fire in these tropical forests, including historical fire regimes, compounding effects of human activities and climate on fire, fauna, and forest resilience, as well as strategies to enhance fire management. Historically, forest fires have occurred at relatively low frequencies. More recently, extensive deforestation and climate change significantly increased the vulnerability of forests to fire. The consequences are extensive and affect biodiversity, carbon storage, and human wellbeing. In the short term, fires kill small, thin-barked trees, but over time, remaining large trees become more susceptible to windstorms and

D. V. Silverio (✉)
Universidade Federal Rural da Amazônia, Capitão Poço, Brazil
e-mail: divino.silverio@ufra.edu.br

E. Lenza · L. Gomes
Universidade do Estado de Mato Grosso, Nova Xavantina, Brazil
e-mail: eddie.lenza@unemat.br

L. Paolucci
Universidade Federal de Viçosa, Viçosa, Brazil
e-mail: lucas.paolucci@ufv.br

P. Brando
Yale School of the Environment, Yale University, New Haven, CT, USA
e-mail: paulo.brando@yale.edu

M. Macedo
Woodwell Climate Research Center, Falmouth, USA
e-mail: mmacedo@woodwellclimate.org

A. Fidelis, V. R. Pivello (eds.), *Fire in the South American Ecosystems*, Ecological Studies 250, https://doi.org/10.1007/978-3-031-89372-8_12

drought stresses. Fire also has immediate and long-term effects on fauna, including reductions in abundance and diversity, particularly among invertebrates. A rapid response to the increased frequency of wildfires is crucial for maintaining the resilience of tropical forests, safeguarding biodiversity, and mitigating climate change. We emphasize the importance of proactive conservation and management strategies that integrate scientific understanding with traditional knowledge to effectively address these challenges.

Resumen

Los bosques tropicales de Sudamérica, incluidas la Amazonia y la Mata Atlántica, destacan por su biodiversidad y capacidad de asimilación de carbono, pero se ven cada vez más amenazados por el cambio climático y los incendios forestales. Analizamos aspectos importantes de la ecología del fuego en estos bosques, desde el régimen histórico de incendios y el aumento de la inflamabilidad hasta los impactos de las actividades humanas y del cambio climático en la biota y su capacidad de recuperación. Históricamente, los incendios eran infrecuentes, pero la reciente deforestación y el cambio climático han aumentado la inflamabilidad de los bosques, provocando incendios, con consecuencias a la biodiversidad y al bienestar humano. Después del incendio, los árboles más pequeños y de corteza fina son los primeros en morir, pero a medio y largo plazo, aumenta la vulnerabilidad de los árboles grandes a fuertes vientos y al intenso estrés por sequía. La fauna, especialmente de invertebrados, muestra una reducción a corto plazo de la abundancia y diversidad. La situación actual exige una respuesta rápida a los incendios para preservar la resiliencia de los bosques, proteger la biodiversidad y mitigar los efectos del cambio climático, priorizando estrategias de conservación y gestión que integren conocimientos científicos y tradicionales.

Resumo

As florestas tropicais da América do Sul, abrangendo a Amazônica e Mata Atlântica, são notáveis pela biodiversidade e capacidade de assimilação de carbono, mas estão cada vez mais ameaçadas pelas mudanças climáticas e incêndios florestais. Neste capítulo discutimos alguns aspectos importantes da ecologia do fogo nessas florestas, desde o regime histórico de incêndios e aumento da inflamabilidade até os impactos das atividades humanas e mudanças climáticas sobre a biota e resiliência das florestas. Historicamente, incêndios eram raros, mas os recentes desmatamentos e alterações climáticas aumentaram a inflamabilidade das florestas, levando a incêndios. As consequências são extensas, afetando a biodiversidade e o bem-estar humano. Árvores menores e com casca fina são as primeiras a morrer em decorrência do fogo, mas no médio e longo prazo, o fogo também aumenta a vulnerabilidade das grandes árvores aos ventos fortes e estresses por secas intensas. Os efeitos para a fauna também são complexos, havendo, no curto prazo, redução da abundância e diversidade, especialmente para invertebrados. A situação atual requer uma resposta rápida aos incêndios para preservar a resiliência das florestas, proteger a biodiversidade e mitigar os efeitos das mudanças climáticas, priorizando estratégias de conservação e gestão que integrem os conhecimentos científico e tradicional.

12.1 Introduction

South American (SA) tropical forests (Fig. 12.1) are among the most biologically diverse and carbon-rich ecosystems on the planet, providing essential ecological services and supporting the livelihoods of millions of people. With an area of 8.8 million square kilometers (Olson et al. 2001), tropical forests cover ~50% of the continent,[1] and are concentrated mostly in Amazonia (74%) and the Atlantic Forest (12%; Fig. 12.1). These biomes harbor tremendous species diversity, housing more than 26% of all named tree species (ter Steege et al. 2013) and 10% of all named plant and vertebrate species on Earth (Albert et al. 2023). Forest ecosystems in this region are also an immense global carbon pool, containing approximately 180 Pg of carbon stored in the vegetation biomass and soil (Jung et al. 2021). However, these forest functions are increasingly threatened by changing fire regimes, often associated with changes in climate and land use, including logging, forest fragmentation, and edge effects (Brando et al. 2019; Berenguer et al. 2021; Lapola et al. 2023).

Land use changes across SA have increased pressure on natural forest ecosystems over the course of centuries and are a primary driver of changing fire regimes in the region (Fig. 12.2). The Brazilian Atlantic Forest has experienced high

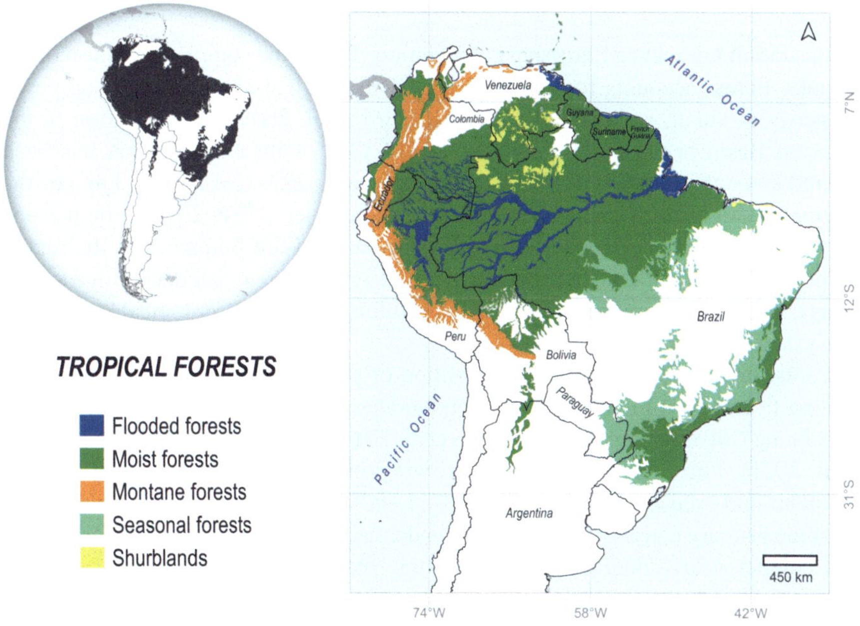

Fig. 12.1 Distribution of humid tropical forests in South America, based on a reclassification analysis according to vegetation types, as defined by Olson et al. (2001). (Produced by Júlia Abrantes Rodrigues)

[1] Although other vegetation types in South America can be defined as forests, here we focus on fire impacts on tropical humid forests of South America, as defined by Olson et al. (2001).

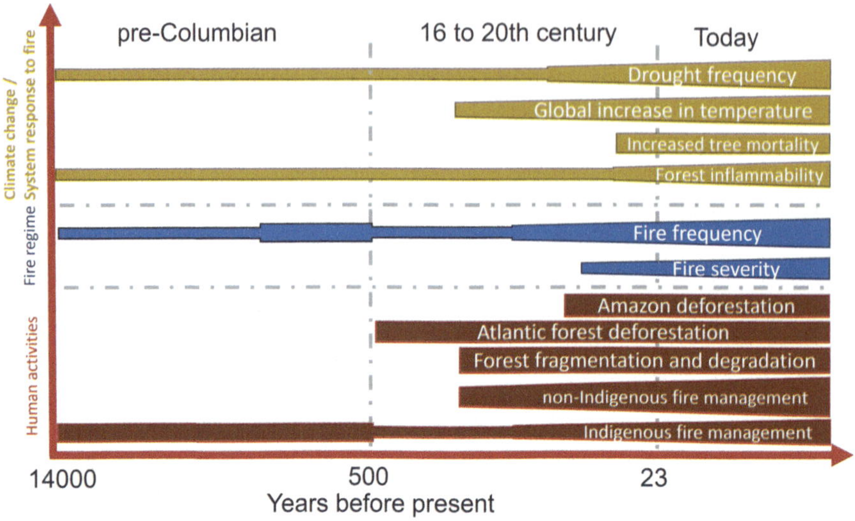

Fig. 12.2 Timeline of human activities, changes in fire regime, and system responses to fire in South American tropical forests. The bar width indicates the intensity of the factor, which may remain similar or widen over time, indicating an intensification of the factor over time

deforestation rates since European colonization. Today, the natural vegetation of the Atlantic Forest accounts for just 11–16% of its original extent and is composed primarily of small, isolated fragments (Ribeiro et al. 2009). Deforestation in the Amazon Basin began much later, but nearly 13% of the Pan-Amazon has been cleared since the 2000s (GFW 2021). Today, nearly 27% (2.4 million km^2) of the original extent of SA tropical forests has been deforested (GFW 2021), which is one of the main factors associated with recent increases in fire frequency in the region (Lapola et al. 2023; Aragão et al. 2018). Fire ignition associated with deforested areas can escape natural vegetation areas, resulting in several ecological impacts on ecosystem functioning.

Forest fires alter the species composition of plants and the structure of forests, leading to losses of biodiversity, carbon storage, ecosystem functions, and human well-being (Silvério et al. 2013; Brando et al. 2014; Nóbrega et al. 2019; Berenguer et al. 2021). Fires can also cause severe air pollution, resulting in serious health problems and even death (Butt et al. 2020). Landscape fires in general are estimated to be the primary cause of ~10,000 human deaths every year in South America, and this number nearly doubles during El Niño years (Johnston et al. 2012). Thus, understanding the ecological, social, and economic impacts of forest fires in SA is crucial for developing effective conservation and management strategies. In this chapter, we conduct a broad, systematic, and integrated review of fire ecology in SA tropical forests by analyzing the historical fire regimes and vegetation cover in the region. We then examine the new paradigm of fire in these forests, discussing the responses of vegetation and animal communities to novel fire regimes and exploring future scenarios and adaptation options. Finally, we identify important unanswered research questions and call for more concerted efforts to protect these critical ecosystems from the devastating effects of fires.

12.2 Historical Fire Regime and Vegetation

Historical records indicate that fires were rare events in SA tropical forests, often taking hundreds of years to reoccur (Sanford et al. 1985; Hammond and ter Steege 1998). Soil charcoal samples dated using C_{14} have revealed a median fire return interval of 500–3000 years for the Amazon region during the period from 900 to 10,000 years before the present (Feldpausch et al. 2022). Wildfire occurrence depends on the presence of ignition sources, fuel availability, oxygen, and climatic conditions (e.g., temperature, humidity, and wind) (de Dios 2020). Although oxygen was abundant in the past and lightning could function as an ignition source, higher rates of precipitation and lower seasonality maintained relatively high fuel moisture and low flammability of forests compared with the present. The use of fire as a management tool by Indigenous populations also represented an important source of ignition for forest fires in many tropical landscapes of South America. However, these fires tended to occur more often near riverbanks in southern Amazonia, where human population densities were higher (Bush et al. 2008).

Tropical forests have historically acted as natural firebreaks, and their normal fire regime is characterized by low frequency and severity (Feldpausch et al. 2022), which is related to relatively high annual precipitation, humidity, and shaded understories (Ray et al. 2005; Fig. 12.2). The 'fire regime' describes the long-term pattern of fire in a given area (Keeley 2009; Archibald et al. 2012), and the low frequency of fires in tropical forests is due to a combination of abundant moisture and infrequent droughts (Satoh et al. 2022), limited fuel availability, and reduced sunlight reaching the forest floor, which contributes to relatively low forest flammability (Keeley and Pausas 2022). Understanding how human activities are changing fire regimes and their interactions with vegetation dynamics is crucial for assessing the possible impacts of novel disturbances on forest ecosystems.

The rarity of forest fires in most tropical regions explains the lack of tree adaptations to survive fires (Pivello 2011). For example, thicker bark is a common trait in fire-adapted ecosystems as it insulates the cambium layer of the tree and helps prevent it from being damaged by heat from the fire (Brando et al. 2012). In the Amazon and Atlantic Forests, trees generally have thinner bark than in other forests, making them more vulnerable to fires (Staver et al. 2020). The capacity to resprout after a fire event is also an important adaptation to fire. Although some Amazon and Atlantic Forest species can resprout after a fire, the resprouts tend to die in the years following disturbance and do not become mature plants (Silvério et al. 2019). Collectively, these patterns indicate that the Amazon's stable climate (historically), coupled with its vegetation dynamics and their interactions, have maintained a long fire return interval for hundreds of years. This low fire frequency likely played an insignificant role in the evolution of plants in the biome, which tend to be highly vulnerable to fire.

Fire management by human activities over the last 14,000 years has also been an important factor driving species composition in SA tropical forests (Fig. 12.2). Indigenous Peoples (IP) played a crucial role in shaping plant species distribution through landscape management practices (Neves et al. 2021). Archaeological evidence has revealed that early Indigenous Peoples engaged in various activities, such as agroforestry and crop cultivation (Levis et al. 2018). Over the last 12,000 years, human habitation of the Amazon region has thus created complex agroecosystems that support high plant and animal diversity (Heckenberger et al. 2008; Neves et al. 2021). Indigenous Peoples have also used fire as a tool for land management, burning some forest areas to create open spaces for crops, and facilitating the growth of valuable plants such as Brazil nuts and palms (Levis et al. 2018). These practices not only left traces of anthropogenic dark earth soils at archaeological sites but also created new ecological niches that facilitated the establishment of plant species requiring higher nutrient levels (Rebellato et al. 2009; Arroyo-Kalin 2010). Although large forest fires occurred only during sporadic droughts (Feldpausch et al. 2022), ancient human activities have shaped the structure and composition of the Amazonian forests we see today, contributing to areas of high plant diversity, including species of great importance for food, medicine, and cultural practices (Clement et al. 2015). Moreover, the influence of ancient human practices extends beyond immediate ecological changes, as they also influenced the distribution of plant species through alterations in local environmental conditions, species dispersal, and biological interactions (Levis et al. 2018), influencing the trajectory of Amazonian forests over millennia (Fig. 12.2).

12.3 The New Paradigm of Fire in Tropical Forests

In recent decades, the occurrence and behavior of fires in tropical forests have shifted dramatically, influenced by changes in key factors that were less important in the past, including anthropogenic disturbances, extreme weather events, and climate change (Fig. 12.2). Although fire frequency may be reduced in landscapes where land use is consolidated and intensified (Andela et al. 2017; Ribeiro et al. 2024), the early phases of land-use change in tropical agricultural frontiers have been associated with pronounced increases in fire frequency (Alencar et al. 2015; Aragão et al. 2018; Oliveira et al. 2022). The clearing of forests for agriculture creates fragmented forest patches that lead to changes in the remaining forest structure and species composition, ultimately increasing the flammability of forest edges (Silvério et al. 2013; Brando et al. 2014; Alencar et al. 2015). Additionally, fire is a common tool for cleaning recently deforested areas and managing pastures (Barlow et al. 2020). These changes can affect fire dynamics, because fragmented forests may be more susceptible to fire ignition due to edge effects, higher fuel loads, and altered microclimatic conditions.

12.3.1 Drivers of Tropical Forest Flammability

Fuel moisture plays a critical role in determining the likelihood and severity of a forest fire. In undisturbed forests, litter is the primary fuel source, but it normally retains high moisture levels owing to the natural closed canopy (Ray et al. 2005; Oliveira et al. 2022). In contrast, secondary or previously disturbed forests have more open canopies that allow solar radiation to reach the forest floor, reducing fuel moisture levels and increasing flammability (Silvério et al. 2013; Brando et al. 2014; Balch et al. 2015). Additional fuel sources, such as grasses and herbaceous plants that colonize along forest edges, may also contribute significantly to increased fuel loads, as they generally have a lower moisture content and are highly flammable (Nepstad et al. 1999; Silvério et al. 2013). Recent studies have indicated that forest degradation, driven by various factors, now affects an area greater than that affected by deforestation (Matricardi et al. 2020; Lapola et al. 2023), which significantly increases the risk of fires in these landscapes (Oliveira et al. 2022). Understanding how this process works on the ground is crucial for implementing effective fire-management strategies under this novel fire regime.

The interplay between fuel characteristics and climate is key for assessing forest vulnerability to fire and predicting possible future trajectories of tropical forests (Fig. 12.3). There is a low probability of fire spreading into intact forests if the vapor pressure deficit (VPD) is lower than 0.75 kPa, which prevents fuel moisture from falling below 23% — the threshold for fire spreading into forests (Ray et al. 2005). As temperatures rise and the air becomes drier, however, the VPD increases (Ray et al. 2005; Brando et al. 2014; Ribeiro et al. 2022). These conditions are associated with a higher fire frequency and a reduction in the amount of biomass as fuels burn, releasing greenhouse gases into the atmosphere (Fig. 12.3). Resulting changes in the forest microclimate, such as increases in temperature and lower humidity, can further increase fire risk in the following years (Drüke et al. 2019; Gomes et al. 2020). These factors trigger a positive feedback loop that increases the likelihood of subsequent fires (Fig. 12.3), leading to forest degradation, as disturbances exceed the resilience of forest ecosystems (Blanco et al. 2014). If left unchecked, these processes can push forests towards a more open, degraded state with reduced capacity for carbon storage or provision of other ecosystem services (Nobre et al. 2016; Uribe et al. 2023).

12.3.2 Climate Change and Fire Occurrence

Increasing temperatures caused by ongoing global climate change have increased the frequency of drought events in SA tropical forests (Duffy et al. 2015; Satoh et al. 2022), making them more vulnerable to fires. Together, these processes increase nighttime VPD, thereby facilitating the spread of fires even during the night (Balch et al. 2022), resulting in larger burned areas. Furthermore, extensive deforestation (as in South America over the last four decades) can reduce precipitation (Smith et al. 2023), a process expected to worsen in the coming decades (IPCC 2021; Satoh

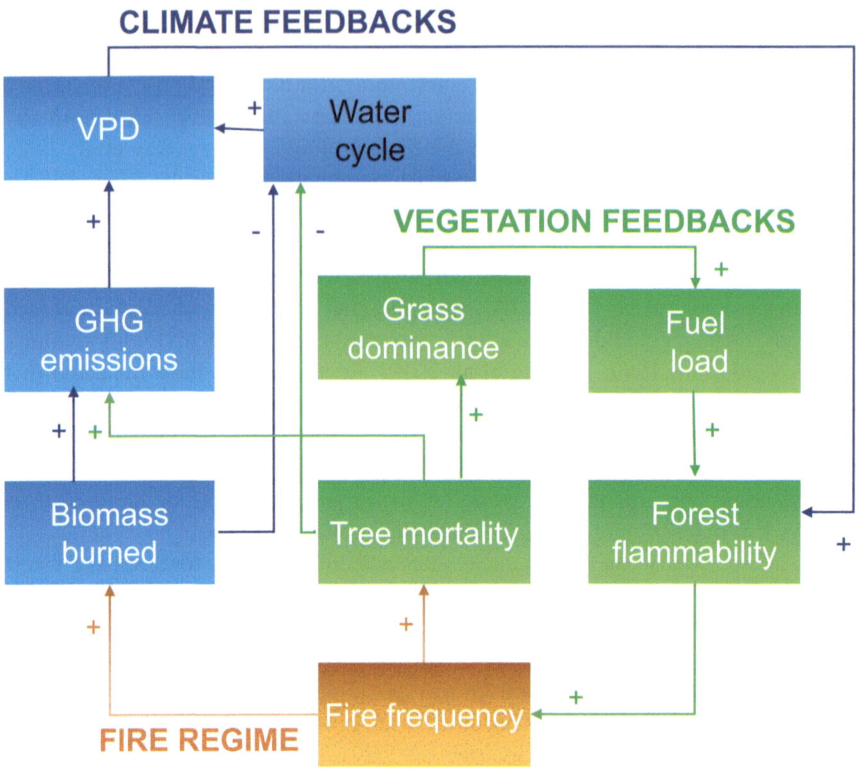

Fig. 12.3 Diagram showing feedback among climate, vegetation, and fire regimes. Arrows originate from the causes and end at the consequences, which can either increase (+) or decrease (−), *VPD* = vapor pressure deficit, GHG = greenhouse gases

et al. 2022), potentially increasing fire frequency and burned areas, especially in protected areas where there are continuous fuel loads (Brando et al. 2020; Flores et al. 2024). In this context, it is imperative to devise strategies to mitigate the impacts of climate change and forest fires on biodiversity and ecosystem functioning.

Historically, protected areas and Indigenous lands have helped to slow deforestation rates and fire activity (Nepstad et al. 2006; Walker et al. 2020; Almada et al. 2024). However, they may not be immune to the effects of climate change and forest fragmentation (Brando et al. 2014). As forest flammability increases, the risk of fire also increases over the next few years. The increasing frequency of droughts and fires is intricately linked to a heightened likelihood of irreversible forest loss (Silvério et al. 2022; Flores et al. 2024). This escalating landscape susceptibility facilitates the rapid propagation of fires and presents compounding challenges for fire management and forest conservation (Brando et al. 2020). Additionally, precipitation seasonality in the Amazon region is predicted to increase significantly over the near-term (IPCC 2021), potentially disrupting the temporal and spatial characteristics of the fire season. Consequently, the combination of climate and fire is expected to exert far-reaching impacts on the ecosystem services provided by these forests (Flores et al. 2024), which include mitigating greenhouse gas emissions (through biomass accumulation) and modulating climate across South America.

During drought years, the amount of dry fuel on the forest floor increases, contributing to a higher probability of wildfires. Many trees in the forest have adapted to shed their leaves to cope with lower soil moisture, which further increases the amount of litter on the forest floor (Lloyd and Farquhar 1994; Nepstad et al. 2004). Higher fuel loads on the ground increase fire spread rates, fuel consumption, and burned area—variables that drive fireline intensity and correlate positively with tree mortality (Brando et al. 2014). The more open canopy allows for a higher incidence of solar radiation in the understory (Trenberth et al. 2014; McDowell and Allen 2015), drying out fine fuel and increasing forest flammability (Malhi et al. 2009), particularly in edge environments (Silva Junior et al. 2020). As expected, the Amazon experienced a large number of forest fires during the regional droughts of 2005, 2010, and 2015 (Aragão et al. 2018). Zero deforestation and forest restoration are key interventions to reduce fire occurrence in tropical forests and will continue to be crucial for effective fire management and conservation policies. Integrated approaches to land-use planning and climate change adaptations will also be essential for promoting greater resilience to fires in tropical forest ecosystems.

12.3.3 Compounding Effects of Human Activities and Climate on Forest Fire Regimes

Human activities have profoundly modified the fire regime in South America (Fig. 12.4). The use of fire as a tool for land clearing is widespread, particularly for pasture management and clearing biomass in recently deforested areas. Given the right climatic conditions, these fires can easily get out of control and spread into unintended areas, causing damage to natural ecosystems (Nepstad et al. 1999; Oliveira et al. 2021). As deforestation rates increase, so does the likelihood of forest fires (Aragão et al. 2018). Recent estimates indicate that nearly 8% of SA tropical forests (743 thousand of km^2) have burned at least once in the last two decades (Fig. 12.4a). This burned area was most concentrated in the southern Amazon and Atlantic forests (Fig. 12.4a). In addition to deforestation, other human activities such as logging, and infrastructure development can increase the risk of fire occurrence by creating more forest edges and introducing new ignition sources (Cochrane and Laurance 2008). These human-induced changes to the fire regime have significant negative impacts on biodiversity, carbon emissions, human health, and livelihoods.

12.4 Vegetation Responses to Fire

The change of the fire regime, especially increased fire frequency, has had significant effects on various aspects of tropical forest vegetation (Figs. 12.2 and 12.4). The effect of fire is closely linked to climate and other compounding environmental

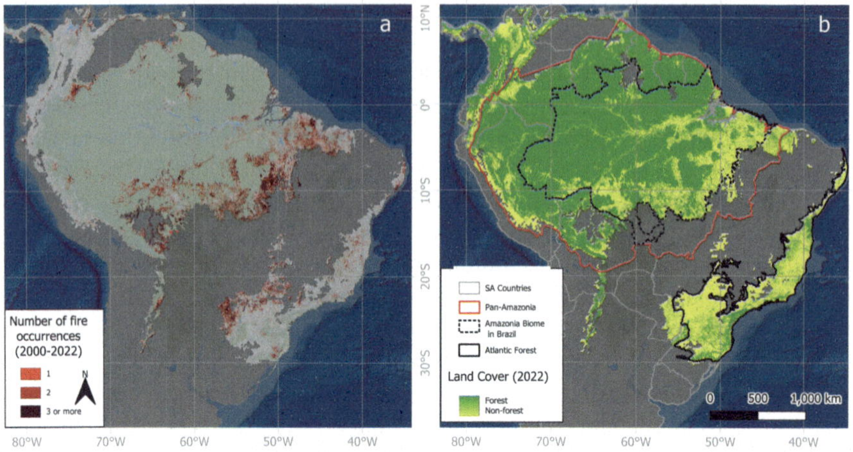

Fig. 12.4 (**a**) Burned area and magnitude of fire occurrences in tropical forests of South America calculated using the MCD64A1 product for the period from 2000 to 2022. (**b**) The current land cover (2022) in the area originally occupied by SA tropical forests. Remaining forests are shown in green and deforested areas in yellow. Also shown are the extents of the Amazon biome in Brazil (black dotted outline); the Atlantic Forest in Brazil (black continuous outline); and the entire Amazon watershed (red outline), spanning nine countries and referred to here as "Pan-Amazonia". (Produced by Eduardo Queiroz Marques)

factors, making it difficult to distinguish the individual effects of stressors on vegetation (Flores et al. 2024). Fires can have profound effects on the vertical and horizontal structure of tropical forests, while altering functional traits, reducing biomass, and species richness of forest communities (Nóbrega et al. 2019; Silvério et al. 2013). The impact of fires on vegetation may also vary across diverse forest types, vegetation layers, and ecological associations between vegetation and fauna (McLauchlan et al. 2020). Unraveling these relationships requires comprehensive, interdisciplinary research.

12.4.1 Forest Resilience to Fire Occurrence

The capacity of a forest to recover from burn events depends, in part, on its inherent resistance to fires, but also on the intensity of a given fire event and the fire frequency. Unlike savanna vegetation, which can recover quickly after a fire, few trees in Amazonian primary forests have adapted to fire (Hoffmann et al. 2012; Brando et al. 2012). Tropical forest resilience to fire is thus relatively low, as forests take a long time to recover biomass, vegetation structure, and plant diversity following a fire event. This slow recovery has far-reaching consequences for critical ecosystem services provided by forests (Trumbore et al. 2015). The long-lasting effect of fires on the provision of essential forest functions (e.g., biodiversity, climate regulation) stress the need for effective fire management to maintain the ecological integrity of these valuable ecosystems (Brando et al. 2020).

Tree mortality after forest fires can range from 5% to 100%, tending to be lower in upland forests burning for the first time and higher in flooded forests, which have little to no resistance to fire (Brando et al. 2014; Flores et al. 2017). In upland forests, tree mortality in response to fire is typically lower than 15% under normal climatic conditions (Balch et al. 2008). Subsequent fires under drought conditions increase fire intensity and severity, especially in edge environments, where tree mortality can exceed 70% (Brando et al. 2014). Seasonally flooded forests, on the other hand, can burn only under extreme drought conditions, but tree mortality can approach 100% when fire occurs (Flores et al. 2017). Recovery of burned flooded forests is very slow, limited by seed, nutrients, and an intense annual flooding regime (Flores and Holmgren 2021). Understanding the relationship between fire and forest resilience is critical for effective forest management and conservation (Aragão and Shimabukuro 2010; Flores et al. 2024).

Forest resistance to fire depends on traits that help plants to tolerate high temperatures, such as bark thickness and wood density. Individuals with thick bark and dense wood have greater thermal protection of their internal tissues (Brando et al. 2012) and are thus less affected by fire. Given that some of these traits are related to tree size, larger trees tend to have thicker bark and are less likely to be damaged by fire. On the other hand, large trees are more likely to be affected by drought, wind, and lightning (Silvério et al. 2019; Gora and Esquivel-Muelbert 2021). The combined effects of fires and other disturbances can make forests highly vulnerable to future fires. For example, large trees that survive fires typically have fire scars that make them more vulnerable to snapping during intense windstorms (Silvério et al. 2019). Thus, while forest fires kill mostly small individuals, the compounding effects of fire and other disturbances have lasting impacts on large trees, which makes forests highly vulnerable to future fires and degradation (Silvério et al. 2019; Gora and Esquivel-Muelbert 2021).

12.4.2 Impacts of Fire on Forest Biodiversity, Growth, and Reproduction

Fires in tropical forests profoundly modify the composition and richness of woody species and, consequently, the taxonomic and functional diversity of these forests (Nóbrega et al. 2019). This is because fire not only affects the survival of plants but also their growth and reproduction. Over the short- to medium-term, these changes can negatively impact animal and human populations that depend on forest resources (e.g., flowers, fruits, seeds, and construction material) (Kelly et al. 2020). High temperatures generated by flames can lead to immediate leaf loss and damage to flowers and fruits, particularly during the reproductive stage. Consequently, forest fires can significantly reduce the availability of resources during the months following fires.

The phenology (e.g. flowering, fruiting, and leaf dynamics) of Amazonian tree species is often poorly understood, as are the effects of anthropogenic disturbances on these processes. Together, fire and drought events can generate immense physiological stresses, with the potential to fundamentally change flowering and fruit production. For instance, previous studies (Brando et al. 2024) have reported lower fruit and flower production in areas where fires were most severe, and a reduction in the number of species producing fruits and flowers, particularly when fires co-occurred with drought and windthrow events. Where forests burned multiple times, the species composition of tree communities producing fruits and seeds became dominated by species with relatively faster growth, thinner leaves and bark, and lower height (Brando et al. 2024). Ultimately, such changes in reproductive phenology in response to fires can have a significant effect on faunal species that depend on forest resources, particularly during periods of food scarcity, leading to changes in faunal species abundance and diversity. This in turn can affect ecosystem processes that are essential for the proper functioning of ecosystems and related benefits to human wellbeing (Cardinale et al. 2012).

12.5 Fire Effects on Animal Communities

Unlike vegetation, fauna is often highly mobile, which facilitates immediate and short-term responses to fires. Nevertheless, forest fires can directly or indirectly affect associated fauna over several timescales. The direct and short-term effects of fires on fauna include mortality, forced emigration, and reduced reproductive success (Kaufman et al. 1990). Indirect and long-term effects may result from habitat changes, as fire occurrence tends to open the forest canopy and reduce understory vegetation. Although larger animals manage to escape from fires (Kinnaird and O'Brien 1998), small mammals, birds, snakes, and tortoises can suffer direct mortality from heat stress, carbon dioxide asphyxiation, physiological stress, or predation during the escape process (Kaufman et al. 1990).

Forest fires may also affect invertebrate species associated with leaf litter, often leading to decreased post-fire abundance and diversity. These organisms may experience increased temperatures and exposure during combustion (Swengel 2001), which increases the risk of desiccation and direct burning (New 2014). However, many arthropods manage to escape the direct effects of fire because of their greater mobility, nest location, or ability to take shelter in the lower layers of the soil (Swengel 2001). Indirect fire effects are the main cause of insect decline, and include changes in habitat structure, microclimate, resource availability, and competitive interactions (Andersen 2019). In some cases, insects may benefit from reductions in predators and competitors, greater food availability for detritivores, and regrowth of seedlings that offer more nutrition for herbivores (Coleman and Rieske 2006).

The indirect impacts of burning are also important for vertebrates. Some families of fruit plants are especially vulnerable to fires (Brando et al. 2024), which can reduce the abundance of large frugivorous vertebrates and even the local extinction of species associated with primary forests (Barlow and Peres 2006). Such indirect effects of fires may significantly alter the structure of faunal communities. As forest canopies become more open following fires, several studies have documented cascading effects on the associated fauna, with local extinctions of forest specialists and colonization by open-habitat specialists seen in the case of both invertebrate

(Andrade et al. 2017; Paolucci et al. 2017), and vertebrate (Sales et al. 2020) assemblages.

Interactions between plants and animals—including both mutualistic (e.g., seed dispersers) and antagonistic (e.g., herbivores) animal species—play a critical role in shaping the diversity and composition of tropical tree communities, particularly through their effects on regeneration (Jakovac et al. 2021). Animals mediate key ecological processes that are directly related to natural forest recovery, including seed dispersal, the top-down control of herbivores, nutrient cycling, and decomposition. By altering animal communities, fires may also affect these plant-animal interactions, triggering feedbacks that may decrease forest resistance and resilience to fire. For example, seed dispersal by invertebrates, such as ants and dung beetles, which frequently disperse early successional species, may decrease after forest fires França et al. 2020. With reduced dispersal, seedling density and diversity may decline, diminishing the ability of disturbed areas to recover (i.e., resilience).

Burned forests may also suffer from higher herbivory due to lower top-down control of insect herbivores; therefore, indirect effects mediated by animals can further increase forest degradation by fire (Queiroz et al. 2022). On the other hand, some animal groups can increase their contribution to resilience across burned forests. The lowland tapir (*Tapirus terrestris*), for example, defecates more often in burned forests and can disperse up to three times more seeds (including an abundance of seeds from large, climax tree species) in these areas than in undisturbed forests (Paolucci et al. 2019), potentially contributing to forest recovery. By understanding how different animal species respond to understory fires, we can better evaluate ecosystem functioning in the face of a changing climate and fire regimes. More research is needed to explore the long-term effects of understory fires on faunal communities and to devise effective conservation and management strategies to mitigate the negative effects of fires in tropical forest landscapes.

12.6 Future Scenarios of Fire and Adaptation

12.6.1 Land Tenure Regularization and Fire

Land tenure plays a crucial role in shaping present-day and future fire regimes, with the potential to either promote or discourage anthropogenic use of fire in tropical landscapes. Land ownership remains unclear in many different regions, which creates challenges for identifying and fining people who use fire inappropriately in the landscape. It also hinders implementation of sustainable land management practices and measures for fire prevention and control. For instance, Undesignated Public Forests occupy ~60 million hectares (15%) (SFB 2023) of the Brazilian Amazon—creating landscapes that are prone to land grabbing and illegal exploitation of natural resources and pose significant challenges for preventing deforestation and forest fires (Moutinho and Azevedo-Ramos 2023a). Recognizing the importance of land tenure regularization as a key component of future fire management strategies is essential for fostering community resilience, reducing fire-related risks, and promoting long-term sustainability in tropical forest ecosystems.

Land tenure insecurity is a key problem that must be addressed before adopting sustainable land use models in Amazonia (Azevedo-Ramos et al. 2020). Carbon emissions inside Indigenous territories and protected areas remain well below the level of private property (Walker et al. 2020; Almada et al. 2024). On the other hand, a large fraction of deforestation has occurred inside Undesignated Public Forests (Alencar et al. 2021), which today accounts for a large proportion of carbon losses in the Brazilian Amazon (Kruid et al. 2021).

Fires have significant and long-lasting effects on the biodiversity, carbon storage, ecosystem services, and livelihoods of local communities. It is imperative to improve the land-tenure regime and prioritize sustainable land management practices. This is essential to promote fire-resilient landscapes, and enhance the capacity for rapid response to wildfires. Additionally, integrating traditional knowledge, scientific research, and community engagement is essential for developing holistic fire management approaches that consider both ecological and socioeconomic dimensions. By addressing these challenges, we can strive to maintain the integrity and resilience of tropical forests and ensure their vital role in mitigating climate change, supporting biodiversity, and sustaining regional community well-being.

12.6.2 Fire Management

As climate changes, effective fire management will be critical for continued ecosystem functioning and sustainable development of tropical forests in South America. Because climate change has intensified wildfire regimes, this will require a paradigm shift in the fire culture. Climate-induced increases in fire frequency and severity have increased the complexity and cost of fire management in tropical forest landscapes. Whereas natural fires in the past have occurred under conditions of lower temperatures and higher air humidity, recent fires are often associated with higher temperatures, more severe droughts, and abundant anthropogenic ignition sources (Aragão et al. 2018; Brando et al. 2014). Given this scenario, a focus on fire prevention and suppression is essential. The use of fire as a tool for clearing deforested land and managing pasture areas in the Amazon has become ecologically inappropriate and socially costly. If changes in fire management are not implemented, the impact on natural ecosystems will be extensive. Below, we highlight three main changes that can help improve fire management in tropical forests.

1. *Early warning systems and long-term monitoring of fire impacts on ecosystems.* Controlling forest fires is a high-risk task that requires substantial financial resources. Early warning of fires is key as it helps reduce the time required to detect fires and increases the efficiency of command-and-control actions. The long distances and poor road conditions in the Amazon, for example, make it difficult to access active fire fronts. Although several early warning initiatives exist, timely access to information by firefighters and environmental agencies on the ground is hindered by field conditions with little connectivity; limited training in the use of geospatial data; and poor integration of data streams for use in real-time fire suppression activities. This is particularly important under the

extreme climatic conditions that enable large wildfires, when timely information could help prioritize actions and maximize the use of limited (financial and human) resources to manage fires. At the same time, the creation of local firefighting brigades and stricter fire management protocols can help distribute these resources to remote regions before fires get out of control. Creating and training professional and volunteer firefighting brigades at local scales is, thus, as essential as expanding requirements for fire management and firefighting plans for large companies.

2. *Investing in new technologies that can serve as alternatives to fire.* Despite the urgent need to reduce fires in tropical forests, the use of fire as a tool for land management is deeply rooted in the culture of South American farmers, ranchers, and Traditional and Local Communities (LC). Traditional and Local Communities and Indigenous Peoples living in the Amazon have used fire for millennia. Historically, this fire management occurred at specific locations and times of year, accounting for climatic and ecological conditions to minimize impacts on intact forests. These communities have a deep understanding of local ecology and use fire as a tool to promote biodiversity, renew soils, and facilitate sustainable agricultural practices (Clement et al. 2015; Levis et al. 2018). However, the new climatic reality (Silvério et al. 2022; Flores et al. 2024) requires new methods of fire management, grounded in participatory and collaborative approaches that enable these communities to produce food and access natural resources, while limiting fire use and minimizing cultural and ecological impacts. Reducing the use of fire while minimizing impacts on the cultures and livelihoods of Indigenous Peoples and other Traditional and Local Communities may be the greatest challenge to successful climate adaptation in tropical forest regions.
3. *Effective implementation of fire management laws.* Laws regarding fire management vary among South American countries. Most governments have established regulations that control or prohibit the use of fire, with the aim of avoiding uncontrolled fires and protecting sensitive ecosystems. However, the effective implementation of these laws can be challenging, especially in remote areas such as the Amazon. The Brazilian Federal Constitution establishes that any activities considered harmful to the environment are crimes and are subjected to sanctions (Brasil 1988). Building on this foundation, several laws have been established to protect native ecosystems. For example, the Brazilian Forest Code (Federal Law N°. 12,651/2012), the main legislation governing conservation on private properties, prohibits fire use in areas with natural vegetation (Brasil 2012). While the use of fire to manage agricultural lands is permitted with prior approval by environmental agencies, in most cases fires are illegal and enforcement is lacking. Thus, neither prohibition nor legal authorization has been sufficient to inhibit tropical degradation by fires. In contrast, the recently enacted Brazilian law for Integrated Fire Management (Federal Law No. 14.944/2024) significantly enhances fire management by allowing regulated prescribed burns across diverse biomes (Brasil 2024). This law aligns fire management with ecological requirements, a tool that helps to address the need for effective mitigation of fire impacts throughout the country.

12.7 Important Unanswered Research Questions

Despite significant advancements in our understanding of fire ecology in tropical forests, knowledge gaps remain, necessitating further investigation (McLauchlan et al. 2020). In this chapter, we discuss the historical and underlying environmental drivers of forest fires, as well as their links with climate, vegetation dynamics, and impacts on fauna. We also discuss the changing fire regime associated with contemporary human activities and consequences for species diversity, carbon storage, ecosystem services, and human well-being. Several important and intriguing questions remain, ranging from the ecological resilience of forests to fires to how the integration of scientific and traditional knowledge could help to enhance tropical forest resilience. By answering these questions, we can learn more about controlling fires, protecting biodiversity, and making the world a better place for future generations.

How does water limitation stress forest trees and influence understory fires? We know that in extreme droughts, for example, forest fires are more intense and kill more trees (Brando et al. 2014). Nevertheless, it remains unclear when water stress in the soil causes plants to shed their leaves. It is energetically expensive to produce new leaves in poor soil; therefore, trees usually retain their leaves for a long time. However, we do not know how nutrients and soil moisture work together with climate conditions to control litter production. Ongoing climate change is causing an increase in air temperature and precipitation patterns in South America (IPCC 2021), which may lead to an increase in fuel production and forest fire susceptibility. However, an increase in the CO_2 levels is expected to enhance the water use efficiency by plants. The interaction between climate, soil nutrients, and land-use changes requires further investigation to understand when water limitation stresses forests, increasing forest flammability.

How are animal communities and food webs affected by fire and how do they recover over time after disturbances? Fire modulates food webs in many different ecosystems and its effects across all trophic levels can be positive or negative (Bowman et al. 2016). The impact of fire on the food web is expected to have more negative effects in tropical forests, where biomass is constrained by climate and soil nutrients, but herbivory is less important for controlling biomass and fire regimes (Bond 2005). Indeed, there is a need to better understand how plant-animal interactions are affected by fire, particularly in Amazonia (Ballarin et al. 2023), as well as how animal communities and interactions recover from forest fires, and what strategies can be employed to facilitate the overall recovery of these vital ecosystems. Uncovering the intricacies of post-fire ecological resilience, particularly in terms of fauna, is essential for developing conservation and management strategies and that promote resilience and sustain healthy ecosystems.

How does the expansion and intensification of agriculture affect fire occurrence in tropical landscapes? The emergence and intensification of agriculture has been closely linked to alterations in fire occurrence patterns (Andela et al. 2017; McLauchlan et al. 2020; Ribeiro et al. 2024). However, understanding dynamics and feedbacks between agriculture and fire in South America remains a significant unanswered question, particularly regarding their variability across diverse biomes and the land use following deforestation. Investigate the interaction between these factors and fire occurrence, particularly in rapidly transitioning landscapes, is key to formulate informed land-use policies and fire management practices. Answering these questions will provide crucial insights into how fire patterns in the tropical forests of South America are likely to evolve over the next few decades.

Can monitoring forest degradation and innovative fire management help slow the adverse effects of changing climate and fire regimes on South America's tropical forests? Understanding the effectiveness of fire prevention, control, and combat strategies in these regions is crucial for promoting the sustainable coexistence of people with fires (Bowman et al. 2016). Because climate change, land use, and invasive species are shifting fire regimes in tropical forests, suppressing fires has become essential for preventing ecosystem degradation and negative social outcomes (Wu et al. 2022). A comprehensive understanding of fire regimes can guide various objectives, from reducing carbon emissions to enhancing biodiversity and human wellbeing. This multifaceted question spans both the scientific and policy domains, directly influencing landscape planning. Furthermore, integrating different knowledge systems (Alves et al. 2022) and clarifying land tenure (Moutinho and Azevedo-Ramos 2023b) may be prerequisites for the effective management of extensive tropical forest areas, particularly those overseen by Indigenous Peoples (Silvério et al. 2022). The incorporation of traditional knowledge related to fire and scientific information can lead to more holistic and culturally appropriate fire management approaches that benefit both the environment and local populations.

References

Albert JS, Carnaval AC, Flantua SGA, Lohmann LG, Ribas CC et al (2023) Human impacts outpace natural processes in the Amazon. Science 379:1–10. https://doi.org/10.1126/science.abo5003

Alencar AA, Brando PM, Asner GP, Putz FE (2015) Landscape fragmentation, severe drought, and the new Amazon forest fire regime. Ecol Appl 25:1493–1505. https://doi.org/10.1890/14-1528.1

Alencar A, Castro I, Laureto L, Guyot C, Stabile M, Moutinho P (2021) Amazon on Fire—deforestation and fire in undesignated public forests: Technical Note nº 7. https://ipam.org.br/wp-content/uploads/2021/04/Amazon-on-Fire-7-en.pdf. Accessed 21 Apr 2024

Almada HK, Macedo MN, Lenza E, Maracahipes L, Silvério DV (2024) Indigenous lands and conservation units slow down non-GHG climate change in the Cerrado-Amazon ecotone. Perspect Ecol Conserv 22:1–8. https://doi.org/10.1016/j.pecon.2024.03.002

Alves RP, Levis C, Bertin VM, Ferreira MJ, Cassino MF et al (2022) Local forest specialists maintain traditional ecological knowledge in the face of environmental threats to Brazilian Amazonian protected areas. Front For Glob Change 5:1028129. https://doi.org/10.3389/ffgc.2022.1028129

Andela N, Morton DC, Giglio L, Chen Y, Van Der Werf GR et al (2017) A human-driven decline in global burned area. Science 356:1356–1362. https://doi.org/10.1126/science.aal4108

Andrade RB, Balch JK, Carreira JYO, Brando PM, Freitas AVL (2017) The impacts of recurrent fires on diversity of fruit-feeding butterflies in a south-eastern Amazon forest. Journal of Tropical Ecology, 33, 22–32. https://doi.org/10.1017/S0266467416000559

Andersen AN (2019) Responses of ant communities to disturbance: five principles for understanding the disturbance dynamics of a globally dominant faunal group. J Anim Ecol 88:350–362. https://doi.org/10.1111/1365-2656.12907

Aragão LEOC, Shimabukuro YE (2010) The incidence of fire in Amazonian forests with implications for REDD. Science 328:1275–1278. https://doi.org/10.1126/science.1186925

Aragão LEOC, Anderson LO, Fonseca MG, Rosan TM, Vedovato LB et al (2018) 21st Century drought-related fires counteract the decline of Amazon deforestation carbon emissions. Nat Commun 9:1–12. https://doi.org/10.1038/s41467-017-02771-y

Archibald S, Staver AC, Levin SA (2012) Evolution of human-driven fire regimes in Africa. Proc Natl Acad Sci USA 109:847–852. https://doi.org/10.1073/pnas.1118648109

Arroyo-Kalin M (2010) The Amazonian formative: crop domestication and anthropogenic soils. Diversity 2:473–504. https://doi.org/10.3390/d2040473

Azevedo-Ramos C, Moutinho P, Arruda VLS, Stabile MCC, Alencar A, Castro I, Ribeiro JP (2020) Lawless land in no man's land: the undesignated public forests in the Brazilian Amazon. Land Use Policy 99:104863. https://doi.org/10.1016/j.landusepol.2020.104863

Balch JK, Brando PM, Nepstad DC, Coe MT, Silvério DV et al (2015) The susceptibility of southeastern Amazon forests to fire: insights from a large-scale burn experiment. Bioscience 65:893–905. https://doi.org/10.1093/biosci/biv106

Balch JK, Abatzoglou JT, Joseph MB, Koontz MJ, Mahood AL et al (2022) Warming weakens the night-time barrier to global fire. Nature 602:442–448. https://doi.org/10.1038/s41586-021-04325-1

Balch JK, Nepstad DC, Brando PM, Curran LM, Portela O, Carvalho O, Lefebvre PP (2008) Negative fire feedback in a transitional forest of southeastern Amazonia. Global Change Biology, 14, 2276–2287. https://doi.org/10.1111/j.1365-2486.2008.01655.x

Ballarin CS, Mores GJ, Alcarás de Goés G, Fidelis A, Cornelissen T (2023) Trends and gaps in the study of fire effects on plant–animal interactions in Brazilian ecosystems. Austral Ecol 49:1–16. https://doi.org/10.1111/AEC.13420

Barlow J, Peres CA (2006) Effects of single and recurrent wildfires on fruit production and large vertebrate abundance in a central Amazonian forest. Biodivers Conserv 15:985–1012. https://doi.org/10.1007/s10531-004-3952-1

Barlow J, Berenguer E, Carmenta R, França F (2020) Clarifying Amazonia's burning crisis. Glob Chang Biol 26:319–321. https://doi.org/10.1111/gcb.14872

Berenguer E, Lennox GD, Ferreira J, Malhi Y, Aragão LEOC et al (2021) Tracking the impacts of El Niño drought and fire in human-modified Amazonian forests. Proc Natl Acad Sci USA 118:1–8. https://doi.org/10.1073/pnas.2019377118

Blanco CC, Scheiter S, Sosinski E, Fidelis A, Anand M, Pillar VD (2014) Feedbacks between vegetation and disturbance processes promote long-term persistence of forest–grassland mosaics in South Brazil. Ecol Model 291:224–232. https://doi.org/10.1016/j.ecolmodel.2014.07.024

Bond WJ (2005) Large parts of the world are brown or black: a different view on the 'Green World' hypothesis. J Veg Sci 16:261–266. https://doi.org/10.1111/J.1654-1103.2005.TB02364.X

Bowman DMJS, Perry GLW, Higgins SI, Johnson CN, Fuhlendorf SD, Murphy BP (2016) Pyrodiversity is the coupling of biodiversity and fire regimes in food webs. Philos Trans R Soc Lond Ser B Biol Sci 371:1–11. https://doi.org/10.1098/RSTB.2015.0169

Brando PM, Nepstad DC, Balch JK, Bolker B, Christman MC, Coe M, Putz FE (2012) Fire-induced tree mortality in a neotropical forest: the roles of bark traits, tree size, wood density and fire behavior. Glob Chang Biol 18:630–641. https://doi.org/10.1111/j.1365-2486.2011.02533.x

Brando PM, Balch JK, Nepstad DC, Morton DC, Putz FE et al (2014) Abrupt increases in Amazonian tree mortality due to drought-fire interactions. Proc Natl Acad Sci USA 111:6347–6352. https://doi.org/10.1073/pnas.1305499111

Brando PM, Paolucci L, Ummenhofer CC, Ordway EM, Hartmann H et al (2019) Droughts, wildfires, and forest carbon cycling: a pantropical synthesis. Annu Rev Earth Planet Sci 47:555–581. https://doi.org/10.1146/annurev-earth-082517-010235

Brando PM, Soares-Filho B, Rodrigues L, Assunção A, Morton D et al (2020) The gathering firestorm in southern Amazonia. Sci Adv 6:1–10

Brando PM, Silvério D, Maracahipes L, Benzi R, Paolucci L et al (2024) Legacies of multiple disturbances on fruit and seed patterns in southeast Amazonia: implications for forest functional traits. Ecosphere 15:e4780. https://doi.org/10.1002/ecs2.4780

BRASIL (1988) Constituição da República Federativa do Brasil, de 5 de outubro de 1988. Diário Oficial da República Federativa do Brasil, Brasília, DF, 05 October 1988. https://www.planalto.gov.br/ccivil_03/constituicao/constituicao.htm. Accessed 13 Apr 2024

BRASIL (2012) Lei de proteção da vegetação nativa—Lei 12.727, 17 October 2012. www.planalto.gov.br/ccivil_03/_Ato2011-2014/2012/Lei/L12727.htm. Accessed 13 Apr 2024

BRASIL (2024) Política Nacional de Manejo Integrado do Fogo – Lei Nº 14.944, 31 de julho de 2024. https://www.planalto.gov.br/ccivil_03/_ato2023-2026/2024/lei/L14944.htm. Accessed 26 Mar 2025

Bush MB, Silman MR, McMichael C, Saatchi S (2008) Fire, climate change and biodiversity in Amazonia: a Late-Holocene perspective. Philos Trans R Soc Lond Ser B Biol Sci 363:1795–1802. https://doi.org/10.1098/rstb.2007.0014

Butt EW, Conibear L, Reddington CL, Darbyshire E, Morgan WT et al (2020) Large air quality and human health impacts due to amazon forest and vegetation fires. Environ Res Commun 2:1–18. https://doi.org/10.1088/2515-7620/abb0db

Cardinale BJ, Duffy JE, Gonzalez A, Hooper DU, Perrings C et al (2012) Biodiversity loss and its impact on humanity. Nature 486:59–67. https://doi.org/10.1038/nature11148

Clement CR, Denevan WM, Heckenberger MJ, Junqueira AB, Neves EG, Teixeira WG, Woods WI (2015) The domestication of Amazonia before European conquest. Philos Trans R Soc Lond Ser B Biol Sci 282:20150813. https://doi.org/10.1098/rspb.2015.0813

Cochrane MA, Laurance WF (2008) Synergisms among fire, land use, and climate change in the Amazon. Ambio 37:522–527. https://doi.org/10.1579/0044-7447-37.7.522

Coleman TW, Rieske LK (2006) Arthropod response to prescription burning at the soil–litter interface in oak–pine forests. For Ecol Manag 233:52–60. https://doi.org/10.1016/j.foreco.2006.06.001

da Silva Junior CA, Teodoro PE, Delgado RC, Teodoro LPR, Lima M et al (2020) Persistent fire foci in all biomes undermine the Paris Agreement in Brazil. Sci Rep 10:16246. https://doi.org/10.1038/s41598-020-72571-w

de Dios VR (2020) Fire as an earth system process. In: Plant-fire interactions: applying ecophysiology to wildfire management. Springer, Lleida, Spain, pp 31–51

Drüke M, Forkel M, von Bloh W, Sakschewski B, Cardoso M et al (2019) Improving the LPJmL4-SPITFIRE vegetation-fire model for South America using satellite data. Geosci Model Dev 12:1–27. https://doi.org/10.5194/gmd-2019-92

Duffy PB, Brando P, Asner GP, Field CB (2015) Projections of future meteorological drought and wet periods in the Amazon. Proc Natl Acad Sci USA 112:13172–13177. https://doi.org/10.1073/pnas.1421010112

Feldpausch TR, Carvalho L, Macario KD, Ascough PL, Flores CF et al (2022) Forest fire history in Amazonia inferred from intensive soil charcoal sampling and radiocarbon dating. Front For Glob Change 5:1–14. https://doi.org/10.3389/ffgc.2022.815438

Flores BM, Holmgren M (2021) Why forest fails to recover after repeated wildfires in Amazonian floodplains? Experimental evidence on tree recruitment limitation. J Ecol 109:3473–3486. https://doi.org/10.1111/1365-2745.13769

Flores BM, Holmgren M, Xu C, van Nes EH, Jakovac CC et al (2017) Floodplains as an Achilles' heel of Amazonian forest resilience. Proc Natl Acad Sci USA 114:4442–4446. https://doi.org/10.1073/pnas.1617988114

Flores BM, Montoya E, Sakschewski B, Nascimento N, Staal A et al (2024) Critical transitions in the Amazon forest system. Nature 626:555–564. https://doi.org/10.1038/s41586-023-06970-0

França FM, Ferreira J, Vaz-de-Mello FZ, Maia LF, Berenguer E et al (2020) El Niño impacts on human-modified tropical forests: consequences for dung beetle diversity and associated ecological processes. Biotropica 52:252–262. https://doi.org/10.1111/btp.12756

GFW (2021) Monitoramento de Florestas Projetado para a Ação. https://www.globalforestwatch.org/. Accessed 01 Sept 2023

Gomes L, Miranda HS, Silvério DV, Bustamante MMC (2020) Effects and behaviour of experimental fires in grasslands, savannas, and forests of the Brazilian Cerrado. For Ecol Manag 458:117804. https://doi.org/10.1016/j.foreco.2019.117804

Gora EM, Esquivel-Muelbert A (2021) Implications of size-dependent tree mortality for tropical forest carbon dynamics. Nat Plants 7:384–391. https://doi.org/10.1038/s41477-021-00879-0

Hammond DS, ter Steege H (1998) Propensity for fire in Guianan rainforests. Conserv Biol 12:944–947

Heckenberger MJ, Russell JC, Fausto C, Toney JR, Schmidt MJ et al (2008) Pre-Columbian urbanism, anthropogenic landscapes, and the future of the Amazon. Science (1979) 321:1214–1217. https://doi.org/10.1126/science.1159769

Hoffmann WA, Geiger EL, Gotsch SG, Rossatto DR, Silva LCR et al (2012) Ecological thresholds at the savanna-forest boundary: how plant traits, resources and fire govern the distribution of tropical biomes. Ecol Lett 15:759–768. https://doi.org/10.1111/j.1461-0248.2012.01789.x

IPCC (2021) IPCC AR6 WGI technical summary. In: Masson-Delmotte BZ, Zhai P, Pirani A, Connors SL, Péan C (eds) The physical science basis. Contribution of working group I to the sixth assessment report of the Intergovernmental Panel on Climate Change. Cambridge University Press

Jakovac CC, Junqueira AB, Crouzeilles R, Peña-Claros M, Mesquita RCG, Bongers F (2021) The role of land-use history in driving successional pathways and its implications for the restoration of tropical forests. Biol Rev 96:1114–1134. https://doi.org/10.1111/brv.12694

Johnston FH, Henderson SB, Chen Y, Randerson JT, Marlier M et al (2012) Estimated global mortality attributable to smoke from landscape fires. Environ Health Perspect 120:695–701. https://doi.org/10.1289/ehp.1104422

Jung M, Arnell A, de Lamo X, García-Rangel S, Lewis M et al (2021) Author correction: Areas of global importance for conserving terrestrial biodiversity, carbon and water. Nat Ecol Evol 5:1557–1557. https://doi.org/10.1038/s41559-021-01569-y

Kaufman DW, Finck EJ, Kaufman GA (1990) Small mammals and grassland fires. In: Fire in North American tallgrass prairies. University of Oklahoma Press, Norman, OK, pp 46–80

Keeley JE (2009) Fire intensity, fire severity and burn severity: a brief review and suggested usage. Int J Wildland Fire 18:116–126. https://doi.org/10.1071/WF07049

Keeley J, Pausas J (2022) Evolutionary ecology of fire. In: Evolutionary ecology of lizards. MDPI Books. https://doi.org/10.3390/books978-3-0365-4052-8

Kelly LT, Giljohann KM, Duane A, Aquilué N, Archibald S et al (2020) Fire and biodiversity in the Anthropocene. Science 370:eabb0355. https://doi.org/10.1126/science.abb0355

Kinnaird MF, O'Brien TG (1998) Ecological effects of wildfire on lowland rainforest in Sumatra. Conserv Biol 12:954–956. https://www.jstor.org/stable/2387568

Kruid S, Macedo MN, Gorelik SR, Walker W, Moutinho P et al (2021) Beyond deforestation: carbon emissions from land grabbing and forest degradation in the Brazilian Amazon. Front For Glob Change 4:1–10. https://doi.org/10.3389/ffgc.2021.645282

Lapola DM, Pinho P, Barlow J, Aragão LEOC, Berenguer E et al (2023) The drivers and impacts of Amazon forest degradation. Science 379:1–11. https://doi.org/10.1126/science.abp8622

Levis C, Flores BM, Moreira PA, Luize BG, Alves RP et al (2018) How people domesticated Amazonian forests. Front Ecol Evol 5:1–21. https://doi.org/10.3389/fevo.2017.00171

Lloyd J, Farquhar GD (1994) 13C discrimination during CO_2 assimilation by the terrestrial biosphere. Oecologia 99:201–215. https://doi.org/10.1007/BF00627732

Malhi Y, Aragão LEOC, Metcalfe DB, Paiva R, Quesada CA et al (2009) Comprehensive assessment of carbon productivity, allocation and storage in three Amazonian forests. Glob Chang Biol 15:1255–1274. https://doi.org/10.1111/j.1365-2486.2008.01780.x

Matricardi EAT, Skole DL, Costa BO, Pedlowski MA, Samek JH, Miguel EP (2020) Long-term forest degradation surpasses deforestation in the Brazilian Amazon. Science 369:1378–1382. https://www.science.org/doi/10.1126/science.abb3021

McDowell NG, Allen CD (2015) Darcy's law predicts widespread forest mortality under climate warming. Nat Clim Chang 5:669–672. https://doi.org/10.1038/nclimate2641

McLauchlan KK, Higuera PE, Miesel J, Rogers BM, Schweitzer J et al (2020) Fire as a fundamental ecological process: research advances and frontiers. J Ecol 108:2047–2069. https://doi.org/10.1111/1365-2745.13403

Moutinho P, Azevedo-Ramos C (2023a) Author correction: Untitled public forestlands threaten Amazon conservation. Nat Commun 14:1425. https://doi.org/10.1038/s41467-023-37218-0

Moutinho P, Azevedo-Ramos C (2023b) Untitled public forestlands threaten Amazon conservation. Nat Commun 14:1425. https://doi.org/10.1038/s41467-023-37218-0

Nepstad DC, Verssimo A, Alencar A, Nobre C, Lima E et al (1999) Large-scale impoverishment of Amazonian forests by logging and fire. Nature 398:505–508. https://doi.org/10.1038/19066

Nepstad D, Lefebvre P, Da Silva UL, Tomasella J, Schlesinger P et al (2004) Amazon drought and its implications for forest flammability and tree growth: a basin-wide analysis. Glob Chang Biol 10:704–717. https://doi.org/10.1111/j.1529-8817.2003.00772.x

Nepstad D, Schwartzman S, Bamberger B, Santilli M, Ray D et al (2006) Inhibition of Amazon deforestation and fire by parks and indigenous lands. Conserv Biol 20:65–73. https://doi.org/10.1111/j.1523-1739.2006.00351.x

Neves EG, Junqueira AB, Levis C, Clement CR, Schaan DP et al (2021) Biodiversidade e agrobiodiversidade como legados de povos indígenas. In: Cunha MC, Magalhães SB, Adams C (eds) Contribuições dos povos indígenas, quilombolas e comunidades tradicionais para a biodiversidade, políticas e ameaças. SBPC, São Paulo, pp 1–85

New TR (2014) Insects, fire and conservation. Springer International Publishing, Cham

Nobre CA, Sampaio G, Borma LS, Castilla-Rubio JC, Silva JS, Cardoso M (2016) Land-use and climate change risks in the Amazon and the need of a novel sustainable development paradigm. Proc Natl Acad Sci USA 113:10759–10768. https://doi.org/10.1073/pnas.1605516113

Nóbrega CC, Brando PM, Silvério DV, Maracahipes L, de Marco P (2019) Effects of experimental fires on the phylogenetic and functional diversity of woody species in a neotropical forest. For Ecol Manag 450:1–7. https://doi.org/10.1016/j.foreco.2019.117497

Oliveira U, Soares-Filho B, de Souza Costa WL, Gomes L, Bustamante M, Miranda H (2021) Modeling fuel loads dynamics and fire spread probability in the Brazilian Cerrado. For Ecol Manag 482:118889. https://doi.org/10.1016/j.foreco.2020.118889

Oliveira U, Soares-Filho B, Bustamante M, Gomes L, Ometto JP, Rajão R (2022) Determinants of fire impact in the Brazilian biomes. Front For Glob Change 5:1–12. https://doi.org/10.3389/ffgc.2022.735017

Olson DM, Dinerstein E, Wikramanayake ED, Burgess ND, Powell GV et al (2001) Terrestrial ecoregions of the world: a new map of life on earth. Bioscience 51:933–938. https://doi.org/10.1641/0006-3568(2001)051[0933:TEOTWA]2.0.CO;2

Paolucci LN, Schoereder JH, Brando PM, Andersen AN (2017) Fire-induced forest transition to derived savannas: Cascading effects on ant communities. Biological Conservation, 214, 295–302. https://doi.org/10.1016/j.biocon.2017.08.020

Paolucci LN, Pereira RL, Rattis L, Silvério DV, Marques NCS, Macedo MN, Brando PM (2019) Lowland tapirs facilitate seed dispersal in degraded Amazonian forests. Biotropica 51:245–252. https://doi.org/10.1111/btp.12627

Pivello VR (2011) The use of fire in the Cerrado and Amazonian rainforests of Brazil: past and present. Fire Ecol 7:24–39. https://doi.org/10.4996/fireecology.0701024

Queiroz EA, Schoereder JH, Cornelissen TG, Brando PM, Maracahipes L, Paolucci LN (2022) Reduced predation by arthropods and higher herbivory in burned Amazonian forests. Biotropica 54:1052–1060. https://doi.org/10.1111/btp.13129

Ray D, Nepstad D, Moutinho P (2005) Micrometeorological and canopy controls of fire susceptibility in a forested Amazon landscape. Ecol Appl 15:1664–1678. https://doi.org/10.1890/05-0404

Rebellato L, Woods W, Neves E (2009) Pre-Columbian settlement dynamics in the Central Amazon. In: Amazonian dark earths: Wim Sombroek's vision. Springer, Netherlands, Dordrecht, pp 15–31

Ribeiro MC, Metzger JP, Martensen AC, Ponzoni FJ, Hirota MM (2009) The Brazilian Atlantic Forest: how much is left, and how is the remaining forest distributed? Implications for conservation. Biol Conserv 142:1141–1153. https://doi.org/10.1016/j.biocon.2009.02.021

Ribeiro AFS, Brando PM, Santos L, Rattis L, Hirschi M et al (2022) A compound event-oriented framework to tropical fire risk assessment in a changing climate. Environ Res Lett 17:065015. https://doi.org/10.1088/1748-9326/ac7342

Ribeiro AFS, Santos L, Randerson JT, Uribe MR, Alencar AAC et al (2024) The time since land-use transition drives changes in fire activity in the Amazon-Cerrado region. Commun Earth Environ 5:1–11. https://doi.org/10.1038/s43247-024-01248-3

Sales LP, Galetti M, Pires MM (2020) Climate and land-use change will lead to a faunal "savannization" on tropical rainforests. Glob Chang Biol 26:7036–7044. https://doi.org/10.1111/gcb.15374

Sanford RL, Saldarriaga J, Clark KE, Uhl C, Herrera R (1985) Amazon rain-forest fires. Science 227:53–55. https://doi.org/10.1126/science.227.4682.53

Satoh Y, Yoshimura K, Pokhrel Y, Kim H, Shiogama H, Yokohata T et al (2022) The timing of unprecedented hydrological drought under climate change. Nat Commun 13:1–11. https://doi.org/10.1038/s41467-022-30729-2

SFB (2023) Serviço Florestal Brasileiro. https://www.gov.br/mma/pt-br/composicao/servico-florestal-brasileiro/cadastro-nacional-de-florestas-publicas/cnfp-2022. Accessed 21 Apr 2024

Silvério DV, Brando PM, Balch JK, Putz FE, Nepstad DC et al (2013) Testing the Amazon savannization hypothesis: fire effects on invasion of a neotropical forest by native cerrado and exotic pasture grasses. Philos Trans R Soc Lond Ser B Biol Sci 368:20120427. https://doi.org/10.1098/rstb.2012.0427

Silvério DV, Brando PM, Bustamante MMC, Putz FE, Marra DM, Levick SR, Trumbore SE (2019) Fire, fragmentation, and windstorms: a recipe for tropical forest degradation. J Ecol 107:656–667. https://doi.org/10.1111/1365-2745.13076

Silvério D, de Oliveira RS, Flores B, Brando P, Almada HK et al (2022) Intensification of fire regimes and forest loss in the Território Indígena do Xingu. Environ Res Lett 17:045012. https://doi.org/10.1088/1748-9326/AC5713

Smith C, Baker JCA, Spracklen DV (2023) Tropical deforestation causes large reductions in observed precipitation. Nature 615:270–275. https://doi.org/10.1038/s41586-022-05690-1

Staver AC, Brando PM, Barlow J, Morton DC, Paine CET et al (2020) Thinner bark increases sensitivity of wetter Amazonian tropical forests to fire. Ecol Lett 23:99–106. https://doi.org/10.1111/ele.13409

Swengel AB (2001) A literature review of insect responses to fire, compared to other conservation managements of open habitat. Biodivers Conserv 10:1141–1169. https://doi.org/10.1023/A:1016683807033

ter Steege H, Pitman NCA, Sabatier D, Baraloto C, Salomão RP et al (2013) Hyperdominance in the Amazonian tree flora. Science 342:1243092. https://doi.org/10.1126/science.1243092

Trenberth KE, Dai A, van der Schrier G, Jones PD, Barichivich J, Briffa KR, Sheffield J (2014) Global warming and changes in drought. Nat Clim Chang 4:17–22. https://doi.org/10.1038/nclimate2067

Trumbore S, Brando P, Hartmann H (2015) Forest health and global change. Science 349:814–818. https://doi.org/10.1126/science.aac6759

Uribe MR, Coe MT, Castanho ADA, Macedo MN, Valle D, Brando PM (2023) Net loss of biomass predicted for tropical biomes in a changing climate. Nat Clim Chang 13:274–281. https://doi.org/10.1038/s41558-023-01600-z

Walker WS, Gorelik SR, Baccini A, Aragon-Osejo JL, Josse C et al (2020) The role of forest conversion, degradation, and disturbance in the carbon dynamics of Amazon indigenous territories and protected areas. Proc Natl Acad Sci USA 117:3015–3025. https://doi.org/10.1073/pnas.1913321117

Wu C, Sitch S, Huntingford C, Mercado LM, Venevsky S, Lasslop G, Archibald S, Staver AC (2022) Reduced global fire activity due to human demography slows global warming by enhanced land carbon uptake. Proc Natl Acad Sci USA 119:e2101186119. https://doi.org/10.1073/pnas.2101186119

Fire in the Anthropocene

13

Patrícia S. Silva, Renata Libonati, Jose A. Marengo, Mabel Calim Costa, Lincoln Muniz Alves, and Isabel Belloni Schmidt

Abstract

Fire has been a natural disturbance for millions of years, influencing and shaping most terrestrial ecosystems on Earth. The dawn of the Anthropocene, an epoch in which humans represent a driving force within the Earth system, raises new questions and challenges for fire science. Humankind has been disrupting natural fire regimes in fire-prone regions and inducing fires in fire-sensitive ecosystems through increased ignition sources, fire suppression, and anthropogenic climate change. In South America, these dynamics are fundamental to understand current fire regimes, as anthropogenic burning now dominates vast regions across

P. S. Silva (✉)
Instituto Dom Luiz, Faculdade de Ciências, Universidade de Lisboa, Lisboa, Portugal

Yale School of the Environment, Yale University, New Haven, CT, USA
e-mail: pssilva@fc.ul.pt

R. Libonati
Departamento de Meteorologia, Universidade Federal do Rio de Janeiro, Rio de Janeiro, Brazil
e-mail: renata.libonati@igeo.ufrj.br

J. A. Marengo · M. C. Costa
Coordenação Geral de Pesquisa e Desenvolvimento, Centro Nacional de Monitoramento e Alertas de Desastres Naturais, São José dos Campos, Brazil
e-mail: jose.marengo@cemaden.gov.br

L. M. Alves
Instituto Nacional de Pesquisas Espaciais (INPE), São José dos Campos, Brazil
e-mail: lincoln.alves@inpe.br

I. B. Schmidt
Departamento de Ecologia, Instituto de Ciências Biológicas, Universidade de Brasília, Campus Universitário Darcy Ribeiro, Brasília, Brazil
e-mail: isabels@unb.br

A. Fidelis, V. R. Pivello (eds.), *Fire in the South American Ecosystems*, Ecological Studies 250, https://doi.org/10.1007/978-3-031-89372-8_13

the continent. Associated with rising temperatures, changing rainfall patterns, and the increased frequency of extreme events, the heightened risk of wildfires poses significant challenges for various sectors and the overall well-being of the region. Events such as the Pantanal 2020 fires, in Brazil, or the 2024 Chilean fires, should not be evaluated through a single meteorological or anthropogenic perspective, and have been shown to be an interplay of both factors. In this chapter, we will delve on the fire regimes of the Anthropocene, and how humans directly and indirectly influence fire activity in South America.

Resumen

El fuego es una perturbación natural que lleva millones de años influyendo en los ecosistemas terrestres. El Antropoceno—época que tiene el ser humano como fuerza transformadora del sistema terrestre—plantea nuevas preguntas y desafíos para la ciencia. La presencia humana ha deteriorado los regímenes naturales de incendios en ecosistemas donde el fuego ocurre de forma natural, e introduciéndolo en ecosistemas sensibles a esta perturbación, mediante un mayor número de igniciones, políticas de supresión y el cambio climático antropogénico. En Sudamérica, estas dinámicas son esenciales para estudiar los regímenes de incendios actuales, pues los incendios provocados por el hombre constituyen la gran mayoría de los incidentes. El aumento del riesgo de incendios asociado al aumento de temperaturas, la alteración de los regímenes de precipitaciones y la mayor frecuencia de fenómenos extremos plantea nuevos desafíos para la región, en diversos sectores de la sociedad. Incendios como los de 2020 en el Pantanal (Brasil) y de 2024 en Chile no deben valorarse desde una perspectiva puramente antropogénica o climática, sino como la acción combinada de los dos factores. Exploramos aquí los regímenes de incendios del Antropoceno y cómo los seres humanos influyen directa e indirectamente en el fuego en Sudamérica.

Resumo

O fogo é uma perturbação natural que vem influenciando os ecossistemas terrestres há milhões de anos. O começo do Antropoceno—época em que os humanos constituem uma força transformadora no sistema terrestre—levanta novas questões e desafios para a ciência. A presença humana vem deteriorando os regimes de fogo naturais em ecossistemas onde este ocorre naturalmente, e introduzindo fogo em ecossistemas sensíveis a esta perturbação, através de um maior número de ignições, políticas de supressão e alterações climáticas antropogênicas. Na América do Sul, estas dinâmicas são essenciais para o estudo dos atuais regimes de fogo, dado que fogos de origem humana constituem a vasta maioria das incidências. O aumento no risco de fogo associado a temperaturas mais elevadas, regimes de precipitação alterados e maior frequência de fenômenos extremos, acarreta novos desafios para a região, em vários setores da sociedade. Eventos como os fogos de 2020 no Pantanal (Brasil) e os de 2024 no Chile não devem ser avaliados numa perspectiva puramente antrópica ou climática, mas sim, como a ação combinada de ambos estes fatores. Neste capítulo, exploramos os regimes de fogo do Antropoceno e como os humanos influenciam direta e indiretamente o fogo na América do Sul.

13.1 Introduction

Humans have impacted the Earth for thousands of years. Anthropogenic activity is a key characteristic and distinguishing feature of the Holocene (*holos* meaning "whole" and *cene* "new"), the interglacial period of the past 10–12 millennia. According to the Early Anthropogenic Hypothesis, humans may have been responsible for rises in carbon dioxide and methane levels as early as 7000 years ago, leading to anomalous carbon dioxide and methane increasing trends that severely contrast with the natural downward trends of previous interglacial periods (Ruddiman et al. 2020). Linked with changes in climate, humans have also precipitated the mass extinction of large-bodied fauna between 50,000 and 10,000 years ago (Koch and Barnosky 2006). However, these pre-industrial human impacts developed relatively slowly compared to the exponential and near-globally synchronous changes of a post-industrial time (Zalasiewicz et al. 2019). Ocean acidification due to anthropogenic carbon dioxide input into the oceans is happening at rates possibly unseen over the last 300 million years (Hönisch et al. 2012). Current anthropogenic-driven carbon emissions are also occurring at a rate that is unprecedented over the last 66 million years (Zeebe et al. 2016) and may possibly postpone the next ice age for 100,000 years (Ganopolski et al. 2016). Recent extinction rates are 100–1000 times their pre-human levels (Pimm et al. 1995), with a potential mass extinction currently unfolding with impacts that may last over geological timescales (Williams et al. 2015).

Given the profound impacts of human activity on the Earth system, scientists have put forward the term Anthropocene (from *anthropo* = human and *cene* = new) to define this human-dominated epoch (Crutzen and Stoermer 2000). The Anthropocene defines a period where humans are a predominant "force of nature", capable of discernibly affecting the Earth system beyond natural variability (Steffen et al. 2007). While extensively used, the Anthropocene is not formalised as an epoch in the Geological Time Scale. It is currently being examined as a potential new time unit, but its definition poses novel and difficult challenges. The Anthropocene relies on a paradigm shift from traditional environmental science to Earth system science, in which the Earth is considered to be a complex planetary-level system, not entirely describable by the sum of its parts (Hamilton 2015). The Anthropocene, as initially proposed (Crutzen and Stoermer 2000), is not merely the effects of human activity on ecology or the environment pinpointed by stratigraphic markers, but rather a detectable impact on functioning of the Earth system. From the mid-twentieth century onwards, there is clear evidence for fundamental shifts in the state and functioning of the Earth system driven by human activities. This period, known as the "Great Acceleration", is characterised by the acceleration of the human imprint on the Earth system across the socio-economical and biophysical spheres, and has been proposed as the starting date of the Anthropocene (Steffen et al. 2015). Other authors consider that the Anthropocene began with the Industrial Revolution (Crutzen 2002), or with the advent of nuclear bombings and testing (Zalasiewicz et al. 2015; Waters et al. 2015). Although there is ongoing scientific debate, the formal establishment of this new epoch would mark the end of the Holocene and the beginning of a new era fundamentally impacted by humankind.

One of the very first impacts of humans on the Earth system was the use of fire. Fire has been a natural feature of the Earth system for over 420 million years, impacting all of its components: the biosphere, hydrosphere, geosphere, cryosphere and atmosphere (Bowman et al. 2020). As seen in the previous chapters, fire is a fundamental part of the ecosystem in the fire-dependent ecosystems, sustaining native and evolutionarily adapted fauna and flora while occurring at the natural bounds, set by seasonal climates, fuel accumulation, ignitions and other natural factors (Hardesty et al. 2005). These natural fires are often triggered by lightning and incandescent fallout from volcanic eruptions (Glikson 2013). Nevertheless, exposure to anomalous fires in fire-prone ecosystems, such as savannas, increases the exposure and vulnerability of adjacent ecosystems not adapted to fire, such as seasonally flooded forests (Bilbao et al. 2020). Likewise, in fire-sensitive ecosystems, fire severely damages ecosystem structure and the relative abundance of species, as they lack adaptations that allow them to respond positively to, or rapidly rebound after fire (Hardesty et al. 2005) (see Chap. 12).

Around 1 million years ago, fire-wielding *hominids* started to disrupt these natural fire regimes (Bowman et al. 2020). Albeit long-term datasets of fire activity are lacking, proxies such as charcoal records indicate that there has been a global increase on biomass burning since the last glacial maximum, around 21,000 years ago, and a shift from low to high spatial heterogeneity around 12,000 years ago (Marlon et al. 2016; Power et al. 2008). Initially, fire was used for warmth, cooking and hunting (Glikson 2013). More recently, as the transition from subsidence to industrial economies unfolded, forests turned into agricultural or pastoral landscapes through the use of fire (Bowman et al. 2009), and with increased landscape fragmentation the use of fire became utilitarian, for uses such as crop enhancement, charcoal production, and smelting activities (Gillson et al. 2019). Humans are now directly influencing fire regimes by modifying landscapes and fuel loads, by suppressing natural ignitions whilst also being an ignition source, and indirectly by affecting the climate. The beginning of the twentieth century saw a decrease in global fire activity, mostly due to fire suppression and elimination, and highly modified and fragmented landscapes (Mouillot and Field 2005). In many fire-prone ecosystems, fire suppression and elimination has homogenised and simplified landscapes, eroding their ability to recover from disturbance, and setting the stage for larger and more intense fires (Gillson et al. 2019). Accordingly, the latter half of the twentieth century is marked by an increase in global burned areas that are not spatially homogeneous, with regional fire regimes showing great differences in the magnitude and direction of change (Mouillot and Field 2005). Many ecosystems where fire is historically absent or rare have burned recently, such as the Amazon

(Barlow et al. 2019). Tropical forests saw an exponential increase in biomass burning over the last few decades, due to deforestation associated with agricultural expansion (Mouillot and Field 2005).

Currently, there are very few environments that are not influenced by anthropogenic fires: around 93% of fire-sensitive ecosystems and 77% of fire-dependent ecosystems worldwide show altered fire regimes where fire is thought to be occurring beyond natural bounds (Hardesty et al. 2005). Changes in historical fire activity threaten biodiversity worldwide, particularly savannas, grasslands, shrublands and forests (Kelly et al. 2020), ecosystems that are present in South America. Fires also are a substantial source of carbon dioxide, releasing 2.2 Pg of carbon per year in the period of 1997–2016, representing almost a quarter of global carbon emissions from fossil fuel combustion (Van Der Werf et al. 2017). Around a fifth of this value is due to deforestation and agricultural practices (Van Der Werf et al. 2017), estimated to contribute up to ~19% of the total increased radiative forcing since pre-industrial times (Bowman et al. 2009). While vegetation fires may see some of the carbon re-captured due to post-fire recovery, permanent deforestation or burning of organic deposits—such as peatlands—are net sources of carbon to the atmosphere (Bowman et al. 2020). These emissions represent a substantial emission of greenhouse gases to the atmosphere, conducive to changes in global and regional climate.

The Anthropocene is undoubtedly connected to climate change, as it was initially proposed to have begun in the late eighteenth century, based solely on the measured increase in atmospheric carbon dioxide and methane (Crutzen 2002). It is known since the nineteenth century that changes in greenhouse gas concentration in the atmosphere lead to an increase in global temperature. Palaeoclimate studies also show that the Earth's climate has naturally changed on all time scales reflecting variations in its radiation balance (IPCC 2021). These variations may be due to natural changes in the incoming solar radiation (or orbital forcing, through variations in Earth's orbit around the Sun or in the Sun itself), changes in the fraction of solar radiation that is reflected (the planet's albedo), or in the longwave energy radiated back to space (by means of greenhouse gas concentrations in the atmosphere). However, ongoing changes in climate are not attributed to orbital forcing as the Earth is not thought to naturally enter another Ice Age for at least another 30,000 years (IPCC 2021). Current atmospheric concentration of carbon dioxide and methane exceed considerably the natural range of the last 650,000 years, and anthropogenic carbon dioxide emissions are increasing around 250 times faster than those from natural sources since the Last Ice Age (IPCC 2021). The increasing rate of radiative forcing due to greenhouse gases over the past century is possibly unseen over the last 16,000 years (IPCC 2021), and current atmospheric carbon dioxide levels, above 400 parts per million (ppm), have not been seen in ice core measurements over the last 800,000 years (Lüthi et al. 2008). The Intergovernmental Panel on Climate Change (IPCC), the leading international body for the assessment of climate change, has deemed undisputed that human activities are currently disturbing the climate system, mostly through the emission of greenhouse gases, and may have impacted the Earth for millennia going forward (Calvin et al. 2023; IPCC 2021).

Changes in historical fire activity over the last 21,000 years have been linked to changes in climate, especially temperature variations (Daniau et al. 2012; Osmont et al. 2019). Wildfires, defined as unusual or extraordinary free-burning vegetation fires with negative impacts, have a synergic relationship with climate change in which they are "mutually exacerbating": meteorological conditions favourable to wildfires are increasing over many regions of the globe, while wildfires further contribute to climate change by emitting massive amounts of greenhouse gases to the atmosphere and accelerating ecosystem degradation (United Nations Environment Programme 2022). Moreover, as climate is changing and ecological thresholds are crossed, the relative dominance of species will alter, both responding to and driving changes in fire conditions (Gillson et al. 2019).

Understanding the underlying factors that influence current fire activity is crucial for effective ecosystem management and informed policymaking. In this chapter we compile state-of-the-art science on how anthropogenic activity is influencing fire regimes in South America, and how future climate change may influence future fire activity.

13.2 How Anthropogenic Activity Affects Fire Regimes

Anthropogenic activity influences fire over multiple temporal and spatial scales. Considering the fire behaviour and regime triangles, humans now directly or indirectly influence all its components by altering the amount, type, and connectivity of fuel, the frequency of ignitions, and disrupting climate patterns (Bowman et al. 2020).

Fire regimes across vast areas of the South American continent are undergoing significant changes, primarily driven by anthropogenic pressure, land use conversion, and climate change. Although South America vegetation fires are influenced by the interplay of climate and natural vegetation dynamics, the prevalence of anthropogenic burning now dominates extensive regions across the continent, not only altering natural fire regimes in fire-prone regions but inducing fires in fire-sensitive ecosystems (Pereira et al. 2022). Indeed, South America is one of the regions with the highest fire activity in the world, with most fires having an anthropogenic origin (Andela et al. 2017) and spanning across the entire continent (Bella et al. 2006). Satellite data for Argentina, Brazil and Bolivia indicates a very large fire incidence (Fig. 13.1), and Brazil has larger burned areas and number of fires than Europe and the US combined (Manzello et al. 2018). In terms of burned area, Brazil and Bolivia are the most affected countries with 4% of their territories burning annually, in most part savannas and the Amazon rainforest (Bilbao et al. 2020). Paraguay is the fourth country in South America with the highest number of active fires during the 2003–2022 period (Fig. 13.1) and the first in terms of the average number of active fires per area (Bilbao et al. 2020), even though there is no official record of the incidence of fire in its territory.

Although there has been extensive research on naturally induced fires in boreal and temperate ecosystems of the North Hemisphere, relatively less attention has been devoted to other regions of the globe due to their lower occurrence compared

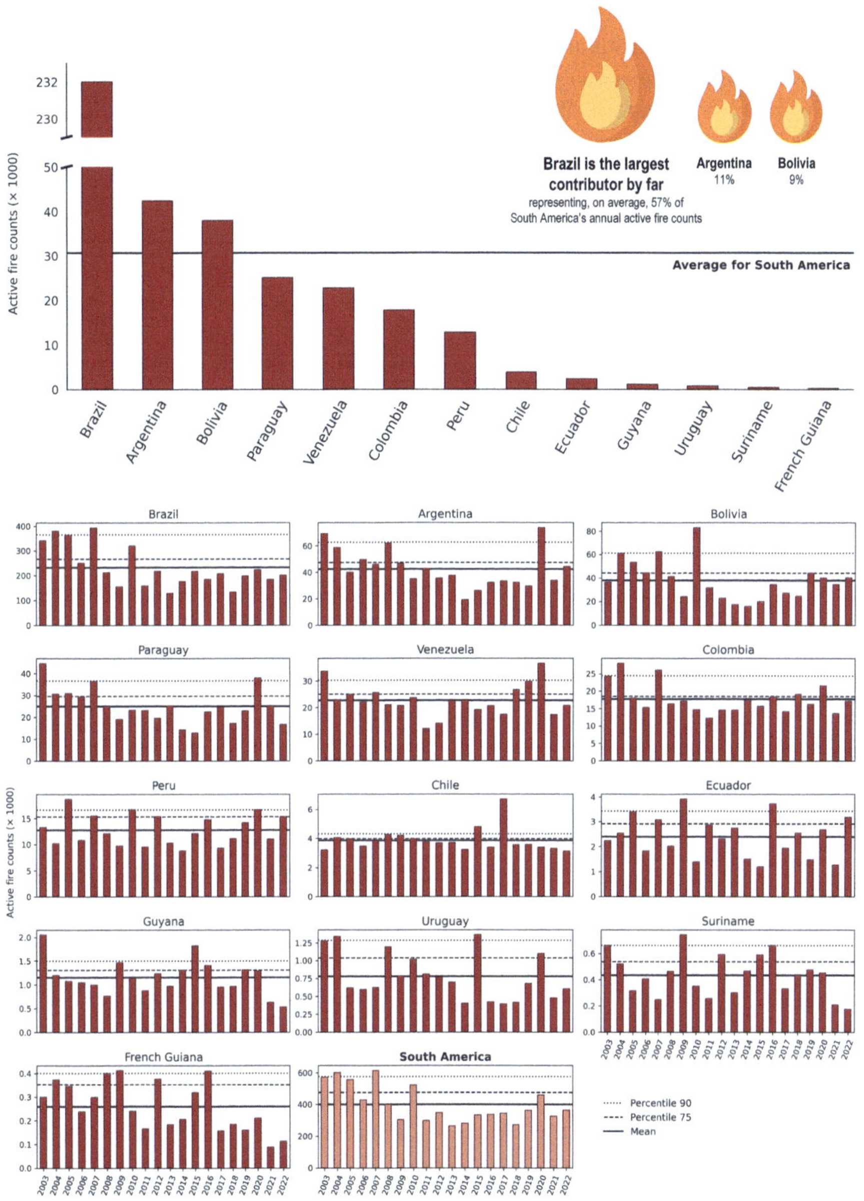

Fig. 13.1 Top panel: Annual average active fire counts as detected by MODIS/AQUA sensor from 2003 to 2022 per South American country. The average of all countries in South America is represented in a solid black line. Bottom panels: Interannual variability of active fire counts per country in South America from 2003 to 2022, in descending order of annual averages. South America totals are also included in lighter red bars. Dotted lines represent the percentile 90 of the time series, dashed lines represent the percentile 75, and solid lines represent the mean. (Data from BDQUEIMADAS/INPE (https://terrabrasilis.dpi.inpe.br/queimadas/bdqueimadas/), using icons from Freepik (Flaticon))

to fires caused by human activities in South America, Asia, and Africa (Moris et al. 2023). Fires induced naturally by lightning are infrequent and typically limited in size within South American ecosystems, such as wet tropical forests (Gora et al. 2020; Gora and Yanoviak 2020), Mediterranean regions of Chile (Corporación Nacional Forestal 2023), the savannas of Brazil (Schumacher et al. 2022) and wetlands (Menezes et al. 2022). In the Cerrado, lightning-induced ignitions primarily occur during the transition period between the rainy and dry seasons (Ramos-Neto and Pivello 2000; Schumacher et al. 2022), producing low-intensity fires that trigger a positive feedback loop by favouring the growth of grass species that is not to the detriment of tree species (Hoffmann et al. 2012). In the Pantanal wetlands, while the huge majority of large fires are not naturally caused, lightning-induced fires are an important feature from the summer fire regime (Menezes et al. 2022). While lightning may not be a primary cause of significant fires in most tropical forests, the direct damage from lightning strikes plays an important ecological role in this ecosystem (Yanoviak et al. 2020).

Throughout the continent, the vast majority of ignitions are of anthropogenic origin: in south-central Chile, 98.5% of fire ignitions are associated with human activities, either by accident or intent (Pozo et al. 2022); and in the Pantanal, natural-caused fires account for only 1% of the total scars (Menezes et al. 2022). Even in fire-dependent ecosystems, where natural occurring fires are a common and essential disturbance, anthropogenic fires are dominating the current fire regime. For instance, the vast majority of ignitions in the Cerrado are anthropogenic, leading to more intense fires that persist for a greater duration and occur mainly in the dry season. Anthropogenic ignitions during the dry season occur for a variety of reasons, including species and landscape management (Schmidt et al. 2007), cattle rearing on native or exotic pasturelands, subsistence agriculture, and industrial agriculture (Eloy et al. 2019b). These anthropogenic fires trigger changes to the floristic composition and community structure of tree-shrub vegetation, affect the accumulation of aerial woody biomass and cause a substantial reduction in the biomass of fire-sensitive species, which may transform the Cerrado from a carbon sink to a source of CO_2 emissions (da Silva Junior et al. 2020; de Azevedo et al. 2020; Hoffmann et al. 2012).

Human activity further influences fire extent and intensity by land conversion. Around 40% of South American ecosystems (approximately 430 million hectares), have already been impacted by human activity (Zalles et al. 2021): since 1985, natural cover has decreased by 16% whereas pasture, croplands and plantations have increased by 23%, 160% and 288%, respectively. Moreover, around 55 million hectares of disturbed land has no particular use, representing degraded ecosystems with no economic value (Zalles et al. 2021). The ongoing economic development has been at the expense of native cover and associated ecosystem services, including fire, severely disturbing and threatening South American unique landscapes. However, the dynamics of these factors and their influence on fire differ greatly depending on the regional context and socio-economical aspects. In areas experiencing agricultural expansion and native vegetation clearance, fires tend to be larger and more intense. Conversely, in highly fragmented landscapes, fires tend to be smaller and less intense (Campagnolo et al. 2021; Rosan et al. 2022; Silva et al. 2021). The Cerrado, for example, exhibits both these dynamics between fire extent and fuel continuity. The northeastern part of the biome, known as MATOPIBA (the confluence of states Maranhão, Tocantins, Piauí, and Bahia), represents Brazil's latest agricultural frontier and has witnessed significant deforestation rates in recent decades (Trigueiro et al. 2020). In this region, larger fire scars and higher fire intensity have been associated with land conversion activities (Silva et al. 2020). Conversely, in the southern Cerrado, where agricultural and pasture fields dominate, fire scars are smaller, and fires are less intense (Silva et al. 2021).

Similarly, in Chile, fire extent and probability has been shown to be linked, among other factors, with vegetation type and population density (McWethy et al. 2018). Most fires occur over savanna, pastures and croplands (51%), while only 31% of fires occur in native forests or forest plantations (Sarricolea et al. 2020). Forested regions in Chile show increasing trends of burned areas, while grasslands and croplands have decreasing trends, possibly due to effective firefighting or prevention facilitated by easier access (Sarricolea et al. 2020). The large wildfires of 2017 in Chile have been associated with large-scale plantations of non-native species *Pinus* and *Eucalyptus* that provided extensive and continuous fuel loads (Bowman et al. 2019), and plantations show a positive burned area trend over the country (Sarricolea et al. 2020). Over several regions of Argentina, including the Pampa, Chaco and Patagonia, land-use changes, fire suppression strategies, and agricultural expansion, have been pinpointed as important causes of large wildfires (Bilbao et al. 2020). Recent studies have also linked the degradation of wetlands in South America as a driver for larger and more severe fires (Kumar et al. 2022). Fire-related human activities by means of illegal land-use changes and use of the fire to prepare the land for agriculture (Correa et al. 2022; Ferreira Barbosa et al. 2022), associated with favourable meteorological conditions, led to the devastating events of 2020 when approximately one-third of the Pantanal was burned (Libonati et al. 2020).

Likewise, this is an ongoing problem in the Amazon rainforest, a fire-sensitive ecosystem where fire is not a natural occurrence. It is estimated that around 3% of the Amazon has fire return intervals shorter than the time required for grass exclusion by canopy recovery, putting these regions at a high risk of irreversible shifts to a fire-maintained degraded state (De Faria et al. 2021). While specific meteorological conditions may promote fires spreading, in particular extreme events such as droughts, fire in the Amazon is ultimately influenced by deforestation practices (Libonati et al. 2021). The process of deforestation, driven by intense anthropogenic pressure over the past decades, has played a pivotal role in shaping land use changes and consequently fire activity in the Amazon. This includes the conversion of forests into agricultural and pasture lands, selective logging, and forest fragmentation (Lapola et al. 2014; Ometto et al. 2016; Venter et al. 2016). Notably, between 2000 and 2013, agro-industrial activities for pasture accounted for 63% of total deforestation in the Brazilian Amazon, with an additional 9% attributed to cropland (Tyukavina et al. 2017). Moreover, the recent decade fires are very unlikely to have been caused by meteorological conditions alone, and have been linked with changes in land use, land cover and management (Cano-Crespo et al. 2021; Libonati et al. 2021; Kelley et al. 2021). In the 2020–2022 period, increasing deforestation rates were observed within the context of the relaxation of environmental legislation which resulted in the opening of the Amazon rainforest to agribusiness activities, as well as the weakening of licensing requirements for the construction of dams, roads, and mining permits (Abessa et al. 2019). The significant levels of deforestation in Brazil and Colombia, along with the increasing occurrence of agricultural fires in Venezuela, Guyana, and Bolivia, could be identified as a major contributing factor to the enhancement of fire activity in the Amazon during 2020–2022 (Silveira et al. 2022).

In response to large wildfires and deforestation, many countries in South America tried to impose zero-fire policies for decades (Moura et al. 2019). In many instances, such policies led to loss of traditional knowledge and fire management practices, conflicts between local communities and environmental authorities and actually led to increased wildfire occurrences (Fidelis et al. 2018; Durigan and Ratter 2016). Aiming to decrease such negative effects, fire management policies have been implemented in South American countries in the recent decades (Mistry et al. 2019; López-Mársico et al. 2021). Integrated Fire Management (IFM) became the main management framework in federal protected areas in Brazil with positive results in decreasing the occurrence of wildfires and fire related GHG, as well as social conflicts and firefighting expenses (Berlinck and Batista 2020; Mistry et al. 2019; Santos et al. 2021; Franke et al. 2024). (More details on fire management policies and implementation are reviewed in Chap. 14.)

South America is also home to some of the most destructive wildland-urban interface (WUI) fires in the world. The WUI is defined as a transitional area between unoccupied lands and human development, where natural and (possibly) fire prone landscapes meet human communities (United States Fire Administration (USFA) 2022). There have been recent efforts to characterise and study WUI communities in South America where WUI fires are recurrent, particularly in Chile and Argentina, resulting in severe economic losses and human casualties (Manzello et al. 2018). Recent estimates point to central Argentina having a WUI fire problem: around 15% of Sierras Chicas of Córdoba are WUI areas, which include most of the buildings exposed to wildfires and most of the buildings located in areas of higher burn probabilities (Argañaraz et al. 2017). In the city of Bariloche, located in the Argentinean Patagonia, fires have increased from less than 50 to more than 1800 per season between 1984 and 2004, especially in the urban-forest interface (Bilbao et al. 2020). Wildfires in Mediterranean regions of central Chile have been increasing due to increased WUI areas and it is estimated that around 25% of the Chilean population inhabits WUI areas (over three million people), where 40% of its inhabitants are exposed to burned areas (Sarricolea et al. 2020). The city of Valparaiso in central Chile is particularly affected by WUI fires with five documented events between 2013 and 2017 (Manzello et al. 2018). In 2014, the city saw one of the most catastrophic fire events in Chile, and possibly South America (Manzello et al. 2018), in the now called "Great Valparaiso Fire", where more than 1000 hectares burned displacing around 12,500 people, injuring more than 500 and claiming the lives of 15

(Reszka and Fuentes 2015). The fire started on the outskirts of Valparaiso, then spread through the WUI towards the city, where over 2900 homes were destroyed. Most of the affected areas were dense and low-income neighbourhoods with substandard buildings of easily ignitable materials, and only one road to both evacuate people and provide access to fire brigades and appliances (Reszka and Fuentes 2015). These conditions, associated with challenging topography (most of the population lives in steep hills that overlook the Valparaiso Bay) and poor water availability, hindered firefighting efforts (Reszka and Fuentes 2015). Recently, in 2024, Chile experienced its deadliest wildfires, which resulted in at least 133 fatalities, 7000 homes destroyed, and hundreds injured. Those events occurred in Viña del Mar located at the WUI of Valparaíso (Libonati 2024) where a growing of informal settlements in forest zones is observed. It was the worst natural disaster in the country since the 2010 earthquake and tsunami which killed about 500 people. Like previous catastrophic fire events, the 2024 Chilean wildfires resulted from a complex interplay between a long and severe drought and extreme weather conditions, and poor land management being characterised as 1 in 30-year event in today's climate (Kimutai et al. 2024).

13.3 How Climate Change Affects Fire Regimes

During the twenty-first century, human activities linked to landscape fragmentation led to a decline in global fire activity, including South America (Andela et al. 2017). However, compensatory increases in fire occurrence have been observed in tropical regions since the 1960s, primarily attributed to deforestation and land use and cover changes (Mouillot and Field 2005). The impact of global warming, alongside socio-economic and environmental changes, has contributed to a rise in extreme fires worldwide, driven by alterations in temperature and precipitation patterns (Bowman et al. 2020). In turn, extended and severe fire seasons contribute to higher levels of atmospheric aerosol concentrations, triggering feedback mechanisms between fire and climate (Bevan et al. 2009; Cox et al. 2008). Furthermore, varying vegetation responses to different durations of drought can also disrupt fire activity (Vicente-Serrano et al. 2013). The strong association and feedback mechanisms between anomalous fire weather and extreme fire behaviour suggests that anthropogenic climate change will impact future fire regimes (Bowman et al. 2020; Flannigan et al. 2013). Although there are large uncertainties, it is likely that the economic and environmental impacts of vegetation fires will worsen as a result of anthropogenic climate change (Bowman et al. 2020).

Because of the continent's large latitudinal (10°N to 55°S) and orographic variations, there is a high north-south and west-east variability in climate conditions in South America. According to the Köppen–Geiger climate classification (Beck et al. 2018), South America presents 17 different climates. The diversity of climate occurrence, induces diverse fire regimes and different relationships between fire and ecosystems around the continent, encompassing fire-prone, fire-sensitive and fire independent regions (Pivello 2011).

13.3.1 Ongoing Climate Change

Climate change is already having visible effects on the world. The IPCC Sixth Assessment Report (AR6; Castellanos et al. 2022) concluded that several regions in the world are warming, including South America, and rainfall patterns are changing. The scientific evidence since the IPCC's Fifth Assessment Report (AR5) (IPCC 2014) increased the confidence in the synergy among fire, land use change, particularly deforestation, and climate change. Regional increases in temperature, aridity and drought increased the frequency and intensity of fire. On average, people in the region were more exposed to high fire danger between 1 and 26 additional days depending on the sub-region for the years 2017–2020 compared to 2001–2004 (Castellanos et al. 2022).

Climate change has been increasing the likelihood of aggravated fire seasons and higher burned areas worldwide, and South America is no exception. Historical records of extreme meteorological fire danger reveal a systematic increase over many regions of South America, in response to higher regional temperatures and lower relative humidity (Fig. 13.2). The number of days of extreme meteorological fire danger has been increasing over South America's main fire hotspots, namely Brazil, Argentina, Bolivia, and Paraguay. The rapid and widespread deforestation of

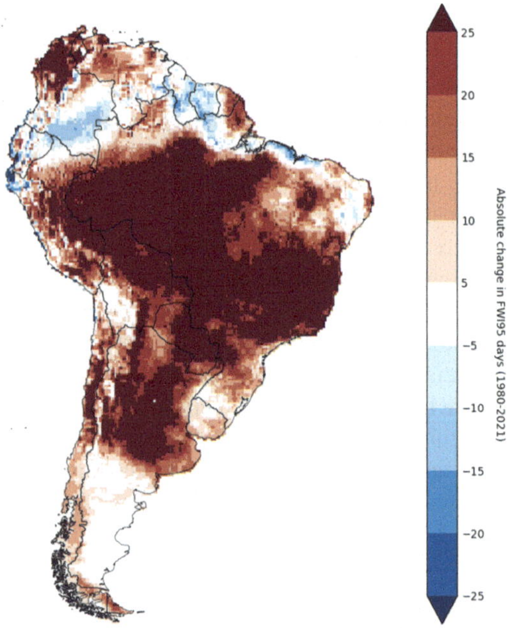

Absolute change in FWI95 days	
Brazil	25
Argentina	17
Bolivia	24
Paraguay	27
Venezuela	10
Colombia	7
Peru	16
Chile	11
Ecuador	-2
Guyana	-5
Uruguay	8
Suriname	-7
French Guiana	11

Fig. 13.2 Absolute change in the number of days per year where the Fire Weather Index (FWI) surpassed the local 95th percentile over the 1980–2021 period (FWI95). Estimated as the linear trend (days/year) multiplied by the 42 years of the study period. The computation of FWI95 followed that of Bowman et al. (2020). Table shows the average absolute change in FWI95 days per country. (Data source: ECMWF's ERA5 FWI version 4.0 (Vitolo et al. 2020))

savannas and tropical rainforests, compounded by droughts/heatwaves and zero-fire policies, which lead to biomass accumulation in savannas, have led to disastrous wildfires in recent decades (Eloy et al. 2019a; Holz et al. 2017). Over the last few years, South America has seen extremely severe fire events, including the 2017 megafires in Mediterranean Chile (Bowman et al. 2019), the 2005, 2010 and 2015 extreme fire season in the Amazon rainforest (Libonati et al. 2021), and the 2020 outstanding fire outbreaks in the Pantanal wetlands (Libonati et al. 2020, 2022a; Marengo et al. 2021). In 2019, more than 6 million hectares burned in the Chiquitania, Cerrado and Amazon regions in Bolivia, Brazil, Colombia, Paraguay and Peru, mostly within protected areas of native vegetation (Vargas-Cuentas and Roman-Gonzalez 2021). During the dry season of 2020, another long and destructive wave of wildfires swept through the area (Garcia et al. 2021).

Considering the last few decades, many parts of South America have increased temperatures, altered precipitation patterns, and an increase in extreme events. A consistent increase in temperature is found across global and regional models' projections (Coppola et al. 2021) and the number of heatwave days has increased significantly around the South American continent (Ceccherini et al. 2016; Geirinhas et al. 2018; Hannart et al. 2015; Libonati et al. 2022a; Monteiro dos Santos et al. 2024; Rusticucci et al. 2016). Many regions of Brazil have seen an increase in the number of consecutive dry days, particularly in the southern portions of the Amazon region, the Northeast region, Midwest, and Southeast (Regoto et al. 2021), and an increase in the number of days with extremely high temperatures, particularly in the central, northern, and north-eastern regions. Precipitation extremes, on the other hand, show more spatially heterogeneous signals. Recent global studies have highlighted South America as a potential hotspot for the occurrence of compound events, specifically those associated with droughts and heatwaves (Mukherjee and Mishra 2021; Raymond et al. 2022). Compounded drought-heatwave events have been increasing over the Northeast and Southeast regions of Brazil, Amazon, and Pantanal, and affecting larger areas for longer periods (Geirinhas et al. 2021; Libonati et al. 2022a, b; Marengo et al. 2022). In the Amazon rainforest, the observed strengthening of the hydrological cycle (Jimenez et al. 2018) is connected to the escalation of surface temperature, with the most severe heatwave events occurring simultaneously during exceptionally dry years (Costa et al. 2022).

Such extreme climate has been shown to influence a large part of interannual fire activity in the Cerrado (Li et al. 2022; Silva et al. 2019), the Amazon (Libonati et al. 2021), and emerging studies show a similar relationship for Pantanal (Libonati et al. 2022a; Silva et al. 2022). The tragic events of 2020 in the Pantanal, where around a third of the biome was burnt (Libonati et al. 2020), have been connected with extreme climate conditions, including an unprecedented summertime drought (Marengo et al. 2021; Thielen et al. 2021) concurrent with a heat-wave event in spring 2020 (Libonati et al. 2022a; Marengo et al. 2022), which led fire danger levels to unseen values over the last 40 years (Libonati et al. 2022a). Recent results show that the occurrence of compound dry and hot events exacerbate fire activity in both Cerrado and the Amazon over the last decades (Libonati et al. 2022b). Studies also observed changes in Amazon forest productivity linked to increased drought frequencies and warming trends (Machado-Silva et al. 2021). Fire risk related to concurrent droughts and heatwaves at Xingu forests and Pantanal's grassland/wetland has increased by 8.5% and 9.8% since 1981–2000, respectively (Ribeiro et al. 2022). In the Cerrado region, wet grassland areas have been shown to be drying due to lowering level of underground water, which leads to changes in vegetation structure and composition (de Souza et al. 2022). Such changes in wet grasslands may turn the gallery forests (fire-sensitive vegetation) more vulnerable and prone to be hit by wildfires.

The Amazon forest was highly impacted by the unprecedented droughts and higher temperatures observed in 1998, 2005, 2010 and 2015/2016, which are attributed partly to climate change. From 2000 to 2015, the decadal occurrence of droughts in the Amazon was nearly three times more frequent compared to the previous century, posing a significant threat to the forest ecosystem (Panisset et al. 2018). Severe drought events, such as the one in 1997–1998 linked to an intense El Niño-Southern Oscillation (ENSO) episode, made a significant portion of the Amazon susceptible to fires (Nepstad et al. 2004). Reduced rainfall in 2005, influenced by the warming tropical North Atlantic Ocean, led to an extended and intense fire season in western Amazon (Marengo et al. 2008). In 2010, the co-occurrence of positive phases of ENSO and the Atlantic Multidecadal Oscillation (AMO) resulted in record-breaking drought and fire activity in western and southern Amazon (Marengo et al. 2011). The extreme temperatures and precipitation levels, further exacerbated by the strong ENSO event of 2015, caused unprecedented drought conditions in the rainforest, though fire activity remained relatively low due to reduced deforestation (Libonati et al. 2021). Ongoing climate anomalies associated with ENSO and AMO are expected to continue impacting the Amazon with more frequent extreme droughts (Duffy et al. 2015). Combined with human-induced land cover changes, these hydroclimatic extremes are likely to intensify fire activity. Overall, addressing the complex interplay of these factors is crucial for the preservation of the Amazon rainforest.

Other terrestrial ecosystems in Central South America have been impacted by climate change, through persistent drought or extreme climatic events. In central Chile, the last 30 years were marked by a significant decrease in precipitation, along with a reduced river streamflow (Boisier et al. 2016). In fact, from 1946 to 2017, the number of fires and areas burned have increased significantly in Chile (González et al. 2011). Observations from 363 rain gauge stations in central Chile show a decrease of 65 mm per decade from 1979 (Boisier et al. 2016), and 2010–2015 saw the longest continuous dry spell since 1915, with only one analogue event in tree-ring-based data (Garreaud et al. 2017). Although intense albeit short (1–2 years) droughts are common in this region, the 2010–2015 megadrought reached farther south than previous droughts and its duration and severity were unprecedented considering records from the mid-twentieth century onwards (Garreaud et al. 2017). A precipitation deficit of approximately 30% in central Chile affected snow cover in the Andes and river flow across central Chile and western Argentina (Garreaud et al. 2017). This megadrought coincided with above-normal-temperatures, triggering changes in vegetation productivity (Garreaud et al. 2017). It is unlikely that these rainfall deficits were solely driven by the Pacific Decadal Oscillation (PDO) and, from 2010 onwards, it is estimated that around a quarter of the rainfall deficit in central Chile can be attributed to anthropogenic activity (Boisier et al. 2016).

Mean annual temperatures, along with agricultural use and human impacts, were found to have the highest direct positive effects on burned areas and number of fires in central Chile, while topography had multiple indirect effects (Gómez-González et al. 2019). Interannual burned area in Chile has also been positively linked to summer temperature, vapor pressure deficit, extreme fire danger days, and total precipitation in the preceding year; and negatively correlated with total precipitation in the previous winter (Bowman et al. 2019). This suggests that fuel abundance and fuel dryness are important controls of fire activity in the region. This was also the case in the 2017 Chilean fires, the "*tormenta de fuego*" event, where hundreds of fires burned more than 5000 km^2 (Bowman et al. 2019). These fires are amongst the most intense fires in the world, and were linked to unprecedented meteorological conditions, involving a complex interaction between drought, high temperatures, and the persistence of high pressures on the Antarctic periphery exacerbated by global climate change processes (Bilbao et al. 2020). The all-time Chilean national record for fire danger was reached on the 25th of January, when daily maximum temperatures went above the climatological 95th percentile for several days and reached maximum values at two local long-term meteorological stations in the region (Bowman et al. 2019). In the height of the Las Máquinas fires, on the 26th and 27th of January, record temperatures of 35.2–41.5 °C were reported, along with warm high-speed winds (Bilbao et al. 2020). Moreover, these fires occurred after the 2010–2015 multiannual drought, following a warm and moist growing season in 2016. The 2010–2015 megadrought extended the duration of fire season which, associated with relatively simple and poorly diversified silviculture schemes applied to large and continuous areas with highly flammable species, contributed to increasing magnitude of the incidents (Bilbao et al. 2020). This entails that interaction between climatic variables and vegetation cover should be considered for fire prevention (Gómez-González et al. 2019). Emissions from the 2017 Chilean fires affected regional atmospheric composition and climate and increased the concentration of pollutants over the capital city, with far-reaching effects over the Pacific Ocean and the Andes cordillera (Lapere et al. 2021).

In Bolivia the years of severe drought favoured unusually intense wildfires due accumulation of forest fuels, such as fallen woody material and leaves (Bilbao et al. 2020). In 2010, Bolivia had the most extreme fire season ever recorded (with approximately 6 million hectares burned) during an ENSO year. Climate trends observed over the past 50 years show more pronounced changes during the dry season, increasing the risk of fire and warmer and drier general conditions throughout the country (Bilbao et al. 2020).

In Colombian Orinoco savannas, which have been steadily transformed into pastures for more than a century (Jimenez et al. 2022), satellite measurements show that 7% of its area burned every year (during 1997–2016) and up to 40% of the burned area interannual variability is due to irregular rainfall and drought. In Orinoco Llanos rapid changes in land-use due to overgrazing, frequent and extensive burning and conversion of native vegetation into agricultural lands resulted in progressively increasing atmospheric carbon emissions (San José and Montes 2001). The sink capacity of Orinoco vegetation is limited by climatic conditions, soil type, soil pH, soil minerals and drainage (San José and Montes 2001). The Colombian natural savanna's net ecosystem exchange was found to be highly dependent on drought duration after ~10 days of nominal drought, i.e. with less than 5 mm/day of precipitation (Morales-Rincon et al. 2021). Morales-Rincon et al. (2021) noted that a shift to a climate with similar total precipitation but split into extreme dry and wet seasons might eventually suppress the savanna carbon uptake capacity. In the Orinoco savannas, the frequency of disturbance is comparable to the lifespan and recolonization capability of the native species, which creates an impression of an ecosystem in equilibrium; however, arid savannas are always recovering from disturbance and are in non-equilibrium condition (San José et al. 2008). A digital soil map that measures and predicts dynamic soil properties such as soil organic carbon (SOC) in Eastern Plains of Colombia might provide insight to conserve and manage natural savanna's ecosystems to face agricultural and pastureland expansions (Rainford et al. 2021).

A recent analysis of 4500 meteorological droughts globally found increased drought frequency over the Amazon, north-eastern Brazil and Patagonia (Spinoni et al. 2019). The study also found drought intensity has become more severe in parts of Patagonia and South Chile (Jia et al. 2022; Spinoni et al. 2019). In northern Patagonia modelling results showed an increase in predicted fire activity in less productive ecosystems (shrublands and steppes), although fuel-rich productive forests (subalpine deciduous and evergreen *Nothofagus*-dominated rainforests) are more vulnerable mainly due to shorter fire intervals and lack of fire-adapted traits in the dominant trees that compose these forests (Kitzberger et al. 2022).

13.3.2 Future Climate Change

Future climate assessments rely on global and regional climate models, and a set of possible future pathways, to evaluate future climate parameters such as temperature and precipitation. A climate model is a numerical representation of physical, chemical and biological processes (Schoeman et al. 2023) translated in a computer simulation of Earth's climate system, including the atmosphere, ocean, land and ice. A global climate model (GCM) is a complex mathematical representation divided into three-dimensional grid cells representing specific geographic locations and elevations. Each climate component (atmosphere, ocean, land, and sea ice) is calculated on the global grid and their interactions in terms of exchange fluxes of heat, water and momentum is tracked over time. The size of these grid cells depends on the power of the computer that is available to solve these equations. Finer resolution implies bigger and faster computer capacity to perform the simulation.

The World Climate Research Programme (WCRP) Coupled Model Intercomparison Project (CMIP) gathers output from global climate models in order to detect anthropogenic effects in the climate record of the past century and to project future climatic changes due greenhouse gas concentration. For the IPCC AR5 published in 2017, future projections were based on the four Representative Concentration Pathways (RCPs) that represent classes of climate effects (radiative forcing targets for 2100) pertaining to the CMIP5 phase.

Following the AR5 report, scientific publications on climate model performance and projections in South America have increased, mainly focusing on regional climate models (RCMs) (Boulanger et al. 2016; Ambrizzi et al. 2019). Overall, global and regional models demonstrate good skill in representing the climatological features of South America. Nonetheless, significant biases persist, mainly at regional scales (Blázquez and Nuñez 2013; Gulizia et al. 2013; Joetzjer et al. 2013; Jones and Carvalho 2013). For instance, most models tend to underestimate precipitation during the dry season in the Amazon and misrepresent climate features in regions with complex topography like the Andes. RCMs, while effective in simulating temperature and precipitation patterns, also manifest systematic biases, notably precipitation overestimation and temperature underestimation in specific regions (Díaz et al. 2021; Rivera and Arnould 2020; Sierra et al. 2015; Zazulie et al. 2017). It is also important to mention that regions with sparse observational data potentially exacerbate these biases. CMIP5 models have shown some improvements over CMIP3 models in representing certain South American regions, but whether CMIP6 models offer further improvements remains inconclusive. The models' performance depends on multiple factors, including the region, time scale, and the specific variables analysed. In summary, while progress has been made in climate modelling for South America, no single model excels across all climate aspects, and persistent biases underscore the need for continued improvement and evaluation.

For the IPCC AR6 published in 2021 (IPCC 2021), results pertaining to the latest CMIP phase, version 6 (CMIP6), used a predefined subset of scenarios that represent different socio-economic developments as well as different pathways of atmospheric greenhouse gas concentration (O'Neill et al. 2016). These scenarios, called Shared Socioeconomic Pathways (SSPs), consider new future pathways of societal development and are relevant to societal concerns regarding climate change mitigation, adaptation and impacts (O'Neill et al. 2016). The IPCC AR6 WGI report focused on five core SSP scenarios: SSP1-1.9, SSP1-2.6, SSP2-4.5, SSP3-7.0 and SSP5-8.5. The first number in each label corresponds to the set of socioeconomic assumptions driving the emissions and other climate forcing inputs taken up by climate models showcased in Fig. 13.3, and the second number corresponds to the radiative forcing level reached in 2100 (Rose et al. 2022). The IPCC AR6 further employs the climate reference regions described in Iturbide et al. (2020), which, for South America, include: North-West South America (NWS); Northern South America (NSA); North-East South America (NES); South American Monsoon (SAM); South-West South America (SES); and Southern South America (SSA).

IPCC AR6 projections (IPCC 2021) point to increases in mean temperatures between 2 and 6 °C by the end of the century (Fig. 13.3). Mean temperatures have increased in all South America and will continue to increase at rates greater than the global average. The AR6 also pointed to patterns of changes in precipitation, and changes in the duration of dry spells. Mean precipitation is projected to change,

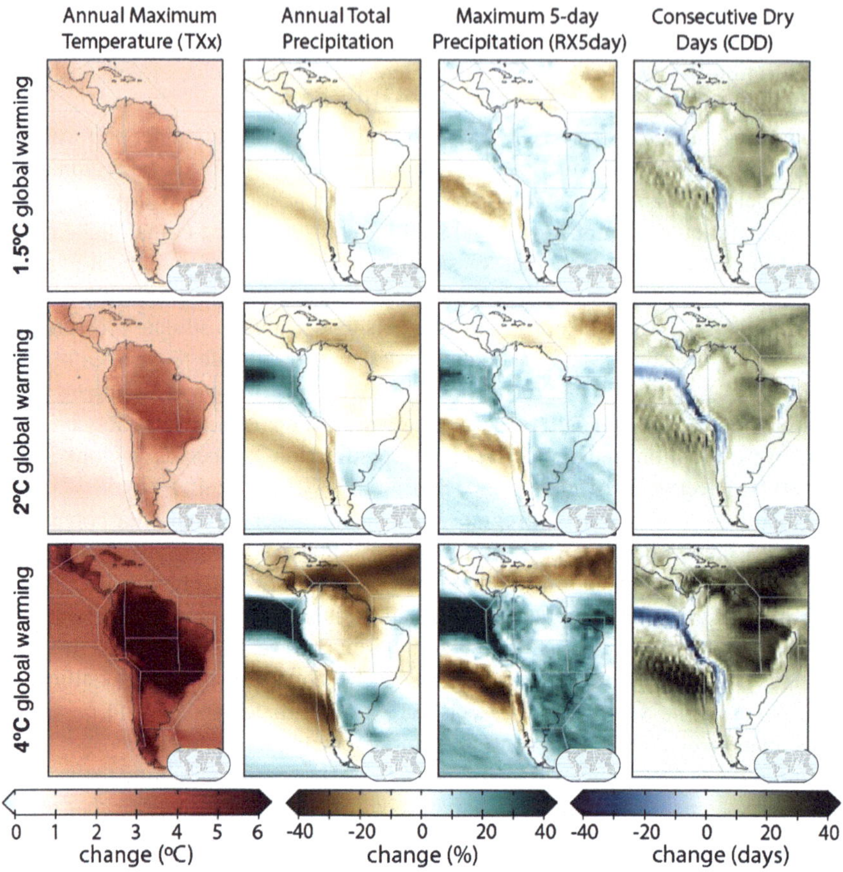

Fig. 13.3 Projected changes in annual maximum temperature (TXx), annual total precipitation, annual maximum 5-day precipitation (RX5day) and annual consecutive dry days (CDD) at 1.5, 2, and 4 °C (in rows) global warming relative to 1850–1900. Results are based on simulations from the CMIP6 multi-model ensemble (32 global climate models) using the SSP5-8.5 scenario to compute the warming levels. (Source: Regional fact sheet—Central and South America Interactive Atlas (Gutiérrez et al. 2021))

with increases in NWS and SES, and decreases in NES and SWS. Mean precipitation will decrease, and heavy precipitation, aridity and drought are projected to increase with medium confidence, whereas mean temperature, extreme heat, fire weather and coastal and oceanic climate impact drivers will all increase with high confidence (Castellanos et al. 2022). This is consistent among model projections by mid- and end of the twenty-first century for RCP 4.5 and RCP 8.5 scenarios. There are projections of increased agricultural and ecological drought for the mid-twenty-first century, for 2 °C of global warming level and above in the NSA, NES, and SAM regions. This will potentially impact a wide range of sectors, including agriculture, forestry, health, and ecosystems. In particular, over the Amazon, the number of days per year with maximum temperatures exceeding 35 °C would increase by more than 150 days by the end of the twenty-first century in the SSP5-8.5 scenario, while it is expected to increase by less than 60 days under the SSP1-2.6 scenario (Gutiérrez et al. 2021).

In addition, climate change projections point to major increases in several heat indices across the South America region for all scenarios. Heatwaves are projected to increase in frequency, intensity and duration in this century, and, under high emission scenarios, will become extremely long (60 consecutive days) and frequent (once every 2 years) (Jia et al. 2022). Largest increases in the frequency of hot days (maximum temperatures, Tx > 35 °C) are projected for the Amazon basin under SSP5-8.5 with more than 200 days per year at the end of the century under SSP5-8.5 relative to 1995–2014, while such increases remain moderate (50–100 days) in SSP1-2.6 (Coppola et al. 2021; Schwingshackl et al. 2021). There is high confidence of an increase in drought duration in NES, and in the number of dry days and drought frequency in NSA and SAM. Dry climatic impact-drivers are projected to increase at the regional level with higher global warming levels. The strongest signal of future increase in agricultural and ecological drought, aridity and fire weather is over the Amazon region (Sun et al. 2019). This is consistent with the large increase in the frequency of joint occurrence of extreme hot and dry days projected for a 2 °C warming level or more (Vogel et al. 2020). Projections of fire weather indices also show an increased risk in SWS and SSA. Globally, Quilcaille et al. (2023) found at a global warming level of 3 °C, the mean fire weather would increase on average by at least 66% in duration and frequency, while associated 1-in-10-year events would approximately triple in duration and increase by at least 31% in intensity. However, wildfires highly depend on land use and appropriate management may help mitigate future increases in fire risk (Fonseca et al. 2019).

Climate change is projected to lead to widespread increases in risk, with at least 30 additional days above critical thresholds for fire activity in forest biomes on every continent by 2100 under rising emissions scenarios (Table 13.1; Clarke et al. 2022). The authors found under the RCP 8.5 scenario of CMIP5 models that an escalating forest fire risk threatens catastrophic carbon losses in the Amazon and major population health impacts from wildfire smoke in southeast Asia and east Africa. However, the role of deforestation as a driver of fire activity in the region should not be underestimated and must be taken into account when implementing measures to protect the Amazon forest (Brando et al. 2020; Libonati et al. 2021).

However, these future projections of climate, and by extent fire risk, are not without uncertainties. The limited availability of station data records particularly regarding precipitation in high latitudes and in tropical regions as the Amazon rainforest (Spinoni et al. 2019) hinder world climate databases. Global datasets based on station-observed data are fundamental to grounded model's simulation, and the lack of reliable data in South America is notable (López-Bermeo et al. 2022). Over areas of poor data availability, interpolation algorithms and long-term climatology are used as strategies to fulfil this void and can cause critical misguided interpretation of data. In general, South American precipitation is not well reproduced by CMIP6 models, and some of the reasons are well known, such as the misrepresentation of

Table 13.1 Overview of studies that evaluate future fire activity in South America

Country	Ecosystem	Projected change in fire parameter/index	Reference(s)
South America		Increase	Burton et al. (2022) Oliveira-Júnior et al. (2021)
Brazil		Increase	Viegas et al. (2022) Oliveira et al. (2022) da Silva Junior et al. (2020) Justino et al. (2011)
	Amazon	Increase	Ribeiro et al. (2022) Fonseca et al. (2019) Le Page et al. (2017)Silvestrini et al. (2011)
	Cerrado	Increase	Silva et al. (2019) Silva et al. (2016)
	Pantanal	Increase	Ribeiro et al. (2022) Silva et al. (2022)
Argentina	Patagonia	Increase	Kitzberger et al. (2022)
Bolivia	Chiquitania	Increase	Devisscher et al. (2016)
Colombia	Orinoquia/ Llanos	Decrease	Jimenez et al. (2022)
Chile	South-Central Chile	Increase	Ciocca et al. (2023)

We only considered peer-reviewed scientific publications that assessed future changes in a fire index and/or parameter over South America and/or its unique biomes/ecosystems, from 2010 onwards

cloud physics (Firpo et al. 2022; Khairoutdinov et al. 2005). Precipitation is underestimated in the Amazon mainly due dipolar bias associated with the wrong position of maximum precipitation centre. The South American Low-Level Jet, a known low-level austral summer circulation pattern associated with the South American monsoon, with trade winds entering in South America through the east coast of the Amazon rainforest, and an important feature for maintaining South American Monsoon System, tends to be misrepresented by CMIP models by weaken the trade winds compared to reanalysis data (Firpo et al. 2022). CMIP6 models also show deficiencies in simulating temperature over the Amazonian region (Firpo et al. 2022).

13.4 Final Remarks

As summarised by Shuman et al. (2022) fire is a fundamental ecological and societal force on our planet. However, we are currently dealing with fire regimes of the Anthropocene—which essentially means a world where fire ignition and spread are exacerbated by human ignitions and global warming, ecosystems are increasingly flammable because of activities such as fire suppression, and fire risks are more severe because there are more people living in areas susceptible to burning.

Throughout this chapter, we have seen how human activity now influences all aspects of fire behaviour and the fire regime, and why wildfire events, such as the Pantanal 2020 fires or the 2024 Chilean fires, can and should not be evaluated through a single meteorological or anthropogenic perspective. These extreme events, with severe economic, ecological and health consequences, are an interplay of both climate and human activity.

Climate change is also visibly impacting South America. Rising temperatures, changing rainfall patterns, increased frequency of extreme events, and the heightened risk of wildfires pose significant challenges for various sectors and the overall well-being of the region. The combined effect of anthropogenic land use change and climate change increases the vulnerabilities of terrestrial ecosystems to extreme climate events and fires (IPCC 2021). Urgent action and appropriate management practices are crucial to mitigate these impacts and build resilience in the face of climate change.

Ecosystem-based adaptation measures can reduce climatic risks to people, including from flood, drought, fire, and heat. Restoring natural vegetation cover and wildfire regimes can reduce risks to people from catastrophic fires (Pörtner et al. 2022). Active fire management policies are being adopted worldwide to decrease the ecological, socioeconomic, and human-health damages related to wildfires. Specific fire management strategies should be considered for each socio ecological context, and thus applying an adaptive management framework is essential (Berlinck and Batista 2020; Myers 2006; Schmidt et al. 2018; Whelan 2002). Firewise policies and management are essential to plan and manage natural and anthropogenic landscapes, considering both climatic and socioeconomic contexts.

Applied research focused on informing managers and decision-makers on how to manage fire and reduce climate change consequences in natural ecosystems is also essential to allow for more effective conservation practices. For science to inform policy and management, clear communication in both directions needs to be effectively established, for scientists to understand the need for rapid management decisions and for managers to use parameters in decision-making that are based on state-of-the-art scientific knowledge.

References

Abessa D, Famá A, Buruaem L (2019) The systematic dismantling of Brazilian environmental laws risks losses on all fronts. Nat Ecol Evol 3:510–511. https://doi.org/10.1038/S41559-019-0855-9

Ambrizzi T, Reboita MS, da Rocha RP, Llopart M (2019) The state of the art and fundamental aspects of regional climate modeling in South America. Ann N Y Acad Sci 1436:98–120. https://doi.org/10.1111/nyas.13932

Andela N, Morton DC, Giglio L, Chen Y, Van Der Werf GR et al (2017) A human-driven decline in global burned area. Science 356(6345):1356–1362. https://doi.org/10.1126/science.aal4108

Argañaraz JP, Radeloff VC, Bar-Massada A, Gavier-Pizarro GI, Scavuzzo CM, Bellis LM (2017) Assessing wildfire exposure in the Wildland-Urban Interface area of the mountains of central Argentina. J Environ Manag 196:499–510. https://doi.org/10.1016/J.JENVMAN.2017.03.058

Barlow J, Berenguer E, Carmenta R, França F (2019) Clarifying Amazonia's burning crisis. Glob Chang Biol 26(2):319–321. https://doi.org/10.1111/gcb.14872

Beck HE, Zimmermann NE, McVicar TR, Vergopolan N, Berg A, Wood EF (2018) Present and future Köppen-Geiger climate classification maps at 1-km resolution. Sci Data 5:180214. https://doi.org/10.1038/sdata.2018.214

Bella CM, Jobbagy EG, Paruelo JM, Pinnock S (2006) Continental fire density patterns in South America. Glob Ecol Biogeogr 15:192–199. https://doi.org/10.1111/j.1466-822X.2006.00225.x

Berlinck CN, Batista EKL (2020) Good fire, bad fire: it depends on who burns. Flora 268:151610. https://doi.org/10.1016/J.FLORA.2020.151610

Bevan SL, North PRJ, Grey WMF, Los SO, Plummer SE (2009) Impact of atmospheric aerosol from biomass burning on Amazon dry-season drought. J Geophys Res Atmos 114(D9):13283–13301. https://doi.org/10.1029/2008JD011112

Bilbao B, Steil L, Urbieta IR, Anderson L, Pinto C et al (2020) Wildfires. In: Adaptation to climate change risks in Ibero-American countries—RIOCCADAPT report. McGraw Hill, Madrid, Spain, pp 435–496

Blázquez J, Nuñez MN (2013) Performance of a high resolution global model over southern South America. Int J Climatol 33:904–919. https://doi.org/10.1002/JOC.3478

Boisier JP, Rondanelli R, Garreaud RD, Muñoz F (2016) Anthropogenic and natural contributions to the Southeast Pacific precipitation decline and recent megadrought in central Chile. Geophys Res Lett 43:413–421. https://doi.org/10.1002/2015GL067265

Boulanger JP, Carril AF, Sanchez E (2016) CLARIS-La Plata Basin: regional hydroclimate variability, uncertainties and climate change scenarios. Clim Res 68:93–94. https://doi.org/10.3354/CR01392

Bowman DMJS, Balch JK, Artaxo P, Bond WJ, Carlson JM et al (2009) Fire in the Earth system. Science 324(5926):481–484. https://doi.org/10.1126/science.1163886

Bowman DMJS, Moreira-Muñoz A, Kolden CA, Chávez RO, Muñoz AA et al (2019) Human-environmental drivers and impacts of the globally extreme 2017 Chilean fires. Ambio 48:350–362. https://doi.org/10.1007/s13280-018-1084-1

Bowman DMJS, Kolden CA, Abatzoglou JT, Johnston FH, van der Werf GR, Flannigan M (2020) Vegetation fires in the Anthropocene. Nat Rev Earth Environ 1:500–515. https://doi.org/10.1038/s43017-020-0085-3

Brando P, Macedo M, Silvério D, Rattis L, Paolucci L, Alencar A, Coe M, Amorim C (2020) Amazon wildfires: scenes from a foreseeable disaster. Flora 268:151609. https://doi.org/10.1016/J.FLORA.2020.151609

Burton C, Kelley DI, Jones CD, Betts RA, Cardoso M, Anderson L (2022) South American fires and their impacts on ecosystems increase with continued emissions. Clim Resil Sustain 1(1):e8. https://doi.org/10.1002/CLI2.8

Calvin K, Dasgupta D, Krinner G et al (2023) IPCC, 2023: Climate change 2023: synthesis report. In: Core Writing Team, Lee H, Romero J (eds) Contribution of working groups I, II and III to the sixth assessment report of the Intergovernmental Panel on Climate Change. IPCC, Geneva, Switzerland. https://doi.org/10.59327/IPCC/AR6-9789291691647

Campagnolo ML, Libonati R, Rodrigues JA, Pereira JMC (2021) A comprehensive characterization of MODIS daily burned area mapping accuracy across fire sizes in tropical savannas. Remote Sens Environ 252:112115. https://doi.org/10.1016/j.rse.2020.112115

Cano-Crespo A, Traxl D, Thonicke K (2021) Spatio-temporal patterns of extreme fires in Amazonian forests. Eur Phys J Spec Top 230:3033–3044. https://doi.org/10.1140/epjs/s11734-021-00164-3

Castellanos E, Lemos MF, Astigarraga L et al (2022) Central and South America. In: Pörtner HO, Roberts DC, Tignor M et al (eds) Climate change 2022: impacts, adaptation and vulnerability. Contribution of working group II to the sixth assessment report of the Intergovernmental Panel on Climate Change. Cambridge University Press, Cambridge, UK, pp 1689–1816. https://doi.org/10.1017/9781009325844.014

Ceccherini G, Russo S, Ameztoy I, Patricia Romero C, Carmona-Moreno C (2016) Magnitude and frequency of heat and cold waves in recent decades: the case of South America. Nat Hazards Earth Syst Sci 16:821–831. https://doi.org/10.5194/nhess-16-821-2016

Ciocca I, Fernández A, Jaque E, Justino F, Shumacher V et al (2023) Increased wildfire hazard along South-Central Chile under the RCP8.5 scenario as revealed by high-resolution modeling. Environ Res Lett 18:034023. https://doi.org/10.1088/1748-9326/acba33

Clarke H, Nolan RH, De Dios VR, Bradstock R, Griebel A et al (2022) Forest fire threatens global carbon sinks and population centres under rising atmospheric water demand. Nat Commun 13:1–10. https://doi.org/10.1038/s41467-022-34966-3

Coppola E, Raffaele F, Giorgi F, Giuliani G, Xuejie G et al (2021) Climate hazard indices projections based on CORDEX-CORE, CMIP5 and CMIP6 ensemble. Clim Dyn 57:1293–1383. https://doi.org/10.1007/s00382-021-05640-z

Corporación Nacional Forestal (CONAF) (2023) Estadísticas históricas [WWW Document]. https://www.conaf.cl/incendios-forestales/incendios-forestales-en-chile/estadisticas-historicas/. Accessed 24 Apr 2023

Correa DB, Alcântara E, Libonati R, Massi KG, Park E (2022) Increased burned area in the Pantanal over the past two decades. Sci Total Environ 835:155386. https://doi.org/10.1016/j.scitotenv.2022.155386

Costa DF, Gomes HB, Silva MCL, Zhou L (2022) The most extreme heat waves in Amazonia happened under extreme dryness. Clim Dyn 59:281–295. https://doi.org/10.1007/s00382-021-06134-8

Cox PM, Harris PP, Huntingford C, Betts RA, Collins M et al (2008) Increasing risk of Amazonian drought due to decreasing aerosol pollution. Nature 453:212–215. https://doi.org/10.1038/nature06960

Crutzen PJ (2002) Geology of mankind. Nature 415:23. https://doi.org/10.1038/415023a

Crutzen PJ, Stoermer EF (2000) The 'Anthropocene'. In: Benner S, Lax G, Crutzen PJ, Pöschl U, Lelieveld J, Brauch HG (eds) Paul J. Crutzen and the Anthropocene: a new epoch in earth's history. The Anthropocene: politik—economics—society—science, vol 1. Springer Cham, pp 19–21. https://doi.org/10.1007/978-3-030-82202-6_2

da Silva Junior CA, Teodoro PE, Delgado RC, Teodoro LPR, Lima M et al (2020) Persistent fire foci in all biomes undermine the Paris agreement in Brazil. Sci Rep 10:16246. https://doi.org/10.1038/s41598-020-72571-w

Daniau AL, Bartlein PJ, Harrison SP, Prentice IC, Brewer S et al (2012) Predictability of biomass burning in response to climate changes. Global Biogeochem Cycles 26:4007. https://doi.org/10.1029/2011GB004249

de Azevedo GB, Rezende AV, de Oliveira Sousa Azevedo GT, Miguel EP, de Gois Aquino F et al (2020) Woody biomass accumulation in a Cerrado of Central Brazil monitored for 27 years after the implementation of silvicultural systems. For Ecol Manag 455:117718. https://doi.org/10.1016/j.foreco.2019.117718

De Faria BL, Staal A, Silva CA, Martin PA, Panday PK, Dantas VL (2021) Climate change and deforestation increase the vulnerability of Amazonian forests to post-fire grass invasion. Glob Ecol Biogeogr 30:2368–2381. https://doi.org/10.1111/geb.13388

de Souza GF, Ferreira MC, Munhoz CBR (2022) Decrease in species richness and diversity, and shrub encroachment in Cerrado grasslands: a 20-year study. Appl Veg Sci 25:e12668. https://doi.org/10.1111/avsc.12668

Devisscher T, Anderson LO, Aragão LEOC, Galván L, Malhi Y (2016) Increased wildfire risk driven by climate and development interactions in the Bolivian Chiquitania, Southern Amazonia. PLoS One 11:e0161323. https://doi.org/10.1371/journal.pone.0161323

Díaz LB, Saurral RI, Vera CS (2021) Assessment of South America summer rainfall climatology and trends in a set of global climate models large ensembles. Int J Climatol 41:e59–e77. https://doi.org/10.1002/joc.6643

Duffy PB, Brando P, Asner GP, Field CB (2015) Projections of future meteorological drought and wet periods in the Amazon. Proc Natl Acad Sci USA 112:13172–13177. https://doi.org/10.1073/pnas.1421010112

Durigan G, Ratter JA (2016) The need for a consistent fire policy for Cerrado conservation. J Appl Ecol 53:11–15. https://doi.org/10.1111/1365-2664.12559

Eloy L, Bilbao B, Mistry J, Schmidt IB (2019a) From fire suppression to fire management: advances and resistances to changes in fire policy in the savannas of Brazil and Venezuela. Geogr J 185:10–22. https://doi.org/10.1111/geoj.12245

Eloy L, Schmidt IB, Borges SL, Ferreira MC, dos Santos TA (2019b) Seasonal fire management by traditional cattle ranchers prevents the spread of wildfire in the Brazilian Cerrado. Ambio 48:890–899. https://doi.org/10.1007/s13280-018-1118-8

Ferreira Barbosa ML, Haddad I, da Silva Nascimento AL, Máximo da Silva G, Moura da Veiga R et al (2022) Compound impact of land use and extreme climate on the 2020 fire record of the Brazilian Pantanal. Glob Ecol Biogeogr 31:1960–1975. https://doi.org/10.1111/geb.13563

Fidelis A, Alvarado ST, Barradas ACS, Pivello VR (2018) The year 2017: megafires and management in the Cerrado. Fire 1:49. https://doi.org/10.3390/fire1030049

Firpo MÂF, Guimarães BS, Dantas LG, Silva MGB, Alves LM et al (2022) Assessment of CMIP6 models' performance in simulating present-day climate in Brazil. Front Clim 4:948499. https://doi.org/10.3389/fclim.2022.948499

Flannigan M, Cantin AS, de Groot WJ, Wotton M, Newbery A, Gowman LM (2013) Global wildland fire season severity in the 21st century. For Ecol Manag 294:54–61. https://doi.org/10.1016/j.foreco.2012.10.022

Fonseca MG, Alves LM, Aguiar APD, Arai E, Anderson LO et al (2019) Effects of climate and land-use change scenarios on fire probability during the 21st century in the Brazilian Amazon. Glob Chang Biol 25:2931–2946. https://doi.org/10.1111/gcb.14709

Franke J, Sena Barradas AC, Borges KMR et al (2024) Prescribed burning and integrated fire management in the Brazilian Cerrado: demonstrated impacts and scale-up potential for emission abatement. Environ Res Lett 19:034020. https://doi.org/10.1088/1748-9326/ad2820

Ganopolski A, Winkelmann R, Schellnhuber HJ (2016) Critical insolation—CO_2 relation for diagnosing past and future glacial inception. Nature 529:200–203. https://doi.org/10.1038/nature16494

Garcia LC, Szabo JK, de Oliveira Roque F, de Matos Martins Pereira A, Nunes da Cunha C et al (2021) Record-breaking wildfires in the world's largest continuous tropical wetland: integrative fire management is urgently needed for both biodiversity and humans. J Environ Manag 293:112870. https://doi.org/10.1016/j.jenvman.2021.112870

Garreaud RD, Alvarez-Garreton C, Barichivich J, Pablo Boisier J, Christie D et al (2017) The 2010–2015 megadrought in central Chile: impacts on regional hydroclimate and vegetation. Hydrol Earth Syst Sci 21:6307–6327. https://doi.org/10.5194/hess-21-6307-2017

Geirinhas JL, Trigo RM, Libonati R, Coelho CAS, Palmeira AC (2018) Climatic and synoptic characterization of heat waves in Brazil. Int J Climatol 38:1760–1776. https://doi.org/10.1002/joc.5294

Geirinhas JL, Russo A, Libonati R, Sousa PM, Miralles DG, Trigo RM (2021) Recent increasing frequency of compound summer drought and heatwaves in Southeast Brazil. Environ Res Lett 16:034036. https://doi.org/10.1088/1748-9326/abe0eb

Gillson L, Whitlock C, Humphrey G (2019) Resilience and fire management in the Anthropocene. Ecol Soc 24:14. https://doi.org/10.5751/ES-11022-240314

Glikson A (2013) Fire and human evolution: the deep-time blueprints of the Anthropocene. Anthropocene 3:89–92. https://doi.org/10.1016/j.ancene.2014.02.002

Gómez-González S, González ME, Paula S, Díaz-Hormazábal I, Lara A, Delgado-Baquerizo M (2019) Temperature and agriculture are largely associated with fire activity in Central Chile across different temporal periods. For Ecol Manag 433:535–543. https://doi.org/10.1016/j.foreco.2018.11.041

González ME, Lara A, Urrutia R, Bosnich J (2011) Cambio climático y su impacto potencial en la ocurrencia de incendios forestales en la zona centro-sur de Chile (33°–42°s). Bosque 32:215–219. https://doi.org/10.4067/S0717-92002011000300002

Gora EM, Yanoviak SP (2020) Lightning-caused disturbance in the Peruvian Amazon. Biotropica 52:813–817. https://doi.org/10.1111/btp.12826

Gora EM, Burchfield JC, Muller-Landau HC, Bitzer PM, Yanoviak SP (2020) Pantropical geography of lightning-caused disturbance and its implications for tropical forests. Glob Chang Biol 26:5017–5026. https://doi.org/10.1111/gcb.15227

Gulizia C, Camilloni I, Doyle M (2013) Identification of the principal patterns of summer moisture transport in South America and their representation by WCRP/CMIP3 global climate models. Theor Appl Climatol 112:227–241. https://doi.org/10.1007/s00704-012-0729-4

Gutiérrez JM, Jones RG, Narisma GT, Alves LM, Amjad M et al (eds) (2021) Climate change 2021: the physical science basis. Contribution of working group I to the sixth assessment report of the Intergovernmental Panel on Climate Change. IPCC. http://interactive-atlas.ipcc.ch

Hamilton C (2015) Getting the Anthropocene so wrong. Anthr Rev 2:102–107. https://doi.org/10.1177/2053019615584974

Hannart A, Vera C, Cerne B, Otto FEL (2015) Causal influence of anthropogenic forcings on the Argentinian heat wave of December 2013. Bull Am Meteorol Soc 96:S41–S45. https://doi.org/10.1175/BAMS-D-15-00137.1

Hardesty J, Myers R, Fulks W (2005) Fire, ecosystems and people: a preliminary assessment of fire as a global conservation issue. Fire Manag 22:78–87. https://doi.org/10.1177/0146167209351886

Hoffmann WA, Geiger EL, Gotsch SG, Rossatto DR, Silva LCR et al (2012) Ecological thresholds at the savanna-forest boundary: how plant traits, resources and fire govern the distribution of tropical biomes. Ecol Lett 15:759–768. https://doi.org/10.1111/j.1461-0248.2012.01789.x

Holz A, Paritsis J, Mundo IA, Veblen TT, Kitzberger T et al (2017) Southern Annular Mode drives multicentury wildfire activity in southern South America. Proc Natl Acad Sci USA 114:9552–9557. https://doi.org/10.1073/pnas.1705168114

Hönisch B, Ridgwell A, Schmidt DN, Thomas E, Gibbs SJ et al (2012) The geological record of ocean acidification. Science 335:1058–1063. https://doi.org/10.1126/science.1208277

IPCC (2014) Climate change 2014: synthesis report. In: Core Writing Team, Pachauri RK, Meyer LA (eds) Contribution of working groups I, II and III to the fifth assessment report of the Intergovernmental Panel on Climate Change. IPCC, Geneva, Switzerland

IPCC (2021) Climate change 2021: the physical science basis. Contribution of working group I to the sixth assessment report of the Intergovernmental Panel on Climate Change. Cambridge University Press

Iturbide M, Gutiérrez JM, Alves LM, Bedia J, Cerezo-Mota R et al (2020) An update of IPCC climate reference regions for subcontinental analysis of climate model data: definition and aggregated datasets. Earth Syst Sci Data 12:2959–2970. https://doi.org/10.5194/essd-12-2959-2020

Jia G, Shevliakova E, Artaxo P, De Noblet-Ducoudré N, Houghton RA et al (eds) (2022) Climate change and land. Cambridge University Press, pp 131–248. https://doi.org/10.1017/9781009157988.004

Jimenez JC, Libonati R, Peres LF (2018) Droughts over Amazonia in 2005, 2010, and 2015: a cloud cover perspective. Front Earth Sci 6:424137. https://doi.org/10.3389/feart.2018.00227

Jimenez R, Ardila AV, Vargas-Burbano AC, Hernandez AJ, Leon-Velasquez E (2022) Biomass burning-agriculture coupling in the Orinoco savannas—particulate matter emission scenarios. Front Environ Sci 10:689844. https://doi.org/10.3389/fenvs.2022.689844

Joetzjer E, Douville H, Delire C, Ciais P (2013) Present-day and future Amazonian precipitation in global climate models: CMIP5 versus CMIP3. Clim Dyn 41:2921–2936. https://doi.org/10.1007/s00382-012-1644-1

Jones C, Carvalho LMV (2013) Climate change in the South American monsoon system: present climate and CMIP5 projections. J Clim 26:6660–6678. https://doi.org/10.1175/jcli-d-12-00412.1

Justino F, de Mélo AS, Setzer A, Sismanoglu R, Sediyama GC et al (2011) Greenhouse gas induced changes in the fire risk in Brazil in ECHAM5/MPI-OM coupled climate model. Clim Chang 106:285–302. https://doi.org/10.1007/s10584-010-9902-x

Kelley DI, Burton C, Huntingford C, Brown MAJ, Whitley R, Dong N (2021) Technical note: Low meteorological influence found in 2019 Amazonia fires. Biogeosciences 18:787–804. https://doi.org/10.5194/bg-18-787-2021

Kelly LT, Giljohann KM, Duane A, Aquilué N, Archibald S et al (2020) Fire and biodiversity in the Anthropocene. Science. https://doi.org/10.1126/science.abb0355

Khairoutdinov M, Randall D, DeMott C (2005) Simulations of the atmospheric general circulation using a cloud-resolving model as a superparameterization of physical processes. J Atmos Sci 62:2136–2154. https://doi.org/10.1175/jas3453.1

Kimutai J, Carrasco-Escaff T, Garreaud RD et al. (2024) Despite known coastal cooling trend, risk of deadly wildfires in central Chile increasing with changing land management in a warming climate. https://doi.org/10.25561/109375

Kitzberger T, Tiribelli F, Barberá I, Gowda JH, Morales JM, Zalazar L, Paritsis J (2022) Projections of fire probability and ecosystem vulnerability under 21st century climate across a trans-Andean productivity gradient in Patagonia. Sci Total Environ 839:156303. https://doi.org/10.1016/j.scitotenv.2022.156303

Koch PL, Barnosky AD (2006) Late quaternary extinctions: state of the debate. Annu Rev Ecol Evol Syst 37:215–250. https://doi.org/10.1146/annurev.ecolsys.34.011802.132415

Kumar S, Getirana A, Libonati R, Hain C, Mahanama S, Andela N (2022) Changes in land use enhance the sensitivity of tropical ecosystems to fire-climate extremes. Sci Rep 12:1–11. https://doi.org/10.1038/s41598-022-05130-0

Lapere R, Mailler S, Menut L (2021) The 2017 mega-fires in central Chile: impacts on regional atmospheric composition and meteorology assessed from satellite data and chemistry-transport modeling. Atmos 12:344. https://doi.org/10.3390/atmos12030344

Lapola DM, Martinelli LA, Peres CA, Ometto JPHB, Ferreira ME et al (2014) Pervasive transition of the Brazilian land-use system. Nat Clim Chang 4:27–35. https://doi.org/10.1038/nclimate2056

Le Page Y, Morton D, Hartin C, Bond-Lamberty B, Pereira JMC, Hurtt G, Asrar G (2017) Synergy between land use and climate change increases future fire risk in Amazon forests. Earth Syst Dynam 8:1237–1246. https://doi.org/10.5194/esd-8-1237-2017

Li S, Rifai S, Anderson LO, Sparrow S (2022) Identifying local-scale meteorological conditions favorable to large fires in Brazil. Clim Resil Sustain 1:e11. https://doi.org/10.1002/cli2.11

Libonati R (2024) Megafires are here to stay—and blaming only climate change won't help. Nature 627:10. https://doi.org/10.1038/d41586-024-00641-4

Libonati R, DaCamara CC, Peres LF, de Carvalho LAS, Garcia LC (2020) Rescue Brazil's burning Pantanal wetlands. Nature 588:217–219. https://doi.org/10.1038/d41586-020-03464-1

Libonati R, Pereira JMC, DaCamara CC, Peres LF, Oom D et al (2021) Twenty-first century droughts have not increasingly exacerbated fire season severity in the Brazilian Amazon. Sci Rep 11:4400. https://doi.org/10.1038/s41598-021-82158-8

Libonati R, Geirinhas JL, Silva PS, Russo A, Rodrigues JA et al (2022a) Assessing the role of compound drought and heatwave events on unprecedented 2020 wildfires in the Pantanal. Environ Res Lett 17:015005. https://doi.org/10.1088/1748-9326/ac462e

Libonati R, Geirinhas JL, Silva PS, dos Santos DM, Rodrigues JA et al (2022b) Drought–heatwave nexus in Brazil and related impacts on health and fires: a comprehensive review. Ann N Y Acad Sci 1517:44–62. https://doi.org/10.1111/nyas.14887

López-Bermeo C, Montoya RD, Caro-Lopera FJ, Díaz-García JA (2022) Validation of the accuracy of the CHIRPS precipitation dataset at representing climate variability in a tropical mountainous region of South America. Phys Chem Earth Parts A/B/C 127:103184. https://doi.org/10.1016/j.pce.2022.103184

López-Mársico L, Lezama F, Altesor A (2021) Heterogeneity decreases as time since fire increases in a South American grassland. Appl Veg Sci 24:e12521. https://doi.org/10.1111/avsc.12521

Lüthi D, Le Floch M, Bereiter B, Blunier T, Barnola JM et al (2008) High-resolution carbon dioxide concentration record 650,000–800,000 years before present. Nature 453:379–382. https://doi.org/10.1038/nature06949

Machado-Silva F, Peres LF, Gouveia CM, Enrich-Prast A, Peixoto RB et al (2021) Drought resilience debt drives NPP decline in the Amazon Forest. Global Biogeochem Cycles 35:e2021GB007004. https://doi.org/10.1029/2021GB007004

Manzello SL, Blanchi R, Gollner MJ, Gorham D, McAllister S et al (2018) Summary of workshop large outdoor fires and the built environment. Fire Saf J 100:76–92. https://doi.org/10.1016/j.firesaf.2018.07.002

Marengo JA, Nobre CA, Tomasella J, Oyama MD, de Oliveira GS et al (2008) The drought of Amazonia in 2005. J Clim 21:495–516. https://doi.org/10.1175/2007JCLI1600.1

Marengo JA, Tomasella J, Alves LM, Soares WR, Rodriguez DA (2011) The drought of 2010 in the context of historical droughts in the Amazon region. Geophys Res Lett 38. https://doi.org/10.1029/2011GL047436

Marengo JA, Cunha AP, Cuartas LA, Leal KRD, Broedel E et al (2021) Extreme drought in the Brazilian Pantanal in 2019–2020: characterization, causes, and impacts. Front Water 4:942068. https://doi.org/10.3389/frwa.2021.639204

Marengo JA, Ambrizzi T, Barreto N, Cunha AP, Ramos AM et al (2022) The heat wave of October 2020 in central South America. Int J Climatol 42:2281–2298. https://doi.org/10.1002/joc.7365

Marlon JR, Kelly R, Daniau A-L, Vannière B, Power MJ et al (2016) Reconstructions of biomass burning from sediment-charcoal records to improve data–model comparisons. Biogeosciences 13:3225–3244. https://doi.org/10.5194/bg-13-3225-2016

McWethy DB, Pauchard A, García RA, Holz A, González ME et al (2018) Landscape drivers of recent fire activity (2001–2017) in south-central Chile. PLoS One 13:e0201195. https://doi.org/10.1371/journal.pone.0201195

Menezes LS, de Oliveira AM, Santos FLM, Russo A, de Souza RAF et al (2022) Lightning patterns in the Pantanal: untangling natural and anthropogenic-induced wildfires. Sci Total Environ 820:153021. https://doi.org/10.1016/j.scitotenv.2022.153021

Mistry J, Schmidt IB, Eloy L, Bilbao B (2019) New perspectives in fire management in South American savannas: the importance of intercultural governance. Ambio 48:172–179. https://doi.org/10.1007/s13280-018-1054-7

Monteiro dos Santos D, Libonati R, Garcia BN et al (2024) Twenty-first-century demographic and social inequalities of heat-related deaths in Brazilian urban areas. PLoS One 19:e0295766. https://doi.org/10.1371/journal.pone.0295766

Morales-Rincon LA, Hernandez AJ, Rodriguez-Hernandez NS, Jimenez R (2021) Carbon exchange and accumulation in an Orinoco high plains native savanna ecosystem as measured by Eddy covariance. Front Environ Sci 9:673932. https://doi.org/10.3389/fenvs.2021.673932

Moris JV, Álvarez-Álvarez P, Conedera M, Dorph A, Hessilt TD, Hunt HGP et al (2023) A global database on holdover time of lightning-ignited wildfires. Earth Syst Sci Data 15:1151–1163. https://doi.org/10.5194/essd-15-1151-2023

Mouillot F, Field CB (2005) Fire history and the global carbon budget: a 1° × 1° fire history reconstruction for the 20th century. Glob Chang Biol 11:398–420. https://doi.org/10.1111/j.1365-2486.2005.00920.x

Moura LC, Scariot AO, Schmidt IB, Beatty R, Russell-Smith J (2019) The legacy of colonial fire management policies on traditional livelihoods and ecological sustainability in savannas: impacts, consequences, new directions. J Environ Manag 232:600–606. https://doi.org/10.1016/j.jenvman.2018.11.057

Mukherjee S, Mishra AK (2021) Increase in compound drought and heatwaves in a warming world. Geophys Res Lett 48:e2020GL090617. https://doi.org/10.1029/2020GL090617

Myers RL (2006) Living with fire—sustaining ecosystems & livelihoods through integrated fire management. Global Fire Initiative, The Nature Conservancy

Nepstad D, Lefebvre P, Da Silva UL, Tomasella J, Schlesinger P et al (2004) Amazon drought and its implications for forest flammability and tree growth: a basin-wide analysis. Glob Chang Biol 10:704–717. https://doi.org/10.1111/j.1529-8817.2003.00772.x

O'Neill BC, Tebaldi C, Van Vuuren DP, Eyring V, Friedlingstein P et al (2016) The Scenario Model Intercomparison Project (ScenarioMIP) for CMIP6. Geosci Model Dev 9:3461–3482. https://doi.org/10.5194/gmd-9-3461-2016

Oliveira U, Soares-Filho B, Bustamante M, Gomes L, Ometto JP, Rajão R (2022) Determinants of fire impact in the Brazilian biomes. Front For Glob Chang 5:20. https://doi.org/10.3389/ffgc.2022.735017

Oliveira-Júnior JF, Mendes D, Correia Filho WLF, Da Silva CA, De Gois G et al (2021) Fire foci in South America: impact and causes, fire hazard and future scenarios. J S Am Earth Sci 112:103623. https://doi.org/10.1016/j.jsames.2021.103623

Ometto JP, Sousa-Neto ER, Tejada G (2016) Land use, land cover and land use change in the Brazilian Amazon (1960–2013). In: Nagy L, Forsberg BR, Artaxo P (eds) Interactions between biosphere, atmosphere and human land use in the Amazon Basin. Springer, Heidelberg, pp 369–383. https://doi.org/10.1007/978-3-662-49902-3_15

Osmont D, Sigl M, Eichler A, Jenk TM, Schwikowski M (2019) A Holocene black carbon ice-core record of biomass burning in the Amazon Basin from Illimani, Bolivia. Clim Past 15:579–592. https://doi.org/10.5194/cp-15-579-2019

Panisset JS, Libonati R, Gouveia CMP, Machado-Silva F, França DA et al (2018) Contrasting patterns of the extreme drought episodes of 2005, 2010 and 2015 in the Amazon Basin. Int J Climatol 38:1096–1104. https://doi.org/10.1002/joc.5224

Pereira JMC, Oom D, Silva PC, Benali A (2022) Wild, tamed, and domesticated: three fire macroregimes for global pyrogeography in the Anthropocene. Ecol Appl 32:e2588. https://doi.org/10.1002/eap.2588

Pimm SL, Russell GJ, Gittleman JL, Brooks TM (1995) The future of biodiversity. Science 269:347–350. https://doi.org/10.1126/science.269.5222.347

Pivello VR (2011) The use of fire in the Cerrado and Amazonian rainforests of Brazil: past and present. Fire Ecol 7:24–39. https://doi.org/10.4996/fireecology.0701024

Pörtner H-O, Roberts DC, Adams H, Adelekan I, Adler C et al (2022) Climate change 2022: impacts, adaptation and vulnerability, Technical summary. Cambridge University Press, Cambridge UK

Power MJ, Marlon J, Ortiz N, Bartlein PJ, Harrison SP et al (2008) Changes in fire regimes since the last glacial maximum: an assessment based on a global synthesis and analysis of charcoal data. Clim Dyn 30:887–907. https://doi.org/10.1007/s00382-007-0334-x

Pozo RA, Galleguillos M, González ME, Vásquez F, Arriagada R (2022) Assessing the socio-economic and land-cover drivers of wildfire activity and its spatiotemporal distribution in south-central Chile. Sci Total Environ 810:152002. https://doi.org/10.1016/j.scitotenv.2021.152002

Quilcaille Y, Batibeniz F, Ribeiro AFS, Padrón RS, Seneviratne SI (2023) Fire weather index data under historical and shared socioeconomic pathway projections in the 6th phase of the Coupled Model Intercomparison Project from 1850 to 2100. Earth Syst Sci Data 15:2153–2177. https://doi.org/10.5194/essd-15-2153-2023

Rainford SK, Martín-López JM, Da Silva M (2021) Approximating soil organic carbon stock in the Eastern Plains of Colombia. Front Environ Sci 9:685819. https://doi.org/10.3389/fenvs.2021.685819

Ramos-Neto MB, Pivello VR (2000) Lightning fires in a Brazilian Savanna National Park rethinking management strategies. Environ Manag 26:675–684. https://doi.org/10.1007/s002670010124

Raymond C, Suarez-Gutierrez L, Kornhuber K, Pascolini-Campbell M, Sillmann J, Waliser DE (2022) Increasing spatiotemporal proximity of heat and precipitation extremes in a warming world quantified by a large model ensemble. Environ Res Lett 17:035005. https://doi.org/10.1088/1748-9326/ac5712

Regoto P, Dereczynski C, Chou SC, Bazzanela AC (2021) Observed changes in air temperature and precipitation extremes over Brazil. Int J Climatol 41:5125–5142. https://doi.org/10.1002/joc.7119

Reszka P, Fuentes A (2015) The great valparaiso fire and fire safety management in Chile. Fire Technol 51:753–758. https://doi.org/10.1007/s10694-014-0427-0

Ribeiro AFS, Brando PM, Santos L, Rattis L, Hirschi M et al (2022) A compound event-oriented framework to tropical fire risk assessment in a changing climate. Environ Res Lett 17:065015. https://doi.org/10.1088/1748-9326/ac7342

Rivera JA, Arnould G (2020) Evaluation of the ability of CMIP6 models to simulate precipitation over Southwestern South America: climatic features and long-term trends (1901–2014). Atmos Res 241:104953. https://doi.org/10.1016/j.atmosres.2020.104953

Rosan TM, Sitch S, Mercado LM, Heinrich V, Friedlingstein P, Aragão LEOC (2022) Fragmentation-driven divergent trends in burned area in Amazonia and Cerrado. Front For Glob Chang 5:15. https://doi.org/10.3389/ffgc.2022.801408

Rose S, Betts R, Wester P, Koutroulis A (2022) Cross-chapter box CLIMATE | Climate reference periods, global warming levels and common climate dimensions. In: Pörtner H-O, Roberts DC, Tignor M et al (eds) Climate change 2022: impacts, adaptation and vulnerability. Contribution of working group II to the sixth assessment report of the Intergovernmental Panel on Climate Change. Cambridge University Press, Cambridge, UK, pp 135–143

Ruddiman WF, He F, Vavrus SJ, Kutzbach JE (2020) The early anthropogenic hypothesis: a review. Quat Sci Rev 240:106386. https://doi.org/10.1016/j.quascirev.2020.106386

Rusticucci M, Kyselý J, Almeira G, Lhotka O (2016) Long-term variability of heat waves in Argentina and recurrence probability of the severe 2008 heat wave in Buenos Aires. Theor Appl Climatol 124:679–689. https://doi.org/10.1007/s00704-015-1445-7

San José JJ, Montes RA (2001) Management effects on carbon stocks and fluxes across the Orinoco savannas. For Ecol Manag 150:293–311. https://doi.org/10.1016/s0378-1127(00)00588-0

San José J, Montes R, Grace J, Nikonova N (2008) Land-use changes alter CO_2 flux patterns of a tall-grass Andropogon field and a savanna–woodland continuum in the Orinoco lowlands. Tree Physiol 28:437–450. https://doi.org/10.1093/treephys/28.3.437

Santos FLM, Nogueira J, de Souza RAF, Falleiro RM, Schmidt IB, Libonati R (2021) Prescribed burning reduces large, high-intensity wildfires and emissions in the Brazilian savanna. Fire 4:56. https://doi.org/10.3390/fire4030056

Sarricolea P, Serrano-Notivoli R, Fuentealba M, Hernández-Mora M, de la Barrera F et al (2020) Recent wildfires in central Chile: detecting links between burned areas and population exposure in the wildland urban interface. Sci Total Environ 706:135894. https://doi.org/10.1016/j.scitotenv.2019.135894

Schmidt IB, Figueiredo IB, Scariot A (2007) Ethnobotany and effects of harvesting on the population ecology of Syngonanthus nitens (Bong.) Ruhland (Eriocaulaceae), a NTFP from Jalapão region, central Brazil. Econ Bot 61:73–85. https://doi.org/10.1663/0013-0001(2007)61[73:EAEOHO]2.0.co;2

Schmidt IB, Moura LC, Ferreira MC, Eloy L, Sampaio AB et al (2018) Fire management in the Brazilian savanna: first steps and the way forward. J Appl Ecol 55:2094–2101. https://doi.org/10.1111/1365-2664.13118

Schoeman DS, Sen Gupta A, Harrison CS, Everett JD, Brito-Morales I et al (2023) Demystifying global climate models for use in the life sciences. Trends Ecol Evol 38:843–858. https://doi.org/10.1016/j.tree.2023.04.005

Schumacher V, Setzer A, Saba MMF, Naccarato KP, Mattos E, Justino F (2022) Characteristics of lightning-caused wildfires in central Brazil in relation to cloud-ground and dry lightning. Agric For Meteorol 312:108723. https://doi.org/10.1016/j.agrformet.2021.108723

Schwingshackl C, Sillmann J, Vicedo-Cabrera AM, Sandstad M, Aunan K (2021) Heat stress indicators in CMIP6: estimating future trends and exceedances of impact-relevant thresholds. Earths Fut 9:e2020EF001885. https://doi.org/10.1029/2020ef001885

Shuman JK, Balch JK, Barnes RT, Higuera PE, Roos CI et al (2022) Reimagine fire science for the anthropocene. PNAS Nexus 1:1–14. https://doi.org/10.1093/pnasnexus/pgac115

Sierra JP, Arias PA, Vieira SC (2015) Precipitation over northern South America and its seasonal variability as simulated by the CMIP5 models. Adv Meteorol. https://doi.org/10.1155/2015/634720

Silva PS, Bastos A, DaCamara CC, Libonati R (2016) Future projections of fire occurrence in Brazil using EC-Earth climate model. Rev Bras Meteorol 31:288–297. https://doi.org/10.1590/0102-778631320150142

Silva PS, Bastos A, Libonati R, Rodrigues JA, DaCamara CC (2019) Impacts of the 1.5°C global warming target on future burned area in the Brazilian Cerrado. For Ecol Manag 446:193–203. https://doi.org/10.1016/j.foreco.2019.05.047

Silva PS, Rodrigues JA, Santos FLM, Pereira AA, Nogueira J et al (2020) Drivers of burned area patterns in Cerrado: the case of MATOPIBA region. Int Arch Photogramm Remote Sens Spat Inf Sci XLII-3/W12:135–140. https://doi.org/10.5194/isprs-archives-xlii-3-w12-2020-135-2020

Silva PS, Nogueira J, Rodrigues JA, Santos FLM, Pereira JMC et al (2021) Putting fire on the map of Brazilian savanna ecoregions. J Environ Manag 296:113098. https://doi.org/10.1016/j.jenvman.2021.113098

Silva PS, Geirinhas JL, Lapere R, Laura W, Cassain D, Alegría A, Campbell J (2022) Heatwaves and fire in Pantanal: historical and future perspectives from CORDEX-CORE. J Environ Manag 323:116193. https://doi.org/10.1016/j.jenvman.2022.116193

Silveira MVF, Silva-Junior CHL, Anderson LO, Aragão LEOC (2022) Amazon fires in the 21st century: the year of 2020 in evidence. Glob Ecol Biogeogr 31:2026–2040. https://doi.org/10.1111/geb.13577

Silvestrini RA, Soares-Filho BS, Nepstad D, Coe M, Rodrigues H, Assunção R (2011) Simulating fire regimes in the Amazon in response to climate change and deforestation. Ecol Appl 21:1573–1590. https://doi.org/10.1890/10-0827.1

Spinoni J, Barbosa P, De Jager A, McCormick N, Naumann G et al (2019) A new global database of meteorological drought events from 1951 to 2016. J Hydrol Reg Stud 22:100593. https://doi.org/10.1016/j.ejrh.2019.100593

Steffen W, Crutzen PJ, McNeill JR (2007) The Anthropocene: are humans now overwhelming the great forces of nature. Ambio 36:614–621

Steffen W, Broadgate W, Deutsch L, Gaffney O, Ludwig C (2015) The trajectory of the Anthropocene: the great acceleration. Anthr Rev 2:81–98. https://doi.org/10.1177/2053019614564785

Sun Q, Miao C, Hanel M, Borthwick AGL, Duan Q, Ji D, Li H (2019) Global heat stress on health, wildfires, and agricultural crops under different levels of climate warming. Environ Int 128:125–136. https://doi.org/10.1016/j.envint.2019.04.025

Thielen D, Ramoni-Perazzi P, Puche ML, Márquez M, Quintero JI et al (2021) The Pantanal under siege—on the origin, dynamics and forecast of the megadrought severely affecting the largest wetland in the world. Water 13:3034. https://doi.org/10.3390/w13213034

Trigueiro WR, Nabout JC, Tessarolo G (2020) Uncovering the spatial variability of recent deforestation drivers in the Brazilian Cerrado. J Environ Manag 275:111243. https://doi.org/10.1016/j.jenvman.2020.111243

Tyukavina A, Hansen MC, Potapov PV, Stehman SV, Smith-Rodriguez K et al (2017) Types and rates of forest disturbance in Brazilian Legal Amazon, 2000–2013. Sci Adv 3:e1601047. https://doi.org/10.1126/sciadv.1601047

United Nations Environment Programme (2022) Spreading like wildfire: the rising threat of extraordinary landscape fires. UNEP, Nairobi

United States Fire Administration (USFA) (2022) What is the WUI?. https://www.usfa.fema.gov/wui/what-is-the-wui.html. Accessed 24 Apr 2023

Van Der Werf GR, Randerson JT, Giglio L, Van Leeuwen TT, Chen Y et al (2017) Global fire emissions estimates during 1997–2016. Earth Syst Sci Data 9:697–720. https://doi.org/10.5194/essd-9-697-2017

Vargas-Cuentas NI, Roman-Gonzalez A (2021) Satellite-based analysis of forest fires in the Bolivian Chiquitania and Amazon region: case 2019. IEEE Aerosp Electron Syst Mag 36:38–54. https://doi.org/10.1109/maes.2020.3033392

Venter O, Sanderson EW, Magrach A, Allan JR, Beher J et al (2016) Sixteen years of change in the global terrestrial human footprint and implications for biodiversity conservation. Nat Commun 7:12558. https://doi.org/10.1038/ncomms12558

Vicente-Serrano SM, Gouveia C, Camarero JJ, Beguería S, Trigo R et al (2013) Response of vegetation to drought time-scales across global land biomes. Proc Natl Acad Sci USA 110:52–57. https://doi.org/10.1073/pnas.1207068110

Viegas LMD, Sales L, Hipólito J, Amorim C, de Pereira EJ et al (2022) We're building it up to burn it down: fire occurrence and fire-related climatic patterns in Brazilian biomes. PeerJ 10:e14276. https://doi.org/10.7717/peerj.14276

Vitolo C, Di Giuseppe F, Barnard C, Coughlan R, San-Miguel-Ayanz J et al (2020) ERA5-based global meteorological wildfire danger maps. Sci Data 7:55. https://doi.org/10.1038/s41597-020-0554-z

Vogel MM, Hauser M, Seneviratne SI (2020) Projected changes in hot, dry and wet extreme events' clusters in CMIP6 multi-model ensemble. Environ Res Lett 15:094021. https://doi.org/10.1088/1748-9326/ab90a7

Waters CN, Syvitski JPM, Gałuszka A, Hancock GJ, Zalasiewicz J et al (2015) Can nuclear weapons fallout mark the beginning of the Anthropocene Epoch? Bull At Sci 71:46–57. https://doi.org/10.1177/0096340215581357

Whelan R (2002) Adaptive management: what does it mean and how can it be used in fire management? Faculty of Science—Papers

Williams M, Zalasiewicz J, Haff PK, Schwägerl C, Barnosky AD, Ellis EC (2015) The Anthropocene biosphere. Anthr Rev 2:196–219. https://doi.org/10.1177/2053019615591020

Yanoviak SP, Gora EM, Bitzer PM, Burchfield JC, Muller-Landau HC et al (2020) Lightning is a major cause of large tree mortality in a lowland neotropical forest. New Phytol 225:1936–1944. https://doi.org/10.1111/nph.16260

Zalasiewicz J, Waters CN, Williams M et al (2015) When did the Anthropocene begin? A mid-twentieth century boundary level is stratigraphically optimal. Quat Int 383:196–203. https://doi.org/10.1016/j.quaint.2014.11.045

Zalasiewicz J, Waters CN, Head MJ, Poirier C, Summerhayes CP et al (2019) A formal Anthropocene is compatible with but distinct from its diachronous anthropogenic counterparts: a response to W.F. Ruddiman's 'three flaws in defining a formal Anthropocene'. Prog Phys Geogr Earth Environ 43:319–333. https://doi.org/10.1177/0309133319832607

Zalles V, Hansen MC, Potapov PV, Parker D, Stehman SV et al (2021) Rapid expansion of human impact on natural land in South America since 1985. Sci Adv 7:e1601047. https://doi.org/10.1126/sciadv.abg1620

Zazulie N, Rusticucci M, Raga GB (2017) Regional climate of the subtropical central Andes using high-resolution CMIP5 models—Part I: Past performance (1980–2005). Clim Dyn 49:3937–3957. https://doi.org/10.1007/s00382-017-3560-x

Zeebe RE, Ridgwell A, Zachos JC (2016) Anthropogenic carbon release rate unprecedented during the past 66 million years. Nat Geosci 9:325–329. https://doi.org/10.1038/ngeo2681

Challenges for a Better Fire Management in South America

14

Livia Carvalho Moura, Vânia R. Pivello, Christian Niel Berlinck, Isabel Belloni Schmidt, and Alessandra Fidelis

Abstract

The use of fire as a management tool by South American Indigenous Peoples is highly connected to their traditional knowledge, culture and governance, serving purposes like landscape management, subsistence agriculture, hunting, cultural rituals and fruiting stimulation. Despite this, many countries worldwide, influenced by European colonialism, adopted zero-fire policies, focused on fire suppression. These command-and-control policies that simply prohibit and criminalize fire tend to be ineffective in fire-prone ecosystems and where local traditions include fire for land management. Instead of inhibiting fires, the zero-

L. C. Moura
Instituto Sociedade, População e Natureza, SHCGN CLR Quadra, Brasília, Brazil
e-mail: livia@ispn.org.br

V. R. Pivello
Departamento de Ecologia, Instituto de Biociências, Universidade de São Paulo, São Paulo, Brazil
e-mail: vrpivel@usp.br

C. N. Berlinck
Centro Nacional de Pesquisa e Conservação de Mamíferos Carnívoros (CENAP), Instituto Chico Mendes de Conservação da Biodiversidade (ICMBio), Estrada Municipal Hisaichi Takebayashi, Atibaia, Brazil
e-mail: christian.berlinck@icmbio.gov.br

I. B. Schmidt
Departamento de Ecologia, Instituto de Ciências Biológicas, Universidade de Brasília, Campus Universitário Darcy Ribeiro, Brasília, Brazil
e-mail: isabels@unb.br

A. Fidelis (✉)
Departamento de Biodiversidade, Instituto de Biociências, Universidade Estadual Paulista, Rio Claro, Brazil
e-mail: alessandra.fidelis@unesp.br

A. Fidelis, V. R. Pivello (eds.), *Fire in the South American Ecosystems*, Ecological Studies 250, https://doi.org/10.1007/978-3-031-89372-8_14

fire policy ended up promoting large wildfires, arson, conflicts, socioeconomic problems, cultural loss and environmental degradation. Alternatively, more adequate, adaptive and participatory policies have increasingly replaced fire suppression, with noticeable results. After long processes of discussion and political persuasion, the Integrated Fire Management approach, that combines techniques and knowledge from different stakeholders, has been implemented in countries like Colombia, Ecuador, Paraguay and Brazil with positive outcomes. However, challenges remain, including the lack of engagement and investment from the private sector, conflicts with political agendas and the absence of state regulations. To improve fire management and achieve long term success, policies should adopt participatory and intercultural approaches, integrate knowledge and emerging technologies, also encourage local, national and international collaborations.

Resumen

El fuego como herramienta de manejo de los pueblos indígenas de Sudamerica está profundamente ligado a su conocimiento tradicional, cultura y gobernanza, como el manejo del paisaje, la agricultura de subsistencia, la caza, los rituales culturales y estimulación de la fructificación. Sin embargo, muchos países, influenciados por el colonialismo europeo, adoptaron políticas de cero-fuego centradas en la supresión. Estas políticas que prohíben y criminalizan el uso del fuego, suelen ser ineficaces en ecosistemas propensos a incendios y donde el uso del fuego es parte de las tradiciones locales. En lugar de reducir los incendios, la política de cero-fuego há fomentado grandes incendios forestales, conflictos, problemas socioeconómicos, pérdidas culturales y degradación ambiental. En cambio, políticas más adaptativas y participativas han reemplazado la supresión del fuego, con resultados positivos. El enfoque de Manejo Integrado del Fuego, que combina conocimientos y técnicas de diversos actores, se implementó con éxito en países como Colombia, Ecuador, Paraguay y Brasil. Sin embargo, persisten desafíos, como la falta de inversión del sector privado, conflictos políticos y la ausencia de regulaciones estatales. Para mejorar el manejo del fuego, las políticas deben adoptar enfoques participativos, integrar conocimientos y tecnologías emergentes, y fomentar la colaboración local, nacional e internacional.

Resumo

O fogo como ferramenta de manejo pelos povos indígenas da América do Sul está fortemente ligado ao conhecimento tradicional, cultura e governança, com objetivos como o manejo da paisagem, a agricultura de subsistência e caça, rituais culturais e estímulo à frutificação. Porém, muitos países, influenciados pelo colonialismo europeu, adotaram políticas de zero-fogo centradas na supressão do fogo. Essas políticas de comando-e-controle, que simplesmente proíbem e criminalizam o uso do fogo, são geralmente ineficazes em ecossistemas propensos ao fogo e onde usar o fogo faz parte das tradições locais. Em vez de reduzir os incêndios, a política de zero-fogo acabou promovendo grandes incêndios florestais, conflitos, problemas socioeconômicos, perdas culturais e degradação ambiental. Por outro lado, políticas mais adaptativas e participativas têm substituído a supressão do fogo, com resultados positivos. A abordagem do Manejo Integrado do Fogo, que combina conhecimentos e técnicas de diversos atores, foi implementado com sucesso em países como Colômbia, Equador, Paraguai e Brasil. No entanto, desafios ainda persistem, como a falta de investimento do setor privado, conflitos políticos e a ausência de regulamentações estatais. Para melhorar o manejo do fogo, as políticas devem adotar abordagens participativas, integrar conhecimentos e tecnologias emergentes e fomentar a colaboração local, nacional e internacional.

14.1 Overview of Fire Management in South American Ecosystems

Fire has long been used as a management tool by Indigenous Peoples in South America (Mistry et al. 2016; see Chap. 3). These communities have employed fire for a variety of purposes, including landscape management, subsistence agriculture, hunting, as well as in cultural rituals and to stimulate fruiting (Welch and Coimbra 2021; Falleiro et al. 2024). Other Traditional and Local Communities and small farmers apply fire for the same purposes and also to raise cattle in native vegetation (Borges et al. 2014; Eloy et al. 2019a). These traditional practices of using fire are highly connected with their knowledge, culture and governance (Mistry et al. 2016). However, traditional fire knowledge used for environmental management is declining, which can be linked to the general loss of knowledge between generations, prohibitive legislations (Moura et al. 2019), conflicts between Indigenous and non-Indigenous stakeholders, and the breakdown of the social cohesion of Indigenous communities (Mistry et al. 2016). For example, the traditional use of fire by Indigenous Peoples has been criticized by governmental institutions in different South American countries. Farmers of European origin, unfamiliar with local ecosystems and Indigenous fire management practices, have influenced policies and environmental educational programmes that focused on fire suppression and exclusion from any type of environment. These approaches disregard traditional Indigenous knowledge, which has long included fire as an integral part of land stewardship (Mistry et al. 2016; Moura et al. 2019).

Zero-fire policies have thus been implemented in many countries worldwide, influenced by European colonialism (Moura et al. 2019). For instance, these policies altered the traditional use of fire by Indigenous Peoples in Brazil and Venezuela (Mistry et al. 2016), overlooking its ecological importance in the ecosystems, as well as the cultural practices of traditional and local communities who had lived in harmony with these ecosystems for centuries or even millennia. In Venezuela for example, a firefighting program has been established in 1981, named Control of Wildfire Programme (PCIV), aiming to detect, prevent and fight wildfires at the headwaters of the Caroní River, in the Gran Sabana. Although the local brigade was composed mostly by Pemón Indigenous People, their traditional knowledge was ignored, and the policy of fire exclusion was implemented. This led to conflicts between the Pemón people and governmental agencies, as fire was an integral part of their daily activities and cultural identity (Mistry et al. 2016).

The imposition of zero-fire policies marginalized the ancient fire practices and led to the loss of traditional ecological knowledge associated with fire use. After years of pressure to replace traditional fire practices with fire suppression, some young members of Indigenous and Traditional and Local Communities came to view fire as harmful and supported substituting it with other technologies (Eloy et al. 2017). These zero-fire policies also restricted access to formal fire permits, creating conflicts between environmental agencies and the local and Traditional Communities (Moura and Viadana 2011). In consequence, illegal and hidden uses of fire increased, often resulting in large wildfires (Moura et al. 2019). Furthermore, the prohibition of fire use led local cattle ranchers—who rely on fire to stimulate native vegetation regrowth for grazing during the dry season—to initiate illegal fires outside their properties, allowing them to spread naturally, as wildfires, to achieve their objectives (Mistry and Bizerril 2011).

Therefore, instead of inhibiting fires, the zero-fire policy ended up generating the opposite effect, promoting large wildfires (Fidelis et al. 2018). In Brazil, for example, before the implementation of fire management policies, huge efforts were made to keep fire away from fire-prone vegetation within protected areas; whether naturally occurring or human-induced, fires were systematically extinguished whenever they happened (Durigan and Ratter 2016). This policy resulted in disastrous wildfires that devastated large areas of native vegetation in several major national parks. Notable examples include the Emas National Park (Central Brazil) in 2010, where 123,200 hectares were burned, representing 93% of the park (Pivello et al. 2021); Serra da Canastra National Park (Southeastern Brazil) in 2016, which lost 91,520 hectares, or 46% of its area (Silva et al. 2018); Araguaia National Park (Northern Brazil) in 2017, where 370,465 hectares, or 67% of the park, were consumed by fire (Hataishi et al. 2019); and Chapada dos Veadeiros National Park (Central Brazil) in 2017, where 85,500 hectares, or 36% of the total area, were burned (Fidelis et al. 2018).

In general, command-and-control policies that simply prohibit and criminalize fire tend to be ineffective in fire-prone ecosystems where local culture and practices include fire as a tool for management and productivity (Carmenta et al. 2019), as these policies commonly generate conflicts and wildfires due to increased arson, protest fires, and/or the accumulation of dry biomass in areas that have not been burned for several years (Moura et al. 2019). In response to these challenges, fire management policies came to replace zero-fire policies in South American countries (Mistry et al. 2019).

Still, over recent decades, climate and land-use changes have increased the likelihood of fire ignition and spread, turning some South American landscapes more flammable. The megafires that occurred in Chile in 2017 and 2023 exemplify this trend, driven by a combination of extreme high temperatures, low humidity and strong winds, associated with highly flammable forests. These conditions made the wildfires exceptionally difficult to control and extinguish (Armenteras and Barrera 2023). Similar scenarios have become increasingly common across the continent, including in the Pantanal region of Brazil (2020, 2024; Tomas et al. 2021; Libonati et al. 2022), the Amazon rainforest (2019, 2024; Oliveira et al. 2023a), the Chiquitano dry forests of Bolivia (2018, 2019; Maillard et al. 2020; Singh et al. 2022) and parts of Argentina (2020, 2024; Di Bitetti et al. 2022).

In response to the significant changes in climate, natural vegetation cover and land uses in South America, new fire policies are being tested and implemented in several countries, particularly those with large savanna regions, such as Brazil, Guyana, Venezuela and Paraguay (Bilbao et al. 2019; Eloy et al. 2019b; Mistry et al. 2016). In places where the use of fire is associated with forest management and timber extraction activities, the REDD++ stimulated the implementation of the low-carbon policy and the creation of a national forest fire management (e.g. Guyana; Mistry et al. 2016). Also in the Uruguayan grasslands, the use of prescribed burns has shown to be a major driver of vegetation change and structural diversity, paving the way for its adoption as a management tool to maintain biodiversity, prevent vegetation homogenization and influence aboveground productivity (López-Mársico et al. 2021).

14.2 Fire Management Legislation and Implementation

14.2.1 Fire Management Policies in South America

Before the last decade, environmental legislations in South America were mostly related to fire in terms of prohibiting its use and trying to exclude it from all landscapes, within the scope of the zero-fire policy (Moura et al. 2019). There were almost no expressive laws or policies that were dedicated exclusively to fire management until 2012, when Argentina and Paraguay launched more specific fire rules and plans, with the Federal Fire Management System in Argentina (Law n. 26.815/2012) and the National Plan for Integrated Fire Management in Paraguay (Table 14.1). Other countries (e.g., Venezuela, Brazil, Bolivia) have also established their national policies or programmes related to the use of fire in natural and rural systems, as well as more specific regulations (Table 14.1), but they were not always implemented or made effective. Legislations and initiatives recognizing the importance of fire as a natural element and its use as a management tool became more common in the last decades, due to greater knowledge over new techniques and technologies, and as a result of new attempts to reduce the occurrence of large wildfires, which have been progressively increasing (Myers 2006; Dowdy et al. 2022).

From 2019, the Integrated Fire Management (IFM) approach began to be implemented as the main national fire strategy in some countries, with some even developing their own policies, such as in Colombia (Bill 221/2019), Ecuador (Conamif/2020), Paraguay (Law 6.818/2021) and Brazil (Law 14.944/2024, Table 14.1). The IFM strategy combines scientific, technical and traditional knowledge into fire management practices. These regulations went through a long process of discussions and political persuasion, challenging the paradigm that all fires are harmful and must be suppressed. However, the increasing frequency and severity of

Table 14.1 Some relevant policies and legislations directly connected to fire management in South American countries

Country	Political and legal instruments related to fire management	Description
Argentina	Law 26815, 28/11/2012—Federal Fire Management System	Focus on fire suppression, with limited attention to fire prevention, and little emphasis on fuel management or active restoration strategies
Bolívia	Law 1171, 25/5/2019—Law on the Use and Rational Management of Fire	Defines guidelines for a comprehensive fire management policy in the national territory; establishes the administrative sanction for unauthorized burning; regulates, on an exceptional basis, a period for the payment of debts and fines related to unauthorized burning
Brazil	Decree 84.017, 21/09/1979—Regulation of Brazilian National Parks	Forbids any practice or act that may start a wildfire in National Parks, and only allows the use of fire as a management technique if it is indicated in the park management plan
	Resolution National Environment Council (CONAMA) 11, 14/12/1988—Fire in Protected Areas	Establishes actions for the planning of management fires in protected areas and the disposal of burnt wood
	Law 12.651, 25/5/2012—Protection of the Native Vegetation	Establishes permissions and prohibitions for the use of fire in protected areas, measures of wildfire control, and predicts the establishment of a national fire policy
	Law 14944, 31/7/2024—National Policy of Integrated Fire Management	Establishes directions and general policies to implement Integrated Fire Management in the Brazilian territory
Chile	Decree 276, 2609/1980—Ministry of Agriculture Supreme	Regulates the use of fire in the vegetation
Colombia	Resolution 532, 26/04/2005—Ministry of Environment, Housing and Territorial Development	Establishes requirements, terms and obligations for controlled open burning in rural areas, agricultural and mining activities
	Law 1575, 21/08/2012—General Firefighters Law of Colombia	Establishes the principle of shared responsibility, between all the authorities and inhabitants of the Colombian territory in relation to the management and risk of wildfires and the preparation and response to rescues during incidents involving hazardous materials
	Bill 388, 16/06/2021—Fire Management Law—Ministry of Agriculture and Rural Development	Bill in progress in the senate to implement Integrated Fire Management in the country

Table 14.1 (continued)

Country	Political and legal instruments related to fire management	Description
Ecuador	Programme Amazon without Fire (PASF)/2017—Ministry of the Environment	Regional project to prevent, control and remediate forest wildfires in Ecuador by strengthening technical capacities in fire management, drawing up public policies, environmental education and promoting good agricultural practices
	National Strategy for Integrated Fire Management and the formation of the National Committee for Integrated Fire Management (Conamif), 2020, Ministry of the Environment and Water	Public policy instrument established in the Regulations of the Organic Code of the Environment. The strategy consolidates an inter-institutional planning to minimize and address the effects of fire, by regulating its use in a controlled, articulated and conditional way
	Decree 752, 12/06/2019—Regulation for the Organic Code of the Environment, Executive. Official Register Supplement 507, 12/06/2019, altered in 18/10/2022	Binds mandatory regulations for all entities, organizations and departments, including the central public sector and the decentralized autonomous sector, natural and legal persons, communities, cities, nationalities and groups, which are permanently or temporarily in the national territory. Integrated Fire Management is mentioned and recognized as a strategy to protect areas from wildfires
	Ministerial Agreement n. MAATE-2023-080, 20/08/2023—Reform of the Organic Statute by Processes of the Ministry of Environment, Water and Ecological Transition (MAATE)	Attributes responsibility to authorize, manage, protect and forbid Integrated Fire Management activities, including the development of mechanisms to promote Integrated Fire Management in the country
Guiana	Act 6, 2009—Forest Act	Contains provisions to prohibit, restrict, and penalize the use of fire in protected areas
	Pilot Fire Management Project under REDD+ in 2019 (Guyana Government)	A pilot project that focuses on how fire can be used by Traditional Communities to receive benefits from the REDD+ for abating emissions. It also established the first Indigenous fire brigade in Guyana (Shulinab Fire Brigade)
Paraguay	National Plan for Integrated Fire Management in Paraguay, 2012–2016	Provides the conceptual, legal and operational basis to address fire management for environmental conservation; emphasizes the need for an ecosystem approach to fire management in order to protect biodiversity and associated environmental services
	Law 6818, 12/10/2021 and Decree 7775/, 8/09/2022—Integrated Fire Management Law and regulation	Establishes actions, rules and procedures for the implementation of Integrated Fire Management in the country, as well as prevention measures necessary for the execution of its activities throughout the national territory

(continued)

Table 14.1 (continued)

Country	Political and legal instruments related to fire management	Description
Peru	Law 29763, 21/07/2011—Forest and Fauna Law	Establishes guidelines to conserve and protect national forest resources
	Law 29664, 19/02/2011—National Disaster Risk Management System Law (SINAGERD), updated in 2021	Establishes guidelines for preventing or reducing disaster risks, preparing for and responding to disasters and minimizing their impact on people, the economy and the environment
Uruguay	Decree 436, 19/11/2007 (modified by Decree 281, 27/08/2021)—General Action Plan for Prevention, Alert and Response to Forest Fires	A plan to prevent wildfire outbreaks, provide early warning and respond effectively so that they do not become major forest wildfires
Venezuela	Official announcement n. 2990, 26/07/1982—Venezuelan Civil Code	Establishes restrictions for forest burnings in the headwaters of rivers and slopes
	Official announcement n. 39913, 2/5/2012—Environmental Criminal Law	Establishes criteria for granting burn permits
	Official announcement n. 40222, 6/8/2013—Forest Law	Creates the National Forest Fire Protection System ensuring that the burning of vegetation complies with the measures and regulations established by the appropriate authorities

wildfires in recent years, exacerbated by climate change, rising arson rates, and unmanaged areas with significant fuel loads (dead and dry biomass), has highlighted the urgent need for effective management policies and guidelines, pressuring governments and decision-makers to adopt new policies. Many of these policies are still in the process of being regulated, and traditional, technical, and scientific knowledge have fostered rules and guided management towards better results since the beginning. One of the key advancements has been the recognition that fire behaves differently and has varied effects in each ecosystem (Hardesty et al. 2005; Myers 2006). This understanding has enabled managers to use prescribed fire as a tool for conserving fire-adapted ecosystems, rather than trying to exclude it from all environments, for example. A resolution of the Brazilian National Environment Council (CONAMA) in 1989 was the first regulation to expressly recognize that fire could be used as an ecological conservation tool in Protected Areas (Resolution CONAMA n. 11 from 14 December 1988) (Table 14.1).

Other countries are still in the process of convincing decision makers that the IFM approach, or other fire management strategies, may be a preferable solution to wildfire problems than the zero-fire policy and are trying to create bills and new fire policies. An example of this is the Guyana Forestry Commission and the Ministry of Natural Resources that have begun discussions to develop a fire management document to minimize the risk of wildfires in forests and savannas (UNESCO 2024).

14.2.2 Integrated Fire Management in Brazil: A Successful Experience

In the early development of Brazil's environmental policies, fire management regulations were largely aimed at preventing economic losses, though they gradually began to incorporate ecological considerations. This can be seen in several early laws, including the Pau-Brasil Regiment (from 1605), the Orders for the Preservation of Naval Timber (from 1795), the Land Law (No. 601/1850), and the law that classified setting fires as a crime (No. 3311/1886) (Berlinck and Lima 2021). It was only after the 1970s that new perspectives on the roles and uses of fire began to emerge, as fire became a subject of scientific studies (Pivello et al. 2021). In 1979, the National Parks Regulation (Decree 84017/1979) marked a turning point by introducing a new perspective on fire management, by using prescribed fires. This approach was subsequently reinforced by various infra-law resolutions, such as CONAMA Resolution n° 11/1988, which authorized the use of fire for ecological management (Berlinck and Lima 2021). The shift was further solidified by the Law for the Protection of Native Vegetation (Lei n° 12651/2012), which formally acknowledged fire as a valuable tool for conserving fire-adapted native vegetation (Pivello et al. 2021).

In this new approach, Brazil moves from the fire-suppression policy (zero-fire policy) to recognizing fire as an ecological factor and including it in the perspective of socio-biodiversity conservation (Garda and Berlinck 2018). Thus, starting in 2011, the foundations for an IFM began to evolve in Emas National Park. The initial aims of the implementation of the IFM were: (1) conservation of the socio-biodiversity; (2) reduction of the burned areas; (3) change in the fire regime (season, frequency, intensity and extent of the burned areas); (4) protection of sensitive species and environments; (5) reduction of the emission of greenhouse gases; and (6) reduction of the firefighting costs (Berlinck and Lima 2021; Schmidt et al. 2018).

The IFM approach was first introduced in Brazilian protected areas in 2014 as a pilot program in Chapada das Mesas National Park, Serra Geral do Tocantins Ecological Station, and Jalapão State Park (Schmidt et al. 2018), all located in the northern Cerrado region. The Cerrado was prioritized for IFM implementation due to its predominantly fire-prone ecosystems (see Chap. 5). Subsequently, the approach expanded to other biomes, including the Atlantic Rainforest, Amazon Forest, Pampas, Pantanal, and Caatinga. In Pantanal, the importance of researches on the ecological and social roles of fire was only recognized in 2020, following catastrophic wildfires that drew national and international attention due to the extensive damage, including massive animal casualties and vast areas burned (Tomas et al. 2021; Berlinck et al. 2022). These events stressed the urgent need for effective fire management strategies in the biome.

Following its initial successful experiences, the IFM approach expanded rapidly, and within less than a decade, it was adopted in over 58 protected areas, including federal and state reserves, 54 Indigenous lands, and one *Quilombola* territory (Eloy et al. 2019b; Falleiro et al. 2021). By 2023, federal environmental agencies had planned to manage more than 30 million hectares under the IFM approach, incorporating prescribed burns as a strategic tool to conserve fire-adapted native vegetation while protecting fire-sensitive ecosystems across Brazil (Berlinck and Batista 2020; Schmidt et al. 2018). Successful results are being documented (Falleiro et al. 2024; Santos et al. 2021a, b) from territories managed with prescribed fire. Overall, IFM has led to notable reductions in late-dry season wildfires and total burned areas across several Brazilian protected areas (Berlinck and Batista 2020; Mistry et al. 2019; Santos et al. 2021b). For instance, at Serra da Canastra National Park, IFM was introduced in 2018, including the use of fire for livestock management. This resulted in a decrease in wildfire-affected areas from 90,000 hectares to 30,000 hectares, with uninhabited areas experiencing the most significant reduction—from approximately 40,000 hectares to 10,000 hectares (Tizianel et al. 2020). In the Amazon savannas, the Campos Amazônicos National Park experienced a 45% reduction in the total burned area since the IFM implementation in 2016, effectively safeguarding the surrounding sensitive tropical forests (Cambraia 2024). Additionally, IFM has substantially lowered firefighting costs and reduced fire-related greenhouse gas emissions (Franke et al. 2024).

However, to achieve such results other important and complementary activities from the IFM approach are also being applied, such as environmental education, local communities sensibilization, participatory monitoring and collaborative planning including meetings to establish burning calendars and minimize conflicts (Schmidt et al. 2018). For example, in the Kalunga Quilombola territory (Traditional Communities from Central Brazil), participatory meetings were held with Quilombola leaders to establish collective burning calendars and authorizations for fire use before implementing prescribed burnings. Similarly, in the Xerente and Araguaia Indigenous lands, participatory meetings were conducted to create collective burning calendars, and Indigenous leaders were consulted to approved fire use before prescribed burns were carried out in collaboration with the Indigenous fire brigade and environmental public institutions. This integrated approach contributed to significantly reducing late-dry-season wildfires, which previously affected over 80,000 hectare and was reduced to just 3,000 hectare in 2021—a decrease of approximately 23.5% (Santos et al. 2021b; ICMBio 2023).

14.2.3 Policy Gaps in the Implementation of Fire Management in South America

Although there have been great advances in the use of fire as a management tool in South American ecosystems, critical policy gaps continue to hinder the effective implementation of fire management across the region. In Brazil, for example, the

National Policy for Integrated Fire Management had been under discussion since 2018 and was only recently approved in July 2024 (Federal Law 14. 944). The delay was due to extended debates over jurisdiction and governance of fire management on federal lands, particularly as local firefighting forces—composed of military personnel—sought to maintain control over fire operations and activities. The approval of this national policy was urgently needed, as its absence was obstructing the governance and coordination with other government sectors beyond environmental agencies, as well as its implementation at state and local levels. Expanding IFM implementation further requires the involvement of private lands, which fall under the jurisdiction of federal states. However, a few Brazilian states have adopted policies aligned with IFM principles, including Mato Grosso (Decree 774, 2024), Mato Grosso do Sul (Decree 15.654, 2021), and Minas Gerais (Decree 47.919, 2020). Currently, up to 113 protected areas—including parks, reserves, Indigenous Territories, and *Quilombola* Territories—have implemented IFM, but greater adherence is needed, especially where the vegetation is more flammable and subjected to greater risks due to neighboring land uses.

Similarly to Brazil, the implementation of the Law on the Rational Use and Management of Fire (Law n°. 1171/2019) in Bolivia finds a significant challenge in integrating actions and responsibilities among various stakeholders and their interests, including government agencies, Indigenous Peoples, local residents, the productive sector, and livestock farmers. This is essential for achieving integrated landscape management and ensuring the respectful and orderly use of fire (MMAA 2024).

In general, there is a pressing need across South American countries to establish a comprehensive fire research agenda, define standardized methodologies, and encourage broader discussions about the perception and use of fire in restoration, conservation, and management planning. Additionally, developing a streamlined and efficient fire management process, including simplified authorization procedures, is essential to support Indigenous Peoples, Traditional and Local Communities, and small farmers who rely on fire to sustain their livelihoods.

14.3 New Perspectives and Challenges in Using Fire as a Management Tool in South America

Fire management remains a complex and controversial issue in South America, largely due to a widespread misunderstanding of the multiple roles of fire, and a key challenge is to differentiate its impacts—positive and negative—on different ecosystems, while also appreciating its importance in rural practices and cultural traditions. Fire has been a tool for maintaining landscapes, promoting biodiversity, and supporting livelihoods in many regions for millenia (see Chap. 2). However, efforts to manage fire often conflict with political agendas or lack the necessary scientific and technical backing.

14.3.1 Integration Among Stakeholders

Wildfire issues in South America involve a wide range of stakeholders, including Indigenous Peoples, local residents, landowners, managers, firefighters, government agencies, and politicians. One of the major challenges is the need to integrate diverse forms of knowledge, by embracing both scientific research and traditional, ancestral practices of fire use, without stigmatizing the latter. Effective fire governance requires cooperation between all these groups to create integrated strategies, protocols, policies, and educational programs. Fire management decisions should be driven by scientific data and technical expertise rather than political and economic motivations, where the goal is to transform fire from a threat into a tool for ecosystem management and sustainable development.

Achieving this requires acknowledging the diverse priorities and interests of stakeholders and leveraging their unique expertise. By setting aside differences and focusing on the shared goal of preventing wildfires from devastating areas of mutual interest, stakeholders can collaborate effectively to implement fire management strategies. Fire policies and management plans shall be developed through participatory and intercultural approaches and encouraged collaboration (Mistry et al. 2019). For example, Venezuela has adopted a participatory and intercultural fire management model that brings together the traditional fire knowledge of the Pémon Indigenous People from Canaima National Park with firefighters' practical knowledge and academic knowledge, involving diverse stakeholders (Bilbao et al. 2010). In 1999, a fire experiment was conducted to simulate the Pémon's traditional use of fire by burning the savanna in different periods of the dry season to create a heterogeneous landscape (Bilbao et al. 2010). The process of bringing the Indigenous community to join ecological experiments resulted in the regain of the traditional ecological knowledge, being an opportunity to consult and learn from the Pémon people more about the traditional use of fire (Mistry et al. 2019).

However, efforts to integrate fire management have often been hindered by conflicting interests, even worsening wildfire problems. As stated above, prohibitive fire policies, such as zero-tolerance approaches to fire use have led to environmental, economic, and social losses, also creating conflicts with local communities who rely on fire for agricultural and cultural practices (Moura and Viadana 2011; Mistry et al. 2016). Furthermore, the organization of firefighting efforts—across military, volunteer, and professional fire brigades—often lacks coordination, with disputes over authority taking precedence over collaboration. This lack of alignment has undermined efforts to implement effective fire management policies that balance environmental, social, and economic needs in many South American countries. Another important point to be raised is that not all fire brigades engage with the local communities to manage their territories, using techniques from their formal technical training and neglecting the traditional fire knowledge (Eloy et al. 2017). In some cases, sacred areas or other areas of interest are burned without consent.

Therefore, to achieve a more participatory approach to fire management and governance, Mistry et al. (2019) suggest some key actions: (1) capacity building and training of decision-makers and local managers in participatory methods, to encourage engagement with local and Indigenous Peoples and their knowledge, (2) legitimizing traditional fire management based on local social-ecological systems as a community-owned solution, and (3) creating common spaces for different stakeholders to share their perspectives and experiences in fire management.

14.3.2 Emerging and Indispensable Technologies

In the last decades, advances in remote sensing, weather monitoring, and fire alert systems have significantly improved fire management in South America. Scientific and technological advancement, together with increased collaboration and experience exchange among researchers, governments, and communities have led to more efficient wildfire prevention and suppression methods (e.g., Anderson et al. 2023; Casallas et al. 2023).

A notable initiative is the Latin-American Network for Remote Sensing and Forest Fires (RedLaTIF), established in 2002, which contributes to global efforts in forest and land cover monitoring. In Brazil, the National Institute for Space Research (INPE) has developed tools for heat detection and wildfire risk monitoring, benefiting all South American countries. NASA's Fire Information for Resource Management System (FIRMS) is another important tool for monitoring fire activity in the region (https://firms.modaps.eosdis.nasa.gov/map/#d:24hrs;@0.0,0.0,3.0z). Additionally, Brazil's Integrated Multi-Agency Operational Coordination Center (CIMAN), created in 2016, coordinates large firefighting national operations by monitoring, organizing and articulating wildfire suppression activities in the Brazilian territory during the dry season.

Despite these advancements, many South American countries still do not effectively address the root causes of wildfire outbreaks, especially the ones based in programs and initiatives focused on fire suppression. In contrast, new approaches such as Community-Based Fire Management (CBFiM) and Integrated Fire Management (IFM) are emerging, recognizing the value of traditional fire knowledge in preventing and managing wildfires (Eloy et al. 2019a; Moura et al. 2019). These approaches emphasize the importance of community participation and flexible methodologies for fire management, but to succeed on a regional or national level will require overcoming established policies that continue to prioritize fire suppression over prevention.

An initiative in Brazil that deserves to be mentioned is the development of a fire spread prediction system for the fire-prone Cerrado (see Chap. 5). The system, named FISC-Cerrado (Fire, Ignition, Spread and Carbon Cycling), automatically uploads hot pixel and satellite data to generate maps of fuel load, vegetation moisture, and burn probability, simulating fire spread three times a day across the entire Cerrado region (Oliveira et al. 2023b). The model results, along with additional data such as fire scars, historical burnt areas and annual CO_2 emissions from fires, are available on an interactive web platform (https://csr.ufmg.br/fipcerrado/en/), which serves as a tool for fire prevention and control (Oliveira et al. 2023b). While currently it is limited to a few protected areas in Brazil, it has the potential to expand

and include new areas and even other countries, improving its accuracy through the use and customization.

In addition to the FISC-Cerrado system, other innovative tools are being developed to enhance fire management and prevention in South America. The *Alarm* system (https://alarmes.lasa.ufrj.br/) identifies heat sources and calculates the area affected by fire, providing real-time data to support decision-making in fire management. Currently available for Pantanal, Cerrado, and Amazon, it offers critical insights into fire behavior and spread across these vulnerable regions. The Mapbiomas Fogo is a collaborative platform supplied by a network of non-governmental organizations, universities and technology companies that provides fire scar information for all the Brazilian territory from 1985 to 2023, based on image mosaics from Landsat satellites with a spatial resolution of 30 meters (https://plataforma.brasil.mapbiomas.org/fogo?). Furthermore, CEMADEM (National Center for Monitoring and Early Warning of Natural Disasters, in Brazil) provides fire propagation risk assessments, helping authorities anticipate and respond to fire outbreaks more effectively (https://produtos.cemaden.gov.br/fogo/). These technologies, along with ongoing advances in remote sensing and predictive modeling, are increasingly fundamental to the region's efforts to combat the growing threat of forest fires.

14.3.3 Integrative Policies

The concept of IFM is gaining traction in South America as a more holistic approach to managing fire, since it integrates the ecological, cultural, and socioeconomic roles of fire. It incorporates strategies for wildfire prevention, suppression, ecosystem conservation and recovery, and the fulfillment of local, Indigenous, and traditional needs (Mistry et al. 2019; Bilbao et al. 2019). This inclusive approach offers a more balanced and sustainable way to manage fire, recognizing that it is not inherently harmful, but rather a natural and sometimes necessary tool for landscape management. By integrating diverse knowledge systems and different stakeholder perspectives, IFM can foster more effective and equitable fire management policies across South America.

Fire management must be adaptive, continuously evolving to respond to changing environmental conditions, and this perspective requires action-based research, continuous learning, and regular monitoring to ensure strategies remain effective over time. Given that the outcomes of adaptive approaches are often long-term and can take years to fully materialize—considering implementation, results gathering and monitoring—fire management policies should not be abandoned or altered abruptly due to shifts in political leadership or changes of decision-makers.

In conclusion, effective fire management in South America requires a comprehensive, adaptive approach. Collaboration among stakeholders—ranging from local communities and Indigenous groups to government agencies and researchers—is essential for developing policies and strategies that balance ecological, cultural, and socioeconomic needs. While technological advancements and new fire management approaches are promising, it is essential to overcome existing rigid policies and encourage long-term collaboration among all stakeholders. By adopting flexible, participatory fire management models, South American countries can more effectively reduce wildfire risks and use fire as a tool for sustainable land management. This will help to protect both fire sensitive and fire dependent ecosystems and support the livelihoods of local communities.

References

Anderson LD, Dix B, Schnell J, Yokelson R, Veefkind JP et al (2023) Analyzing the impact of evolving combustion conditions on the composition of wildfire emissions using satellite data. Geophys Res Lett 50:e2023GL105811. https://doi.org/10.1029/2023GL105811

Armenteras D, Barrera F (2023) Landscape management is urgently needed to address the rise of megafires in South America. Commun Earth Environ 4:305–315. https://doi.org/10.1038/s43247-023-00964-6

Berlinck CN, Batista EKL (2020) Good fire, bad fire: it depends on who burns. Flora 268:151610. https://doi.org/10.1016/J.FLORA.2020.151610

Berlinck CN, Lima LHA (2021) Implementation of integrated fire management in Brazilian Federal protected areas: results and perspectives. Biodivers Bras 11:128–138. https://doi.org/10.37002/biodiversidadebrasileira.v11i2.1709

Berlinck CN, Lima LHA, Pereira AMM, Carvalho-Jr EAR, Paula RC et al (2022) The Pantanal is on fire and only a sustainable agenda can save the largest wetland in the world. Braz J Biol 82:e244200. https://doi.org/10.1590/1519-6984.244200

Bilbao B, Leal AV, Méndez C (2010) Indigenous use of fire and forest loss in Canaima National Park, Venezuela. Assessment of and tools for alternative strategies of fire management in Pemón indigenous lands. Hum Ecol 38:663–673. https://doi.org/10.1007/s10745-010-9344-0

Bilbao B, Mistry J, Millán A, Berardi A (2019) Sharing multiple perspectives on burning: towards a participatory and intercultural fire management policy in Venezuela, Brazil, and Guyana. Fire 2:1–33. https://doi.org/10.3390/fire2030039

Borges SL, Eloy L, Ludewigs T (2014) O gado que circulava: Desafios da gestão participativa e impactos da proibição do uso do fogo aos criadores de gado de solta da Reserva de Desenvolvimento Sustentável Veredas do Acari. Biodivers Bras 4:130–156

Cambraia B (2024) Queima prescrita como técnica de prevenção de incêndios florestais no maior enclave de Cerrado da Amazônia Meridional Brasileira. Dissertação de mestrado. Programa de Pós-Graduação Profissional Biodiversidade em Unidades de Conservação. Jardim Botânico do Rio de Janeiro, Brazil

Carmenta R, Coudel E, Steward AM (2019) Forbidden fire: does criminalising fire hinder conservation efforts in swidden landscapes of the Brazilian Amazon? Geogr J 185:23–37. https://doi.org/10.1111/GEOJ.12255

Casallas A, Castillo-Camacho MP, Sanchez ER, González Y, Celis N et al (2023) Surface, satellite ozone variations in Northern South America during low anthropogenic emission conditions: a machine learning approach. Air Qual Atmos Health 16:745–764. https://doi.org/10.1007/s11869-023-01303-6

Di Bitetti MS, de Angelo C, Paviolo A, Di Giacomo AS, Varela D, Giraudo AR (2022) Argentina: wildfires jeopardize rewilding. Nature 604:246. https://doi.org/10.1038/d41586-022-01006-5

Dowdy A, Purcell L, Boulter S, Moura LC (2022) Wildfires under climate change: a burning issue. In: United Nations Environment Programme (ed) Frontiers 2022: noise, blazes and mismatches—emerging issues of environmental concern. UNEP, Nairobi, pp 24–31

Durigan G, Ratter JA (2016) The need for a consistent fire policy for Cerrado conservation. J Appl Ecol 53:11–15. https://doi.org/10.1111/1365-2664.12559

Eloy L, Carvalho IS, Figueiredo I (2017) Sistemas agrícolas tradicionais no Cerrado: caracterização, transformações e perspectivas. In: Santilli J, Bustamante P, Barbieri RL (eds) Agrobiodiversidade e Agroecologia. ABA/EMBRAPA, Brasília, pp 129–164

Eloy L, Bilbao AB, Mistry J, Schmidt IB (2019a) From fire suppression to fire management: advances and resistances to changes in fire policy in the savannas of Brazil and Venezuela. Geogr J 185:10–22. https://doi.org/10.1111/geoj.12245

Eloy L, Schmidt IB, Borges SL, Ferreira MC, dos Santos TA (2019b) Seasonal fire management by traditional cattle ranchers prevents the spread of wildfire in the Brazilian Cerrado. Ambio 48:890–899. https://doi.org/10.1007/s13280-018-1118-8

Falleiro RM, Steil L, Oliveira MS, Lando I, Machado LOR et al (2021) Histórico, avaliação, oportunidades e desafios do Manejo Integrado do Fogo nas Terras Indígenas Brasileiras. Biodivers Bras 11:75–98. https://doi.org/10.37002/BIODIVERSIDADEBRASILEIRA.V11I2.1742

Falleiro RM, Moura LC, Xerente PP, Pinto CP, Santana MT et al (2024) Using a cultural keystone species in participatory monitoring of fire management in indigenous lands in the Brazilian savanna. Fire 7:231. https://doi.org/10.3390/fire7070231

Fidelis A, Alvarado ST, Barradas ACS, Pivello VR (2018) The year 2017: megafires and management in the Cerrado. Fire 1:49. https://doi.org/10.3390/fire1030049

Franke J, Barradas ACS, Borges KMR, Hoffmann AA, Filho JCO et al (2024) Prescribed burning and integrated fire management in the Brazilian Cerrado: demonstrated impacts and scale-up potential for emission abatement. Environ Res Lett 19:034020. https://doi.org/10.1088/1748-9326/ad2820

Garda AB, Berlinck CN (2018) Histórico do fogo no Brasil com enfoque na gestão de unidades de conservação. In: Lorenzon AS, Brianezi D, Valdetaro EB, Martins MC (eds) Incêndio Florestal: princípios, manejo e impactos. Editora UFV, Viçosa, pp 27–53

Hardesty J, Myers R, Fulks W (2005) Fire, ecosystems and people: a preliminary assessment of fire as a global conservation issue. Fire Manag 22:78–87

Hataishi L, Rodrigues B, Tartar R (2019) Análise espacial dos focos de calor e queimadas no Parque nacional do Araguaia, influências meteorológicas e seus efeitos sobre o NDVI. In: Anais XIX Simposio Brasileiro de Sensoriamento Remoto. Santos, 14–17 April 2019

ICMBio, Instituto Chico Mendes de Conservação da Biodiversidade (2023) Plano de Manejo Integrado do Fogo da Estação Ecológica de Serra Geral do Tocantins: Ciclo 2023 a 2027. Ministério do Meio Ambiente. https://www.gov.br/icmbio/pt-br/centrais-de-conteudo/publicacoes/planos-de-manejo-integrado-do-fogo/PMIF_ESEC_Serra_Geral_do_Tocantins.pdf

Libonati R, Geirinhas JL, Silva PS, Russo A, Rodrigues JA et al (2022) Assessing the role of compound drought and heatwave events on unprecedented 2020 wildfires in the Pantanal. Environ Res Lett 17:015005. https://doi.org/10.1088/1748-9326/ac462e

López-Mársico L, Lezama F, Altesor A (2021) Heterogeneity decreases as time since fire increases in a South American grassland. Appl Veg Sci 24:e12521. https://doi.org/10.1111/AVSC.12521

Maillard O, Vides-Almonacid R, Flores-Valencia M, Coronado R, Vogt P et al (2020) Relationship of forest cover fragmentation and drought with the occurrence of forest fires in the Department of Santa Cruz, Bolivia. Forests 11:910. https://doi.org/10.3390/f11090910

Mistry J, Bizerril M (2011) Por que é importante entender as inter-relações entre pessoas, fogo e áreas protegidas? Why it is important to understand the relationship between people, fire and protected areas. Biodivers Bras 1:40–49. https://doi.org/10.37002/biodiversidadebrasileira.v1i2.137

Mistry J, Bilbao BA, Berardi A (2016) Community owned solutions for fire management in tropical ecosystems: case studies from indigenous communities of South America. Philos Trans R Soc B 317:20150174. https://doi.org/10.1098/rstb.2015.0174

Mistry J, Schmidt IB, Eloy L, Bilbao B (2019) New perspectives in fire management in South American savannas: the importance of intercultural governance. Ambio 48:172–179. https://doi.org/10.1007/s13280-018-1054-7

MMAA—Ministerio de Medio Ambiente y Agua (2024) Plan de acción 2024 Lucho contra los incendios. MMAA, La Paz, p 43

Moura LC, Viadana AG (2011) Contextualização do problema de incêndios florestais atribuídos à causa humana no Parque Nacional da Serra da Canastra MG. In: Anais XI Seminário de Pós-Graduação em Geografia da UNESP, Rio Claro, Brasil

Moura LC, Scariot AO, Scmidt IB, Beatty R, Russell-Smith J (2019) The legacy of colonial fire management policies on traditional livelihoods and ecological sustainability in savannas: impacts, consequences, new directions. J Environ Manag 232:600–606. https://doi.org/10.1016/j.jenvman.2018.11.057

Myers RL (2006) Living with fire: sustaining ecosystems and livelihoods through integrated fire management. The Nature Conservancy, Global Fire Initiative, Tallahassee

Oliveira G, Mataveli G, Stark SC, Jones MW, Carmenta R et al (2023a) Increasing wildfires threaten progress on halting deforestation in Brazilian Amazonia. Nat Ecol Evol 7:1945–1946. https://doi.org/10.1038/s41559-023-02233-3

Oliveira U, Soares-Filhos B, Rodrigues H, Figueira D, Gomes L, Leles W, Berlinck CN, Morelli F, Bustamanete M, Ometto J, Miranda H (2023b) A near real-time web-system for predicting fire spread across the Cerrado biome. Sci Rep 13:4829. https://doi.org/10.1038/s41598-023-30560-9

Pivello VR, Vieira I, Christianini AV, Ribeiro DB, Menezes L, Berlinck CN et al (2021) Understanding Brazil's catastrophic fires: causes, consequences and policy needed to prevent future tragedies. Perspect Ecol Conserv 19:233–255. https://doi.org/10.1016/j.pecon.2021.06.005

Santos AC, Montenegro SR, Ferreira MC, Barradas ACS, Schmidt IB (2021a) Managing fires in a changing world: fuel and weather determine fire behavior and safety in the neotropical savannas. J Environ Manag 289:112508. https://doi.org/10.1016/j.jenvman.2021.112508

Santos FLM, Nogueira J, Souza RAF, Falleiro RM, Schmidt IB, Libonati R (2021b) Prescribed burning reduces large, high-intensity wildfires and emissions in the Brazilian savanna. Fire 4:1–21. https://doi.org/10.3390/fire4030056

Schmidt IB, Moura LC, Ferreira MC, Eloy L, Sampaio AB et al (2018) Fire management in the Brazilian savanna: first steps and the way forward. J Appl Ecol 55:2094–2101. https://doi.org/10.1111/1365-2664.13118

Silva AAC, Vidal JMC, Silva RA, Lacorte GA (2018) Incêndios florestais no Parque Nacional da Serra da Canastra e a implementação de manejo integrado de fogo. ForScience 6:e00404

Singh M, Sood S, Collins CM (2022) Fire dynamics of the Bolivian Amazon. Land 11:1436. https://doi.org/10.3390/land11091436

Tizianel FAT, Bruno SF, Tizianel BZ (2020) Early prescribed fire in the Serra da Canastra National Park as a strategy to contain large fires within the integrated fire management plan. Biodivers Bras 10:34. https://doi.org/10.37002/biodiversidadebrasileira.v10i1.1517

Tomas WM, Berlinck CN, Chiaravalloti RM, Faggioni GP, Strüssmann C et al (2021) Distance sampling surveys reveal 17 million vertebrates directly killed by the 2020's wildfires in the Pantanal, Brazil. Sci Rep 11:1–8. https://doi.org/10.1038/s41598-021-02844-5

UNESCO (2024) Indigenous peoples' traditional knowledge of fire: case studies from the Guiana shield (Guyana, Suriname, Venezuela). UNESCO, Paris, p 88p

Welch JR, Coimbra CEA (2021) Indigenous fire ecologies, restoration, and territorial sovereignty in the Brazilian Cerrado: the case of two Xavante reserves. Land Use Policy 104:104055. https://doi.org/10.1016/J.LANDUSEPOL.2019.104055

Glossary of Terms Related to Fire Ecology

Backfire Fire that spreads against the wind direction.
Combustion efficiency Percentage of biomass effectively burned, indicating the extent to which the fuel is converted into heat or energy.
Crown fire Fire occurring and spreading in the crown of shrubs and trees.
Fire behavior The way a fire starts, develops and spreads.
Fire extent The area affected by a fire.
Fire frequency Number of fire events in a given time period and area.
Fire intensity Energy released in the fire front.
Fire ladder Continuity of fuel from the ground to the tree crowns.
Fire regime Combination of fire frequency, extent, intensity/severity, and season that prevails in an area over time.
Fire residence time Time that flames or heat persist at a specific point.
Fire scar Sign or injury left by fire in aboveground parts of trees and shrubs. It can also refer to a burned patch detected by remote sensed imagery.
Fire season Time of the year when most fires occur.
Fire severity The impact caused by fire on the environment aboveground (vegetation) and belowground (soil components).
Fire triangle The three essential components needed to start and carry a fire: oxygen, heat (ignition source), and fuel.
Flame height Vertical distance from the base to the top of a flame.
Flammability Ease with which a material ignites and sustains combustion.
Fuel bed A distinct layer of combustible material.
Fuel continuity Spatial continuity of fuel in a given area.
Fuel load Amount of biomass available to burn.
Ground fire Fire that occurs belowground, damaging the plant roots and consuming the soil organic matter.
Head fire Fire that spreads towards the wind.
Ignition Initial phase of the combustion process.
Integrated Fire Management (IFM) Holistic fire management approach based on prevention, fire ecology and cultural knowledge.
Megafire Wildfire of large proportions (>100–1000 ha, depending on classification).

A. Fidelis, V. R. Pivello (eds.), *Fire in the South American Ecosystems*, Ecological Studies 250, https://doi.org/10.1007/978-3-031-89372-8

Obligate seeder Plant species that only regenerates by seed germination after a fire event.

Prescribed fire Planned and controlled fire used for specific purposes such as conservation, management, agriculture.

Rate of fire spread Speed of the fire-front progression.

Resprouter Plant species that resprout from above and/or belowground buds after a fire event.

Serotiny Condition of a plant species to retain seeds within cones or closed fruits for long time periods until a disturbance, such as fire, triggers the seeds release.

Smoldering fire Fire that spreads slowly and without flames.

Surface fire Fire that occurs and spreads in surface fuels, such as herbaceous vegetation and litter.

Wildfire Non-planned fire of natural or human origin, which can develop into uncontrolled fire.

The manufacturer's authorised representative in the EU is Springer Nature Customer Service Centre GmbH, Europaplatz 3, 69115 Heidelberg, Germany. If you have any concerns regarding our products, please contact ProductSafety@springernature.com

Printed and bound by CPI Group (UK) Ltd, Croydon, CR0 4YY
30/09/2026
02225584-0015